An Introduction to Economic Geography

In the context of great economic turmoil and uncertainty, the emergent conflict between continued globalisation and growing economic nationalism means that a geographical economic perspective has never been so important. *An Introduction to Economic Geography* guides students through the key debates of this vibrant area, exploring the range of ideas and approaches that invigorate the wider discipline.

This third edition includes new chapters on finance, cities and the digital economy, consumption and the environment. Underpinned by the themes of globalisation, uneven development and place, the text conveys the diversity of contemporary economic geography and explores the social and spatial effects of global economic restructuring. It combines a critical geographical perspective on the changing economic landscape with an appreciation of contemporary themes such as neoliberalism, financialisation, innovation and the growth of new technologies.

An Introduction to Economic Geography is an essential textbook for undergraduate students taking courses in Economic Geography, Globalisation Studies and more broadly in Human Geography. It will also be of much interest to those in Planning, Business and Management Studies and Economics.

Danny MacKinnon is Professor of Regional Development and Governance and Director of the Centre for Urban and Regional Development Studies (CURDS) at Newcastle University, UK.

Andrew Cumbers is Professor in Regional Political Economy at the University of Glasgow, UK.

An Introduction to Economic Geography

Globalisation, Uneven Development and Place

Third Edition

Danny MacKinnon and
Andrew Cumbers

Routledge
Taylor & Francis Group
LONDON AND NEW YORK

Third edition published 2019
by Routledge
2 Park Square, Milton Park, Abingdon, Oxon, OX14 4RN

and by Routledge
711 Third Avenue, New York, NY 10017

Routledge is an imprint of the Taylor & Francis Group, an informa business

First published by Pearson Education Limited 2007
Second edition published by Routledge 2014

British Library Cataloguing-in-Publication Data
A catalogue record for this book is available from the British Library

Library of Congress Cataloging-in-Publication Data
A catalog record has been requested for this book

ISBN: 978-1-138-92450-5 (hbk)
ISBN: 978-1-138-92451-2 (pbk)
ISBN: 978-1-315-68428-4 (ebk)

Typeset in MinionPro
by Apex CoVantage, LLC

Visit the eResources: www.routledge.com/9781138924512

Printed in Canada

Contents

Figures

Figures

Tables

Preface to the third edition

In this third edition of our textbook, our main aim is to provide students with a sense of the diversity and vitality of economic geography as a lens to critically examine the changing economic landscape. The book is intended as an introductory text for undergraduate geography students, preparing them for further more specialist study. The first and second editions were very well received both by students and teachers, regarded as innovative books that, at the same time, strived to keep abreast of a dynamic and volatile global economy. We are gratified by the positive feedback that we have received both from students and the course convenors and tutors that have used the book.

As was the case with the second edition, which appeared in 2011, and was largely written in the midst of the global financial crisis, this third edition was written during a time of continuing economic and political turmoil. Alongside the 'economic' crisis dimension, there is also a sense of ecological crisis as the realities of global warming become increasingly apparent. While the immediate crisis threatening the global economy may have subsided in recent years, the subsequent 'recovery' has been a rather muted affair, with little sign at the time of writing (February 2018) of renewed stability or prolonged growth. In addition to the crisis-prone tendencies of the world economy, widening inequalities between people and places, policies of austerity and the growth of a populist backlash against globalisation are key contemporary trends that are emphasised in this third edition.

In seeking to keep abreast of such changes, this third edition has undergone some substantial revision from the second. We have replaced the threefold structure of the second edition: 'Foundations'; 'Actors and processes'; and 'Contemporary issues' with a four-part text (followed by a conclusion). Part 1 'Foundations' remains, with similar content on key processes underpinning the economy, but modified (as three rather than four chapters) and updated. Part 2, now titled 'Reshaping the economic landscape: dynamics and outcomes' features updated substantive chapters on the state, labour and development with a new Chapter 4 ('Capital unbound? Spatial circuits of finance and investment') that is a substantially revised version of Chapter 9 ('The uneven geographies of finance') from our second edition. Parts 3 and 4, 'Reworking urban and regional economies' and 'Reordering economic life', provide a significant amount of new material. Part 3 ('Reworking urban and regional economies') contains two new chapters (Chapters 8 and 10): 'Connecting cities: transport, communications and the digital economy' and 'Urban agglomeration, innovation and creativity', alongside another, 'Global production networks and regional economic development', that brings together key themes from the second edition's chapters on transnational corporations and commodity chains and global production networks. Part 4 includes new chapters on 'Consumption and retail' (Chapter 11), 'Economic geography and the environment' (Chapter 12) and an updated chapter, 'Alternative economic geographies' (Chapter 13).

We have retained the pedagogic features of boxes, reflective questions and exercises. In addition to the changes we have made to the print version of the book, this edition is novel in featuring accompanying eResource features on the Routledge website with supplementary materials for instructors and students.

We are very grateful to our editors at Taylor & Francis, Andrew Mould and Egle Zigaite, for their support and endless patience. We would also like to thank the six anonymous referees who provided excellent feedback and suggestions for improvement from the previous edition and constructive advice on the new edition.

PART 1

Foundations

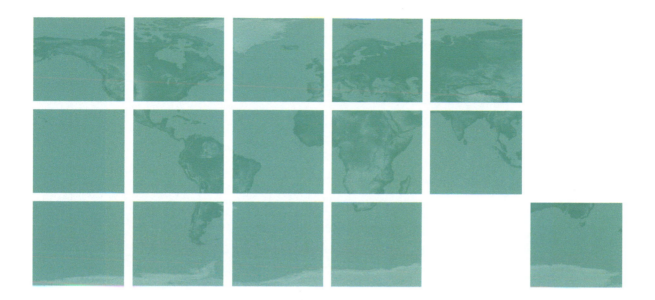

Chapter 1
Introducing economic geography

1.1 Introduction

In late 2013, a wave of local protests and marches broke out in San Francisco against Google's commuter buses (Corbyn 2014). While commuter buses may seem an unlikely trigger for social conflict, they had become symbolic of acute local concerns about the impact of a dramatic process of gentrification (the movement of wealthier groups into an area) on housing affordability in San Francisco and the Greater Bay Area (Schafran 2013), a conflict that is being echoed in global cities like London. In recent years, the renowned urban charms of San Francisco have made it a popular bedroom city for people who work in the booming high-tech industries of 'Silicon Valley' – the shorthand term for the world-leading cluster of electronics and internet industries around San Jose and Palo Alto in the south Bay Area (Figure 1.1). This has fuelled the gentrification of many neighbourhoods, with spiralling house prices making the city increasingly unaffordable to many lower-income residents, leading to rising eviction rates from 2011 (Corbyn 2014). It is the resultant social tension that animates the protests against the commuter buses, laid on by Google and other technology companies to transport their workers to their offices in Silicon Valley.

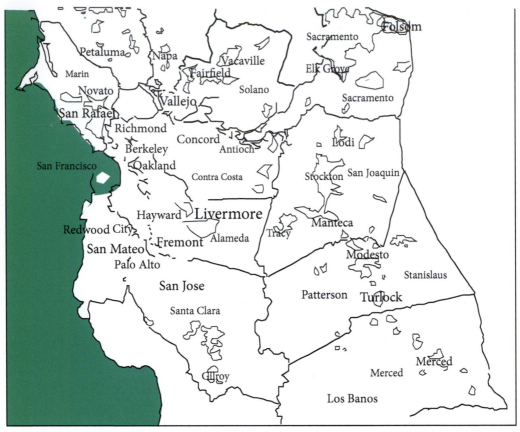

Figure 1.1 The 13-county Bay Area
Source: Schafran 2013: 668, Figure 1.

This collision between the process of economic development based on high-technology industries and the existing social fabric of the urban landscape is at the heart of the economic geography perspective developed in this book, serving to illustrate its main themes. The first of these themes is **globalisation,** which refers to the increased connections and linkages between people, firms and markets located in different places, manifested in flows of goods, services, money, information and people across national and continental borders. Here, the San Francisco Bay Area has attracted huge influxes of investment, becoming the leading centre of venture capital in the United States (US). At the same time, its booming housing market became a key outlet for the money generated on Wall Street prior to the financial crash of 2008 (Walker and Schafran 2015). In addition, economic growth has attracted large numbers of

immigrants, stretching back over many decades, giving the Bay Area its characteristic social and racial diversity.

The second theme is **uneven development**, whereby some countries and regions are more prosperous and economically powerful than others. The Bay Area is the richest major metropolitan area in the US on a per capita (by population) basis, irrespective of whether this is measured by income or wealth, containing more millionaires per capita than any other large metropolitan area (ibid.: 23). Yet, as the protests over gentrification highlight, it is also one of the most socially unequal, symbolising the wider trend of rising inequality in the US. There are millions of ordinary low- and middle-wage workers who are not employed in the high-tech sector in a region with an extremely high cost of living (ibid.). Over the past couple of decades, the spread of urbanisation and the rising costs of housing in San Francisco and the Inner Bay Area more generally have driven many of these groups to live further and further out in the Central Valley counties of San Joaquin, Stanislaus and Merced (Figure 1.1). This process is generally known as ex-urbanisation, representing an extension of the well-established trend of suburbanisation to previously free-standing areas. Yet, reflecting its intense housing boom, the Bay Area has been one of the regions worst affected by the post-2008 recession and foreclosure (housing repossession) crisis. This has been particularly concentrated in the new outer suburbs in the Central Valley, with the city of Stockton being declared bankrupt in 2013.

Third, the social tensions surrounding gentrification in the Bay Area also illustrate the theme of **place**, in terms of how particular areas become entangled in wider economic processes and the consequences of this for their social make-up and identity. From the origins of San Francisco as the supply centre for the Californian gold boom of the late 1840s to the internet boom of the past couple of decades, the region has been transformed by successive waves of investment and immigration. The alternative, bohemian identity of San Francisco was established through the attraction of burgeoning counter-cultures associated with particular neighbourhoods such as hippies in Haight-Ashbury and the gay rights movement in the Castro. For some residents and critics, it is these identities and diverse neighbourhoods that are being threatened by gentrification as their social diversity gives way to the male-dominated, affluent monocultures associated with high-tech workers and entrepreneurs (Corbyn 2014). For others, however, the latter are simply the latest wave of incoming pioneers which the region will be able to accommodate in the same fashion as it accommodated previous migrants.

Reflect

Do you think that San Francisco will be able to accommodate the large-scale in-migration of high-tech workers and entrepreneurs whilst retaining its character and identity?

1.2 Key themes: globalisation, uneven development and place

In this section, we build on the Bay Area example to examine the three main themes of the book – globalisation, uneven development and place – more fully. Our selection of these themes is informed by the basic geographical concepts of location and distance, **scale**, **space** and **place**.

➤ Location is perhaps the most basic geographical concept, referring to the geographical position of people and objects relative to one another (Coe *et al.* 2013), i.e. *where* things are. This is often represented by maps (see Figure 1.1), and systems of grid references have been developed to convey this information in a precise form. It is clearly related to distance which is the area or space between locations, for example cities such as Hong Kong and London. Overcoming what geographers have traditionally called the 'friction' of distance (the effort and cost of moving objects and people between locations) requires time and money, for example the price of a long haul flight between London and Hong Kong. The greatly increased ability of economic actors such as **Transnational corporations (TNCs)** and banks to move money, goods, services, information and people over large distances as a result of the development in transport and communication technologies has been of great significance for the reorganisation of the international economy in recent decades.

➤ Scale refers to the different geographical levels of human activity, from the local to the regional, national and global (Figure 1.2). They are important to the definition and organisation of economies as indicated by the common use of terms such as the local economy, national economy and global economy by policy-makers, media commentators and citizens. It is important to see these different scales of economic organisation as overlapping and interconnected rather than viewing them as entirely separate.

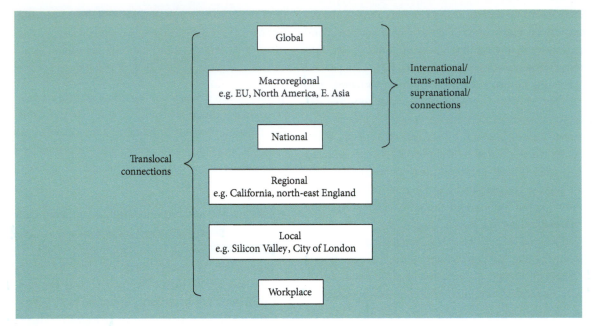

Figure 1.2 Scales of geographical analysis
Source: Castree *et al.* 2004: xvix.

➤ Space is an area of the earth's surface such as that contained within the boundaries of a particular region or country. It is perhaps the most abstract and difficult to grasp of the geographical concepts introduced here, and is best understood by being related to distance and place. Although a more general term, space is related to the more specific notion of distance and can also be expressed partly in terms of the area between two points (locations) in space and the time it takes to move between them. At the same time, it is often contrasted with place, particularly in terms of how spaces can be converted into places through human occupation and settlement

➤ Place refers to a particular area (space) to which a group of people have become attached, endowing it with human meaning and identity. This is evident in how the occupation of San Francisco neighbourhoods by particular counter-cultural groups has defined their character and identity. The geographer Tim Cresswell (2013) illustrates the distinction between space and place by referring to an advertisement in a local furniture shop entitled 'turning space into place', reflecting how people use furniture and interior décor to make their houses meaningful, turning them from empty locations into personalised and comfortable homes. This domestic transformation of space into place is something with which we are all familiar, perhaps from decorating rooms in university halls of residence or shared flats.

1.2.1 Globalisation and connections across space

The first underlying theme which runs through this book is that economic activities are connected across space through flows of goods, money, information and people. The concept of (economic) globalisation can be defined as a process of economic integration on a global scale, creating increasingly close connections between people and firms located in different places. It is manifested in terms of increased flows of goods, services, money, information and people across national and continental borders. These flows are not new as trading relations between distant people and places involving the exchange of goods have existed throughout much of human history. The notion of globalisation, however, emphasises that the volume and scope of global flows has increased significantly in recent decades (Dicken 2015). Increased trade and economic interaction between distant places is dependent on technology in terms of the ease of movement and communication across space.

A new set of transport and communications technologies has emerged since the 1960s, including jet aircraft, shipping containerisation, the internet, email and mobile telephones. The effects of these 'space-shrinking technologies' have brought the world closer together, effectively reducing the distance between places in terms of the time and costs of movement and communication (Figure 1.3).

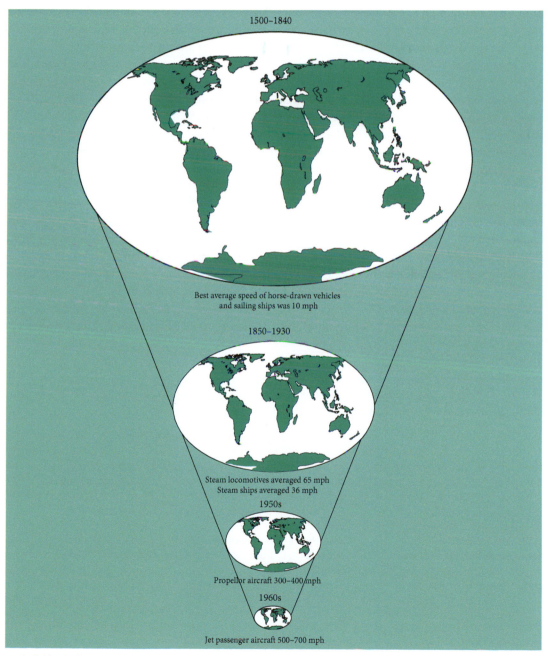

Figure 1.3 'A shrinking world'
Source: Dicken 2003: 92.

Similarly, the rise of new information and communications technologies (ICTs) such as the internet have made it possible for large volumes of information to be exchanged at a fraction of the previous cost, resulting in '**time-space compression**'. The term was introduced by the geographer David Harvey (1989) who argued that the process of 'time-space compression' has been driven by the development of the economy, requiring geographical expansion in search of new markets, supplies of labour and materials. Investments in transport and communications infrastructure have facilitated this process of geographical expansion, reducing the effects of distance as it becomes easier and cheaper to transmit information, money and goods between places. As such, time and space are effectively being compressed through the development of new technologies. This is not an entirely novel process with a previous 'round' of time-space compression occurring towards the end of the nineteenth century through inventions such as railways, steamships, the telegraph and the telephone.

While globalisation has become a key buzzword of the last couple of decades, it should not be viewed as a single, unified phenomenon or thing, but as a group of interrelated processes and activities (Dicken 2015: 6). These include the vastly increased volume of global financial transactions, the impact of internet technologies, the creation of increasingly global consumer markets, the power of TNCs to relocate economic activities to other countries, the rise of new economic powers like China, India and Brazil and the role of international economic organisations in managing the global economy (Box 1.1). This complexity means that globalisation cannot itself be held ultimately responsible as the single cause of other related problems such as social and economic inequality. Such problems are likely to have many causes, and national and local factors are often bound up with the operation of global forces. Furthermore, globalisation is an ongoing set of processes rather than a final condition that has actually been achieved: "although there are undoubtedly globalising forces at work, we do not have a fully globalised world" (ibid.: 7).

In both the academic literature and much popular commentary, the term globalisation is used in two distinct senses (Dicken 2015: 3). The first is empirical or factual, referring to the actual changes that have occurred in the organisation and operation of the world economy. This is often represented in statistics such as figures on the volume of world trade, financial flows or the number of internet users. The second is ideological and refers particularly to the free market, **neoliberal** ideology or world view of globalisation as a project

---

Box 1.1

Managing globalisation: international economic organisations

These organisations were created at the end of the Second World War as part of the Bretton Woods system, alongside the regime of fixed exchange rates. Since the early 1980s, their policies have been shaped by **neoliberalism** (Box 1.2).

➤ The International Monetary Fund (IMF). The role of the IMF is to promote monetary cooperation between countries, and to support economic stability and trade. The provision of financial assistance to countries experiencing budgetary problems allows the IMF to set conditions requiring countries to reform their economies. See www.imf.org/

➤ The World Bank (officially the International Bank for Reconstruction and Development). Its role is to provide development assistance to countries, mainly in the less developed world. The Bank runs a range of programmes and initiatives aimed at reducing poverty and narrowing the gap between rich and poor countries. See www.worldbank.org/

➤ The World Trade Organisation (WTO), established in 1995, taking over from GATT (the General Agreement on Tariffs and Trade). The role of the WTO is to ensure a free and open trading system, working through successive 'rounds' or conferences where member countries come together to negotiate agreements. See www.wto.org/

(Box 1.2) (ibid.: 3). The two are often confused. While they are often linked together in practice – typically, for instance, neoliberal world views emphasise certain factual aspects of globalisation such as increases in international trade, but exaggerate their meaning and significance to support certain political objectives such as reduced government intervention in the economy – they have distinct meanings. Accordingly, drawing an analytical distinction between the empirical and ideological aspects of globalisation is important because it helps to develop a clearer understanding of what can be a complex and confusing concept.

After the collapse of Soviet communism and the broader Cold War order, globalisation became closely tied to claims about the "triumph of free market capitalism" (Jones 2010: 9). This was promoted and justified through a neoliberal project that portrays globalisation as a mutually beneficial and inevitable process which increases economic well-being by enabling a more efficient allocation of resources through the market and free trade, leading to reduced poverty over the longer term.

While the neoliberal project of globalisation encountered little sustained opposition through much of the 1990s, a new wave of opposition and protest emerged with the so-called 'Battle in Seattle' in December 1999. Seattle brought together a large number of protesters to demonstrate against neoliberal globalisation and the WTO's efforts to launch a new round of trade negotiations. It was a key moment in the formation of a **counter-globalisation movement** which espouses a more open, participative 'bottom-up' model of globalisation. The counter-globalisation movement opposes the top-down model of neoliberal globalisation, associating it with increased corporate control, policies of privatisation and liberalisation, and large-scale inequality and poverty in the global South (developing world) particularly. In the wake of the post-2008 economic crisis, the focus of protest and action moved away from globalisation *per se* to focus on issues of inequality and the adoption of austerity policies by governments through movements such as Occupy and the protests in Greece. This underlines the crucial point that globalisation is not only a topic of endless academic discussion, but also the subject of active political contestation and conflict (Routledge and Cumbers 2009).

The real significance of the move towards a more integrated global economy lies not so much in its quantitative extent (for example, the volume of trade or number of countries involved), but in the qualitative transformation of economic relationships across geographical space (Dicken 2015: 6). This emphasises the intensification and spending up of economic relations and economic life as the world economy has become increasingly interconnected. As economic geographers, we are particularly interested in the impact of these changes on urban and regional economies.

Box 1.2

Neoliberalism

A central feature of the period since the 1970s has been a changing economic policy context, with the emergence of a dominant set of policy prescriptions which have exerted a powerful influence on the thinking of many politicians, business leaders, commentators and interest groups. Neoliberalism in simple terms entails a commitment to the promotion of free markets and private property rights and the reduction of state involvement in the economy. It involves the advocacy of more specific reforms to further this objective through measures such as trade liberalisation, financial market deregulation, the elimination of barriers to foreign investment, the privatisation or selling of state-owned enterprises to the private sector and the deregulation of labour markets. Since the 1970s, it has replaced Keynesianism – based on a commitment to full employment and state intervention to stimulate growth during periods of recession – as the dominant mode of thought governing global economic policy-making (section 5.5).

In particular, sub-national economies have become increasingly exposed to a range of global flows and influences. The impact of globalising forces on local, regional and national economies can best be understood in relation to the following key dimensions of change (Coe and Jones 2010).

➤ **Financialisation** which emphasises the increasing role of financial motives, financial markets, financial actors and financial institutions in daily life (Epstein 2005). This is a process that has been developing since at least the 1980s, but was dramatically highlighted by the financial crash of 2008 in terms of economic reliance on the major banks ('too big to fail') and high levels of household indebtedness. The latter reflects the easy availability of credit before the crash as people acquired a range of financial products and services such as mortgages, 'sub-prime' loans and credit cards.

➤ Increased competition which represents a central aspect of globalisation through the liberalisation of trade and the reduction of barriers between national markets. This can have very different impacts on different industries and the cities and regions in which they are located. It has enabled some industries and firms to expand into global consumer markets (think of banks and internet search engines like Google), while exposure to competition and the removal of traditional government protections has rendered other industries and firms less competitive, often leading to plant closures and unemployment. This is characteristic of not only some manufacturing industries like textiles and steel in developed countries but also agricultural and other primary industries in developing countries in the global South, in some cases triggering political protests and demonstrations.

➤ Flexibilisation which refers to the effects of globalisation and increased competition in requiring sub-national and national economies to become more flexible. This involves them responding quickly and efficiently to wider forces of change such as competition, technological change, market trends and broader economic shocks such as the 2008 financial crash and subsequent recession. Here, we are particularly interested in the flexibilisation of employment and labour markets that has occurred since the 1980s, linked to a sectoral shift from manufacturing to services in developed countries and the growth of part-time and contract employment, coupled with reductions in employment stability and security (the end of the job for life). These changes have been most prevalent in Anglo-Saxon countries such as the US, United Kingdom (UK) and Australia, reflecting their strong attachment to neoliberal policies (Box 1.2) which have championed the need to create flexible labour markets, supported by curbs on trade union power. While flexibility is attractive to business, it is associated with greater insecurity and precarity for a significant proportion of the workforce (Coe and Jones 2010: 6).

1.2.2 Uneven development

A basic feature of economic development under **capitalism** is its geographical unevenness. Uneven development is an inherent feature of the capitalist economy, reflecting the tendency for growth and investment to become concentrated in particular locations. These areas may be favoured by a particular set of advantages such as their geographical position, resource base, availability of capital or the skills and capabilities of the workforce. Once growth begins to accelerate in a particular area, it tends to 'suck in' investment, labour and resources from surrounding regions. Capital (investment) is attracted by the opportunities for profit whilst workers are drawn by abundant job opportunities and high wages. Surrounding regions are often left behind, relegated to a subordinate role supplying resources and labour to the growth area.

One key aspect of the process of uneven economic development is that it occurs at different geographical scales (Figure 1.2). This can be illustrated with reference to three key scales of activity: the global, regional and local.

➤ At the global level, there is a marked divergence between the 'core' in North America, Japan and Western Europe and the 'periphery' in the 'global South' of Asia, Latin America and Africa (see Figure 1.4). This pattern reflects the legacy of colonialism, whereby the core countries in Europe and North America produced high-value manufactured goods and the colonies produced low-value raw materials and agricultural products. Whilst a number of East Asian countries, including China, have been able to overcome this legacy, experiencing rapid growth and rising prosperity over the past 25 years, others, particularly in sub-Saharan Africa, have been left behind, experiencing conditions of extreme deprivation and poverty.

➤ Within individual countries too, economic disparities between regions are evident. The rapid economic development of China, for instance, has opened up a growing divide between the booming coastal provinces in the south and east and a poor, underdeveloped interior (Box 1.3). Developed countries are also characterised by regional disparities, such as the persistent north-south divide which has characterised the economic geography of the United Kingdom since the 1930s.

➤ Even on a local level within cities, uneven development is present in the form of social polarisation between affluent middle-class neighbourhoods and poorer inner city areas and public housing schemes or projects. For example, a study of neighbourhood inequality in the eight largest Canadian cities found that it has increased between 1980 and 2005 (Chen et al. 2012). This was due to differential growth in family earnings which rose rapidly for the most affluent families whilst declining or stagnating for the poorer ones. A strong pattern of residential clustering, whereby affluent people tend to live in affluent areas and poorer people in poorer ones, meant that this resulted in increased inequality between neighbourhoods.

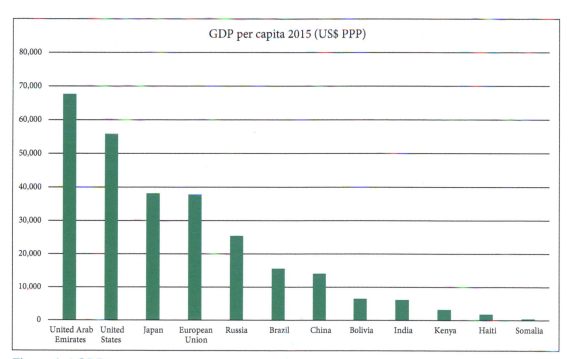

Figure 1.4 GDP by country
Source: www.cia.gov/library/publications/the-world-factbook/rankorder/2004rank.html.

Box 1.3

Rapid economic growth and uneven development in China

Since the Communist regime opened up its economy to attract foreign investment in 1978, China has experienced rapid economic development, becoming a world manufacturing and assembly centre, with the economy growing at an average rate of 10 per cent per year from the early 1980s to 2012. As a result, GDP per capita was 35 times higher in 2011 than 1978, lifting over half a million people out of poverty (Knight 2013). Rapid growth has, however, created considerable strains in terms of environmental degradation, increased inequality and the ruling Communist Party's ability to manage the social pressures generated by people's rising expectations. The rate of GDP growth actually fell to an average of 7 per cent between 2012 and 2015 (although many commentators are sceptical of these official figures), sparking fears about economic slowdown in the context of reduced demand from China for commodities such as oil, copper and iron ore, as well as broader imbalances in the economy.

The experience of China since 1978 is consistent with the argument that uneven geographical development is inherent to the process of economic development. The degree of regional inequality often reflects the speed and magnitude of economic and social transformation, meaning that it is generally highest in countries undergoing rapid industrialisation and development which tends to be concentrated in particular regions (World Bank 2009). Economic growth has created a complex pattern of regional inequality in China, though the underlying divide is between the coastal regions and the interior. The 'open door' policy was initially based on the designation of special economic zones in the coastal provinces to attract foreign investment, encouraging these regions to 'get rich quick' (Wei 1999: 51). Coastal provinces such as Guangdong, Jiangsu, Zhejiang, Fujian and Shandong experienced rapid growth in the 1980s and 1990s, while much of the interior lagged behind. This fuelled the migration of millions of rural workers to the coastal cities in search of employment, generating what has been called the largest mass migration in human history. As a result, 55 per cent of China's population now lives in cities, compared to less than 20 per cent in 1978 (Anderlini 2015). About 275 million workers, more than a third of China's labour force, are migrants from the countryside, often leaving their families behind in their rural villages of origin.

In recent years, the Chinese government has placed more emphasis on the development of the interior, linked to a rebalancing away from export-led growth towards the domestic market, particularly in response to the effects of the financial crisis in 2008 (Yang 2013). This has fostered rapid industrial growth in parts of central and western China with the city of Chongqing in Sichuan province, for instance, becoming a key centre of personal computer manufacturing (Yang 2017). This highlights the dynamic nature of uneven development processes, particularly in conditions of rapid economic growth as new centres of production emerge while established ones undergo periods of restructuring. In recent years, as some interior provinces have experienced rapid development, the flow of migrants to the coastal regions has slowed markedly and some migrant workers have returned to their home areas, with Chongqing, for instance, attracting around 300,000 returning migrant workers in 2014 (ibid.).

The process of economic development is highly dynamic in nature as new technologies are developed, new forms of customer demand emerge and work practices change. Over time, patterns of uneven development are periodically restructured as capital investment moves between different locations, investing in those which offer the highest rate of return (profit). As a result, new growth regions emerge whilst established ones can experience stagnation and decline. As broader market conditions and technologies change, the specialised economic base of formerly prosperous regions can be undermined by reduced demand, rising costs, competition and the invention of new products and methods of production. On a global scale, the most dramatic change in patterns of uneven development is the emergence of East Asia as a dynamic growth region over the last 30 years or so (Box 1.3). Within developed countries, established industrial regions based on nineteenth- and early twentieth-century industries such as coal, steel and shipbuilding have experienced decline whilst new growth centres have emerged in regions such as the south and west of the US (the so-called 'sunbelt'), southern Germany and Cambridge in England.

1.2.3 The importance of place

The role of place in shaping economic activity is a third key theme of this book. As we suggested in the previous section, processes of uneven geographical development have created distinctive forms of production in particular places. During the nineteenth and early twentieth century, specialised industrial regions emerged in Europe and North America. As a result,

> . . . distinct places are associated with sectoral and functional divisions of labour. In the United States for example, "Pittsburgh meant steel, Lowell meant textiles, and Detroit meant automobiles" (Clark *et al.* 1986: 23), while in the United Kingdom, "one finds metal workers in the Midlands, office professionals in London, miners in South Wales, and academics in Oxford" (Storper and Walker 1989: 156).
>
> (Peck 1996: 14)

Although some of these specific associations have been weakened by **deindustrialisation**, the general point about distinctive forms of production being associated with particular places remains important. The City of London continues to be associated with finance and business services, for instance, Los Angeles with movies and Milan with clothing design and fashion. Silicon Valley in California has become the hub of the internet and social media industries, building on its previous success in electronics and computers. This geographical variety of local economies is continually reproduced through the interaction between wider processes of uneven development and local political, social, economic and cultural conditions. These conditions reflect the economic history of a place in terms of the particular industries found there and the institutions and practices associated with them. In general, the interaction between the legacies of established industries and institutions and emerging forces of economic change shapes and moulds the economic landscape (Massey 1984).

It has become increasingly clear in recent years that globalisation is a differentiated and uneven process, generating different outcomes in different places. In particular, it seems to be associated with a resurgence of certain regions as economic units in recent decades. In particular, the success of dynamic growth regions such as the City of London, Silicon Valley, Taipei in Taiwan and Guangdong and Shanghai in China are rooted in the specialised production systems that have flourished there. Geographical proximity seems to encourage close linkages and communication between firms, enabling them to share information and resources. The existence of a large pool of skilled labour is a crucial feature of such regions, allowing firms to recruit easily and workers to move jobs without leaving the local area. These aspects of the local production system encourage innovation and entrepreneurship, enhancing their competitiveness within the global economy.

Whilst globalisation is not leading to the erasure of place in absolute terms, it does undermine traditional notions of places as homogenous and clearly bounded local areas. In response, there is a need to rethink place in terms of connections and relations across space (Massey 1994). It is in this sense that the British geographer Doreen Massey's work on the development of a **'global sense of place'** is of particular interest. Massey develops a new conception of place as a meeting place, a kind of node or point where wider social relations and connections come together:

> . . . what gives place its specificity is not some long internalised history but the fact that it is constructed out of a particular constellation of social relations, meeting and weaving together at a particular locus. . . . Instead . . . of thinking of places as areas with boundaries around them, they can be imagined as articulated moments in networks of social relations and understandings . . . and this in turn allows a sense of place which is extroverted, which includes a consciousness of its links with the wider world, which integrates in a positive way the global and the local . . .
>
> (Massey 1994: 154–5)

From this perspective, place can itself be regarded as a process rather than being seen as a static and unchanging entity. Places are connected and linked through wider processes of uneven development operating through flows of capital, goods, services, information and people. Movement of particular commodities like bananas, for example, links different parts of the UK to the economies of certain Caribbean islands (see Box 1.5).

1.3 The economy and economic geography

1.3.1 The capitalist economy

'The economy' refers to the interrelated processes of production, circulation, exchange and consumption through which wealth is generated (Hudson 2005: 1). It is through such processes that people strive to meet their material needs, earning a living in the form of wages, profits or rent. Production involves combining land (including resources), capital, labour and knowledge – commonly known as the **factors of production** – to make or provide particular commodities. It relies on a supply of resources from nature, meaning that economic activities have a direct impact on the environment. The commodity, defined as any product or service that is sold commercially, is so basic to the workings of the economy that Karl Marx described it as the 'economic cell form' of capitalism. The modern economy involves the production and consumption of a vast array of commodities, spanning everything from smart phones to package holidays (Boxes 1.4, 1.5).

Human societies have tended to organise and structure their economic activities through overarching **modes of production**. These can be defined as economic and social systems which determine how resources are deployed, how work is organised and how wealth is distributed. Economic historians have identified a number of modes of production, principally subsistence, slavery, feudalism, capitalism and socialism. Each of these creates distinctive relationships between the main **factors of production**. Capitalism is clearly the dominant mode of production in the world today, operating at an increasingly global scale. It is defined by private ownership of the means of production – factories, offices, equipment and money capital – and the associated need for most people to sell their labour power to employers in order to earn a wage. This allows them to purchase commodities produced by other firms, creating the market demand that underpins the capitalist

Box 1.4

Global commodity chains

Global commodity chains (GCCs) link together the production and supply of raw materials, the processing of these materials, the production of components, the assembly of finished products, and the distribution, sales and consumption of these products. They involve a range of different organisations and actors, for example farmers, mining or plantation companies, component suppliers, manufacturers, subcontractors, transport operators, distributors, retailers and consumers. Global commodity chains have a distinct geography, linking together different stages of production carried out in different places (Watts 2014). Some parts of the production process add more value or profit, creating tensions between the different participants in the supply chain over who captures this value. Relationships between large supermarkets and their suppliers are a good example of this tension, with farmers in the UK staging protests, including the removal of milk from supermarket shelves, against low prices in August 2015. Exploring the production and consumption of particular commodities helps us to trace and uncover economic connections between places, linking different localities to global trading networks and showing how the range of goods (such as milk) that we consume on a daily basis are produced, distributed and sold.

system. Compared to earlier modes of production, production and consumption are often geographically separate under capitalism, creating a need for extensive transport and distribution networks.

A key underlying point here concerns the fact that the principal features of the modern capitalist economy – such as the role of the market, profits and competition – are not natural and eternal forces that determine human behaviour, as many mainstream economists and business commentators tend to assume. Instead, capitalism is a historically specific mode of production that emerged from its roots in early modern Europe in the sixteenth and seventeenth centuries to encompass virtually the entire globe today. It has been superimposed on a complex mosaic of pre-capitalist societies and cultures, resulting in great regional variation as pre-existing local characteristics interact with broader global processes.

Whilst capitalism is clearly the dominant mode of production in the world today, it does not follow that all economic activity is capitalist in nature. In reality, the formal, capitalist economy based on trying to maximise profits or earnings coexists with a range of other economic activities and motivations such as domestic work, volunteering and the exchange of gifts. In practice, non-capitalist activities interact with capitalism in a variety of ways. Think of the relationship between domestic work and paid employment, for example, or the role of gift-buying in supporting consumption. The geographers Gibson-Graham (1996) emphasise the existence of **diverse economies**, criticising the preoccupation with the formal, capitalist economy amongst economists and economic geographers. This critique has informed a number of studies of 'diverse' or 'alternative' economies such as informal work, local currencies and cooperatives.

1.3.2 An economic geography perspective

Economic geography is concerned with concrete questions about the location and distribution of economic activity, the role of uneven geographical development and processes of local and regional economic development. It asks the key questions of 'what' (the type of economic activity), 'where' (location), 'why' (requiring explanation) and 'so what' (referring to the implications and consequences of particular arrangements and processes). According to one definition:

> Economic geography is concerned with the economics of geography and the geography of economics. What is the spatial distribution of economic activity? How is it explained? Is it efficient and/or equitable? How has it evolved, and how can it be expected to evolve in the future? And what is the appropriate role of government in influencing this evolution?
>
> (Arnott and Wrigley 2001: 1).

Three key themes emerge from this:

➤ The geographical distribution and location of economic phenomena. Figure 1.5 provides an example of such a distribution, showing variations in employment in financial and business services across the UK prior to the financial crash of 2008 and subsequent recession. Describing and mapping distributions of economic activity in this way can perhaps be seen as the first task of economic geography, addressing the basic questions of 'what' and 'where'.

➤ The next stage is to explain and understand these spatial distributions and patterns of economic activity. This is a more demanding task which brings in theory and requires us to have some appreciation of history. It involves addressing the more advanced questions of 'why' and 'how'. In trying to explain why financial and business services employment is clustered in south-east England and one or two other areas, then, we would need to draw on theories of the spatial concentration of economic activity and to have some understanding of the economic history of the UK since at least the 1930s.

➤ A third issue is that of engagement with policymakers in government and the private sector, making recommendations and offering advice about particular geographical issues and problems. As well as describing and explaining the distribution of certain economic phenomena, geographers have also sought to outline how the economic geography of particular countries and regions *should* be organised. This role has sparked periodic debates about the social 'relevance' of the subject.

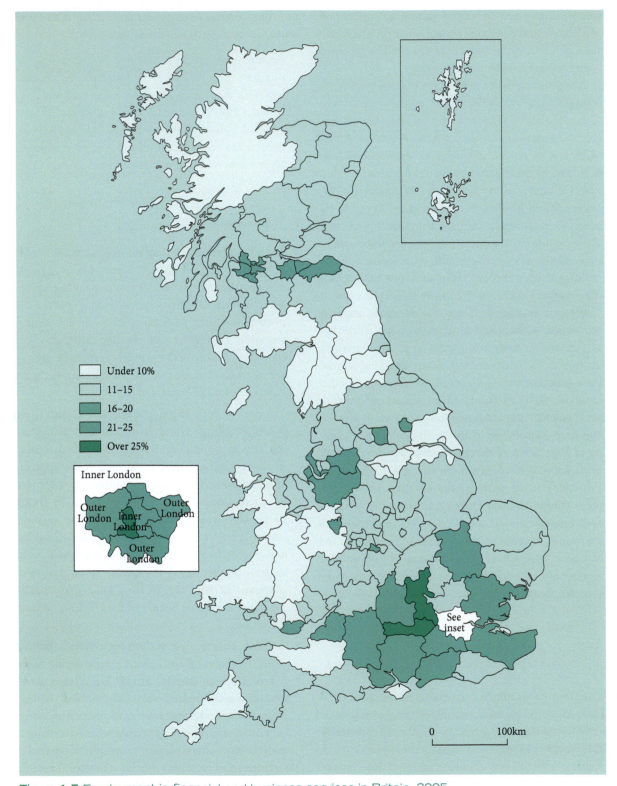

Figure 1.5 Employment in financial and business services in Britain, 2005
Source: ONS 2006.

Reflect

➤ How would economic geographers explain the concentration of financial and business services employment in the south-east of England? Is this pattern of concentration likely to persist in the future?

1.4 A political economy approach

In this section, we build on the above definition of economic geography by introducing our favoured approach to the subject: **political economy**. Political economy analyses the economy within its social and political context, rather than seeing it as a separate entity driven by its own set of rules based on individual self-interest. It is concerned not only with the exchange of commodities through the market, but also with production and the distribution of wealth between the various sections of the population (Barnes 2000). This focus on questions of production and distribution as well as production and exchange (markets) distinguishes political economy from much of mainstream neoclassical economics. We adopt a geographical political economy approach in this book because of this breadth of scope and its analytical power as a framework for understanding the evolution and restructuring of the capitalist economy, particularly in terms of increasing globalisation, associated processes of uneven development and the implications of this for urban and regional economies (place).

Since the 1970s, economic geographers have sought to apply a political economy framework to geographical questions such as regional development and urban restructuring (see section 2.4). This has given rise to what the economic geographer Eric Sheppard (2011) terms **geographical political economy** (GPE), defined by three key characteristics. First, capitalism is just one way of organising the economic activities of a society rather than being historically inevitable or natural. This reflects the historically specific nature of political economy in analysing the transformation of social and economic relations over time. Second, geography is not passive or external to the economy, but is actively produced by the

process of economic development, leading to the periodic restructuring of urban and regional economies. As a result, new forms of economic activity give rise to new growth regions, alongside the decline of older industrial regions. Third, economic processes must be considered in relation to parallel bio-physical, social and cultural processes such as the utilisation of natural resources, social class, gender relations and identity formation. GPE approaches should be seen as open and adaptable, being receptive to insights from other perspectives and evolving in line with capitalism as their main object of analysis (section 2.4).

1.4.1 Social relations

Social relations provide the general link between the economy and society. In contrast to mainstream economics which is underpinned by the assumption of individual rationality, political economists believe that economic activity is grounded in social relations. These simply refer to the relationships between different groups of people involved in the economy such as employers, workers, consumers, government regulators, etc. A key defining social relationship is that between employers and workers, structuring activity within the workplace. The two groups are commonly regarded as different classes within society, a position set out with particular force by Marx, although for some commentators the growth of a large middle class of 'white collar' workers has blurred the distinction considerably. Other important social relations are those between producers and consumers, different firms (for example manufacturers and suppliers), different groups of workers (e.g. supervisors and ordinary employees) and government agencies and firms. As our case study of the banana indicates (Box 1.5), the production of a simple commodity creates complex social relations between people based in different places. Even if some of the parties – for example consumers at large supermarkets in the developed world – are not aware of such relationships, they still exist.

Once we have accepted that the economy is structured by social relations, it is important to recognise that these relations will change over time as society evolves. We have already pointed to a transformation occurring in western societies from around the fifteenth century

17

Box 1.5

The 'secret life of a banana' (Vidal 1999)

The banana is the world's most popular fruit, with consumers spending £10 billion a year on the fruit globally (Fairtrade Foundation 2009). This simple statistic, however, masks a complex geography of production and distribution which links different people and places together (Watts 2014). Whilst individual consumers may remain unaware of such linkages, they create real social relationships between people in different places, for example, consumers in countries such as the UK and banana farmers in tropical countries in Africa, the Caribbean and Latin America.

Bananas are grown by either individual small farmers, sometimes working under contract, or on large industrial plantations run by MNCs. Either way, the fruit must be grown, picked and packed and transported to the nearest port from where it is shipped in special temperature-controlled compartments to another port in the destination country. The fruit is then ripened in special ripening centres before being sent to the supermarket by truck. Bananas exported from the tiny West Indian island of St Vincent to the UK, for instance, are transported to Southampton by the Geest line shipping company, taking roughly two weeks (Vidal 1999).

The complex chain of linkages involved in the production, distribution and exchange of any particular commodity creates real conflicts of interest between groups of people over who captures the most added value from the product (Watts 2014). Figure 1.6 shows how the price of a 30 pence banana is distributed between the various actors in the production chain. In the UK, supermarket price wars have seen the price of bananas fall by nearly half to just 11p between 2004 and 2014 while the costs for farmers have doubled (Butler 2014).

The costs of cheap bananas are typically passed down the supply chain to small farmers and plantation workers, leading to deteriorations in pay and conditions. In 1999, for instance, Del Monte sacked all 4,300 of its workers on a large plantation in Costa Rica, re-employing them on wages reduced by 30–50 per cent and on longer hours with fewer benefits (Lawrence 2009). These cuts underpinned a global deal with Wal-Mart, allowing Del Monte to grant the global retailer a low price. In response, development organisations and activists have promoted Fairtrade which now accounts for one in every three bananas sold in the UK, having been adopted by large retailers like Sainsbury's, the Co-operative and Waitrose.

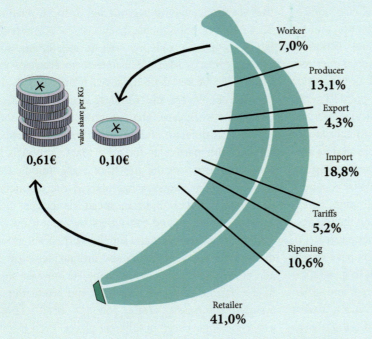

value share per KG

0,61€   0,10€

Worker
7,0%

Producer
13,1%

Export
4,3%

Import
18,8%

Tariffs
5,2%

Ripening
10,6%

Retailer
41,0%

Figure 1.6 Banana split: who gets what in the banana chain (for main countries supplying the EU in 2014)
Source: Banana link website, www.bananalink.org.uk/banana-value-chains-eu.

onwards where a system of market capitalism gradually replaced feudalism as the underlying mode of production. In other words, feudal ties and values between the peasantry and the nobility have given way to market competition, the pursuit of profit and the wage relationship between capital (employers) and labour (workers) as the key social relationship shaping economic and indeed human development. Increasingly intensive competition between firms is also a critical relationship, driven by the relentless pursuit of profit and the subsequent need to eliminate rivals to gain a greater share of the market.

1.4.2 Institutions and the construction of markets

A second crucial element in our political economy perspective is a view of markets as socially constructed entities which require social and political regulation rather than representing naturally occurring phenomena capable of self-regulation. Unlike mainstream economics which holds that markets, if left to their own devices, will return to an equilibrium position where supply equals demand and waste is eliminated, we believe that unregulated ('free') markets are destabilising and socially destructive.

While the notion of the *free* market retains its ideological and political power, in reality virtually all economies are mixed, containing substantial public sectors. Following Karl Polanyi (Box 1.6), we emphasise the institutional foundations of markets, recognising that the economy is shaped by a wide range of institutional forms and practices (Box 1.7). These include cultural rules, habits and norms which structure the social relations between individuals, helping

---

Box 1.6

Karl Polanyi: the economy as an instituted process

The work of Karl Polanyi (1886–1964) on the institutional foundations of economic processes is well known within the social sciences. Polanyi's broad conception of the economy and his insights into the underlying nature of market society have come to inform debates about contemporary processes of globalisation and the social 'embeddedness' of the economy.

In *The Great Transformation*, published in 1944, Polanyi explores the origins and development of market society during the nineteenth century. He identified land, labour and money as 'fictitious commodities' because they are not actually produced for sale on the market like true commodities, a crucial point neglected by orthodox economic analyses which assume that the price mechanism will balance supply and demand in the normal fashion. The identification of these fictitious commodities is important to Polanyi's argument, indicating that a pure self-regulating market economy is

impossible since state intervention is required to match the supply and demand for land, labour and money. The book demonstrates how the construction of competitive markets depended upon state action through the upholding of property rights and contracts, the introduction of labour legislation, the establishment of measures to ensure a stable food supply and the regulation of the banking system. In other words, as Polanyi famously remarked, 'laissez-faire was planned'. This point is just as applicable to the contemporary project of neoliberal globalisation as to the creation of market society in the nineteenth century.

In the 1950s, Polanyi turned away from modern market economies to the analysis of primitive and archaic economies. This substantive focus was closely informed by his enduring interest in the scope of the economy and the role of institutions in shaping economic processes, as demonstrated by his influential characterisation of 'the economy as an

instituted process'. As this phrase suggests, Polanyi views institutions as constitutive of the economy:

> The instituting of the economic process vests that process with unity and stability; it produces a structure with a definite function in society; it shifts the place of the process in society, thus adding significance to its history; it centres interest on values, motives and policy.
>
> (Polanyi [1959] 1992: 34)

Institutions are both economic and non-economic with the inclusion of the latter regarded as vital since religion or government, for example, may be as important in underpinning the operation of the economy as the monetary system or the development of new labour-saving technologies. From this perspective, then, the research agenda is one of examining the manner in which "the economic process is instituted at different times and places" (ibid.).

to generate the trust which underpins legal and contractual relationships, and the direct intervention of the state in managing the economy and in running systems of social welfare.

Key forms of **institution** at the national level include firms, markets, the monetary system, business organisations, the state and a wide range of state agencies and trade unions. These are not merely organisational structures; they also tend to incorporate and embody specific practices, strategies and values which evolve over time. Economic geographers are particularly interested in how urban and regional economies are shaped by distinctive institutional arrangements (Martin 2000). Important forms of institutions in this respect include local authorities and development agencies, employers' organisations, business associations and chambers of commerce, local political groupings, trade union branches and voluntary agencies.

According to Amin and Thrift (1994) institutional 'thickness' or density is an important factor shaping local economic success, referring to the capacity of different organisations and interests to work together in the pursuit of a common agenda. The **industrial districts** of central and north-eastern Italy provide a good example of such institutional 'thickness' (Figure 1.7). Here, the strongly communitarian local political cultures (socialist in Tuscany and Emilia-Romagna, Catholic in Veneto) underpinned a process of renewed economic growth in the 1970s and 1980s through the roles of political parties, trade unions, craft associations, local authorities, chambers of commerce and cooperatives of small entrepreneurs in providing support and a range of services to small firms (Amin 2000). This created a sophisticated reservoir of knowledge and skills and generated high levels of trust between firms, facilitating collaborative forms of innovation.

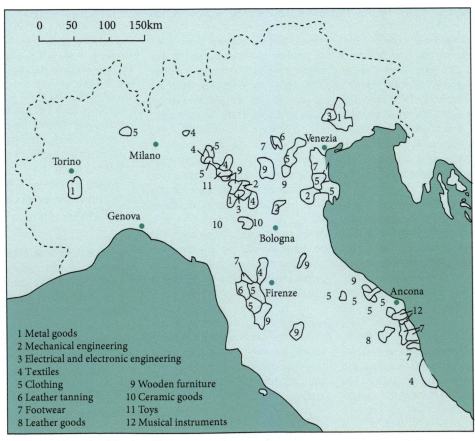

1 Metal goods
2 Mechanical engineering
3 Electrical and electronic engineering
4 Textiles
5 Clothing
6 Leather tanning
7 Footwear
8 Leather goods
9 Wooden furniture
10 Ceramic goods
11 Toys
12 Musical instruments

Figure 1.7 Italian industrial districts
Source: Amin 2000: 155.

Box 1.7

The construction of housing markets in the UK

Markets have traditionally been taken for granted in economic analysis, but this has changed as social researchers (including sociologists, anthropologists and geographers) have challenged conventional approaches. This alternative approach stresses how markets are actively made, not given, by economic actors who construct and perform them in various ways. This process has been examined in a fascinating study of housing markets in the Scottish city of Edinburgh by Susan Smith, Moira Munro and Hazel Christie (2006). The study focused on key 'market intermediaries' – solicitors, estate agents, surveyors and property developers – who shape the flow of information between buyers and sellers. It was conducted in a period of sustained economic growth prior to the financial crash of 2008. As such, the UK housing market was booming with Edinburgh experiencing the second steepest rise in prices after the south-east of England.

One of the key themes of the research was how housing professionals constructed the housing market as a separate and self-contained economic object which was invested with a life of its own (described, variously, as 'hot', 'active', 'exciting', 'amazing') (Smith *et al.* 2006: 86). They performed housing markets through a process of rational economic calculation as prices were set according to their detailed knowledge of market forces. Yet, while market professionals characterised housing markets according to established economic models, this framing became increasingly incompatible with the actual performance of markets during the boom. Increasingly, asking prices and property values became detached from actual sale prices as some prospective buyers were prepared to pay up to 25 to 30 per cent over valuation in order to secure a property. This created great uncertainty, disrupting the smooth operation of the market as decision making became confusing and idiosyncratic. Interestingly, Smith *et al.* (2006) suggest that professionals' detachment from the market may itself be a contributory factor to price instability. In particular, by acting as if they are powerless to influence the market, informed by the powerful economic vision of self-regulating markets, these professionals may actually be helping to place the housing market beyond control (ibid.: 92). As this indicates, "far from *being* the economy, markets have to be made 'economic' through a complex interplay of cultural, legal, political and institutional arrangements" (ibid.: 95; original emphasis).

1.4.3 History and evolution

In contrast to mainstream economics which adopts a timeless purview of universal economic forces, political economy is a historically sensitive approach which stresses that the economy evolves over time. As we have already seen, capitalism is a historically specific mode of production that originated in early modern Europe before spreading to encompass much of the globe. It is a dynamic and unstable economic system which gives rise to periods of both growth and stagnation with booms often followed by crises as markets become overheated and profits crash. Over the past couple of decades, for example, the sustained growth of the late 1990s and early 2000s gave way to a global recession triggered by a financial crisis which originated in housing markets and banking, followed by a slow and geographically uneven recovery in more recent years.

Our GPE approach is concerned with historical processes of economic change and development that have been marginalised by mainstream economics. The key concept here is 'path dependence', which means that the ways in which economic actors respond to wider processes of economic change are shaped and informed by past decisions and experiences. Typically, geographers have 'spatialised' this concept, adapting it to explore patterns of urban and regional growth and decline (section 2.5.3). Our approach emphasises the interaction between pre-existing regional characteristics and wider processes of change. Thus, for example, the San Francisco Bay Area was able to mobilise its existing resources, expertise and entrepreneurial networks to generate and attract new internet and social media companies like Facebook and Twitter in the 2000s as this technology took off, building on its existing concentration of internet and electronics companies.

1.4.4 Power

Having accepted that the economy is structured by social relationships, it is also important to recognise the role of **power** in underpinning these relationships. Ultimately all human social relations are underpinned by power, in the sense of the ability or capacity to take decisions that involve or affect other people (see Allen 2003). Economic relations in this sense are no different. Power percolates through economic relationships at all geographic levels: from that of the household in terms of who makes decisions regarding the domestic budget and who 'goes out to work' and who 'stays at home'; at the level of the firm in terms of the share of the wealth generated in production that accrues to employers rather than employees; in the relationships between firms within particular industries with large retailers and manufacturers often able to dictate prices and terms to their suppliers; and at the international level, in the way that some institutions and governments – the WTO or the United States, for example – have greater power to set the rules of trade than others.

The 'secret life of a banana' (Box 1.5) indicates how the social relationships between different groups of people located in different places are structured by power. At a very basic level, it is clear that some actors in the chain, particularly the supermarkets and the multinational firms who coordinate the production and distribution processes, are in a more powerful position that the small Caribbean farmers or the labourers in the large Central American plantations. Unequal power relations in this sense are central in understanding processes of uneven development, particularly in terms of how they affect different economic actors and places.

Reflect

➤ In what ways do institutions shape the process of economic development?

1.5 Outline of the book

In this first part of the book, entitled 'Foundations', we develop our conceptual approach and highlight some of the main underlying features of capitalism and its geographies. In Chapter 2 we consider the main sets of approaches that have been adopted by economic geographers, focusing particularly on GPE and the cultural, institutional, evolutionary and relational perspectives that have been developed since the 1990s. Chapter 3 provides a broad historical foundation for the remainder of the text, assessing the changing geographies of production from the nineteenth century, covering colonialism and industrialisation, Fordism, flexible accumulation and changing international divisions of labour in the late twentieth and early twenty-first centuries. This is related back to the underlying dynamics of capitalism in terms of competition, the labour process and innovation and technological change.

Informed by these 'Foundations', we turn to explore the geographies of the contemporary economy in Part 2, entitled 'Reshaping the economic landscape: dynamics and outcomes'. Chapter 4 examines the uneven geographies of finance and investment in the context of the broader relations between money, credit and debt. It covers: the changing geographies of money; financial crises and cycles, illustrated by the 2008 crisis, subsequent global recession and recent recovery; and the broader implications of financialisation for everyday life. Chapter 5 focuses on the changing role of the state in the economy, assessing its main economic roles and identifying different types of state. The chapter also considers how the economic role of the state has changed since the 1980s, not least through post-crisis programmes of fiscal austerity. Chapter 6 is concerned with the geography of employment and the changing role of labour within a more integrated world economy. The active role of workers in shaping processes of economic development is stressed in relation to the key shifts and trends that have reshaped employment in recent decades. This is followed by a chapter on development in the global South which identifies the main approaches to development, examines broad patterns of inequality, assesses the effects of the liberalisation of markets and trade on commodity production and livelihoods, and reviews the emergence of post-neoliberal states in Latin America in the late 1990s and 2000s.

The third part of the book is concerned with 'Reworking urban and regional economies', based upon the need to ground the broader themes of the book in specific geographical contexts and issues. Chapter 8 is concerned with the role of transport and ICTs in increasing connectivity between cities. Alongside an assessment of the relationship between transport infrastructure investment and urban and regional development, the chapter emphasises the tendency for key **digital economy** sectors to be concentrated in cities, something that is being reinforced by the urban basis of the '**sharing economy**' and '**smart city**' initiatives. Chapter 9 discusses the organisational geographies of global industries and the relationships between key actors in these industries and regional economies. It focuses particularly on the **global production network (GPN)** approach, processes of '**strategic coupling**' between GPN actors and regional economic actors and broader debates around the attraction of inward investment as a development strategy. Chapter 10 addresses questions of urban **agglomeration**, **innovation** and creativity, emphasising the increased importance of knowledge, skills and creativity in economic development. It discusses the urban economics approach to urban agglomeration, territorial innovation models and Richard Florida's creative cities approach.

The fourth part of the book, entitled 'Reordering economic life', is concerned with changes in the nature and direction of economic relations. Chapter 11 examines debates around consumption and retail, covering the changing geographies of consumption, the emergence of new consumption practices and technologies, and the restructuring and **internationalisation of retail**. Chapter 12 focuses on economic geography and the environment. It reviews work on **sustainability transitions**, addresses questions of energy transition and the **green economy** and outlines the emerging geographies of the renewable energy sector. Chapter 13 is concerned with alternative economic geographies which extend beyond and challenge the capitalist mainstream, covering the counter-globalisation movement, alternative local initiatives and broader geographies of responsibility and justice, as expressed in the notion of fair trade, for instance.

The final part of the book, 'Prospects', consists of a brief conclusion which summarises the main themes of the book, and offers some brief reflections on the key developments that are likely to shape patterns of urban and regional development over the next few years.

Exercise

Think of a commodity that you have recently consumed. This could be something you ate for lunch or breakfast or an item of clothing that you have recently bought. A jar of coffee or a pair of training shoes could be an example.

When you purchased these, were you primarily concerned with the price and physical qualities of these goods? Was there anything, a label, to indicate the geographical origin of this good? Why would it have been produced in that particular region or country? Under what conditions do you think it would have been produced, for example, in a factory, by craft workers, on a farm? What main actors would have been involved in its production, TNCs, small firms, farmers, etc? How might the profits be distributed among these main actors?

Key reading

Daniels, P. and Jones, A. (2012) Geographies of the economy. In Daniels, P., Bradshaw, M., Shaw, D. and Sidaway, J. (eds) *An Introduction to Human Geography 4th edition.* Harlow: Pearson, pp. 292–313.
An accessible and well-illustrated introduction to the geographies of the economy, particularly in relation to the rise of the global economy, processes of uneven development and the changing role of places and localities. Also summarises changes in economic geography and the nature of a geographical approach to the economy.

Dicken, P. (2015) *Global Shift: Mapping the Changing Contours of the World Economy*, 7th edition. London: Sage, pp. 1–9, 74–103.
Quite simply the key text on the geography of economic globalisation. Chapter 1 provides a splendidly concise and up-to-date assessment of the latest debates on economic globalisation following the post-2008 recession while chapter 4 offers an excellent account of the role of 'space-shrinking technologies' in facilitating globalisation, based upon successive advances in transport and communications technologies.

Taylor, P.J., Watts, M. and Johnston, R.J. (2002) Geography/globalisation. In Johnston, R.J., Taylor, P. and Watts, M. (eds) *Geographies of Global Change: Remapping The World*, 2nd edition. Oxford: Blackwell, pp. 1–17, also 21–8.
A stimulating introduction to the relationships between geography as a discipline and the issue of globalisation. Outlines

the rise of globalisation as a key topic of interest, discusses the uneven geography of globalisation and highlights recent political debates about the nature of globalisation.

Watts, M. (2014) Commodities. In Cloke, P., Crang, P. and Goodwin, M. (eds) *Introducing Human Geographies*, 3rd edition. London: Arnold, pp. 391–412.
A review of the commodity as a key topic of interest to geographers. Highlights the economic importance of the commodity within capitalism and the role of commodities in linking together production in different places through commodity chains and networks.

Useful websites

www.polity.co.uk/global/
Provides a very accessible and comprehensive introduction to key aspects of globalisation. The companion website for the *Global Transformations* textbook, edited by David Held, Anthony McGrew, David Goldplatt and Jonathan Perraton (2nd edition, 2004), Polity Press Limited.

www.exchange-values.org/
The website of a social sculptures project by Shelley Sacks in collaboration with the banana growers of the Windward Islands and a range of representative organisations. Contains a range of useful articles under the 'debates and discussion' link. The articles by the geographers Ian Cook and Luke Desforges are particularly recommended along with the ones on 'beyond banana wars', 'banana lives' and 'beyond unfair trade'.

References

Allen, J. (2003) *Lost Geographies of Power*. Oxford: Blackwell.

Amin, A. (2000) Industrial districts. In Sheppard, E. and Barnes, T.J. (eds) *A Companion to Economic Geography*. Oxford: Blackwell, pp. 149–68.

Amin, A. and Thrift, N. (1994) Living in the global. In Amin, A. and Thrift, N. (eds) *Globalisation, Institutions and Regional Development in Europe*. Oxford: Oxford University Press, pp. 1–22.

Anderlini, J. (2015) China's great migration. *Financial Times*, 30 April.

Arnott, R. and Wrigley, N. (2001) Editorial. *Journal of Economic Geography* 1: 1–4.

Barnes, T.J. (2000) Political economy. In Johnston, R.J., Gregory, D., Pratt, G. and Watts, M. (eds) *The Dictionary of Human Geography*, 4th edition. Oxford: Blackwell, pp. 593–4.

Butler, S. (2014) Banana price war requires government intervention, says Fairtrade Foundation. *The Guardian*, 23 February.

Castree, N., Coe, N., Ward, K. and Samers, M. (2004) *Spaces of Work: Global Capitalism and Geographies of Labour*. London: Sage.

Chen, W.H., Myles, J. and Picot, G. (2012) Why have poorer neighbourhoods stagnated economically while the richer have flourished? Neighbourhood income inequality in Canadian cities. *Urban Studies* 49: 877–96.

Coe, N.M. and Jones, A. (2010) Introduction: the shifting geographies of the UK economy? In Coe, N.M. and Jones, A. (eds) *The Economic Geography of the UK*. London: Sage, pp. 3–11.

Coe, N.M., Kelly, P.F. and Yeung, H.W. (2013) *Economic Geography: A Contemporary Introduction*, 2nd edition. Oxford: Wiley Blackwell.

Corbyn, Z. (2014) San Francisco 2.0. *Observer Magazine*, 23 February, pp. 17–27.

Cresswell, T. (2013) *Place: A Short Introduction*, 3rd edition. Oxford: Blackwell.

Dicken, P. (2003) *Global Shift: Reshaping the Global Economic Map in the 21st Century*, fourth edition. London: Sage.

Dicken, P. (2015) *Global Shift: Mapping the Changing Contours of the World Economy*, 7th edition. London: Sage.

Epstein, G. (2005) *Financialisation and the World Economy*. London: Edward Elgar.

Fairtrade Foundation (2009) *Unpeeling the Banana Trade*. Briefing Paper. London: Fairtrade Foundation.

Gibson-Graham, J.K. (1996) *The End of Capitalism (As We Knew It): A Feminist Critique of Political Economy*. Oxford: Blackwell.

Harvey, D. (1989) *The Condition of Postmodernity*. Oxford: Blackwell.

Hudson, R. (2005) *Economic Geographies: Circuits, Flows and Spaces*. London: Sage.

Jones, A. (2010) *Globalisation: Key Thinkers*. Cambridge: Polity.

Knight, J. (2013) Inequality in China: an overview. Policy Research Working Paper 6482. Washington, DC: World Bank.

Lawrence, F. (2009) The banana war's collateral damage is many miles away. *The Guardian*, 13 October.

Martin, R. (2000) Institutionalist approaches to economic geography. In Sheppard, E. and Barnes, T. (eds) *Companion to Economic Geography*. Oxford: Blackwell, pp. 77–97.

Massey, D. (1984) *Spatial Divisions of Labour: Social Structures and the Geography of Production*. London: Macmillan.

Massey, D. (1994) A global sense of place. In Massey, D. (ed) *Place, Space and Gender*. Cambridge: Polity, pp. 146–56.

Peck, J. (1996) *Work-Place: The Social Regulation of Labour Markets*. New York: Guildford.

Polanyi, K. (1944) *The Great Transformation: The Political and Economic Origins of Our Time*. Boston, MA: Beacon Press.

Polanyi, K. [1959] (1992) The economy as an instituted process. In Granovetter, M. and Swelberg, R. (eds) *The Sociology of Economic Life*. Boulder, CO: Westview Press, pp. 29–51.

Routledge, P. and Cumbers, A. (2009) *Global Justice Networks*. Manchester: Manchester University Press.

Schafran, A. (2013) Origins of an urban crisis: the restructuring of the San Francisco Bay Area and the geography of foreclosure. *International Journal of Urban and Regional Research* 37: 663–88.

Sheppard, E. (2011) Geographical political economy. *Journal of Economic Geography* 11: 319–31.

Smith, S.J., Munro, M. and Christie, H. (2006) Performing (Housing) Markets. *Urban Studies* 43: 81–98.

Vidal, J. (1999) Secret life of a banana. *The Guardian*, 10 November.

Walker, R. and Schafran, A. (2015) The strange case of the Bay Area. *Environment & Planning A* 47; published online 16 October 2014. Doi: 10.1068/a46277.

Watts, M. (2014) Commodities. In Cloke, P., Crang, P. and Goodwin, M. (eds) *Introducing Human Geographies*, 3rd edition. London: Arnold, pp. 391–412.

Wei, Y.D. (1999) Regional inequality in China. *Progress in Human Geography* 23: 49–59.

World Bank (2009) *World Development Report 2009: Reshaping Economic Geography*. Washington, DC: World Bank.

Yang, C. (2013) From strategic coupling to recoupling and decoupling: restructuring global production networks and regional evolution in China. *European Planning Studies* 21: 1046–63

Yang, C. (2017) The rise of strategic partner firms and the reconfiguration of personal computer networks in China: insights from the emerging laptop cluster in Chongqing. *Geoforum* 84: 21–31.

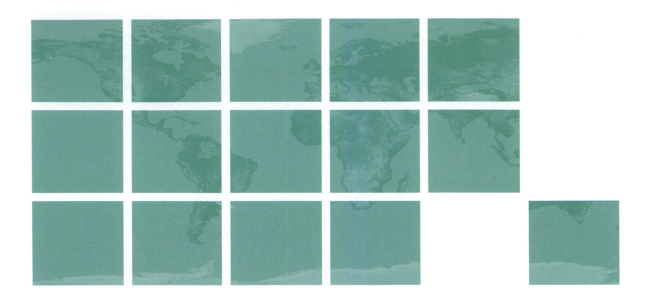

Chapter 2
Approaches to economic geography

2.1 Introduction

A key starting point for this chapter is to recognise that no academic subject has a natural existence. Instead, as Trevor Barnes (2000) argues, subjects must be 'invented' in the sense of being created by people at particular times. Accordingly, the first economic geography course was taught at Cornell University in 1893, the first English-language textbook, George G. Chisholm's *Handbook of Commercial Geography*, was published in 1889 and the journal *Economic Geography* established in 1925. The neighbouring discipline of economics was also established in the late nineteenth century, along with a number of other social sciences. From the start, however, the two disciplines assumed different characteristics.

Economics views the economy as governed by market forces which basically operate in the same fashion everywhere, irrespective of time and space. The market is comprised of a multitude of buyers and sellers – the forces of demand and supply – who dictate how scarce resources are allocated through their decisions about what to produce and consume. Mainstream neoclassical economics is underpinned by the idea of 'economic "man"', assuming that people act in a rational and self-interested manner, continually weighing up alternatives on the basis of cost and benefits, almost like calculating machines. The market is viewed as an essentially self-regulating mechanism, tending towards a state of equilibrium or balance through the role of the price mechanism in mediating between the forces of demand and supply (Figure 2.1).

Whilst economics developed as a theoretical discipline adopting the methods of natural sciences like physics and chemistry, economic geography established

itself as a strongly factual and practical enterprise (see section 2.2):

> As a discipline it [EG] grew less out of concerns by economists to generalise and theorise, than the concerns of geographers to describe and explain the individual economics of different places, and their connections one to another.
>
> (Barnes and Sheppard 2000: 2–3)

The practical character of economic geography was established by **commercial geography** which was highly prominent from the 1880s to the 1930s. This was based on the 'great geographical fact' that different parts of the world yield different products, underpinning the system of global commerce (Barnes 2000: 15). The development of commercial geography was crucial in establishing economic geography as a distinct sub-discipline, helping to define many of its enduring characteristics such as an avoidance of theory, an emphasis on factual detail, a celebration of numbers and a reliance on geographical categories made visible by the map (ibid.: 16). From the 1930s, however, the focus of economic geography "shifted from the general commercial relations of a global system to the geography of narrowly bounded, unique regions, especially those close to home" (ibid.: 18). This was the era of **regional geography**, defined by Hartshorne (1939) as a project of 'areal differentiation' which describes and interprets the variable character of the earth's surface, expressed through the identification of distinct regions.

More generally, a clear contrast can be drawn between the formal and theoretical approach of mainstream neo-classical economics and geography's more open-ended ethos and more substantive concerns. Whilst geography can be seen as synthetic in nature, focusing on the relationships between, rather than the separation of, processes and things, mainstream economics is analytic, seeking to separate the economy from its social and cultural context (Table 2.1). Key features of each of the first three approaches to economic geography examined in this chapter are set out in Table 2.2 (the four new approaches are presented later), providing an important backdrop to the ensuing discussion.

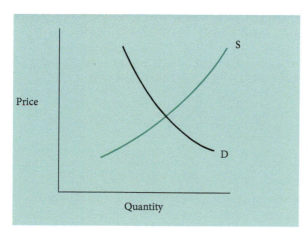

Figure 2.1 Demand and supply curves
Source: Lee 2002: 337.

Table 2.1 Economics and economic geography

	Mainstream neoclassical economics	Economic geography
Nature of discipline	Analytic, separate economy from context and into particular components	Synthetic, concerned with relations between processes and things
Definition of the economy	Autonomous sphere with its own rules and 'laws'	Grounded in wider social, political and cultural relations
Foundations of economic processes	Rational actions of self-interested individuals	Individuals and groups acting in particular geographical contexts. Emphasis depends upon particular approach adopted (Chapter 2)
Substantive concerns	Workings of the economy and its main components	The spatial distribution of economic phenomena
Importance of geography	Ignored, universal economic forces operate in the same way everywhere	Fundamental, purpose of discipline
View of uneven development	A temporary phenomenon which will be eliminated by market forces	Tends to be persistent and deep-rooted

Table 2.2 Approaches to economic geography I

	Spatial analysis	New economic geography (NEG)	Geographical political economy (GPE)
Supporting philosophy	Positivism (Box 2.1)	Implicit positivism	Dialectical materialism[a]
Main source of ideas	Neoclassical economics	Neoclassical economics, revised to accommodate imperfect competition and increasing returns	Marxist economics, sociology and history
Conception of the economy	Driven by rational choices of individual actors	Driven by rational choices of individual actors	Structured by social relations of production. Driven by search for profit and competition
Geographical orientation (place or wider processes)	Wider forms of spatial organisation	Wider forces shaping economic landscape	Wider processes of capitalist development
Geographical focus	Urban regions in North America, Britain and Germany	Urban regions in North America and Europe	Major cities in industrial regions in Europe and North America. Cities and regions in global South
Key research topics	Industrial location; urban settlement systems; spatial diffusion of technologies; land use patterns	Industrial location, agglomeration, urbanisation	Urbanisation processes; industrial restructuring in developed countries; global inequalities and underdevelopment
Research methods	Quantitative analysis based on survey results and secondary data	Mathematical modelling	Reinterpretation of secondary data according to political-economic categories. Interviews and surveys

[a] Dialectical means that social change is seen in terms of a struggle between opposing forces (Box 2.2). Materialism stresses the real social and economic conditions of existence (production, labour, class relations, technology, resources) over ideas and culture, effectively privileging matter over mind.

2.2 Spatial analysis

By the mid-1950s, considerable dissatisfaction was being directed towards regional geography. A new generation of researchers increasingly came to reject the idea that regional synthesis was the proper goal of geography, seeking to develop a more scientific approach. The attack on the traditional establishment found particularly cogent expression in a 1953 paper by Fred K. Schaefer, who called for geographers to employ scientific methods in a search for general theories and laws of location and spatial organisation (Scott 2000). This argument drew directly on the established **positivist** philosophy of science (Box 2.1).

Schaefer's philosophical arguments fitted with a new style of practical research being developed at the Universities of Iowa and Washington, Seattle, where younger geographers were using statistical and mathematical methods to analyse problems of industrial location, distance and movement (Barnes 2000). A vibrant **spatial analysis** research programme developed at Seattle, for example, focusing on issues of industrial location and land use patterns, urbanisation and central place theory, transport networks and the geographical dynamics of trade and social interaction.

The new scientific geography spread to other departments in the US whilst the Universities of Cambridge and Bristol became key centres in the UK in the 1960s. Progress was such that Burton (1963: 151) could proclaim the 'quantitative revolution' complete, defined "as a radical transformation of the spirit and purpose of geography". Economic geography was at the forefront of this movement, viewed as an area that was particularly suited to the application of quantitative methods.

Real world conditions were increasingly favourable to this new approach in the late 1950s and 1960s as policy-makers focused on economic and urban problems in developed countries, providing funds for research and a demand for academic analysis and advice. A period of sustained economic growth and an underlying faith in science and technology created an optimistic 'can do' attitude with urban and regional planning embraced as the means of addressing problems of location, land use management and transportation. In this context, regional geography appeared increasingly backward and anachronistic, with its traditionalist focus on rural areas and its concern with description and classification offering little of practical value to the planner or developer.

Box 2.1

Positivism

This is a philosophy of science originally associated with the French philosopher and sociologist Auguste Comte (1798–1857) and developed further by the Vienna Circle of thinkers in the 1920s and 1930s. It gained some acceptance as an account of the goals of natural science in the post-war period, although specific aspects of it sparked debate. Since the 1970s particularly, some social scientists (including many human geographers) have rejected positivism.

Positivism holds that a real world exists independent of our knowledge of it. This real world has an underlying order and regularity which science seeks to discover and explain. Facts can be directly observed and analysed in a neutral manner. The separation of fact and value is a central tenet of positivism; personal beliefs and positions should not influence scientific research. The aim of science is to generate explanatory laws which explain and predict events and patterns in the real world. In the

classic deductive method (moving from theory to practical research), scientists formulate hypotheses – formal statements of how a force or relationship is thought to operate in the real world – which are then tested against data collected by the scientist through experiment or measurement. Hypotheses that are supported by initial testing must then be verified or proved correct through objective and replicable procedures. If verification is successful, they gain the status of scientific laws.

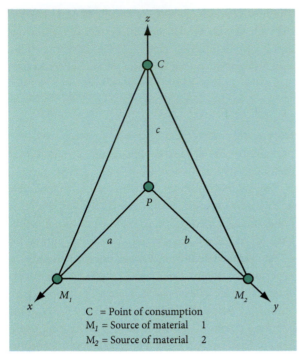

C = Point of consumption
M_1 = Source of material 1
M_2 = Source of material 2

Figure 2.2 Weber's locational triangle
Source: Knox and Agnew 1994: 77.

Neoclassical economic theory provided a ready source of concepts for quantitative economic geography in the 1960s (Box 2.2). Geographers sought to apply the same style of deductive theorising and analysis, moving from simplifying assumptions to the development and testing of hypotheses and models against numerical data representing real world conditions. As one of the pioneers of quantitative economic geography, Harold McCarty of the University of Iowa, put it, "economic geography derives its concepts largely from the field of economics and its method largely from the field of geography" (McCarty 1940, quoted in Barnes 2000: 22).

The tradition of **German location theory** provided a body of economic theory applied to geography that the new economic geography of the 1950s and 1960s could draw upon. The work of theorists such as Von Thunen, Weber, Christaller and Losch was applied to the circumstances of North America and the UK in the 1960s, being used to explain and predict land use patterns, the location of industry and the organisation of settlements and market areas. Weber's theory of industrial location emphasised the importance of transport costs in determining where a factory or plant would be located in relation to the sources of raw materials and the market area, represented in terms of a locational triangle (Figure 2.2). Point P is where the costs of transporting the material to the factory and the finished goods to market are minimised. If the raw materials lost weight during manufacturing, the factory would be drawn towards the material sources. If, on the other hand, distribution costs were higher than the costs of transporting materials, the industry would be drawn towards its market. Weber formulated this model in 1929, a time in which heavy industries based in coalfield regions dominated the economic landscape.

Box 2.2

The neoclassical model of regional convergence

The neoclassical model of regional convergence has been influential in framing both analyses of regional disparities as well as discussion of policy solutions. Like neoclassical economic theory more generally, it presents a simplified model that focuses on the operation of a small number of key variables, based on a number of highly restrictive assumptions. In particular, the model is concerned with the movement of the relevant factors of production, capital and labour (given that land is immobile), between regions. The key simplifying assumptions that it makes include: markets are characterised by perfect competition, constant not increasing returns to scale in production; migration of labour and capital is costless and unrestricted; factor prices (e.g. wages) are perfectly flexible; factors of production are homogeneous rather than comprised of different types and groups; and the owner and bearers of labour and capital are fully informed about factor prices in all regions (Armstrong and Taylor 2000: 141).

Based upon these assumptions, the model predicts that the movement of labour and capital between regions

Box 2.2 (continued)

will iron out regional disparities in output or income and foster convergence in the long run. This occurs because labour and capital move in opposite directions (Pike *et al.* 2017: 63). Initially, labour will flow from low-wage to high-wage regions. Over time, however, high-wage regions will tend to offer lower rates on return to capital (profits). Accordingly, capital will flow out of high-wage regions to low-wage regions where costs are lower and returns are higher. This market adjustment mechanism brings about regional convergence in the long term.

The simplicity and clarity of this model help to explain its influence in informing research and policy. It does identify some of the key forces and trends shaping regional disparities.

The movements of labour and capital are undoubtedly crucial factors in shaping regional difference, whilst there is a discernible tendency for growth and in-migration to generate increasing costs in high-wage regions. Yet the evidence for this bringing about regional convergence is rather mixed. From a historical perspective, a period of strong economic growth and marked regional convergence in Western Europe gave way from the mid-1970s to one of slower growth and greater inequality (Dunford and Perrons 1994). More recently, regional inequalities as measured by GDP per head in the EU fell from 3.8 to 2.5 times higher in the 20 per cent most developed regions compared to the 20 per cent least developed ones between 2000 and 2008, only to rise

again in the period of economic crisis between 2008 and 2011 (European Commission 2014: 3). This suggests that if there is a trend toward regional convergence, it is a slow and discontinuous one.

The disjuncture between the neoclassical model and real world trends can be explained by the unrealistic theories that the model is based on. In practice, migration is not costless, information is not fully available and competition is not perfect. Despite this, the model retains some influence among some economists and policy-makers, informing, for instance, the World Bank's expectations of long-run regional convergence in developing countries after a period of divergence in its *World Development Report 2009* (see Box 2.3).

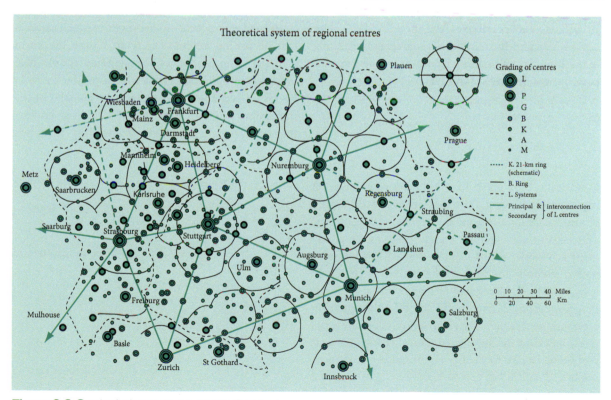

Figure 2.3 Central places in southern Germany
Source: Christaller 1966: 224–5.

Perhaps the best known of the **German locational models** is Christaller's **central place theory**. Based on the assumption of economic rationality and the existence of certain geographical conditions such as uniform population distribution across an area, central place theory offers an account of the size and distribution of settlements within an urban system. The need for shop owners to select central locations produces a hexagonal network of central places, organised into a distinct hierarchy of lower- and higher-order centres (Figure 2.3).

The so-called quantitative revolution transformed the nature of economic geography "from a field-based, craft form of inquiry to a desk-bound technical one in which places were often analysed from afar . . ." (Barnes 2001: 553). Instead of directly observing and mapping regions in the field, economic geographers now tended to use secondary information and statistical methods to analyse patterns of spatial organisation from their desks. While not all practitioners of economic geography adopted the new methods, spatial analysis came to occupy centre-stage with those who refused to follow its approach increasingly relegated to the sidelines (ibid.). By the late 1960s, however, the mood was changing again with a growing number of geographers beginning to question 'the spirit and purpose' of this new quantitative geography (see section 2.4).

2.3 The new economic geography

Historically, economists have not been very interested in geography, viewing it as rather irrelevant to their goal of understanding the general working of the economy. This began to change in the early 1990s as the prominent economist Paul Krugman started to apply economic methods and tools to the analysis of economic geography topics. Krugman termed this new approach the '**new economic geography**' (NEG), although many economic geographers argued that it should more accurately be called the 'new geographical economics' as it represented a new form of geographically oriented economics. The field has grown rapidly as it has attracted many economists to apply mathematical modelling techniques to questions of uneven development,

industrial location and urbanisation. In 2008, Krugman was awarded the Nobel Prize for economics for his work on economic geography and international trade.

The NEG applies the methods of mainstream economics, devising models based on a number of simplifying assumptions, described as "silly but convenient" (Krugman 2000: 51). It retains much of the basic architecture of neoclassical economics, requiring explanations that are based in the rational decisions of individual actors, what economists call microfoundations. For the first time, however, patterns of industrial location could be modelled from these microfoundations. This breakthrough was made possible because economics was now able to incorporate imperfect competition through a model developed by Dixit and Stiglitz in 1977. Rather than assuming perfect competition, this model recognised that some markets can be dominated by a limited number of powerful firms. This is important from an economic geography perspective since this monopoly power creates increasing returns for these firms. Increasing returns essentially refer to economies of scale where additional investment in production capacity generates increased profits. This can again be contrasted with neoclassical economics which assumed that decreasing or constant returns were the norm, equating to no economies of scale (Figure 2.4). Interestingly, the NEG dealt with transportation costs – a key issue in economic geography and regional science – through an 'iceberg' metaphor whereby it is assumed that part of the good simply 'melts away' in transit,

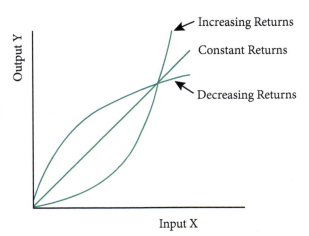

Figure 2.4 Returns to scale
Source: FAO 2010: 10.

simplifying the model by avoiding the need to include transport as an additional industry.

NEG models assess the tension between the forces in the economic landscape that encourage geographical concentration in one or more regions and those favouring the geographical dispersal of production to a large number of locations. Centripetal forces favouring geographical concentration are identified as the effects of market size, a large and specialised labour market and access to information from other firms located there. The main centrifugal forces encouraging dispersal, conversely, are immobile factors of production such as land and, to a considerable extent, labour, and the costs of concentration such as congestion. Much actual research has been devoted to assessing the interaction between these forces under different conditions. Small changes in, for example, transport costs or technology can 'tip' the economy from a pattern of dispersal to one of concentration. The typical outcome of concentration processes will be a simple core-periphery pattern whilst dispersal will create a number of specialised and relatively evenly sized centres of industry. For Krugman and many other proponents of the NEG, this provided a plausible explanation of real world patterns such as the development of the US manufacturing belt in the north-east and Midwest in the nineteenth century (Figure 3.10) and the emergence of a clear core-periphery pattern in China in recent decades (Box 1.3).

The NEG has become increasingly influential with policy-makers and at both the international and national scales, as demonstrated by its wholesale adoption in the World Bank's *World Development Report 2009* (Box 2.3). In part, this reflects the ability of its mathematical, model-based approaches to address 'what if' questions in the sense of how economic outcomes might be changed by altering underlying parameters or decisions, offering potential guidance to policy-makers (Krugman 2011). At the same time, the deductive approach of the NEG provides a rather narrow account of uneven development, concentrating only on a small number of variables that can be measured and modelled. In particular, it concentrates on the most tangible location factors such as economies of scale, transport costs and market size, ignoring the more invisible or intangible sources of geographical concentration, particularly information spillovers, learning and informal relations between firms. This is highly problematic since the latter have become probably the most important sources of spatial concentration and uneven development in the world today as transport costs have fallen and international connections have growth (Storper 2011).

The new economic geography, therefore, focuses on the same basic questions which have long interested economic geographers, but adopts a distinctive approach based on the methods of economics. For many economic geographers, it is simply a more sophisticated

Box 2.3

Reshaping economic geography: the *World Development Report 2009*

In the *World Development Report* (WDR) 2009, the World Bank used the NEG both to frame its analysis of development conditions and trends and identify broad policy directions. It was structured by a 3D perspective of density, distance and division (World Bank 2009). Density refers to the concentration of people and firms in space, best represented by the growth of cities. Distance is related to the distance between the geographical areas in which economic activity is concentrated and the areas that are lagging behind. Division is relevant in terms of the effects of economic borders and divisions in relation to national markets, distinct regulations and separate currencies. While density is most prevalent at the local scale, distance is associated with the (intra) national scale and division with the international.

The 'main message' of the Report is that economic growth will inevitably be unbalanced and that this should be accepted since any attempt to spread out economic activity (for example through regional policies) will discourage it (ibid.: xvi). Three principal drivers of balanced growth are discussed in the second part of the report – the market forces of agglomeration, migration, and specialisation and trade. The WDR 2009

Box 2.3 (continued)

argues that unbalanced growth is still compatible with inclusive development through economic integration to ensure that people who are initially located in places distant from economic opportunity will benefit. This is dependent on the promotion of three main sets of policies: urbanisation which should be encouraged by governments through the provision of effective **institutions**, connected infrastructure and some limited targeted interventions to improve poor neighbourhoods and slums; territorial development that is designed to integrate all areas, particularly by encouraging the mobility of people from lagging to growing areas; and

regional integration between countries (for example Western and Eastern Europe, South America, southern Africa) to increase market size and foster specialisation.

Despite its geographical focus and the wealth of interesting data and analysis it contained, the WDR 2009 was not universally welcomed by geographers. The narrow and definitive picture of development presented in the WDR 2009 echoes the abstract 'bird's eye' view offered by the NEG, isolating a few key forces without any appreciation of geographical complexity and context (Storper 2011). The social and environmental origins and effects of poverty are

omitted entirely from the report while the role of institutions and policy is simply to facilitate the role of the market forces promoting economic integration. Characteristically, the economistic vision of the WDR 2009 ignores the more critical work of geographers and other social scientists (Rigg *et al.* 2009). As such, it is testament to both the power and the limitations of the NEG as a research programme in economic geography, offering a simplified and compelling vision of economic development, but one that fails to learn the broader lessons of development from the European and North American experience (Storper 2011).

version of the spatial analysis popular in the 1960s, sharing its underlying limitations and generating "a dull sense of déjà vu" (Martin 1999: 70). As indicated above, a particular weakness is the characteristic tendency to "focus on what is easier to model rather than on what is probably most important in practice" (Krugman 2000: 59). At the same time, whilst this approach can never hope to capture the complexity and richness of the real economic landscape, its analytical clarity, use of general models and sense of purpose does perhaps carry some lessons for economic geography proper (Storper 2011).

Reflect

Do you think that economic geographers should adopt similar methods and perspectives to economists or should they seek to differentiate themselves?

2.4 Geographical political economy

Political economy should be seen as a broad framework of analysis, encompassing a range of strands and traditions, rather than as a single theory or concept. Its

origins lie in the classical political economy of Adam Smith and David Ricardo in the late eighteenth and nineteenth centuries, becoming most associated with Marx's radical critique and reformulation of their work (Box 2.4). **Geographical political economy** (GPE) emerged in the 1970s as geographers sought to apply the insights of Marxist political economy to questions of urban and regional restructuring and change. It provides a broad and powerful approach to the economy that views capitalism as just one form of economic organisation (albeit a dominant one), sees geography as actively produced through the development and restructuring of the economy, and emphasises the need to link economic relations to parallel bio-physical, social and cultural processes (Sheppard 2011).

2.4.1 The origins of GPE

In the late 1960s and early 1970s, quantitative economic geography was subject to criticism for its lack of social relevance and concern (section 2.2). To a new generation of geographers, it seemed as if the discipline was narrowly focused on technical issues of urban and regional planning to the neglect of deeper questions about how society was organised. The key issues here were racial divisions in US cities, the Vietnam War (symbolising the imperialism of US foreign policy), gender inequalities and the rediscovery of poverty in inner city ghettos. Geography remained largely silent about such questions, leading Harvey (1973) to call for a revolution in geographic thought:

> The quantitative revolution has run its course, and diminishing marginal returns are apparently setting in . . . There is an ecological problem, an urban problem, an international trade problem, and yet we seem incapable of saying anything of depth or profundity about any of them.

(Harvey 1973: 128–9)

In response to these pressing social issues, a group of geographers in the US particularly sought to fashion a new radical geography. This movement began at Clark University in Massachusetts, where a group of postgraduate students, led by Richard Peet, launched *Antipode: A Radical Journal of Geography* in 1969.

This radical new geography turned to political economy for its intellectual foundations. The work of Marx in particular provided a framework for the critical analysis of advanced capitalism. Once again, economic geography was at the forefront of these developments. In general, **Marxism** emphasises processes and relationships rather than fixed things, adopting a dialectical perspective which sees change as driven by the tensions between opposing forces, usually in the form of thesis – antithesis – synthesis (Figure 2.5). Capitalist society in particular is

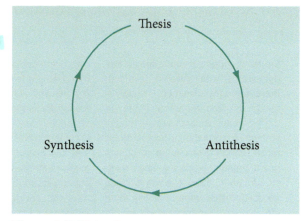

Figure 2.5 Dialectics
Source: Johnston and Sidaway 2004: 231.

characterised by continual change and flux, driven by the search for profits in the face of competition:

> The bourgeoisie [capitalists] cannot exist without constantly revolutionising the instruments of production, and thereby the relations of production, and with them the whole relations of society . . . Constant revolutionising of production, uninterrupted disturbance of all social conditions, everlasting uncertainty and agitation distinguish the bourgeois epoch from all earlier ones. All fixed, fast-frozen relations . . . are swept away, all new formed ones become antiquated before they can ossify. All that is solid melts into air, all that is holy is profaned . . .

(Marx and Engels [1848] 1967: 83)

Box 2.4

The political economy tradition and the origins of a Marxist approach

It is important to set Marx's contribution to political economy within its historical context. Born in 1817, Marx's early academic career was as a student of German philosophy, influenced by Hegel's dialectical approach, in which human relations are conceived in terms of the struggle between opposing forces (Figure 2.5). During the 1840s, however, he became increasingly concerned with economic issues, fascinated by the rise of industrial and urban society. His big theoretical project, culminating in the three volumes of *Capital*, became one of revising the classical political economy of Smith and Ricardo to understand the dramatic changes that were sweeping society in the mid-nineteenth century (Dowd 2000).

Smith, whose great work *The Wealth of Nations* was published in 1776 and Ricardo, whose *Principles of Political Economy and Taxation* appeared in 1817, were great advocates of the market and free trade.

Box 2.4 (continued)

Smith's guiding philosophy was that greater wealth would accrue to the community of nations if governments desisted from intervening in the economy. Ricardo argued that a system of international trade without tariffs (import taxes) would encourage much greater efficiency since each country would specialise and produce goods that they were relatively more efficient at, therefore benefiting the broader commonwealth of nations (Box 3.3).

Both, however, were writing during the very early stages of industrial capitalism when feudalism allied to mercantilism – national protectionism – were viewed as impediments to social progress. For Smith, the development of market relations and competition would produce a more progressive and equal society than feudalism, liberating the individual from traditional social bonds. In this sense, he was very much in tune with the

eighteenth-century enlightenment which sought to overthrow the *ancien regime* and which influenced the French Revolution and American War of Independence. Whilst Marx, writing over half a century after Smith, also recognised that capitalism was more progressive than feudal society, unlike Smith he was able to witness the negative consequences of industrial capitalism. The growth of the factory system, mass urbanisation and the development of a large industrial working class, many living in conditions of appalling poverty and squalor, had transformed Smith's capitalist utopia into William Blake's 'dark satanic mills'.

According to Marx, class struggle is the basic driving force of history, from the relations of master and slave, to feudal lord and serf to capitalist and labourer. Under the capitalist mode of production, class relations are structured by the private ownership

of capital (money, factories and equipment), creating a class of capitalists and a class of labourers who must sell their labour in exchange for a wage. The difference between the value of what the labourer actually produces and what he/she is paid in wages is retained and accumulated by the capitalist as surplus value or profit, representing the basis for class exploitation. The development of factories under capitalism brings large numbers of workers together in industrial cities, providing them with the means of organising against the system that exploits them. When coupled with the inevitable overproduction that would result from the continual expansion of production, driven by competition, the growing power of the working class would eventually bring down the system and usher in a new socialist era. In this way, capitalism acts as its own gravedigger (Dowd 2000: 87).

From this perspective, particular geographical objects, for example cities or a transport system, exist as an expression of wider relationships and are subject to transformation through the movement of broader forces of change.

2.4.2 The development of Marxist theory

Whilst the writings of Marx provided only a few scattered comments and insights into the geography of capitalism, Marxist geographers such as David Harvey have sought to build on these by developing a distinctively Marxist analysis of geographical change (Box 2.5). From this perspective, the economic landscape is shaped by the conflict-laden relationship between capital and labour, mediated by the state, providing a stark contrast to the harmonious equilibrium state of regional balance posited by neoclassical economic theory (Box 2.2).

The first phase of Marxist geography concentrated on establishing how capitalism produces specific geographical landscapes (Smith 2001). In *The Limits to Capital*, Harvey (1982) identified a central contradiction between the geographical fixity and motion of capital. There is a need, on the one hand, for fixity of capital in one place for a sustained period, creating a built environment of factories, offices, houses, transport infrastructures and communication networks that enables production to take place. Such fixity is countered by the need for capital to remain mobile, on the other hand, enabling firms to respond to changing economic conditions by seeking out more profitable locations (Box 2.5). This may require them to withdraw from existing centres of production in which they have invested heavily. Capital is never completely mobile, but must put down roots in particular places to be effectively deployed. Nonetheless, its relative mobility lends capital an important spatial advantage over labour which is more place-bound.

Harvey argues that capital overcomes the friction of space or distance through the production of space in the form of a built environment that enables production and **consumption** to occur. Indeed, such investment in the built environment can act as a **'spatial fix'** to capitalism's inherent tendency towards over-production by absorbing excess capital, performing an important displacement function. As economic conditions change, however, these infrastructures can themselves become a barrier to further expansion, appearing increasingly obsolete and redundant in the face of more attractive investment opportunities elsewhere. In these circumstances, capital is likely to abandon existing centres of production and establish a new 'spatial fix' involving investment in different regions (Box 2.5). The deindustrialisation of many established centres of production in the 'rustbelts' of North America and Western Europe since the late 1970s and the growth of new industry in 'sunbelt' regions and the newly industrialising countries of East Asia can be understood in this light.

Box 2.5

Neil Smith's 'see-saw' theory of uneven development

One of the most notable contributions to Marxist economic geography is Neil Smith's theory of **uneven development**. Processes of uneven development, according to Smith, are the result of a dialectic of spatial differentiation and equalisation that is central to the logic of capitalism, transforming the complex mosaic of landscapes inherited from pre-capitalist systems. Capital moves to the areas which offer highest profits for investors, resulting in the economic development of these areas. The geographical concentration of production in such locations results in differentiation as they experience rapid development whilst other regions are left behind. As a result, living standards and wage rates vary markedly between regions and, especially, countries. At the same time, the tendency towards equalisation reflects the importance of expanding the market for goods and services, implying a need to develop newly incorporated colonies and territories so as to generate the income to underpin consumption.

Over time, the process of economic development in a particular region tends to undermine its own foundations, leading to higher wages, rising land prices, lower unemployment and the development of trade unions, reducing profit rates. In other regions, underdevelopment leads to low wages, high unemployment and the absence of trade unions, creating a basis for profit that attracts capital investment. Over time, capital will 'see-saw' from developed to underdeveloped areas, 'jumping' between locations in its efforts to maintain profit levels. It is this movement of capital that creates patterns of uneven development. In this sense, "capital is like a plague of locusts. It settles on one place, devours it, then moves on to plague another place" (Smith 1984: 152).

According to Smith, the production of space under capitalism leads to the emergence of three primary geographical scales of economic and political organisation: the urban, the national and the global. The dynamic nature of the uneven development process is most pronounced at the urban scale at which capital is most mobile, resulting in, for example, the rapid gentrification (upgrading through the attraction of investment and new middle-class residents) of previously declining inner city areas like Glasgow's Merchant City (Figure 2.6). Conversely, patterns of uneven development exhibit most stability at the global scale where the divide between developed and developing countries remains as wide as ever, although East Asia has risen to the core of the world economy through sustained economic growth since the 1960s.

Figure 2.6 Urban gentrification in Glasgow
Source: RCAHMS Enterprises: Resource for Urban Design Information (RUDI). Licensor www.scran.ec.uk.

Although Marxist and neoclassical economic theories are polar opposites in many respects, there is some overlap in their accounts of uneven regional development (see Box 2.2). They both emphasise the movement of capital and labour to more developed regions and the tendency for rising costs in such regions to trigger subsequent flows of capital towards less developed regions where costs are lower. While the neoclassical model sees this as resulting in regional convergence in the long term, the Marxist approach views it as part of periodic processes of economic restructuring, generating shifting patterns of uneven regional development.

The second phase of Marxist geography began in the early 1980s, concentrating on developing Marxist analyses of specific situations and circumstances. One key research question was how particular places were affected by wider processes of economic restructuring. In a landmark text, Doreen Massey (1995) investigated the changing location of industry in Britain, developing the concept of **'spatial divisions of labour'**. This emphasised how unequal social relations are played out across space, particularly in terms of how capital seeks to exploit geographical differences in the supply, skills and costs of labour. From this perspective, regional inequality reflects the relationships between regions and the roles that they play in the wider capitalist economy, not just their internal characteristics as in traditional theories. At the same time, local conditions also shape the broader process and relations as geography, in the form of regional differentiation, becomes internal to the workings of the economy as part of firms' investment and decision-making criteria.

The theory of spatial divisions of labour reflects a move away from the traditional model of local firms, where all production functions were locally concentrated in the same area (Figure 2.7). The other two models reflect the spatial stretching of production over time, underpinned by the growth of large corporations and advanced transportation and communication systems. These two types of spatial structure involve the separation of production functions over space, particularly in the division between headquarters activities (investment, supervision and control) and routine work. This was evident in the growth of branch plants in the shape of factories performing specific production tasks on behalf of the wider corporation. The first type is the cloning branch plant, where the whole process of production takes place at each site and only management functions are stretched over space. Second, part process production is defined by the separation of the production process as well as management, such that different functions are carried out in different places. Only one specific part is concentrated in each plant.

Each one of these two types of industrial organisation results in a different pattern of uneven development and regional inequality. Typically, headquarters functions were located in core regions such as south-east England or New York where there were supplies of qualified 'white collar' labour, whilst more skilled production work tended to be concentrated in established manufacturing regions like the West Midlands of England or Midwest of the US, and routine assembly was relocated to peripheral regions such as north-east England or the US south in the 1970s on the basis of their lower-cost supplies of labour. More recently, of course, it has become clear that production is not just reorganised within firms on a national basis, but also between firms on a global basis with many leading electronics firms, for example, concentrating routine production and assembly in developing countries like China (section 3.5).

2.4.3 The regulation approach

Another strand of Marxist-informed research which was very influential in economic geography during the 1990s is the **regulation approach**. Derived from the work of a group of French economists in the 1970s, the regulation approach stresses the important role that wider processes of social regulation play in stabilising and sustaining capitalist development. These wider processes of regulation find expression in specific institutional arrangements which mediate and manage the underlying contradictions of the capitalist system (see Chapter 3), expressed in the form of periodic crises, enabling renewed growth to occur. This occurs through the coming together and consolidation of specific **modes of regulation**, referring to the institutions and conventions which shape the process of capitalist development. Regulation is focused on five key aspects of

1. *The locationally-concentrated spatial structure*

| all administration and control |
| total process of production |

| all administration and control |
| total process of production |

| all administration and control |
| total process of production |

(No intra-firm hierarchies)

2. *The cloning branch-plant spatial structure*

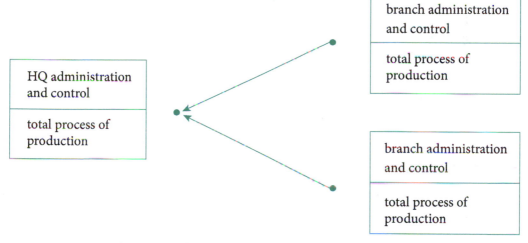

(Hierarchy of relations of ownership and possession only)

3. *The part-process spatial structure*

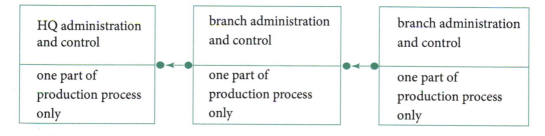

(Plants distinguished and connected by place both in relations of ownership and possession and in the technical division of labour)

Figure 2.7 Examples of possible spatial structures
Source: Massey 1995: 75.

capitalism in particular: labour and the wage relation, forms of competition and business organisation, the monetary system, the state and the international regime (Boyer 1990). When these act in concert, a period of stable growth, known as a **regime of accumulation**, ensues.

Two regimes of accumulation can be identified since the 1940s (Table 2.3). After the global economic depression of the 1930s, a new regime of **Fordism** emerged, named after the American car manufacturer, Henry Ford, who pioneered the introduction of mass production techniques. Fordism was based on a crucial link between mass production and mass consumption, provided by rising wages for workers and increased productivity in the workplace. The state adopted an interventionist approach, informed by the theories of the British economist John Maynard Keynes which stressed the important role of government in managing the overall level of demand in the economy to secure full employment.

Fordism experienced mounting economic problems from the early 1970s, however, and elements of a new 'post-Fordist' regime of accumulation based on flexible production became prominent in the 1980s. Theories of **post-Fordism** emphasise the role of small firms, advanced information and communication technologies and more segmented and individualised forms of consumption (Chapter 11). Whether these features amount to a distinctive and coherent regime of accumulation remains questionable, however, with mass production remaining important in some sectors. One key trend has been the abandonment of Keynesian policies since the 1970s in favour of a **neoliberal** approach which seeks to reduce state intervention in the economy and embrace the free market, stressing the virtues of enterprise, competition and individual self-reliance.

2.4.4 The relevance and value of Marxist political economy

By the late 1980s, Marxist geography was becoming subject to increasing criticism, informed by the emergence of postmodern thought (Box 2.7). Marxism

Table 2.3 Fordist and post-Fordist modes of regulation

	Fordism	Post-Fordism
The wage-relation	Rising wages in exchange for productivity gains. Full recognition of trade union rights. System of national collective bargaining	Flexible labour markets based on individual's position in market. Limited recognition of trade unions
Forms of competition and business organization	Dominance of large corporations. Nationalization of key sectors such as utilities	Key role of small- and medium-sized enterprises (SMEs) alongside large corporations. Privatization of state enterprises and general liberalization of the economy
The monetary system	Monetary policy focused on demand management and maintaining full employment. Use of interest rates to facilitate economic expansion and contraction	Focus on reducing inflation, relying on high interest rates if necessary. International monetary integration and coordination, for example the European single currency
The state	Highly interventionist. Adoption of Keynesian policies of demand management. Provision of social services through welfare state. Goals of full employment and modest income redistribution	Reduce state intervention in the economy. Abandonment of Keynesian policies for neoliberalism. Efforts to reduce welfare expenditure and privatize services
The international regime	Cold War division into capitalist and communist blocs. Bretton Woods system of fixed exchange rates, anchored to the US dollar. Promotion of free trade, monetary stabilization and 'Third World' development	Increased global economic integration. System of floating exchange rates. Renewed emphasis on free trade and openness to foreign investment. Imposition of neoliberal adjustment policies on developing countries

seemed to have also become rather out of touch with the 'new times' of the 1980s, marked by the dominance of neoliberal ideas, particularly in the UK and US. In the realm of left-wing politics too, the focus was shifting from the traditional 'politics of distribution' concerned with work, wages and welfare to a 'politics of identity' promoting the rights of various groups such as women, ethnic minorities and gay people to recognition and justice (Crang 1997). Such claims were channelled through broader social movements rather than the traditional labour movement.

Three main criticisms of Marxism in geography can be identified:

➤ Its apparent neglect of human agency in terms of an impoverished view of individuals and a failure to recognise human autonomy and creativity. Instead, Marxists tend to privilege wider social forces such as class and see people as bearers of class powers and identities, reading off their behaviour from this.

➤ Its emphasis on economic forces and relations. Whilst the Marxist concept of production is, as we have seen, much broader than conventional notions of the economy, Marxists have been criticised for stressing the determining role of economic forces. Culture and ideas are often viewed as products of this economic base.

➤ Its overwhelming emphasis on class, and neglect of other social categories like gender and race. Leading Marxists such as Harvey were attacked by feminist geographers in the early 1990s for their neglect of gender issues, which they were accused of subsuming within a class-based Marxist analysis.

These are important criticisms, although it is questionable whether Marxist geography really was as economically determinist as its critics allege (Hudson 2006). We do not think that they mean that GPE is no longer relevant or useful and should be abandoned (Box 2.6). It remains influential in economic geography, with some of its key arguments around the unstable and crisis-prone character of capitalism as a mode of production, the inseparability of the economy from the social, political and ecological spheres, and the need to relate regional problems to wider processes of uneven development continuing to be accepted by many researchers in the field. In many respects, the relevance of GPE has been reinforced by the post-2008 financial crisis and recession, emphasising the unequal impact of these episodes on particular groups of people and places.

Partly in response to the above criticisms and reflecting its encounters with other contemporary perspectives such as the cultural 'turn', institutionalism, and evolutionary economic geography, GPE has become increasingly diffuse and plural in the 1990s and beyond. This draws attention to the diverse strands of research that GPE continues to inform, often alongside other theoretical influences, making it harder to identify a single school of GPE compared to the 1980s and early 1990s (Jones 2015). These other theories and approaches have helped to raise important new questions about knowledge, identity, evolution and practice, whilst also shedding new light on older economic geography questions such as agglomeration and the sources of urban and regional growth. As such, the approach that we adopt in this book is best characterised as an 'open' and plural version of GPE that is responsive to the evolution of capitalism as its main object of analysis and receptive to the insights offered by other perspectives (Box 2.6).

Box 2.6 (continued)

of ideas over socialism. The spirit of market triumphalism associated with globalisation in the 1990s has, however, subsequently been punctured as the limitations of global capitalism have been highlighted by the anti-globalisation movement and the 2008 financial crisis and subsequent recession.

As part of the more sober climate of the late 1990s and early 2000s, Marx was rediscovered. Triggered by the 150th anniversary of the publication of *The Communist Manifesto* in 1998 and the financial crises then engulfing East Asia and Russia, "impeccably bourgeois magazines" such as the *Financial Times* and *New Yorker* published articles heralding Marx's thought (Smith 2001: 5). Writing in the *New Yorker*, John Cassidy praised Marx as the 'next big thinker', citing his relevance to the workings of the global financial system and stating that his analyses will be worth reading "as long as capitalism endures" (quoted in Rees 1998). Although Wall Street's discovery of Marx was predictably

short-lived, listeners voted Marx the greatest philosopher of all time in a BBC Radio 4 poll in 2005. The late British Marxist historian, Eric Hobsbawm, explained this in terms of the "stunning prediction of the nature and effects of globalisation" found in *The Communist Manifesto* (quoted in Seddon 2005).

We believe that GPE is still relevant because of its value as a holistic framework for understanding the evolution of the global capitalist system. Marx's primary contribution to knowledge was as an analyst of capitalism not as an architect of communism. Whilst he offered only a few scattered comments about geography, this has been rectified by Marxist geographers like Harvey and Smith who have developed theories of uneven development. In the absence of other approaches which can match its historical-geographical reach and analytical purchase, GPE provides the most suitable framework for analysing the 'big questions' concerning the economic geography of global capitalism (Swyngedouw 2000). GPE also

retains a sense of social and political commitment, emphasising issues of inequality and injustice and the need to change the world as well as interpret it.

At the same time, we recognise that GPE does not hold all the answers, containing several limitations and weaknesses. This points to the need for a modified approach which can incorporate insights from other sources, particularly the cultural, institutional and evolutionary perspectives that economic geographers have turned towards over the past couple of decades (section 2.5). In particular, these approaches can provide a stronger sense of agency, recognition of the social and institutional construction of markets, sensitivity towards difference and concern with the evolution of cities and regions. In short, we favour an open and plural GPE which does not claim to have a monopoly on truth, is receptive to insights from other perspectives and which evolves in line with capitalism as its object of analysis.

Reflect

Do you agree that GPE is still relevant to the analysis of the uneven development of the capitalist economy?

2.5 New approaches in economic geography

A new set of approaches have emerged in economic geography since the early 1990s, emphasising the cultural, institutional and evolutionary foundations of

economic processes. These approaches derive theoretical inspiration from various sources, including postmodernism, **post-structuralism**, cultural studies, anthropology, economic sociology and institutional and evolutionary economics. In general, they have served to direct attention towards questions of difference, embeddedness, evolution and practice that were previously marginalised and neglected within economic geography. These 'new' approaches remain distinct from the NEG by emerging within the sub-discipline of economic geography proper rather than economics.

To a considerable extent, these 'new' approaches in economic geography have been defined against the 'old' economic geography of spatial analysis and

Marxian political economy. These perspectives have been variously criticised for ignoring the wider social and cultural practices which shape economic life and for viewing economic actors in essentialist and reductionist ways – a critique that could be extended to the NEG. These new approaches have not always remained separate from work in GPE, however, despite their initial counter-posing, with important areas of overlap and cross-fertilisation developing as indicated above, reflecting the increasingly hybrid and fast-moving nature of the field. They are often characterised as 'turns' in economic geography that involve a reorientation of the field (or, at least, significant parts of it), towards a fresh set of concepts and research questions. Four such 'turns' are discussed in the remainder of this section: cultural economic geography, institutional economic geography, evolutionary economic geography and relational economic geography (Table 2.4).

Table 2.4 Approaches to economic geography II			
	Cultural economic geography	**Institutional economic geography**	**Evolutionary economic geography (EEG)**	**Relational economic geography**
Supporting philosophy	Postmodernism, post-structuralism, feminism and post-colonialism	Institutionalism	Evolutionary approaches in natural and social sciences	Post-structuralism, institutionalism, actor-network approaches
Main source of ideas	Cultural studies, philosophy, literary studies	Institutional economics, economic sociology	Evolutionary economics, biology	Philosophy, economic sociology, science and technology studies
Conception of the economy	Shaped by wider social and cultural relations	Importance of social context. Informal conventions and norms shape economic action	Historically evolving sets of activities and relations; driven by innovation and subject to path dependence	Embedded in wider social and economic relations and grounded in practice
Geographical orientation (place or wider processes)	Emphasis in individual places and sites in context of wider relations	Emphasis on individual places in context of globalisation	Concerned with both the broader evolution of economic landscapes and adaptation of cities and regions	Concerned with particular places and sites in context of wider networks
Geographical focus	Key sites of consumption, global financial centres, workplaces	Growth regions in developed countries	Largely confined to global North; strong European orientation	Diverse, case studies from global North and South
Key research topics	Identity, performance, discourse, industrial cultures	Social and institutional foundations of economic development; territorial embeddedness; geographies of knowledge	Path dependence and lock-in; path branching; the spatial evolution of industries; industrial clusters	Everyday activities through actor-networks and communities of practice; relational cities and regions
Research methods	Interviews, focus groups, textual analysis, ethnography	Case studies (interviews, surveys, documentary analysis)	Varied, covering both quantitative analysis and case studies	Case studies (interviews, surveys, ethnography)

2.5.1 Cultural economic geography

Cultural approaches in economic geography were inspired by the wider **cultural 'turn'** in human geography and the social sciences in the late 1980s and 1990s, creating a 'new cultural geography'. This viewed culture as a process through which individuals and social groups make sense of the world, often defining their identity against 'other' groups regarded as different according to categories such as nationality, race, gender and sexuality (Jackson 1989). Meaning is generated through language which, instead of simply reflecting an underlying reality, actively creates that reality through **discourses** – networks of concepts, statements and practices that produce distinct bodies of knowledge. The cultural turn has been closely tied to the rise of **postmodernist** and **post-structuralist** philosophies which stress the fractured nature of individual identities, the social construction of meaning, the importance of difference and variety and the effects of broader social categories and discourses (Box 2.7).

In response to the broader cultural 'turn', some economic geographers have sought to adapt their interests, approaches and methods, incorporating notions of difference, identity and language into their research. The links between economy and culture are of central importance here with many observers agreeing that the economy has become increasingly cultural in terms of the growing importance of sectors such as entertainment, retail and tourism whilst culture has become increasingly economic, viewed as a set of commodities that can be bought and sold in the market.

The adoption of cultural approaches in economic geography has focused attention on the links between economic action and social and cultural practices in different places. As Wills and Lee (1997: xvii) put it, "the point is to contextualise rather than undermine

---

Box 2.7

Postmodernism

Postmodernist ideas have attracted widespread interest since the 1980s, coming to exert considerable influence in architecture, the humanities and the social sciences. Ley (1994: 466) defines postmodernism as "a movement in philosophy, the arts and social sciences characterised by scepticism towards the grand claims and grand theory of the modern era, and their privileged vantage point, stressing in its place openness to a range of voices in social enquiry, artistic experimentation and political empowerment".

As this quote indicates, pluralism is a key characteristic of postmodern thought in terms of embracing the knowledge claims of different social groups. Grand theories or 'meta-narratives' (big stories) claiming to uncover the changing organisation of society are rejected as a product of the privileged position and authority of the observer rather than being accepted as objective representations of the realities that they purport to explain. Instead of functioning as a set of universal truths, then, knowledge should be regarded as partial and situated in particular places and times. Postmodernists reject conventional notions of scientific rationality and progress, favouring an open interplay of multiple local knowledges.

Rather than assuming that social life has an underlying order and coherence, postmodernists celebrate difference and variety. Difference and variety are held to be a basic characteristic of the world, applied to a range of different phenomena including human groups and cultures, buildings, urban neighbourhoods, texts and artistic products. This basic attention to difference quickly attracted the interest of geographers, reflecting the fact that "the discipline has always . . . displayed a sensitivity to the specific kinds of differences to be found between different (and 'unique') places, districts, regions and countries" (Cloke et al. 1991: 171). For Gregory ([1989] 1996), postmodernism provides an opportunity for geographers to return to the notion of areal differentiation, emphasised by traditional regional geography, but armed with a new 'theoretical sensitivity' derived from work in cultural studies in particular.

the economic, by locating it within the cultural, social and political relations through which it takes on meaning and direction". This has involved a reframing of the economic, demonstrating how economic relations are saturated with issues of identity, meaning and representation. Four main strands of culturally informed research in economic geography can be identified:

➤ Consumption, with studies focusing, for instance, on the creation and experience of particular landscapes of consumption such as shopping malls, supermarkets and heritage parks (see Chapter 11).

➤ Gender, performance and identity in the workplace and labour market. This work focuses on how employees perform particular roles and tasks at work, often informed by cultural and gendered norms. One of the most notable studies on this is Linda McDowell's research on work cultures in merchant banks in the City of London (see Box 2.8).

➤ Research on the importance of personal contact and interpretative skills in financial and business services. This has focused on the cultural practices which underpin communication and interaction, the sites in which this takes place and the consequences for our understanding of financial markets. Research on financial centres such as the City of London, for example, has shown the importance of **social networks** and trust, encouraging geographical concentration through the need for regular face-to-face contact.

➤ Corporate cultures and identities. Research has focused on how managers and workers create distinctive corporate cultures through particular discourses and day-to-day practices. Of particular interest here is Erica Schoenberger's work on large American corporations such as Xerox, DEC and Lockheed, showing the limitations of such cultures in dealing with a turbulent and unpredictable economic environment.

Box 2.8

Capital culture: gender and work in the City

Linda McDowell's research on the construction and performance of gender relations in the City of London represents one notable example of culturally informed economic geography research, linking economic issues to wider social and cultural relations and focusing on questions of discourse, identity and power. McDowell aims to understand how an international financial centre like the City of London actually operates, "viewing it through the lens of the lives and careers of individual men and women working in the City's merchant banks at the end of this period of radical change [the late 1980s and early 1990s] both in the global economy and in the city" (McDowell 1997: 4). In focusing on gender segregation and identity at

work, the research assesses men's experiences in addition to women's. At the same time, the extent to which the City has become more open to women and the career prospects it offers them remain key issues.

Capital Culture illustrates another key feature of culturally oriented economic geography by employing a range of research methods, going beyond the conventional reliance on questionnaires and statistics to employ qualitative methods involving direct fieldwork. McDowell focused on the merchant bank sector, utilising postal questionnaires sent out to all such banks in the City and detailed face-to-face interviews with employees of three.

This multi-method approach meant that information on changing

employment trends and relations within companies could be combined with first-hand accounts of how individual employees have managed and negotiated processes of employment change within the workplace. Through the case study interviews in particular, the research examined people's everyday work experience, assessing how they assumed and performed particular gender roles. The importance of image, bodies and the presentation of self are major themes. A key conclusion is that changing work relations and power structures in the City still favour men as "the cultural construction of the banking world remains elitist and masculinist" (ibid.: 207).

2.5.2 Institutional economic geography

Economic geographers have drawn upon concepts from institutional economics which emphasise the social context of economic life and the role of institutions in shaping and 'embedding' economic action. Here, institutions are broadly defined as the 'rules of the game', incorporating both formal and tangible rules and laws and more informal and intangible conventions and norms. According to one comprehensive definition, institutions incorporate:

> Formal regulations, legislation, and economic systems as well as informal societal norms that regulate the behaviour of economic actors: firms, managers, investors, workers. They govern the workings of labour markets, education and training systems, industrial relations regimes, corporate governance, capital markets, the strength and nature of domestic competition, and associative behaviour. . . . Collectively, they define the system of rules that shape the attitudes, values, and expectations of individual economic actors. Institutions are also responsible for producing and reproducing the conventions, routines, habits, and 'settled habits of thought' that, together with attitudes, values, and expectations, influence actors' economic decisions.
>
> (Gertler 2004: 7–8).

This understanding of institutions as sets of rules, habits and values is informed by the 'old' institutional economics of the late nineteenth and early twentieth centuries, rather than the 'new' neoclassically oriented institutionalism developed by mainstream economists. It maintains that economic life cannot be understood simply by reference to the rational actions of individual actors, paralleling cultural approaches in extending the definition of the economic beyond the scope of conventional models. Indeed, institutions are important because they link 'the economic' and 'the social' through a set of habits, practices and routines which help to structure and stabilise economic activity.

A related set of ideas from economic sociology have also been influential in stressing that economic processes are grounded or 'embedded' in social relations. In contrast to the orthodox economic conception of the economy as a separate domain driven by the rational decisions of individual actors, economic sociologists, following Polanyi, argue that the economy is socially constructed with social norms and institutions playing a key role in shaping economic action (see Box 1.6). Rather than representing an external interference with the 'free' operation of markets, as portrayed by neoclassical economics, institutions serve to enliven, stabilise and regulate economic relations. An important distinction is that between the institutional environment, referring to both the informal conventions and formal rules that enable and control the behaviour of economic actors, and institutional arrangements, referring to particular organisational forms (such as markets, firms, local authorities, trade unions), that reflect and embody this wider environment (Martin 2000).

Characteristically, institutionally inclined economic geographers have 'spatialised' the sociological concept of 'embeddedness', emphasising how particular forms of economic activity are not only grounded in social relations, but also rooted in particular places through the concept of territorial 'embeddedness'. In a study of advanced manufacturing technologies in southern Ontario, for instance, Gertler (1995) shows the importance of 'being there' in facilitating adoption of the technologies by ensuring close links between producers (organisations who actually develop, distribute and sell aspects of the technology) and users (manufacturing firms). A shared 'embeddedness' in a distinctive industrial culture enabled users and producers to develop appropriate training regimes and industrial practices, involving the sharing of information and knowledge. This does not mean, however, that industrial cultures are always defined geographically or that learning and interaction cannot occur over longer distances, utilising information and communication technologies. An awareness of the importance of this issue has fostered an interest in the development and organisation of knowledge within large firms, raising questions about how such global and local knowledges are combined.

Institutionalist ideas encouraged the rise of a 'new regionalism' in economic geography which emphasises the importance of social and cultural conditions within regions in helping to promote or hinder economic growth. In particular, inherited institutional frameworks and routines are held to be of considerable importance in influencing how particular regions respond to the challenges of globalisation. Individual places therefore have attracted renewed attention, as institutional economic geographers have attempted to identify the social and cultural foundations of economic growth and prosperity in successful regions such as 'Silicon Valley' in California, the industrial districts of central and north-eastern Italy and Baden-Wurttemberg. This is associated with a focus on the role of local and regional institutions in mobilising resources for development, particularly Amin and Thrift's (1994) concept of 'institutional thickness'. Four levels of 'institutionalisation' are identified: the number of institutions present; the degree of inter-institutional interaction; the formation of coalitions; and the development of a common agenda between key institutions and actors.

Institutional approaches emphasise the context-specific nature of economic development, stressing the need to avoid 'one size fits all' policy prescriptions that are simply parachuted into a particular city or region from outside. As the economic geographer Andres Rodríguez-Pose (2013: 1042) argues, a well-designed regional economic development strategy can be compared to a bicycle, requiring "two well rounded wheels: a back institutional wheel with efficient formal and informal institutions propelling the bicycle forward and a front development strategy wheel tailor-made to match the institutional environment in which the development interaction takes place" (Figure 2.8). All too frequently, however, development strategies are simply transferred from place to place whilst paying little attention to local institutional conditions. This gives rise to three contrasting scenarios (Figure 2.9). In the first, an imbalanced 'penny farthing' scenario is apparent, whereby a huge strategy front wheel is undermined by a tiny institutional rear wheel (Figure 2.9a). In the second 'square wheels' case, poorly rounded strategies are matched by inadequate institutional conditions, meaning that the process of economic development cannot move forward (Figure 2.9b). In the third, worst-case, scenario, a basic 'bicycle frame' situation indicates a lack of any real strategy and weak local institutions, offering little basis for economic development (Figure 2.9c).

Whilst institutional approaches and the 'new regionalism' have been highly influential in economic geography since the 1980s, they have attracted some criticism. In particular, they have been criticised for remaining rather descriptive and failing to adequately specify the mechanisms which link institutions and territorial embeddedness as conditions for economic development to regional growth and learning as outcomes. While much attention has been devoted to successful growth regions and industrial districts at particular points in time (the 1980s and 1990s), there are examples of both institutionally 'thin' growth regions – for

Figure 2.8 The regional development bicycle

47

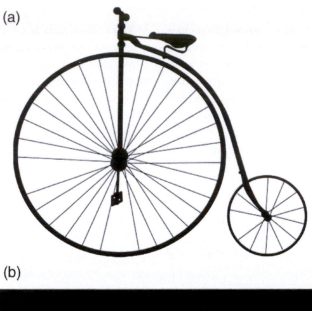

(a)

(b)

(c)

Figure 2.9 Typical mismatch between development strategies and institutions: (a) 'penny farthing' equilibrium; (b) 'square wheels'; and (c) 'bicycle frame' situation
Sources: (a) Graves & Green Engravings, Boston; (b) Michael Vroegop; (c) Rodríguez-Pose, A. (2013).

example 'Motor Sport Valley' in south-east England – and institutionally 'thick' but unsuccessful regions such as the Ruhr in Germany (see Box 2.9) and north-east England. More fundamentally perhaps, the underlying focus on regions as bounded spatial units has been called into question by the 'relational turn' in economic geography which emphasises the importance of the broader networks and relations that link different places in an increasingly globalised world (section 2.5.4).

2.5.3 Evolutionary economic geography

In recent years, **evolutionary economic geography** (**EEG**) has become one of the most vibrant sub-fields of economic geography, focusing attention on how the economic landscape is transformed over time. Informed by concepts from evolutionary economics and biology, EEG is principally concerned with processes of path dependence and lock-in (see below), the clustering of industries in space and the role of innovation and knowledge in shaping economic development. Evolution provides a particular way of thinking about change, inspired by the Darwinian concepts of variety, selection and retention, although not crude notions such as the 'survival of the fittest', or the genetic determination of behaviour.

One source of inspiration for EEG is the famous Austrian economist, Joseph Schumpeter's, definition of capitalism as a process of 'creative destruction'. This is the result of innovation and technological change, driven by the search for profit and wealth creation. It means that new products, technologies, firms, industries and jobs are constantly being added to the economy, while old firms, technologies, products, industries and jobs become uncompetitive and disappear (Boschma and Martin 2007: 537).

In addition, EEG is informed by Nelson and Winter's (1982) evolutionary theory of economic change. Firm routines are central to economic evolution, being regarded as analogous to genes in biological theory. They can be seen as organisational skills, relying on a stock of tacit (practical) knowledge. Routines provide the basis for competition between firms, with the 'fittest' routines being selected through the market and retained and passed on as a firm grows.

Another influence is the economist Paul David's concept of **path dependence**. David defines this as a process of economic development in "which important influences can be exerted by temporally remote events, including happenings dominated by chance elements rather than systemic forces" (David 1985: 332). The process is illustrated by the example of the QWERTY keyboard which emerged as the industry standard following the introduction of the typewriter in the late nineteenth century. The underlying point here is that the decisions of economic actors are shaped and informed by past decisions and experiences. As Walker (2000) puts it:

> One of the most exciting ideas in contemporary economic geography is that industrial history is literally embodied in the present. That is, choices made in the past – technologies embodied in machinery and product design, firm assets gained as patents or specific competencies, or labour skills acquired through learning – influence subsequent choices of methods, design and practices.
>
> (Walker 2000: 126)

The term 'lock-in' is frequently employed to describe how path-dependent economic activities can become overly rigid and trapped as a result of past choices. This has been the focus of one key strand of work in EEG, particularly in the context of old industrial regions (Box 2.9).

Box 2.9

Regional path dependency and lock-in: the case of the Ruhr, Germany

Probably the best-known account of regional path dependence in EEG is Gernot Grabher's work on the coal, iron and steel complex of the Ruhr in the 1970s and 1980s. Located in the north-western *lander* of North Rhine-Westphalia, the Ruhr is the traditional economic heartland of Germany, operating as a polycentric

Box 2.9 (continued)

industrial region of over 5 million people, comprised of a number of large cities such as Dortmund, Bochum and Essen. The industrial decline of the region began in the 1960s, resulting in an intense structural crisis in the late 1970s and early 1980s. This reflected a process of lock-in, based upon the trap of 'rigid specialisation' that emanated from the close intra-regional interdependence of the coal, iron and steel complex. Coal supplied the iron and steel industries, while individual steel producers specialised in different products and used a network of dedicated suppliers, including machine-building, electronics and commercial services.

Grabher distinguishes between three main dimensions of lock-in in the Ruhr:

a) Functional lock-in refers to the close links between the core firms, the major iron and steel manufacturers, and their suppliers, based upon the long-term stability and predictability of demand for iron and steel. As a result, suppliers did not undertake their own research and development or marketing.

This meant that any downturn in demand for iron and steel would have immediate effects on the supplier base as well as the core firms.

b) Cognitive lock-in which relates to how firms thought and operated, pointing to a kind of world view and 'groupthink'. In particular, they assumed that the pattern of stable demand would persist, meaning that the slump of the early 1970s was interpreted as a short-term 'blip' rather than as the start of long-term decline. Cognitive lock-in implies a failure to develop appropriate collective learning mechanisms that allow firms not only to experiment and innovate, but also to be able to read the signs of external change and act appropriately. They responded to the slump of the 1970s by investing heavily in existing technology and process innovation (more of the same) rather than diversifying into new markets.

c) Political lock-in reflects the failure of regional political actors – for example local government,

trade union branches, chambers of commerce – to change policy mechanisms to encourage innovation and learning. In the Ruhr, such lock-in was underpinned by cooperative relations between industry, regional government (the *land* of North Rhine-Westphalia), unions and local authorities. This conservative culture of consensus was expressed in terms of strong political support for the heavy industries, prompting several programmes of investment in the 1970s.

Ultimately, however, the crisis of the late 1970s and 1980s broke these lock-ins, resulting in the substantial destruction of the region's heavy industrial path. Firms responded not only through large-scale plant closures and job losses, but also by reorganising their business and moving into related production sectors and markets. This can be understood as a process of 'path branching' (see below). Firms moved into sectors such as environmental technology which accounted for over 600 firms and 100,000 jobs by the early 1990s (Grabher 1993: 269).

In an effort to move beyond the restrictive theory of path dependency derived from the work of David, the economic geographer Ron Martin (2010) develops a more open model of local industrial evolution which usefully distinguishes between a *preformation phase* dominated by pre-existing economic and technological conditions, together with the resources, competences, skills and experiences inherited from previous local patterns of economic development, and a *path creation* phase where experimentation and competition between different economic agents leads to the emergence of a new path (Figure 2.10). This gives way to a subsequent path development phase based on the development of local increasing returns and network linkages. Path

development, in turn, leads to either a dynamic process of path renewal or 'lock-in' (Box 2.9). In addition, Martin distinguishes between an enabling and constraining institutional environment for the emergence of new technologies and industries.

The process of regional diversification or branching is viewed as a key mechanism of path creation in EEG, involving the development of new industrial growth paths in a city or region. Branching is underpinned by the concept of related variety, defined in terms of regions possessing a number of complementary sectors with overlapping knowledge bases and technological capabilities. A key research finding here, originally based on Swedish data, is that industries are more

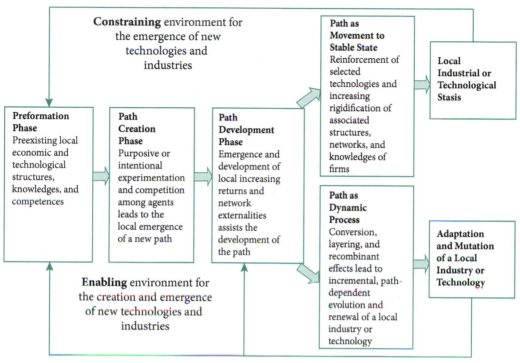

Figure 2.10 Towards an alternative model of local industrial evolution
Source: Martin 2010: 21.

likely to grow in a region if they were technologically related to pre-existing industries in that region (Neffke *et al.* 2011). The processes by which branching actually occurs at the firm level have received less attention in EEG, however, particularly in terms of how firms and other organisations identify opportunities and transfer their knowledge and competencies into new sectors and markets.

Informed by earlier work on industry life cycles, another strand of research in EEG has focused on the spatial evolution of industries. An analysis of the UK automotive (car) industry, for example, shows how it emerged out of the pre-existing bicycle and coaching industries between 1898 and 1922 (Boschma and Wenting 2007). Through this process of regional branching from related industries, car manufacture became concentrated in the West Midlands around Coventry and Birmingham. The process of spin-off was a very important mechanism of diversification with many firms founded by people who had left existing firms, bringing their knowledge and experience with them. This suggests that industrial clusters can emerge from the industrial dynamics of spin-off, whereby successful

routines and capabilities are passed from parent firms to their offspring, challenging the traditional explanation of clustering processes in terms of factors external to individual firms, principally specialised pools of knowledge, labour and suppliers (Boschma and Frenken 2015).

The development of EEG as an exciting field of research has been accompanied by periodic debates about its scope and direction. While Boschma and Frenken's (2006) strong founding statement drew some sharp distinctions between EEG and institutional economic geography and the NEG, others have called for closer links to other approaches, namely institutionalism and GPE, on the one hand, and **relational economic geography,** on the other (Hassink *et al.* 2014; MacKinnon *et al.* 2009). EEG remains diverse, being characterised by the deployment of a range of theories and methods. The distinction between quantitative and qualitative methods is particularly pronounced, with some researchers relying on statistical and formal modelling techniques that are similar to mainstream economics and the NEG, whilst others espouse more qualitative, case study-based approaches.

2.5.4 Relational economic geography

In recent years, economic geography has been influenced by the rise of relational thinking in geography. This builds on Doreen Massey's 'global sense of place' which defines place as constructed out of the coming together of wider social relations (section 1.2.3). According to Massey (2005: 9–12), the relational approach to space is grounded in three basic propositions. First, space should be seen as the product of interrelations, meaning that places and their identities are created through such relations, rather than pre-existing in an *a priori* fashion. Second, space emphasises the possibility of multiplicity, created out of the different relationships that link places together, resonating with the postmodernist concern for difference. Third, space is always changing and becoming rather than being static or fixed. From this perspective, globalisation marks a new era of space/place relations, focusing attention on economic networks and wider circuits of knowledge generation and exchange rather than bounded regions.

Relational economic geography is concerned with economic action and interaction which it views as being embedded in wider social and economic relations. It is particularly associated with the study of networks, defined as "socio-economic structures that connect people, firms and places to one another and that enable knowledge, commodities and capital to flow within and between regions" (Aoyama *et al.* 2011: 181). In this respect, it overlaps significantly with the recent emphasis on economic practices, defined as the diverse and mundane actions by which diverse actors and communities 'make do' and 'get by' to meet their material needs and ensure their social reproduction (Stenning *et al.* 2010: 64). According to Jones and Murphy (2011), this concern with practice should not be seen as a major theoretical 'turn' in its own right, but as a way of understanding broader processes, for instance consumption, through an emphasis on 'ordinary' actions and seemingly 'mundane' activities. For example, Al James's (2007) study of cultural embeddedness in Utah's high-tech economy found that Mormon religious values were upheld by a range of day-to-day practices within firms. These include the reinforcement of group norms through routine association with other employees, observation of fellow workers, the

group ratification of culturally informed corporate decisions, and the recruitment of Mormon employees.

Relational conceptions of space have fostered a new concern with 'the relational region' which reimagines regions as open and discontinuous spaces, defined by the wider social relations in which they are situated (Box 2.10). They are created for particular purposes by, for instance, policy-makers, social movements or academic analysts, meaning that they have no essential character or identity outside of these acts of creation and definition. John Allen, Massey and Allan Cochrane (1998), based at the Open University, illustrate these arguments by an analysis of south-east England, the emblematic growth region of neoliberal Britain since the 1980s. Within the Southeast, areas of economic decline and deprivation exist, complicating and confounding the overarching image of growth and prosperity, while the boundaries of the region can be seen as open and porous (Figure 2.11). This relational theory of

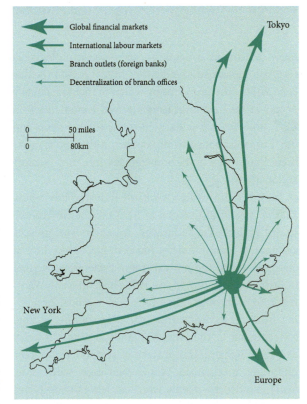

Figure 2.11 International financial links of south-east England
Source: Allen *et al.* 1998: 49.

Box 2.10

Transnational communities and regional development

Anna-Lee Saxenian's work on transnational communities of engineers and entrepreneurs provides a useful illustration of relational approaches to regional development. Saxenian is renowned for her previous work on Silicon Valley in California (Box 3.8), and she has maintained this focus in her recent research, while modifying her approach to take account of the transnational links and flows.

Saxenian (2006) argues that the past couple of decades have seen the rise of what she calls the 'new Argonauts' in the form of engineers and entrepreneurs who operate internationally, often having moved overseas for education or employment and having gone on to launch new firms in the countries they moved to. The main example of such Argonauts discussed by Saxenian are first-generation immigrants to the US from Asian countries such as India, China and Taiwan who gained technical skills in

engineering and who became entrepreneurs in the US. Increasingly, these individuals formed communities of technically skilled immigrants with work experience and connections to Silicon Valley and related American technology centres.

In the 1960s and 1970s, the relationship between the home countries of these Asian engineers and the US as the receiving country conformed to the classic pattern of migration between the rich and poor worlds. From the late 1980s, however, Saxenian argues, a new trend emerged as US-educated engineers began to return to their home countries. This reflected the economic growth of these countries and the development of information and communication technologies (ICTs) which created new opportunities for firms and individuals to operate internationally. Initially, this reverse flow was confined to Taiwan – and other more

developed countries like Israel which had traditionally exported engineers to the US – before spreading to India and China from the late 1990s. This transnational movement of entrepreneurs and engineers in the IT sector underpins Saxenian's key argument that the traditional pattern of 'brain drain', for the originating countries, is being replaced by a mutually beneficial one of 'brain circulation'. As such, the 'new Argonauts' are replacing the old pattern of one-way flows of skilled people from the periphery to the core with new, two-way linkages. This is fuelling growth and employment creation in the (former) periphery through the infusion of skills, capital and technology incubated in the core. It remains unclear, however, how widespread such Argonaut-based 'brain circulation' is compared to traditional 'brain drain' beyond the specific countries she discusses.

Reflect

Which of the new approaches discussed in this section do you consider to offer the most promising perspective for economic geography? Justify your answer.

the region has been accompanied by a critique of 'new regionalist' thinking, rejecting its portrayal of regions as internally coherent and externally bounded.

2.6 Summary

Prevailing approaches to economic geography have changed considerably over time, as we have demonstrated, mirroring the development of geography more

broadly. The traditional framework based on regional classification and description gave way to a more quantitative approach in the late 1950s and early 1960s, based on the development of concepts and techniques of spatial analysis. This was, in turn, superseded by GPE in the 1970s and 1980s before the introduction of a range of new approaches in the 1990s and 2000s. These periodic shifts in intellectual orientation should be seen as changes in broad focus and direction rather than as wholesale transformations, with research in the earlier spatial analysis and GPE remaining important.

Our favoured approach in this book can be described as 'new' GPE, signalling that it has moved beyond the rather clunky and deterministic theories of the 1970s and early 1980s to become more flexible and open to the importance of context, difference and identity, partly as a result of encountering the new approaches outlined in the previous sections. In particular, our approach is a plural GPE that combines the breadth and analytical power of GPE thinking with an evolutionary focus on historical transformation and change and a sensitivity to context and differences associated with the cultural and institutional 'turns'. It represents, as such, a "culturally-sensitive political economy" rather than a "politically-sensitive cultural economy" (Hudson 2005: 15). Our approach can be described as broadly regulationist in nature, stressing the important role that wider processes of social regulation play in stabilising and sustaining capitalism.

From a base in GPE, we examine the uneven development of capitalism, "the individual economics of different places, and their connections one to another" (Barnes and Sheppard 2000: 2–3) in this book. In adopting a broad, synthetic conception of the economy, we recognise the importance of the links between the economy, institutions and culture, and assess some connections between them in specific geographical settings. Our approach also takes difference and variety seriously, emphasising the distinctive character of individual places, but insisting that we need to consider how geographical difference is produced and reproduced through wider processes of economic development and state intervention.

Exercise

A major supermarket chain is proposing to open a new out-of-town superstore near to the town or city where you live.

How would spatial analysis, GPE *and* the cultural *or* institutional approaches examine and understand this issue? What aspects of the development would each approach focus attention on: for example, people's experience of shopping, the broader policies and practices of the corporation in question, finding the optimum location, the links between consumption and identity, the impact on local shops, analysing the characteristics of the local market, assessing why customers shop in major superstores, competition amongst supermarket chains, relationships with suppliers?

On this basis, assess the strengths and weaknesses of each approach. Which, if any, offers the 'best' understanding of the issue? Why? Is it appropriate to try and bring elements of the different approaches together? If so, which ones?

Key reading

Barnes, T.J. (2000) **Inventing Anglo-American economic geography**. In Sheppard, E. and Barnes, T.J. (eds) *A Companion to Economic Geography*. Oxford: Blackwell, pp. 11–26.
An engaging account of the formation and early development of economic geography. Stresses how academic subjects are invented as specific projects at particular times and in particular places by groups of people. Covers the traditional approach and spatial analysis.

Barnes, T.J. (2012) **Economic geography**. In Johnston, R.J., Gregory, D., Pratt, G. and Watts, M. (eds) *The Dictionary of Human Geography*, 5th edition. Oxford: Blackwell, pp. 178–81.
Provides a concise summary of economic geography's concerns and methods. Outlines how the subject has evolved over time and highlights a range of examples of contemporary economic geography research.

Gertler, M.G. (2010) **Rules of the game: the place of institutions in regional economic change**. *Regional Studies* 41: 1–15.
Offers a comprehensive and accessible definition of institutions and discusses several strands of institutionalist research in economic geography and regional studies, arguing that they have failed to take institutions seriously. Sets out an alternative approach to studying institutions for economic geography.

Scott, A.J. (2000) **Economic geography: the great half century**. In Clark, G., Feldmann, M. and Gertler, M. (eds) *The Oxford Handbook of Economic Geography*. Oxford: Oxford University Press, pp. 18–44.
An upbeat review of the development of economic geography in the post-war period. Focuses particularly on the spatial analysis and political economy approaches, highlighting specific research topics such as localities and the rediscovery of regions since the 1980s.

Sheppard, E., Barnes, T.J., Peck, J. and Tickell, A. (2004) **Introduction: reading economic geography**. In Barnes, T., Peck, J., Sheppard, E. and Tickell, A. (eds) *Reading Economic Geography*. Oxford: Blackwell, pp. 1–9.
A concise and authoritative summary of the field from the introduction to a selection of key journal articles in economic geography. Introduces the subject matter of economic geography before providing an essential history of it in two pages.

This is followed by an explanation of the immediate context for the companion and a useful guide on critical reading techniques.

Sheppard, E., Barnes, T.J. and Peck, J. (2012) The long decade: economic geography unbound. In Barnes, T.J., Peck, J. and Sheppard, E. (eds) *The Wiley-Blackwell Companion to Economic Geography*. Oxford: Blackwell, pp. 1–24. An important introduction to economic geography as an academic subject area from a recent handbook which provides a comprehensive guide to the subject. This introduction highlights some of the key themes and recent debates in economic geography, as well as introducing the contents of the handbook.

Useful websites

www.egrg.rgs.org/

The website of the Economic Geography Research Group of the Royal Geographical Society (with the Institute of the British Geographers). This site is mainly used by academic researchers in the field, so much of the material is likely to be difficult. It is worth exploring the site, however, to get a feel for the types of issues and topics that economic geographers conduct research on.

References

Allen J., Massey D. and Cochrane A. (1998) *Rethinking the Region*. London: Routledge.

Amin, A. and Thrift, N. (1994) Living in the global. In Amin, A. and Thrift, N. (eds) *Globalisation, Institutions and Regional Development in Europe*. Oxford: Oxford University Press, pp. 1–22.

Aoyama, Y., Murphy, J.T. and Hanson S. (2011) *Key Concepts in Economic Geography*. London: Sage.

Armstrong, H. and Taylor, J. (2000) *Regional Economics and Policy*, 3rd edition. London: Blackwell.

Barnes, T.J. (2000) Inventing Anglo-American economic geography. In Sheppard, E. and Barnes, T.J. (eds) *A Companion to Economic Geography*. Oxford: Blackwell, pp. 11–26.

Barnes, T.J. (2001) Retheorising economic geography: from the quantitative revolution to the 'cultural turn'. *Annals of the Association of American Geographers* 91: 546–65.

Barnes, T.J. and Sheppard, E. (2000) The art of economic geography. In Sheppard, E. and Barnes, T.J. (eds) *A Companion to Economic Geography*. Oxford: Blackwell, pp. 1–8.

Boschma, R. and Frenken, K. (2006) Why is economic geography not an evolutionary science? Towards an evolutionary economic geography. *Journal of Economic Geography* 6: 273–302.

Boschma, R. and Frenken, K. (2015) Evolutionary economic geography. *Papers in Evolutionary Economic Geography* 15.18. Utrecht: Utrecht University, Urban and Regional Research Centre. At http://econ.geog.uu.nl/peeg/peeg.html.

Boschma, R.A. and Martin, R. (2007) Constructing an evolutionary economic geography. *Journal of Economic Geography* 7: 537–48.

Boschma, R.A. and Wenting, R. (2007) The spatial evolution of the British automobile industry: does location matter? *Industry and Corporate Change* 16: 213–38.

Boyer, R. (1990) *The Regulation School: A Critical Introduction*. New York: Columbia University Press.

Burton, I. (1963) The quantitative revolution and theoretical geography. *Canadian Geographer* 7: 151–62.

Christaller, W. (1966) *Central Places in Southern Germany*, translated by Baskin, C.W. Englewood Cliffs, NJ: Prentice-Hall.

Cloke, P., Philo, C. and Sadler, D. (1991) *Approaching Human Geography: An Introduction to Contemporary Theoretical Debates*. London: Paul Chapman.

Crang, P. (1997) Introduction: cultural turns and the (re)constitution of economic geography. In Lee, R. and Wills, J. (eds) *Geographies of Economies*. London: Arnold, pp. 3–15.

David, P.A. (1985) Clio and the economics of QWERTY. *American Economic Review* 75: 332–7

Dowd, D. (2000) *Capitalism and its Economics: A Critical History*. London: Pluto.

Dunford, M. and Perrons, D. (1994) Regional inequality, regimes of accumulation and economic development in contemporary Europe. *Transactions, Institute of British Geographers* NS 19: 163–82.

European Commission (2014) *Investment for Jobs and Growth: Promoting Development and Good Governance in EU Regions and Cities*, 6th Report on Economic, Social and Territorial Cohesion. At http://ec.europa.eu/regional_policy/sources/docoffic/official/reports/cohesion6/6cr_en.pdf.

FAO (Food and Agriculture Organisation) (2010) Measuring and assessing capacity in fisheries: 2 issues and methods. *FAO Fisheries Technical Paper* 433/2. Rome: Food and Agriculture Organisation of the United Nations.

Gertler, M.S. (1995) 'Being there': proximity, organisation and culture in the development and adoption of advanced manufacturing technologies. *Economic Geography* 71: 1–26.

Gertler, M.S. (2004) *Manufacturing Culture: The Institutional Geography of Industrial Practice*. Oxford: Oxford University Press.

Grabher, G. (1993) The weakness of strong ties: the lock-in of regional development in the Ruhr area. In Grabher, G. (ed) *The Embedded Firm: On the Socio-economics of Industrial Networks*. London: Routledge, pp. 255–77.

Gregory, D. [1989] (1996) Areal differentiation and post-modern human geography. In Agnew, J., Livingstone, D.N. and Rogers, A. (eds) *Human Geography: An Essential Anthology*. London: Blackwell, pp. 211–32.

Hartshorne, R. (1939) *The Nature of Geography: A Critical Survey of the Present in the Light of the Past*. Lancaster, PA: The Association.

Harvey, D. (1973) *Social Justice and the City*. London: Arnold.

Harvey, D. (1982) *The Limits to Capital*. Oxford: Blackwell.

Hassink, R., Klaerding, C. and Marques, P. (2014) Advancing evolutionary economic geography by engaged pluralism. *Regional Studies* 48: 1295–307.

Hudson, R. (2005) *Economic Geographies: Circuits, Flows and Spaces*. London: Sage.

Hudson, R. (2006) On what's right and keeping left: Or why geography still needs Marxian political economy. *Antipode* 38: 374–95.

Jackson, P. (1989) *Maps of Meaning: An Introduction to Cultural Geography*. London: Unwin Hyman.

James, A. (2007) Everyday effects, practices and causal mechanisms of 'cultural embeddedness': learning from Utah's high tech regional economy. *Geoforum* 38: 393–413.

Johnston, R.J. and Sidaway, J.D. (2004) *Geography and Geographers: Anglo-American Human Geography Since 1945*, 6th edition. London: Arnold.

Jones, A. (2015) Geographies of production I: political economic geographies: a pluralist direction? *Progress in Human Geography*. OnlineFirst, doi 10.1177/0309132515595553.

Jones, A. and Murphy, J.T. (2011) Theorising practice in economic geography: foundations, challenges and possibilities. *Progress in Human Geography* 35: 266–92.

Knox, P. and Agnew, J. (1994) *The Geography of the World Economy*, 2nd edition. London: Arnold.

Krugman, P. (2000) Where in the world is the 'new economic geography'?. In Clark, G., Feldmann, M. and Gertler, M. (eds) *The Oxford Handbook of Economic Geography*. Oxford: Oxford University Press, pp. 49–60.

Krugman, P. (2011) The new economic geography, now middle-aged. *Regional Studies* 45: 1–8.

Lee, R. (2002) Nice maps, shame about the theory? Thinking geographically about the economic. *Progress in Human Geography* 26: 333–55.

Ley, D. (1994) Postmodernism. In Johnston, R.J., Gregory, D. and Smith, D.M. (eds) *The Dictionary of Human Geography*, 3rd edition. Oxford: Blackwell, pp. 466–8.

MacKinnon, D., Cumbers, A., Birch, K., Pike, A. and McMaster, R. (2009) Evolution in economic geography: institutions, political economy and regional adaptation. *Economic Geography* 85: 129–50.

Martin, R. (1999) The new 'geographical turn' in economics: some critical reflections. *Cambridge Journal of Economics* 23: 65–91.

Martin, R. (2000) Institutionalist approaches to economic geography. In Sheppard, E. and Barnes, T. (eds) *Companion to Economic Geography*. Oxford: Blackwell, pp. 77–97.

Martin, R. (2010) Rethinking regional path dependence: beyond lock-in to evolution. *Economic Geography* 86: 1–27.

Marx, K. and Engels, F. [1848] (1967) *The Communist Manifesto*. London: Penguin.

Massey, D. (1995) *Spatial Divisions of Labour: Social Structures and the Geography of Production*, 2nd edition. London: Macmillan.

Massey, D. (2005) *For Space*. London: Sage.

McDowell, L. (1997) *Capital Culture: Gender at Work in the City*. Oxford: Blackwell.

Neffke, F., Henning, M. and Boschma, R. (2011) How do regions diversify over time? Industry relatedness and the development of new growth paths in regions. *Economic Geography* 87: 237–65.

Nelson, R.R. and Winter, S.G. (1982) *An Evolutionary Theory of Economic Change*. Cambridge, MA: Harvard University Press.

Pike, A., Rodríguez-Pose, A. and Tomaney, J. (2017) *Local and Regional Development*, 2nd edition. London: Routledge.

Rees, J. (1998) The return of Marx? *International Socialism* 79. Available at http://pubs. socialistreviewindex.org.uk/isj79.rees.htm. Accessed 2 November 2005.

Rigg, J., Bebbington, A., Gough, K.V., Bryceson, D.F., Agergaard, J., Fold, N. and Tacoli, C. (2009) The World Development Report 2009 'reshapes' economic geography: critical reflections. *Transactions, Institute of British Geographers* NS 34: 128–36.

Rodríguez-Pose, A. (2013) Do institutions matter for regional development? *Regional Studies* 47: 1034–47.

Saxenian, A.L. (2006) *The New Argonauts: Regional Advantage in a Global Economy*. Cambridge, MA: Harvard University Press.

Scott, A.J. (2000) Economic geography: the great half century. In Clark, G., Feldmann, M. and Gertler, M. (eds) *The Oxford Handbook of Economic Geography*. Oxford: Oxford University Press, pp. 18–44.

Seddon, M. (2005) Kapital gain: Karl Marx is the Home Counties favourite. *The Guardian*, 14 July.

Sheppard, E. (2011) Geographical political economy. *Journal of Economic Geography* 11: 319–31.

Smith, N. (1984) *Uneven Development: Nature, Capital and the Production of Space*. Oxford: Blackwell.

Smith, N. (2001) Marxism and geography in the Anglophone world. *Geographishce Revue* 3: 5–22. Available at www.geographische-revue.de/gr2-01.htm.

Stenning, A., Smith, A., Rochovska, A. and Swiatek, D. (2010) *Domesticating Neo-Liberalism: Spaces of Economic Practice and Social Reproduction in Post-Socialist Cities*. Oxford: Wiley-Blackwell.

Storper, M. (2011) From retro to avant-garde: a commentary on Paul Krugman's 'the new economic geography, now middle-aged'. *Regional Studies* 45: 9–16.

Swyngedouw, E. (2000) The Marxian alternative: historical-geographical materialism and the political economy of capitalism. In Sheppard, E. and Barnes, T.J. (eds) *A Companion to Economic Geography*. Oxford: Blackwell, pp. 40–59.

Walker, R. (2000) The geography of production. In Sheppard, E. and Barnes, T.J. (eds) *A Companion to Economic Geography*. Oxford: Blackwell, pp. 111–32.

Wills, J. and Lee, R. (1997) Introduction. In Lee, R. and Wills, J. (eds) *Geographies of Economies*. London: Arnold, pp. xi–xviii.

World Bank (2009) *World Development Report 2009: Reshaping Economic Geography*. Washington, DC: World Bank.

Chapter 3

From regional specialisation to global integration: changing geographies of production

3.1 Introduction

As we indicated in section 1.3.1, the contemporary economy is dominated by a capitalist mode of production in which the pursuit of profit has been the key driving force behind development. Over the last 400 or so years, this has led to the emergence of a single world economy as **capitalism** has spread geographically from its origins in Western Europe across the globe to incorporate virtually all countries and regions (Knox *et al.* 2003). This process of expansion has transformed pre-capitalist societies and systems, not least through the process of **colonialism** which reached its high point in the late nineteenth century. Despite periodic setbacks and reverses, including serious challenges from alternative social systems such as forms of Communism in the Soviet Union and China during the twentieth century, capitalism has continued to expand. Indeed, the latest stage is one of **globalisation** defined by the increasing global integration of markets and production, supported by the latest advances in information and communications technologies (ICTs).

As this chapter demonstrates, the geography and organisation of production has evolved radically over the past 200 or so years, driven by the interactions between profit-seeking capital, the labour process, technological change and geographical expansion. We outline this changing geography of production in order to provide clear historical foundations for the remainder of the book, as well as serving to ground the broader observations that have already been made about capitalism (sections 1.3, 1.4, 2.4, 2.5) in the actual historical geography of production. The process of industrialisation, occurring in successive waves of growth and contraction across the nineteenth century, was characterised by the growth of highly specialised industrial regions in Europe and North America, while the colonies of the global South supplied raw materials and agricultural commodities to the core countries. The leading position of certain core industrial regions in Europe and North America was reinforced by the post-war system of mass production (**Fordism**), while other, more specialised and peripheral industrial regions experienced protracted economic decline from the 1960s. This was followed by a more radical period of transition in the 1970s and 1980s involving the large-scale **deindustrialisation** of developed economies and a concomitant 'global shift' of manufacturing investment to certain lower-cost locations in the global South (Dicken 2015). This has been followed by an extension of the process of globalisation through the geographical relocation or '**offshoring**' of some service activities to lower-cost locations. As a result of this logic of global integration, production is now increasingly geographically dispersed and 'stretched' across different countries and regions. At the same time, however, elements of regional specialisation persist, not least through the geographical concentration of higher-value functions such as advanced financial services, research and development and corporate headquarters in world cities such as London and New York and high-tech centres like Silicon Valley in California.

3.2 The dynamics of capitalist production

Informed by our **geographical political economy (GPE)** approach, this section sketches the driving forces of capitalist production in broader terms, moving from the broad framing arguments that were introduced in Chapter 1 to provide greater historical context and depth. We begin with capital as a key driving force of the economy, moving from the abstract treatment of the **circuit of capital** to consider more specific organisational forms. This is followed by a discussion of the labour process, emphasising the nature of labour under capital as a 'fictitious commodity' and the importance of the **division of labour** in industrial production. The basic dynamic of capital is reflected in the process of **innovation** and technological change which is deeply implicated in the restructuring of the economic landscape, generating new industries and technologies and products whilst eroding the competitiveness of existing ones. At the same time, successful firms have also sought to expand their operations geographically, driven by competition and the pursuit of profit, and providing them with access to new markets, raw materials and supplies of labour. In simplified terms, capital and, to a lesser extent, labour can be seen as the key agents of change, shaped by the labour process, technological change and geographical expansion, among a range of other, more specific factors.

3.2.1 Capital

Capital is a complex term, used in different ways by different people. For our purposes here, however, it can simply be defined as money that is invested in production or financial markets. Capitalists are those people who have acquired capital, allowing them to own the means of production: land, materials, factories, offices and machines. The pressures of competition mean that the capitalist has to seek to expand his/her stock of capital by reinvesting it in production to generate higher profits. This reinvestment in order to generate more profits – which are, in turn, reinvested – is often referred to as the process of **capital accumulation** (Barnes 1997). It lies at the heart of the capitalist system with profit-seeking representing the basic driving force for economic growth and expansion.

The basic economic process under capitalism can be understood in terms of the circuit of capital through which profits are generated (Harvey 1982). The first stage is that capital in its money form (M) is transformed into commodity form (C) by purchasing the means of production (MP) – factories, machines, materials, etc. – and labour power (LP) (Figure 3.1). The means of production (MP) and labour power (LP) are then combined in the production process, under the supervision of the owners of capital or their managers and representatives, to produce a commodity for sale (C*) – for example a car, house or, even, a haircut. This commodity is sold for the initial money outlay plus a profit (Δ), representing what some economists call surplus value.

Part of the money (M) realised from the sale of the commodity is reinvested back into the production process, which recreates the circuit anew. Following each circuit of capital, there is an expansion in the total amount of capital, forming the basis of capital accumulation. Clearly, the distribution of income at the end of each circuit becomes an important political question and governments play a regulatory role in diverting capital into welfare spending and other state functions (investment in transport infrastructure, housing, etc.) through taxation.

The main organisational form that capital takes is **the firm**. Firms are legal entities owned by individual capitalists or, more commonly now, a range of shareholders. Along with the commodity, the firm could be described as a basic 'economic cell form' of capitalism (see section 1.3.1).

Our approach in this book is informed by the **competence or resource-based theory of the firm**. Derived from the work of the economist Edith Penrose (1959), the competence perspective views firms as bundles of assets and competencies that have been built up over time. A firm distinguishes itself from its competitors through the specific way in which it combines the resources of land, labour and capital in the production of certain commodities. Competencies can be seen as particular sets of skills, practices and forms of knowledge. The need for firms to focus on identifying and developing their 'core competencies' is a key theme of recent management theories.

The knowledge and skills developed by a firm become embedded within its organisational rules and routines (Taylor and Asheim 2001: 323). Firms acquire and develop this knowledge and skills through processes of learning associated with the repetition of particular tasks. As such, the competence-based approach draws attention to the dynamic processes of learning and knowledge creation that occur within firms, providing much of the basis for understanding firm strategy (Nelson and Winter 1982). It is well suited to economic geography, indicating that firms can derive some of their competitiveness from factors present in the broader regional environment, combining these with internal resources to create distinctive competencies.

In addition to this abstract theory of what firms are and how they operate, it is important to appreciate the diversity of firms in terms of size, ownership and structure. In practice, the nature of business organisation is very dynamic, taking new forms over time. The heroic small firms and entrepreneurs of the nineteenth century have given rise to large transnational corporations or what some Marxist economists have termed monopoly capital (Baran and Sweezy 1966). Whilst mainstream economics had assumed the existence of a world of small firms with no power to influence prices in the marketplace, the emergence of these large multi-plant corporations required more sophisticated analyses. This work looked at the internal organisation of business, emphasising how the growth of the large joint stock company resulted in a separation of ownership (shareholders) from control (managers) and the emergence of vast internal labour markets through which employment is organised (Doeringer and Piore 1971).

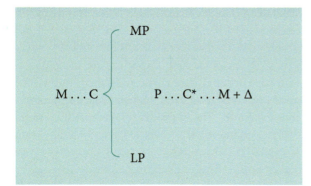

Figure 3.1 The process of production under capitalism

Source: Castree *et al.* 2004: 28.

In order to grow and expand, firms need capital for investment. Some of this capital can be generated internally through the profits made from selling commodities in the marketplace, but many firms require access to external sources of finance such as loans from banks, the financial markets and venture capital. For large public corporations the stock market provides an additional source of funds through the issuing of shares which are purchased by individuals and institutional investors, referred to as **equity finance**. In the large public limited firm or joint stock company, shares are typically distributed between a diverse range of interests that represent different sections of society. **Venture capital** is equity finance that investors provide to firms not quoted on the stock exchange. Such firms are generally small and have a high growth potential, attracting venture capitalists who aim to make high returns by selling their stake in the firm at a later date. Such investment is high risk with investors typically expecting returns of over 30 per cent (Tickell 2005: 249).

Capital is the most mobile of the factors of production, in contrast to labour and land which are relatively place-bound. As we have seen, it tends to become concentrated in regions which offer a high rate of return or profit on money invested, leaving underdeveloped countries and regions bereft of the capital required to invest in production facilities and development projects. At the same time, investment is not contained indefinitely within core regions, with firms and investors also responding to investment opportunities in other regions. Over time, there is a 'see-saw' effect as capital flows back and forth between different sectors and regions (Smith 1984). Financial deregulation in most of the major economies seems to have exacerbated this situation in recent decades, unleashing a far more rapid and volatile phase of accumulation, and culminating in the financial crisis and 'Great Recession' of 2008–09 (Harvey 2010). **Capital switching** has important geographical dimensions, often being transferred from regions dominated by declining sectors to '**new industrial spaces**' in distant regions offering more attractive conditions for investment. An example of the geographical effects of capital switching and **uneven development** is provided by the experience of the north-east of England over the twentieth century (Box 3.1).

Box 3.1

Capital switching and its geographical effects: the case of the north-east of England

As the first country to industrialise, the UK enjoyed an advantage over the rest of the world in the production of heavy engineering and manufacturing goods throughout the nineteenth century. The northeast of England became a heartland region of industrial capitalism during the second half of the nineteenth century on account of its iron ore and coal reserves (Figure 3.2). It became particularly dominant in the growing shipbuilding market – producing around 40 per cent of the world's ships at one point – but also became an important centre for rail and bridge engineering, and arms production. This growth was associated with the emergence of a local bourgeoisie, the 'coal combines' (capitalist class), a group of families that had originally made their fortunes through the coal trade but had diversified into the new industries to take advantage of the investment opportunities.

With the growth of competition from other countries – especially the United States and Germany – and a worldwide downturn in the 1920s, the region's industries faced an economic crisis reflected in declining markets and rising unemployment. At the peak of the Depression in 1933, over 80 per cent of the region's shipbuilding and repair workforce was unemployed. Faced with this situation, the region's capitalists had two alternatives: reinvesting in the traditional industries, introducing new production methods and technologies, or finding alternative avenues for investment. For the most part they chose the latter, switching capital from what were increasingly perceived as declining industries to new growth sectors such as car and aircraft production, the public utilities and financial services. The nationalisation of the coal industry in the 1940s provided

Box 3.1 (continued)

further opportunities for the region's capitalists when "the coal owners' capital, locked in the fixed assets of mines and machinery, became suddenly transformed under the compensation terms into highly liquid government bonds" (Benwell Community Project 1978: 58).

Diversification out of the heavy industrial sector was accompanied by a geographical expansion of the 'coal combines' interests both at the national and international level. Through the establishment of special investment trusts in the inter-war years, local capitalists were able to exploit the growing market in trading and financial transactions to the extent that by 1930, one of the largest of these trusts, the 'Tyneside', had only 23 per cent of its investment in north-eastern industry. Despite efforts to modernise the region by both state-led restructuring and the attempt to encourage new inward investment, the legacy of this withdrawal of capital from the Northeast has been the region's transition from a core region of capitalism in the nineteenth and early twentieth centuries to an increasingly peripheral position from the 1930s. This is manifest in lower levels of economic development and higher levels of social deprivation than the national average.

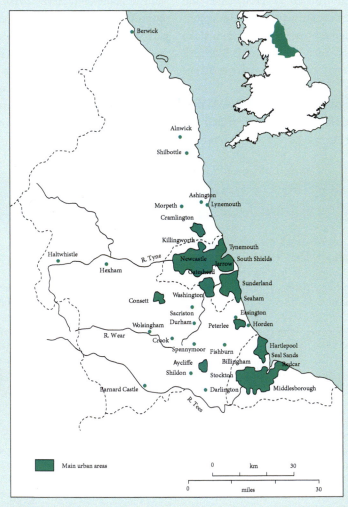

Figure 3.2 North-east England, regional setting and settlements
Source: Hudson 1989: 4.

3.2.2 The labour process

The position of labour is a central defining feature of the capitalist mode of production. Unlike previous and more primitive societies where the great mass of the population, as slaves or peasants, were in thrall and effectively owned by their masters, under capitalism workers are released from feudal ties and are free to sell their own labour. Indeed, labour has to sell its own 'labour power' to earn wages to pay for the essentials of life (e.g. food, clothing, shelter) because it does not own the means of production and therefore the means for its own sustenance. Capital certainly needs labour in order to produce things to sell, but labour's needs are more urgent; it has to engage in production to secure a wage to sustain itself. In this sense, there is a fundamental imbalance at the heart of capitalist social relations. Whilst labour can withdraw its labour power (i.e. strike) as a sanction against capital, its immediate needs are greater than those of capital – which can afford an empty factory/office for a certain amount of time, although not of course for a sustained period.

Labour can be regarded as a 'fictitious commodity' which takes the appearance of a commodity, but cannot be reduced to one (Polanyi 1944). In the sense that it is purchased by capital, and combined with the means of production to produce commodities for sale, labour functions as a commodity. As such, there is a labour market where labour is purchased by capital, and sold by workers, for a certain price (wage). Yet, as Storper and Walker (1989: 154) argue, echoing Marx,

> Labour differs fundamentally from real commodities because it is embodied in living, conscious human beings and because human activity (work) is an irreducible, ubiquitous feature of human existence and social life.

Labour is not produced for sale on the market like other commodities, as Polanyi observed (Box 1.6), but emerges into the labour market from society, through the education and training system. In this sense, the supply of labour is relatively autonomous from demand. Unlike other commodities, when capital buys labour it does not buy a specified amount but rather the time and labour power of the worker. After the working day is over, workers are free to pursue their lives beyond the confines of the factory gates or office.

This brings us to another crucial point; labour must be reproduced outside the market (Peck 1996). Social reproduction, in this context, refers to the daily processes of feeding, clothing, sheltering and socialising which support and sustain labour. These processes rely on family, friends and the local community. As Harvey (1989: 19) has memorably expressed it, "unlike other commodities, labour power has to go home every night". The notion of labour reproduction focuses attention on the connection between work, home and community. In many conventional accounts of work, the focus on full-time waged employment, usually that of men, has obscured the importance of domestic labour in the home, often performed by women (Gregson 2000). Domestic labour includes activities such as childcare, cooking, cleaning, washing, ironing, etc. It is typically unpaid, but vital to the reproduction of the paid workforce.

The key geographical expression of the nature of labour as a 'fictitious commodity' is its relative immobility. Labour is reproduced in particular places, something which is determined by the need for work and home to be located in fairly close proximity for the vast majority of people. The result is a patchwork of local labour markets, the geographical range of which is determined by the distance over which people are able to commute. While this has expanded over time, with the growth of suburbs and the dominance of the private car, most people still live and work within local labour market areas. According to Storper and Walker (1989: 157),

> It takes time and spatial propinquity for the central institutions of daily life – family, church, clubs, schools, sports teams, union locals, etc. – to take shape . . . Once established, these outlive individual participants to benefit and be sustained by generations of workers. The result is a fabric of distinctive, lasting local communities and cultures woven into the landscape of labour.

The relative immobility of labour, and its concern with sustaining and defending its communities, contrasts with the mobility of capital (although this distinction should not be taken too far as firms' investment in particular places does constrain their mobility to a certain extent). This point can be broadened to suggest that economic landscapes are formed out of the interaction between the conflicting forces of capital seeking profits and labour seeking to defend and promote its interests (Peck 1996).

A key feature of the industrial revolution of the late eighteenth and early nineteenth centuries was the reorganisation of production into the factory system, bringing large numbers of workers together under the control of capitalists. A central principle of the factory system and industrial society more broadly is the division of labour which has technical, social and geographical dimensions (Sayer 1995). The technical division of labour can be defined as the process of dividing production into a large number of highly specialised parts, so that each worker concentrates on a single task rather than trying to cover several (Box 3.2).

An increased division of labour results in the **deskilling of labour** as more rewarding aspects of work such as design, planning and variation are removed. The aim of the eighteenth-century pottery owner, Josiah Wedgwood, for instance, was to increase the division of labour so as to convert his employees into "such machines of men that cannot err" (quoted in Bryson and Henry 2005: 315). The subdivision and fragmentation of the labour process increases employers' control of production, reducing workers to small cogs in the system.

The increased technical division of labour in the workplace is matched by what is termed the social division of labour in society. This refers to the vast

Box 3.2

Adam Smith and the division of labour

The concept of the division of labour was first formalised and developed by Adam Smith, writing in the late eighteenth century, just as the first shoots of industrialisation were starting to appear. Smith argued that the division of labour in production is limited by the extent of the market. In pre-industrial times, localised markets were associated with small-scale domestic industry, employing craftsmen and artisans who undertook a number of different tasks. The rapid extension of markets on a global scale that took place from the late eighteenth century (section 3.4), by contrast, created the conditions for large-scale industrial production to flourish, employing an elaborate technical division of labour.

Smith famously used the example of a pin factory to demonstrate that it is far more efficient for an individual worker to concentrate on one particular task rather than to try to perform a number of activities:

One man draws out the wire, another straightens it, a third cuts it, a fourth points it, a fifth grinds the top for receiving the head; to make the head requires two or three distinct operations . . . and the important business of making a pin is, in this manner, divided into about eighteen distinct operations.

(Smith 1991: 14–15)

The key principle here is specialisation. In a small factory employing 10 workers, an untrained worker could make less than 20 pins a day. If workers became specialised through an increased division of labour, though, Smith observed that 10 people could make 48,000 pins in a day, even when the small size of the workforce meant that one worker had to perform two or three operations. In this way, then, an increased

technical division of labour resulted in huge rises in productivity.

There are three specific ways in which an enhanced technical division of labour increases productivity and efficiency in the workplace (Smith 1991: 13):

➤ It improves the dexterity of workers who become highly adept in performing the same routine task thousands of times in the same day.

➤ Lost time, referring to the time wasted by workers moving between tasks, tools and machines, is sharply reduced.

➤ It facilitates the replacement of labour with machines. This is due to the fact that specialisation involves breaking the labour process down into a large number of standardised and routine tasks which can be performed by machines (fixed capital).

array of specialised jobs that people perform in society, from doctors and lawyers to plumbers, painters and construction workers (Sayer and Walker 1992). Modern, industrial societies are characterised by a highly complex division of labour in this respect. In the course of their work, individuals enter into a range of social relations with other people occupying roles such as colleagues, supervisors and clients or competitors (section 1.4). Students, for example, enter into social relations with academics in their role as university teachers during their degree programmes. The different jobs that people do have acquired different values in society, using 'value' in its broadest sense to incorporate the social status and prestige which particular jobs confer on people (compare, for example, an investment banker and a hairdresser). Such varying levels of status and prestige play an important part in determining pay rates for different occupations, alongside patterns of supply and demand in the labour market (Peck 1996).

3.2.3 Innovation and technological change

Capitalism is a highly dynamic economic system, based upon innovation and the development of new technologies. As the economic geographers Ron Boschma and Ron Martin (2007: 537) observe,

> Change is one of capitalism's constants. As a mode of economic organization, capitalism never stands still. Its central imperative – the search for profit and wealth creation – drives a perpetual process of economic flux. Every day new firms, new products, new technologies, new industries and new jobs are added to the economy, whilst old firms, products, technologies, industries and jobs disappear. Joseph Schumpeter once famously described this constant flux as a process of 'creative destruction' that 'incessantly revolutionizes the economic structure from within, incessantly destroying the old one, incessantly creating a new one' (1943: 82).

This process of '**creative destruction**' means that the emergence of new products and technologies often renders existing products and industries obsolete, leaving them unable to compete on the basis of quality or price. In particular, innovation often involves the harnessing of the latest technology to better serve, and sometimes even create, customers' needs and wants. Contemporary examples include the victory of Apple's iPhone and, more recently, Google's Android platform, over the products of rival firms such as BlackBerry and Nokia, the early leaders in the smart phone market, and the threat which the growth of Uber poses to existing taxi operations in many cities (Rogers 2015; Yueh 2014). As part of this process of creative destruction, capital is withdrawn from unprofitable products and industries and invested in new industries, often located in different centres of production (see Box 3.1).

The tendency for major innovations to 'swarm' or cluster together in distinct cycles or waves was emphasised by Schumpeter and other commentators. These are sometimes known as **Kondratiev cycles** after the Soviet economist Kondratieff who first identified them in the 1920s. Lasting for some 50–60 years in length, each cycle is associated with a distinctive system of technology, incorporating the key propulsive industries, transport technologies and energy sources (Figure 3.3). Five Kondratiev cycles are usually distinguished since the late eighteenth century. Each cycle consists of two distinct phases; one of growth (A) and one of stagnation (B) (Taylor and Flint 2000: 14).

The notion of Kondratiev cycles is based on the analysis of price trends which show a characteristic pattern of steady increases for about 20 years, culminating in a rapid inflationary spiral, followed by a collapse with prices reaching a trough some 50–55 years after the start of the cycle. This has sparked much debate about the specific mechanisms behind this pattern. At a basic level, though, each cycle starts with the bunching together of key innovations which create new economic opportunities for firms and entrepreneurs. At this stage, technology is expensive and demand high, resulting in rising prices; after the technology has matured, becoming routine and standardised, prices fall. An inherent problem is **over-production**, reflecting a tendency for the volume of output to grow more rapidly than market demand.

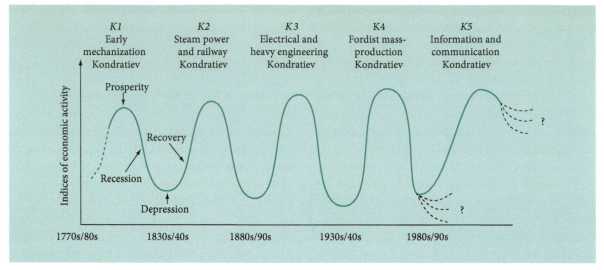

Figure 3.3 Kondratiev cycles
Source: Dicken 2003: 88.

Driven by the search for profits and the pressures of competition, a large number of firms invest in new technologies and products during periods of growth. Accordingly, output increases rapidly until it reaches a point when it can no longer be absorbed by the market. As a result, prices drop and profits and wages are reduced, leading to bankruptcies and unemployment. Even during periods of growth, new technologies make established products and skills obsolete, leading to marginalisation and unemployment for groups of workers. In the boom of the late 1990s and early 2000s, the tendency towards over-production spread from manufacturing to financial services, as banks and other financial institutions invested in complex new products and instruments, borrowing heavily in the markets to do so (Blackburn 2008). High levels of debt and the pattern of complex inter-bank trading ultimately led to a crisis of confidence in credit ('the credit crunch'), prompted by defaults on 'sub-prime' loans in the US.

3.2.4 Colonialism and the geographical expansion of capitalism

The growth of industrial production from the late eighteenth century was supported and facilitated by the geographical expansion of capitalism. The great increases in output and productivity associated with the application of new technology and the development of an elaborate division of labour within factories required, as Adam Smith observed, a corresponding increase in the extent of the market (Box 3.2). At the same time, industrialisation was dependent on an increased volume and range of raw materials, drawn from different parts of the globe. Through the growth of capitalist production and trade, an **international division of labour** was created during the nineteenth century. This involved the developed countries of Europe and North America producing manufactured goods whilst the underdeveloped world specialised in the production of raw materials and foodstuffs. This global trading system was supported and justified by the doctrine of **comparative advantage**, classically expressed by David Ricardo, the great English economist, in 1817 (Box 3.3). It emphasises the benefits of international trade, stating that countries should export the goods which they can produce with greater relative efficiency. In recent years, the theory of comparative advantage has provided an important intellectual foundation for the dominant neoliberal ideology of globalisation.

While the growth of a world economy can be traced back to the voyages of exploration and discovery in the sixteenth century, geographical expansion

Box 3.3

Trade and comparative advantage

Comparative advantage is the principle that a country should specialise in producing and exporting goods in which it has a comparative or relative cost advantage over others and import goods in which it has a cost disadvantage. For example, whilst the developed country in Table 3.1 has an absolute advantage in producing both wheat and cloth (it can produce them more efficiently), the developing country has a comparative advantage in wheat, and the developed country in cloth. Since wheat is relatively cheaper to produce than cloth in the developing country, it only needs to sacrifice 1 metre of cloth to produce 2 kilos of wheat. Conversely, cloth is relatively cheaper to produce than wheat in the developed country with the production of 8 metres entailing the sacrifice of 4 kilos of wheat. In the developing country, by contrast, producing 1 metre of cloth involves giving up 2 kilos of wheat. Thus, the principle of comparative advantage states that countries should specialise in the goods that lead them to give up least in terms of the production

of other goods. Through specialisation, both countries gain by focusing on the good which they can produce most efficiently and by importing the other.

One key question concerns the sources of comparative advantage. How is it that certain countries can produce some goods more efficiently than others? Ricardo explained this in terms of countries' different endowments of the factors of production: land, labour and capital. For example, Canada has a lot of land, China a lot of labour and the US is rich in capital. This means that relative costs of producing goods vary between countries, providing a basis for trade. Basically, countries should specialise in producing goods which use the factors that they have in abundance, enabling them to produce these goods cheaply (grain in Canada, textiles and footwear in China, pharmaceuticals in the US). In this sense, Ricardo believed that trade patterns had a natural basis. The doctrine of comparative advantage supported the colonial trading system where the

European countries exported capital-intensive manufactured goods and the colonies produced raw materials and agricultural goods which were labour- (mines, plantations, etc.) and land-intensive.

In recent decades, however, it has become apparent that most trade takes place between developed countries with similar factor endowments. This has helped to stimulate the development of a **new trade theory** by the Nobel Prize-winning economist Paul Krugman (section 2.3) and others which recognises that comparative advantage does not simply reflect pre-existing factor endowments. Rather, it is actively created by firms through the development of technology, human skills and **economies of scale**, something which is often referred to in terms of **competitive advantage**. This helps to account for patterns of trade and regional specialisation at a more detailed level, for example aircraft in Seattle, cars in southern Germany and finance and business services in London or New York.

Table 3.1 Quantities of wheat and cloth production

	Kilos of wheat	Metres of cloth
Developing country	2	1
Developed country	4	8

Source: Sloman 2000: 659–60.

gained new momentum as the industrial revolution took off, culminating in the 'age of empire' between 1875 and 1914 (Hobsbawm 1987). The inherently expansive nature of capitalism as an economic system has underpinned a search for new markets, new sources of raw materials and new supplies of labour, forging economic relationships between territories on a global scale. As Marx and Engels ([1848] 1967: 83–4) wrote,

The need of a constantly expanding market for its products chases the bourgeoisie over the whole surface of the globe. It must nestle everywhere, settle everywhere, establish connections everywhere . . . All old-established national industries . . . are dislodged by new industries, . . . that no longer work up indigenous raw material, but raw material drawn from the remotest zones; industries whose products are consumed not only at home, but in every corner of the globe . . . In place of the old local and national seclusion and self-sufficiency, we have intercourse in every direction, universal interdependence of nations.

It is this continuing search for new markets, raw materials and labour supplies that underpins recent processes of globalisation, which can be viewed as the latest chapter in the relentless geographical expansion of capitalism.

Colonialism entailed a forcible transformation of pre-capitalist societies in Asia, Africa and Latin America in the eighteenth and nineteenth centuries. As a result, these societies "were no longer locally oriented but had now to focus on the production of raw materials and foodstuffs for the 'core' economies" (Knox et al. 2003: 250). The raw cotton for the mills of Lancashire was supplied by slave plantations in the West Indies and, from the 1790s, the southern US, ensuring that the "most modern centre of production thus preserved and extended the most primitive form of exploitation" (Hobsbawm 1999: 36).

At the same time, the significance of the colonies as outlets for manufactured goods increased greatly. Cotton exports, for instance, multiplied by ten times between 1750 and 1770 (ibid.: 35–6) with the vast majority exported to colonial markets, originally in Africa before India and the Far East took over from the middle of the nineteenth century. The early development of the modern cotton industry relied on Britain's effective monopoly of colonial markets in this period, reflecting both its commercial superiority and naval supremacy.

In many cases, the export of manufactured goods from the core countries resulted in active deindustrialisation in the periphery as traditional local industries were undercut and destroyed by modern factory-based production. Thus, the Indian textile industry was decimated by the much cheaper products of the Lancashire mills. The resultant distress amongst the native hand-loom weavers prompted Marx's ([1867] 1976: 555) comment that "the bones of the cotton-weavers are bleaching the plains of India".

The geographical expansion of capitalism was facilitated by new transport and communications technologies during the nineteenth century. On the transport side, the development of railways and steamships resulted in a dramatic 'shrinking' of space, transporting a growing volume of goods and people over long distances. In 1870, for instance, 336.5 million journeys by rail were made in Britain (Leyshon 1995: 23). The rapid construction of the English railway network in the 1840s had far-reaching consequences:

In every respect this was a revolutionary transformation . . . it reached into some of the remotest areas of the countryside and the centres of the greatest cities. It transformed the speed of movement – indeed of human life – from one measured in single miles per hour to one measured in scores of miles per hour, and introduced the notion of a gigantic, nation-wide, complex and exact interlocking routine symbolised by the railway timetable . . . it revealed the possibilities of technical progress as nothing else had done.

(Hobsbawm 1999: 88)

Railways were subsequently built across much of continental Europe and North America as well as in European colonies overseas, opening up these territories to large-scale investment and trade and creating strong demand for coal, iron and steel. In North America, the railways were crucial to the development of the economy during the nineteenth century, linking the Great Plains to the ports of the Great Lakes and the markets of the east coast and Europe (Leyshon 1995: 28). Huge cities such as Chicago grew as transportation hubs and agricultural markets and processing centres, linking the resources of the American interior to the wider world economy (Figure 3.4) (Cronon 1991). The steamship also played an important role in both transforming the speed of movement between continents and providing a market for the heavy engineering and shipbuilding industries.

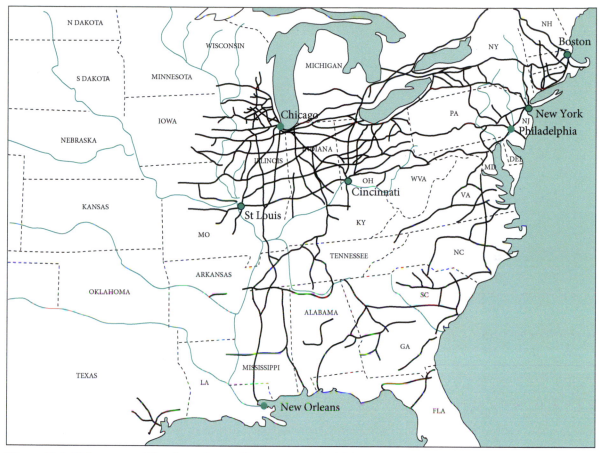

Figure 3.4 Chicago and the American railroad network, 1861
Source: Cronon 1991: iii.

On the communications side, the late nineteenth century saw the invention of the telegram (the 1850s), the telephone (1870s) and the radio (1890s). In terms of facilitating communication over distance and connecting up the realm of everyday life with a geographically dispersed 'out there', these developments were of great significance, paving the way for later developments such as television, the internet and satellite communications. As the German philosopher Martin Heidegger noted in 1916:

> I live in a dull, drab colliery village . . . a bus ride from third rate environments and a considerable journey from any educational, musical or social advantages of a first class sort. In such an atmosphere life becomes rusty and apathetic. Into this monotony comes a good radio set and my little world is transformed.
>
> (Quoted in Urry 2000: 125)

A telegraph cable was laid across the Atlantic Ocean in 1858, enabling rapid communication between North America and Europe for the first time (Leyshon 1995: 24). By 1900, the global telegraph system was complete, creating a communications system which dramatically reduced the time and costs of sending information between places (Figure 3.5). The telegraph was particularly important, supporting the growth of trade and the growing integration of financial markets in particular. For example, foreign currencies could now be easily and rapidly exchanged between markets in London and New York as the telegraph enabled traders to inform one another of the rate at which they would sell pounds for dollars and vice-versa (leading to the sterling–dollar exchange rate becoming known as 'cable') (ibid.: 25).

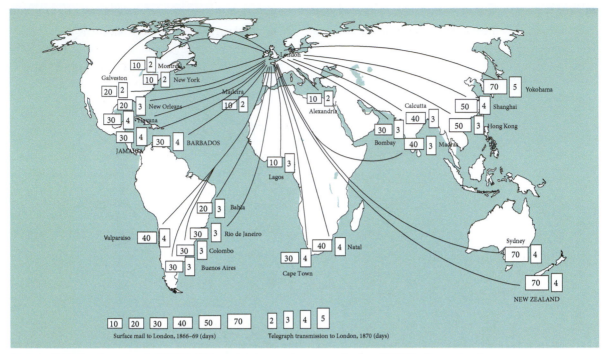

Figure 3.5 Surface mail (1866–69) and telegraph transmission (1870), times in days
Source: Leyshon 1995: 'Map: Nature's Metropolis with American railroads, 1861'.

3.3 Industrialisation and regional specialisation

3.3.1 Industrialisation and regional development

Rather than representing a single abrupt change – the industrial revolution – industrialisation should be seen as an uncertain and uneven process, occurring in successive waves of transformation (Figure 3.3). In the narrow sense, industrialisation refers to a series of changes in manufacturing technology and organisation which radically transformed the economic landscape of capitalism. It can be defined as the "application of power-driven machinery to manufacturing" (Remple, undated: 1). Whilst technological innovations provided a base for large rises in productivity and output, it was the development of the factory system which enabled this possibility to be exploited to the full (Knox *et al.* 2003: 244). The factory was at the core of industrialisation, replacing the existing system of domestic labour and independent craftsmen, representing perhaps its most enduring legacy. Basically, it allowed capitalists to gather and order a large workforce under one roof, enabling them to control the labour process and to exploit the new technologies and emerging division of labour to the full (section 3.3.2).

Waves or cycles of industrialisation are not only a historical phenomenon; they also gave rise to distinct geographies as certain countries and regions assumed technological leadership, leaving others behind. Industrialisation was a regional phenomenon during the nineteenth century, originating in northern and central England before spreading to parts of continental Europe and North America (Pollard 1981). It gave rise to a distinct pattern of **regional sectoral specialisation**, involving particular regions becoming specialised in certain sectors of industry. Characteristically, all the main stages of production from resource extraction to final manufacture were carried out within the same region. Examples include the concentration of industries such as textiles, coal mining, iron and steel in shipbuilding in regions such as north-east England, South Wales, central Scotland, north-east France, the Ruhr and the north-east of the US. These heavy industries were often inter-linked with coal and steel, for instance, providing energy and materials for heavy engineering and shipbuilding. Such regions typically had available reserves of capital, often accumulated from pre-industrial trade, a knowledge of industrial processes and methods among the burgeoning class of entrepreneurs and industrialists, supplies of low-cost labour and relatively advanced transport networks that linked production to local sources of raw materials and international markets.

This pattern of the spatial concentration or **agglomeration** of production in industrial regions (Box 3.4) contrasted with the system of proto-industrialisation which had emerged in seventeenth- and eighteenth-century England where local merchants distributed or 'put out' raw materials to be manufactured by smallholders and artisans in their cottages and workshops. Such domestic industry was small scale and widely dispersed across the countryside, often based on the exploitation of local mineral resources and water power.

Based upon his observation of specialised industrial districts in Britain, Alfred Marshall, the renowned Cambridge economist of the late nineteenth and early twentieth centuries, identified the economic advantages of spatial agglomeration. Marshall's traditional approach emphasised three main factors based on cost reduction (Malmberg and Maskell 2002):

➤ The scope of firms clustered in the same location to share certain collective resources, particularly infrastructure, including transport, communications, electricity and other collective goods, as well as links to the education system and some of the costs of training. This reduces the costs of providing a dedicated infrastructure and other collective resources for individual firms.

➤ The growth of various intermediate and subsidiary industries which provide specialised inputs. This refers to the development of close linkages between manufacturers and suppliers of particular components and services with their co-location serving to reduce transportation costs.

➤ The development of a pool of skilled labour as workers acquire the skills required by local industry. Thus, workers can readily find suitable employment and employers can find skilled labour locally, reducing the search costs for both parties.

These three factors generate **agglomeration economies**, which can be defined as cost advantages that accrue to individual firms because of their location within a cluster of industrial growth (Knox *et al.* 2003: 242). These advantages are sometimes also known as external economies because they stem from circumstances beyond a firm's own practices, reflecting broader features of the local environment. Agglomeration economies can be divided into **localisation economies** such as those listed above, stemming from the concentration of firms in the *same* industry, and **urbanisation economies**, derived from the concentration of firms in *different* industries in large urban areas. Informed by these agglomeration economies, the concept of **cumulative causation** offers a more dynamic theory of the spatial concentration of industry as part of the broader process of uneven development (Box 3.4) (section 1.2.2, Box 2.5).

Box 3.4

Myrdal's model of cumulative causation

A useful model of the process of uneven regional development is that of cumulative causation, derived from the work of the Swedish economist Gunnar Myrdal (1957). This explains the spatial concentration of industry in terms of a spiral of self-reinforcing advantages that build up in a particular area (Figure 3.6) and the adverse

Box 3.4 (continued)

effect this has on other regions, creating a core-periphery pattern. Once an industry starts to grow in a region, for whatever reason, it will tend to attract ancillary (supporting) industry made up of firms supplying it with various inputs and services. At the same time, the expansion of employment and population created by the growth of industry creates a large market which draws in further capital and enterprise. The expansion of industry and the growth in population, moreover, creates increased revenues for local government, resulting in the provision of an improved infrastructure for industrial development (Figure 3.6).

The process of cumulative causation in a growing region is linked to the fate of surrounding ones through flows of capital and labour. Myrdal identified two contrasting types of effects. The first he termed **'backwash' effects**

when investment and people are sucked out of surrounding regions into the growth region, which offers higher profits and wages. In this situation, the virtuous circle of growth in the latter is matched by a vicious circle of decline in the former, leaving such regions suffering from classic symptoms of underdevelopment such as a lack of capital and depopulation. The prevalence of 'backwash' effects, then, means that industry becomes concentred in growth regions, creating an entrenched pattern of core-periphery differentials.

The second set of effects, however, are **'spread' effects** where surrounding regions benefit from increased growth in the core region. One important mechanism here is increased demand in the core region for food, consumer goods and other products, creating opportunities for firms in peripheral

regions to supply this growing market (Knox *et al.* 2003: 243). At the same time, rising costs of land, labour and capital in the core region, together with associated problems like congestion, can push investment out into surrounding regions. Rising costs reflect increased demand and the tendency for growth to outstrip the capacity of the underlying infrastructure to support it. This problem, which has periodically affected the economies of major core regions like south-east England, is often referred to as 'over-heating'. As a result, capital flows out into lower-cost regions, followed by labour. This is a process of **spatial dispersal**, where industry moves out of existing centres of production into new regions. It will create a geographically balanced economy if it is the dominant process over a long period of time.

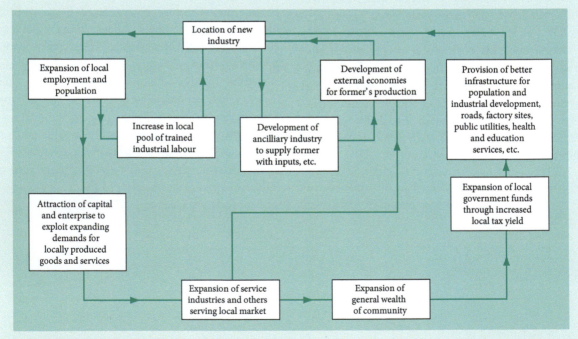

Figure 3.6 The process of cumulative causation
Source: Chapman and Walker 1991: 74.

When we look at patterns of regional development across different Kondratiev cycles, a more complicated and dynamic picture emerges. Many formerly leading regions (for example, nineteenth-century industrial areas in Europe and North America) have been challenged and eclipsed by 'rising' regions such as the US 'sunbelt' or East Asia since the 1970s. At the same time, a select number have been able to maintain their position – particularly metropolitan core regions around cities such as New York and London – whilst much of the developing world outside East Asia has remained peripheral to the world economy. In broad terms, the economic geography of the world can be explained in terms of the interaction between the patterns of investment associated with successive Kondratiev cycles (Massey 1984).

3.3.2 The development of industrial regions in the nineteenth century

The first Kondratiev cycle involved early mechanisation based on water power and steam engines from the 1770s. This was focused on cotton textiles, coal and iron working, facilitated by the development of river systems, canals and turnpike roads for transporting raw materials and finished products. The cotton textiles industry was the main focus of the industrial revolution, acting as the "pacemaker" of change and providing the basis for the first industrial regions (Hobsbawm 1999: 34). The industrial revolution greatly increased output whilst reducing production costs massively. This occurred through the creation of economies of scale, referring to the tendency for firms' costs for each unit of output to fall when production is carried out on a large scale, reflecting greater efficiency. By 1812, for example, the cost of making cotton yarn had dropped by nine-tenths since 1760 and by 1800 the number of workers needed to turn wool into yarn had been reduced by four-fifths (Rempel, undated: 2). By the 1830s and 1840s, however, markets were not growing quickly enough to absorb output, leading to reduced profits, falling wages and unemployment (Box 3.5) (Hobsbawm 1999: 54–5).

This first wave of industrialisation took off in certain regions of Britain, principally Lancashire, the West Riding of Yorkshire, the West Midlands, west-central Scotland and South Wales (Figure 3.7). Possession of substantial coal reserves gave these regions important advantages as industry became increasingly dependent on coal for energy from the 1820s. The development and improvement of canal systems was crucial in enabling raw materials and finished goods to be transported economically.

The textiles industry – at the heart of the industrial revolution – was concentrated in two main regions: Lancashire which specialised in cotton products and the West Riding of Yorkshire which focused on woollens. The city of Manchester, for example, experienced explosive growth as the industrial metropolis of the early nineteenth century ('cottonopolis'), with its population multiplying tenfold between 1760 and 1830 (Hobsbawm 1999: 34).

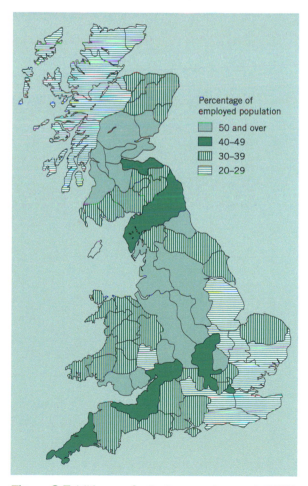

Figure 3.7 UK manufacturing employment, 1851
Source: Lee 1986: 32.

Percentage of employed population

- 50 and over
- 40–49
- 30–39
- 20–29

Box 3.5

The decline of the handloom weavers

The changing fortunes of the hand-loom weavers of England in the late eighteenth and early nineteenth centuries illustrate how the processes of 'creative destruction' associated with industrialisation actually operate, particularly in terms of the impact of new technologies on labour.

The cotton textiles industry – based in Lancashire and West Yorkshire in northern England (Figure 3.8) – in particular was transformed by a sequence of innovations in the late eighteenth century. These included Arkwright's water frame (1771), Hargreaves's spinning jenny (1778) and Compton's mule (1779) which combined the jenny and the water frame. The effect was to revolutionise the spinning process, allowing yarn to be produced in large quantities, feeding into weaving which had been sped up by the flying shuttle (1733).

Weaving expanded hugely in this period, absorbing the growing quantity of yarn through the multiplication of handloom weavers. In retrospect, the years from 1760 to 1810 appear as something of a 'golden age' for handloom weaving with the number of handloom weavers in cotton rising from about 30,000 to over 200,000 (Thompson 1963: 327). Demand for cloth was high and wages rose steadily, attracting many new entrants to the trade. Yet the general climate of prosperity disguised a loss of status as the weavers' ancient craft protections were eroded, with all restrictions on entry to the trade collapsing (ibid.: 305–6).

The removal of such protections exposed the weavers to savage wage cuts after 1800 as markets became glutted through the expansion of output, with a surplus of cheap goods and fierce competition driving wages down. The decline actually pre-dated the spread of power-looms in the cotton textile industry, but this became a key force which rendered handloom weaving uncompetitive. As a result, the average weekly wage of the handloom weaver in Bolton fell from 33s in 1795 to 14s in 1815 to 5s6d in 1829–34 (Hobsbawm 1962: 57). Production moved from the cottage to the factory as the way of life and culture of the handloom weavers were undermined.

Widespread distress and desperation saw the weavers resort to expressions of impotent 'rage against the machine' by smashing the new

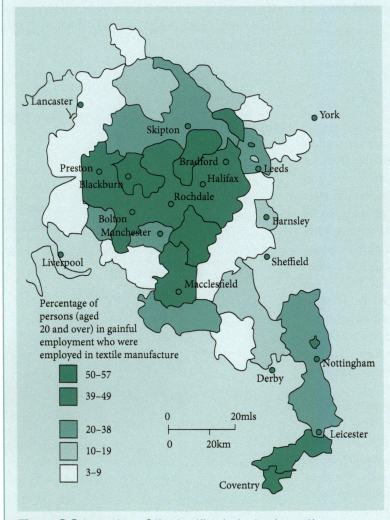

Figure 3.8 Location of the textiles industry in northern England, 1835
Source: Lawton 1986: 109.

Percentage of persons (aged 20 and over) in gainful employment who were employed in textile manufacture

- 50–57
- 39–49
- 20–38
- 10–19
- 3–9

Box 3.5 (continued)

power-looms and factories. Along-side such Luddite acts, strikes were organised and petitions to the government launched. The doctrines of political economy that held sway at the time, however, prohibited any interference with the 'natural' operation of market forces, and the government responded with repression.

By the 1820s, the great majority of weavers were living on the borders of starvation (Thompson 1963: 316). In the town of Blackburn, at the heart of Lancashire's textile industry, over 75 per cent of handloom weavers were unemployed in the early 1820s. Seventy-six people entered the Blackburn Workhouse

in one week in April 1826, bringing the total 'crammed together' to 678 (Turner, undated). The labour of the handloom weavers had been rendered obsolete by the industrial revolution, demonstrating how technological innovation could displace skilled labour from established craft industries.

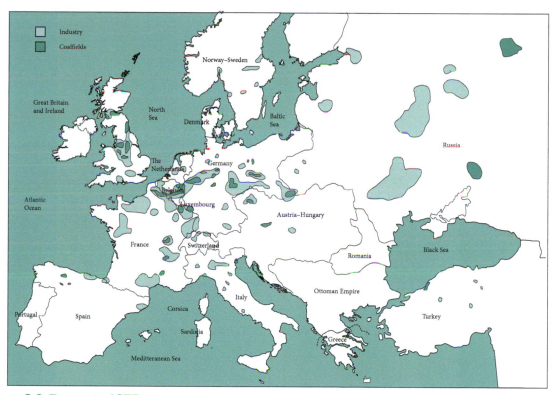

Figure 3.9 Europe in 1875
Source: Pollard 1981.

A second cycle of industrialisation began in the 1840s, lasting until the 1890s, just as the first textile-based phase had reached its limits. It was based on the industries of coal, iron and steel, heavy engineering and shipbuilding, creating a much firmer foundation for economic growth. Advances in transportation were of central importance here, particularly the development of the railways. The industrial regions of Britain maintained their success during the second Kondratiev cycle from the 1840s to the 1890s, based on the railways, iron and steel and heavy engineering. At the same time, other regions of continental Europe such as southern Belgium, the German Ruhr, parts of northern, eastern and southern France (Figure 3.9), together with the north-east of the United States, experienced rapid industrialisation. Again, however, a phase of expansion from the 1840s to the early 1870s gave way to depression as markets became saturated and prices fell.

The development of the internal combustion engine and electricity was central to the third wave, lasting from the 1880s to the 1920s, associated with automobiles, oil, heavy chemicals and plastics. Britain lost its lead to Germany and the US who pioneered the development of the new technologies. The characteristic pattern of a growth phase being succeeded by stagnation was again apparent with the boom years from the mid-1890s to 1914 matched by the inter-war depression.

This period saw industrialisation spread to 'intermediate Europe', parts of Britain, France, Germany and Belgium not directly affected by the first two waves as well as northern Italy, the Netherlands, southern Scandinavia, eastern Austria and Catalonia (Pollard 1981). The position of the US manufacturing belt in the northeast and Midwest was reinforced by new rounds of

cumulative causation (Figure 3.10) whilst Japan also began to experience industrial growth.

Peripheral Europe, on the other hand – encompassing most of Spain and Portugal, northern Scandinavia, Ireland, southern Italy, east-central Europe and the Balkans – was left behind, becoming specialised in a subordinate role supplying agricultural products and labour to the core regions (Knox *et al.* 2003: 148). The relationship between these industrial cores and the surrounding territories was defined by 'backwash' rather than 'spread' effects as industrialisation sucked in capital and labour, creating an uneven economic landscape of urban-industrial cores surrounded by extensive rural peripheries (Box 3.4).

By the early twentieth century, the spatial concentration of industry in a small number of specialised industrial regions in Britain was readily apparent. These regions

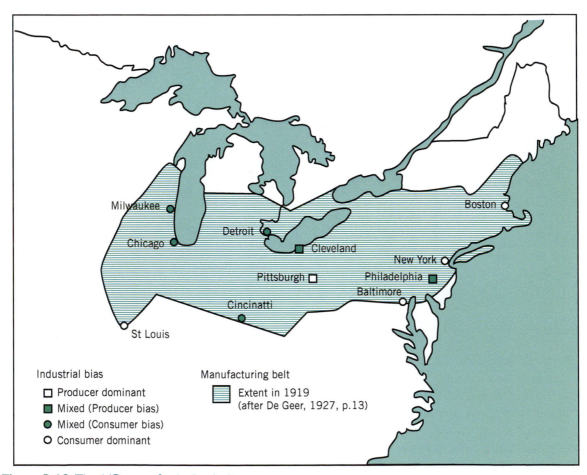

Figure 3.10 The US manufacturing belt
Source: Knox *et al.* 2003: 156.

were built on a coalfield base, supplying the energy for the heavy industries that emerged from the 1840s. In addition to textiles, the iron and steel and shipbuilding industries became highly concentrated. In shipbuilding, for instance, north-east England and west-central Scotland accounted for 94 per cent of employment in Britain in 1911 (Figure 3.11). South Wales was another leading centre of heavy industry by this stage, focused around coal, iron and steel, whilst the West Midlands became specialised in engineering and metal industries. These industrial regions developed a certain '**structured coherence**' as centres of heavy industry (Harvey 1982), becoming working-class regions with strong Labour Party and trade union traditions.

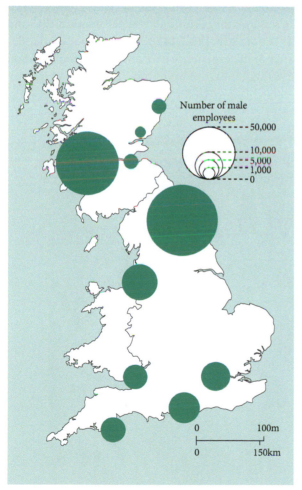

Figure 3.11 Shipbuilding employment in Britain, 1911
Source: Slaven 1986: 133.

3.4 From Fordism to flexible production

3.4.1 Fordism and mass production

Following the initial development of the factory system, further developments took place in the early twentieth century. These can be described as Fordism, after the American automobile manufacturer Henry Ford who introduced the key innovations (section 2.4.3). Fordism is based on an intensification of the labour process, developing techniques of mass production. This was to be balanced by mass consumption with increased wages for workers giving them additional purchasing power in the market, creating the consumer demand needed to underpin mass production. In the 1930s and 1940s, Fordism became consolidated into a new **regime of accumulation** in Europe and North America, based on this link between mass production and mass consumption, supported by Keynesian economic policies of demand management and the development of welfare states (sections 2.4.3, 5.3). A key moment in the development of Fordism as labour process was Ford's introduction of the $5 day at his Highland Park automobile plant in Detroit, Michigan in January 1914. This succeeded in reducing labour turnover and allowed workers themselves to buy the cars they were producing.

The first key element of the Fordist organisation of production occurred with the introduction of **scientific management or Taylorism** – after its key advocate F.W. Taylor – which involved the reorganisation of work according to rational principles designed to maximise productivity. Three key elements of scientific management can be identified (Meegan 1988):

➤ A greatly increased technical division of labour, based on the complete separation of the design and planning of work, undertaken by management, from its execution by workers who became increasingly focused on simplified and repetitive tasks.

➤ The subdivision of operations was matched by the reintegration of the production process, involving increased coordination and control by management who were to exercise complete authority over the planning and direction of work, removing the power of foremen and workers.

➤ The performance and organisation of workers was subject to very close monitoring and analysis by management, employing techniques such as time-and-motion studies.

This organisational revolution was matched by experimentation with new techniques to increase productivity. The most famous and widely adopted of these is the moving assembly line, first established at Ford's Highland Park plant between 1911 and 1913 (Figure 3.12). This revolutionised production methods in the automobile industry. Rather than the worker assembling cars by moving around a factory to pick up different parts, the parts now came to the worker who would be placed at a fixed position with typically just one dedicated task:

'The man who places the part does not fasten it', said Henry Ford. 'The man who puts in a bolt does not put on the nut; the man who puts on the bolt does not tighten it'. Average chassis assembly time fell to ninety-three minutes. The lesson was obvious. Within months Highland Park was a buzzing network of belts, assembly lines, and sub-assemblies . . . The entire place was whirled up into a vast, intricate and never-ending mechanical ballet.

(Lacey 1986, cited in Meegan 1988: 142)

The saving of time and restructuring of work tasks had a dramatic effect on productivity; between 1911 and 1914 production of cars quadrupled from 78,000 to around 300,000 whilst the workforce only doubled in size over the period and even fell between 1913 and 1914 (Meegan 1988: 143). Box 3.6 gives a brief flavour of the experience of 'working for Ford' (Beynon 1984).

Figure 3.12 Ford assembly line
Source: Mary Evans Picture Library.

Box 3.6

Working for Ford

The industrial sociologist Huw Beynon (1984) undertook a detailed study of working conditions at Ford's Halewood plant in Liverpool, north-west England in the late 1960s and early 1970s, conveying workers' experience of assembly line work:

> It's the most boring job in the world. It's the same thing over and over again. There's no change in it, it wears you out. It makes you awful tired. It slows your thinking right down. There's no need to think. It's just a formality. You must carry on. You just endure it for the money. That's what we're paid for, to endure the boredom of it all.
>
> If I had a chance to move I'd leave right way. It's the conditions here. Ford class you more as machines than men. They're on top of you all the time. They expect you to work every minute of the day.
>
> (Ford workers at Halewood, quoted in Beynon 1984: 129)

According to another worker, whilst the white-collar staff that worked in the 'office' were part of Ford's, the men on the shop floor were just regarded as 'numbers'. Getting used to the incessant demands of the production process was difficult for many new workers with no experience of working in a car plant. The effects of assembly line work could sometimes extend into domestic life as well. According to one night-shift worker,

> My wife always used to insist that I had my breakfast before I went to bed. And I would get into such a state that I would sit down to bacon and egg and the table would be moving away from me. I thought, 'crikey, how long am I going to have to put up with this?' But the pay was good. It was a case of really getting stuck in and saying 'to hell with it, get it while it's here'.

And this is the way it went, but the elderly chaps couldn't stand the pace.

(Joe Dennis quoted in Meegan 1988: 144)

It is important to recognise that the workers were not just passive, meekly accepting the dictates of management. The role of the trade unions was central, and in particular the plant-level shop stewards committee in representing the interest of the workers. One of the key things shaping workers' day-to-day experience was the speed of the line, with Beynon (1984: 148) stating that "the history of the line is a history of conflict over speed-up". After considerable disruption and struggle at Halewood, the workers won an important victory when management conceded the right to the shop stewards to hold the key that locked the assembly line (ibid.: 149), giving them more control over their work conditions.

3.4.2 Fordism and regional change

Under the fourth Kondratiev cycle based on the mass production of consumer durables, certain established manufacturing regions such as the US manufacturing belt and the Midlands of England further consolidated their position. Proximity to the main centres of population became important for these market-oriented Fordist industries. In Britain, for example, the new mass production industries were drawn to the Midlands and south-east England.

In the late 1920s and 1930s particularly, the Soviet Union experienced very rapid state-led industrialisation, although this was very much focused on heavy capital goods like iron and steel and heavy engineering rather than consumer goods. This was focused on a core manufacturing belt stretching south and east from St Petersburg and the eastern Ukraine through the Moscow and Volga regions to the Urals (Knox *et al.* 2003: 164) (Figure 3.13). The Japanese economy experienced very rapid growth, averaging around 10 per cent a year, in the 1950s and 1960s, driven by automobiles and electronics as well as more traditional heavy industries like shipbuilding and iron and steel.

The pattern of regional sectoral specialisation created by successive waves of industrialisation in Europe and North America began to break down from the 1920s. As the economic difficulties of the 1920s crystallised into a major depression in the 1930s, the output of industries such as coal, cotton and shipbuilding slumped. The effects were devastating in terms of unemployment, dereliction and poverty. At the peak

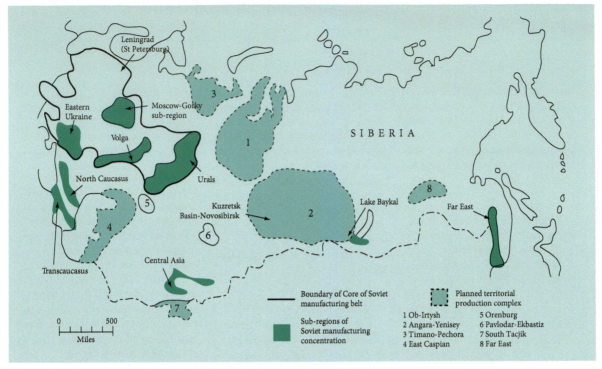

Figure 3.13 The manufacturing belt in the former Soviet Union
Source: Knox *et al.* 2003: 164.

of the slump in 1931–32, 34.5 per cent of coalminers, 43.2 per cent of cotton operatives, 43.8 per cent of pig iron workers, 47.9 per cent of steelworkers and 62 per cent of shipbuilders and ship repairers in Britain were out of work (Hobsbawm 1999: 187).

Such unemployment was concentrated in the traditional industrial regions of northern England, Scotland and Wales, prompting the official recognition of the 'regional problem' by government in 1928 as central Scotland, north-east England, Lancashire and South Wales were designated as 'depressed regions' requiring special assistance (Hudson 2003). As such, a growing north-south divide in economic and social conditions in Britain is apparent from the 1930s, although its roots stretch back further (Box 3.7). Rearmament brought a gradual economic recovery from the late 1930s, reinforced by war and subsequent reconstruction. By the 1960s, however, the underlying problems of nineteenth-century industrial regions in northern Britain, the US manufacturing belt, north-eastern France, southern Belgium and the German Ruhr were becoming increasingly severe.

By the 1960s and 1970s, a new phase of 'neo-Fordism' was apparent – broadly corresponding to the decline phase of this Kondratiev wave – as mass production technologies became increasingly routine and standardised. This was creating a new pattern based on the spatial dispersal of industry to peripheral regions. As Massey (1984) demonstrated, a new '**spatial division of labour**' was emerging in which different parts of the production process were carried out in different regions, reflecting underlying geographical variations in the cost and qualities of labour.

Increasingly, corporations were separating the higher-level jobs in areas like senior management and research and development from the lower-level and more routine jobs such as the processing or final assembly of products (Table 3.2). Through the emergence of large, multi-plant corporations, this division of labour takes on an explicitly spatial form, with companies locating the higher-order functions in cities and regions where there are large pools of highly educated and well-qualified workers, with lower-order functions locating increasingly in those regions and places where

Table 3.2 The spatial division of labour in manufacturing

Functions	Characteristics of the division of labour	Location
Research and development	Conceptualization/mental labour; high level of job control	South-east England
Complex manufacturing and engineering	Mixed; some control over own labour process	Established manufacturing regions like West Midlands
Assembly	Execution, repetition, manual labour, no job control	Peripheral regions such as Cornwall or north-east England

Source: Adapted from Massey 1984. Reproduced with permission of Palgrave Macmillan.

costs (especially wage rates) are lowest. As a result, the organisation of particular industries becomes spatially 'stretched' through corporate hierarchies with different stages of production carried out in different regions (Massey 1988). This contrasts with the nineteenth-century pattern of regional sectoral specialisation where all the main stages of production were concentrated within the same region (section 3.3.).

When applied to the UK, Massey's analysis reveals a sharp divide between London and the south-east of England where a disproportionate amount of company headquarters and R&D facilities are located compared to the outlying regions of the UK such as Scotland, Wales and the north of England which are dominated by 'branch plant' activities. Concentrations of professional and managerial labour could be found in the Southeast whilst peripheral regions contained significant surpluses of lower-cost labour, particularly women, to perform routine work in factories. This spatial dispersal of routine production has also occurred on an international scale through the **'new international division of labour' (NIDL)** (Froebel *et al.* 1980) as TNCs based in western countries have shifted low-status assembly and processing operations to developing countries where costs are much lower (section 3.5). Partly as a result, the newly industrialised countries of East and Southeast Asia have become important centres of industrial production.

3.4.3 Deindustrialisation and 'new industrial spaces'

The geography of the 'fifth Kondratiev cycle' is based on the rise of new 'sunrise' industries such as advanced electronics, computers, financial and business services and biotechnology, employing **flexible production methods**. At the same time, many traditional manufacturing regions have experienced serious deindustrialisation since the late 1960s, as manufacturing industry has declined in the face of competition, overproduction and reduced demand. The legacy for old industrial regions has been one of high unemployment, poverty, industrial dereliction and decay. The West Midlands of England, for example, part of the industrial heartland of the UK in the 1950s and 1960s, lost over half a million manufacturing jobs between 1971 and 1993, 50 per cent of its total manufacturing employment (Bryson and Henry 2005: 358).

The process of 'creative destruction' associated with the rise of a new technology system based upon information technology has led to some dramatic geographical shifts, in terms of, for example, the US 'rustbelt' (the north-east and Midwest) and 'sunbelt' (the south and west), the north and south of Britain (Box 3.7) and north-western and southern Germany. For example, whilst the mid-Atlantic region of the US (New Jersey, New York, Pennsylvania) experienced a net loss of over 175,000 jobs between 1969 and 1976, the south Atlantic region experienced a net gain of over 2 million jobs in the same period (Knox *et al.* 2003: 230). New investment in high-technology industries has been attracted to 'new industrial spaces' distinct from the old industrial cores which offered attractive environments and a high quality of life for managerial and professional workers. This is part of a broader spatial division of labour with industries like computers and semiconductors organised on a global basis. Typically, for instance, R&D functions might be based in Silicon Valley, skilled production carried out in the central belt of Scotland (the so-called 'Silicon Glen'), assembly and testing in the likes of Hong Kong and Singapore and routine assembly in low-cost locations in the Philippines, Malaysia and Indonesia (ibid.: 235–6).

Box 3.7

Britain's north-south divide

Concerns about a north-south divide in levels of wealth and prosperity in Britain have been periodically expressed since the 1930s (Massey 2001). This has sparked intense political debate in the 1980s, the late 1990s, and more recently following the financial crash and recession of 2008–09, reflecting a widening of the regional economic divide as London and the south of England have grown much faster than the rest of the country since the mid-1990s (Martin *et al*. 2016: 344). On a number of measures, the UK has some of the most pronounced regional disparities in the industrialised world (McCann 2016). It also has about the most centralised system of government of any major industrialised country, fuelling recurring proposals for the decentralisation of power to local and regional authorities (Lee 2016).

In the 1980s, the combined impact of the neoliberal reforms of the Thatcher government and wider processes of deindustrialisation seemed to have resulted in the emergence of 'two nations': a prosperous and dynamic south in which most of the growth industries were located and a stagnant and impoverished north, scarred by industrial dereliction, poverty and unemployment (Martin 1988). The late 1990s saw evidence emerge that the divide had widened under New Labour (Massey 2001), creating political problems for a government which drew many of its Ministers and Members of Parliament (MPs) from the north. More recently, the Conservative–Liberal Democratic Coalition Government of 2010–15 sought to rebalance the economy sectorally and geographically away from the dominance of finance and business services in south-east England to support a revival of manufacturing in the north. This concern also informs the post-2015 Conservative Government's plans to create a 'Northern Powerhouse' of growth to match London and to devolve more powers to northern city-regions (Lee 2016).

As Martin (1988) argues, the roots of the north-south divide stretch beyond the 1930s to the nineteenth century, creating a deeply entrenched pattern of regional inequality. Whilst the image of a dynamic, industrial north and sleepy agricultural south in this period remains powerful, the real situation was more complex. For one thing, the South-east was actually the most affluent region throughout the nineteenth century, reflecting the diversified nature of its economy, and the gap with other regions widened from the 1850s. At the same time, the economies of many northern regions remained somewhat precarious, exhibiting a heavy dependence on a narrow range of export-oriented staple industries. This was particularly the case for what Martin calls the 'industrial periphery' of Scotland, the northern region and Wales (Figure 3.14).

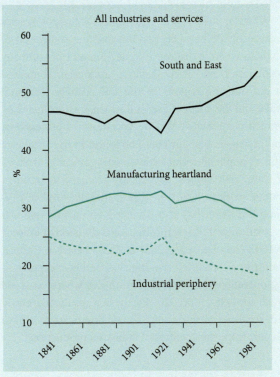

Figure 3.14 The regional distribution of employment in Britain, 1841–1981
Source: Martin 1988: 392.

Box 3.7 (continued)

These industries collapsed in the difficult economic climate of the 1920s, plunging the north into an economic and social crisis. As we have indicated, moreover, the new consumer industries of the inter-war years were drawn to the large market of the South-east and Midlands.

The post-war period saw the economic dominance of the South-east consolidated through the expansion of modern, lighter industries and the service sector, compared to the sluggish growth of much of the north. The deindustrialisation of the 1970s and 1980s decimated the northern regions whereas the financial and business service and high-tech manufacturing sectors which grew rapidly in the 1980s and 1990s were predominantly located in the south and east of the country. Whilst the recession of the early 1990s was most severe in the South-east, leading to a temporary narrowing of the divide, this was soon overtaken by the geography of economic recovery after 1993, which generated a sharp increase in regional inequality with London and the South-east growing more quickly than the remainder of the country (Martin et al. 2016). This reflects the tendency for regional disparities to widen more in the recovery and growth phases of the economic cycle, particularly between 1982 and 1989 and between 1992 and 2007 (Gardiner et al. 2013: 897). This tendency seems to be reoccurring as part of a rather weak and uneven post-2010 economic recovery with Townsend and Champion (2014) finding that the London region experienced a strong economic rebound in 2010–12. By contrast, recovery is taking longer in many of the cities and regions in the north and Midlands of England which were hit hardest by the 2008–09 recession (Industrial Communities Alliance 2015).

Three different kinds of 'new industrial spaces' have been identified in Europe and North America (Knox et al. 2003: 237):

➤ Craft-based industrial districts containing clusters of small and medium-sized firms producing products such as textiles, jewellery, shoes, ceramics, machinery, machine tools and furniture (Figure 1.7). Examples include the districts of central and north-eastern Italy, Jura in Switzerland, parts of southern Germany and Jutland in Denmark. This production system is based on high levels of subcontracting and **outsourcing**, often relying on family labour and artisan skills.

➤ Centres of high-technology industries such as advanced electronics, computer design and manufacturing, pharmaceuticals and biotechnology. Examples include Silicon Valley in California (see Box 3.8), Route 128 around Boston in the US; the M4 corridor and Cambridge region in the UK; and Grenoble and Sophia-Antipolis in France. Such areas are characterised by rapid growth and high levels of innovation, often based around small firm networks, although they have spawned some important TNCs.

They are often close to major cities but offer a high quality of life for workers with an integrated local labour market that enables workers to switch jobs without leaving the area. Links with universities are often significant in terms of providing the research and development infrastructure that supports innovation and learning.

➤ Clusters of advanced financial and producer services, often in central districts of large world cities such as London, New York and Tokyo. The City of London is a good example. The concentration of corporate headquarters functions and major financial institutions found in such metropolitan regions is supported by specialised networks of firms in activities such as accountancy, legal services, management consultancy and advertising. These regions are characterised by high levels of specialisation and large pools of skilled, white-collar labour, although low-wage female and ethnic minority labour is also important. Geographical proximity is important for specialised service firms which are often putting together packages for firms, requiring face-to-face communication and trust (Thrift 1994).

Box 3.8

The development of 'Silicon Valley'

Over the last half century or so, Silicon Valley in California has become probably the most renowned centre of high-tech industry in the world. As such, it has attracted attention from a range of government and development agencies seeking to emulate its success. Whilst the development of Silicon Valley reflects some specific historical circumstances, its experience does highlight some of the factors which explain why particular types of high-tech industry tend to become concentrated in particular places, including links with universities, the availability of skilled labour and the crucial role of social networks (Saxenian 1994).

Silicon Valley is a strip of land in Santa Clara County to the south of San Francisco, stretching from Palo Alto to San Jose (Figure 3.15). In the 1940s and 1950s, this was a sparsely populated agricultural area focused on fruit production. Building on the work of Castells and Hall (1994), its subsequent development and transformation occurred in seven main stages:

➤ The historical roots of technological innovation, dating back to the early twentieth century, based upon a research tradition in electronics which saw the invention of the vacuum tube in 1912 by a firm, the Federal Telegraph Company, founded by a Stanford University graduate with support from the university. Stanford maintained a tradition of excellence in electronics in the 1920s, with many students staying to establish their own firms in the area (ibid.: 15).
➤ The growth of high-tech industry in the 1950s around the Stanford

Industrial Park. Here, the links with Stanford University were crucial. An engineering professor, Frederick Terman, was instrumental in the creation of the Stanford Industrial Park which attracted a growing number of innovative firms (Saxenian 1994). Another key figure was William Shockley, a leading physicist who had won the Nobel Prize for physics with two colleagues in 1955 for their invention of the transistor at Bell Laboratories in New Jersey in 1947. Shockley left Bell in 1954 to set up his own company and moved to Palo Alto, California where his ailing mother lived, establishing Shockley Semiconductor there. Shockley recruited eight talented young researchers from the eastern US to work with him, including Robert Noyce, who invented the integrated circuit in 1957. As such, the recruitment of skills and labour from outside the region was central to the growth of Silicon Valley from the beginning. These eight researchers found Shockley increasingly difficult to work with and left to form Fairchild Semiconductors in 1957.
➤ The growth of innovative electronics companies in the 1960s through spin-offs from the original firms. Fairchild Semiconductors played a particularly significant role as the incubator for the launch of a number of spin-off firms, which became known as 'fairchildren', including Intel and National Semiconductors. This phase of growth was underpinned by strong demand for electronics devices from the military which provided around 50 per cent of

the market for semiconductors in the 1960s (Castells and Hall 1994: 17).
➤ The consolidation of semiconductor firms and the launching of the personal computer (PC) era in the 1970s. While the PC was first produced in Albuquerque, New Mexico in 1974, this event stimulated the activities of a network of computer hobbyists, the Home Brew Computer Club in the San Francisco Bay Area, whose members included Bill Gates and Steve Wozniak. These activities spawned the establishment of firms such as Apple Personal Computer, founded by Wozniak and Steve Jobs in 1976, and Sun Microsystems in the early 1980s as the Valley became increasingly specialised in advanced microelectronics and computers.
➤ The growing dominance of the computer industry, the internationalisation of production and a new round of innovative spin-offs in the 1980s and early 1990s. The region experienced an economic downturn in 1984–86 as a result of a global contraction in the industry and growing Japanese competition, resulting in the layoff of more than 21,000 workers (ibid.: 20). Partly in response, Silicon Valley became increasingly specialised in research and development (R&D), design and advanced manufacturing, whilst the higher-volume and more standardised part of the production process were increasingly relocated to lower-cost regions in the US initially and subsequently internationally as part of an evolving spatial division of labour in the industry. At the same time, the

Box 3.8 (continued)

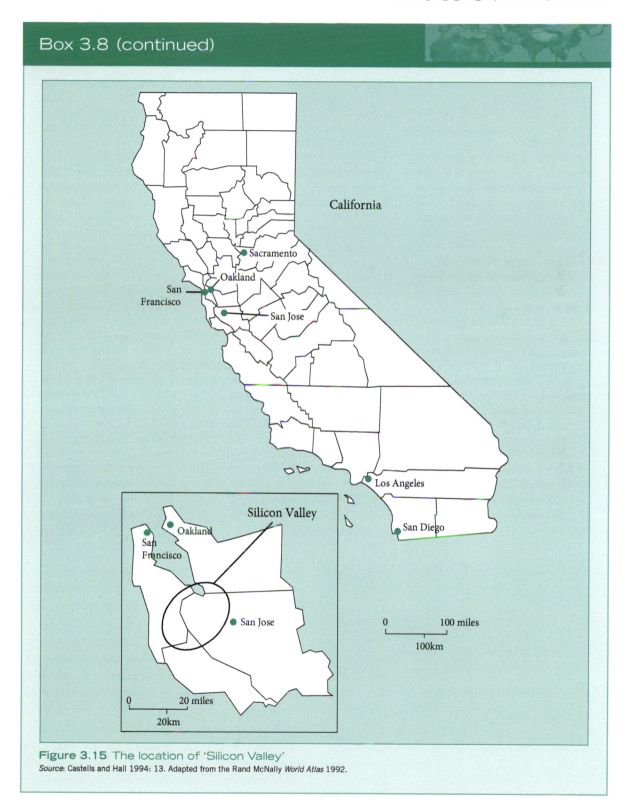

Figure 3.15 The location of 'Silicon Valley'
Source: Castells and Hall 1994: 13. Adapted from the Rand McNally *World Atlas* 1992.

Box 3.8 (continued)

social networks, information and capital of the region continued to generate new firm start-ups and spin-offs into the late 1980s and early 1990s.

➤ Building upon its accumulated success in electronics and computing, Silicon Valley experienced a new wave of growth from the mid-1990s based upon the explosion of internet technology. This gave rise to "an economic boom of unparalleled ferocity [that] hit the place like a bombshell", being likened to a frenzied new high-tech gold rush (Walker 2006: 121). It was fuelled by soaring stock market values with the floating of Netscape in 1995 symbolising this trend. Other start-ups in this period included Yahoo, while Cisco Systems, makers of internet hardware, overtook Microsoft to become the most highly valued company in the world in March 2000. The process of growth expanded to San Francisco itself, particularly through the concentration of 'multimedia' operations incorporating movie special effects, video games and electronic publishing in the area south of Market Street. Reflecting the broader tendency towards boom and bust patterns of development, however, this phase of frenzied growth was followed by a sharp recession triggered by the 'dot.com crash' of stock markets in 2000 with the NASDAQ losing more than 77 per cent of stock values by 2002 (ibid.: 129). As the centre of the internet economy, the Bay Area was the worst hit of any region in the US by this crash with Silicon Valley losing a fifth of its jobs in three years (*The Economist* 2004).

➤ The most recent period of growth since the mid-2000s has been based on more dot.com start-ups, alongside the growth of social media and electronic arts companies. Leading firms based in Silicon Valley in this period include Google, Facebook, ebay, Twitter and Netflix.

In explaining the success of Silicon Valley, Saxenian (1994) has drawn attention to its distinctive **social networks**. These have facilitated high levels of informal communication and cooperation between individual engineers and entrepreneurs, often working in different firms, allowing technical information and ideas to be rapidly circulated and shared. High levels of labour mobility represent a key mechanism for the diffusion of technology as people move between firms or leave to establish their own spin-off ventures. The success of the early pioneers provided a ready source of venture capital and created a culture of entrepreneurialism and individualism which has spurred subsequent generations of innovators. The great advantage of this industrial system is that it enabled the region to maintain competitive advantage through continuous innovation, something which, somewhat paradoxically, required cooperation between firms (ibid.: 46). In particular, close geographical proximity fostered face-to-face communication and trust between key individuals, providing the region with the adaptability and responsiveness that has allowed it to maintain its competitive advantage.

Reflect

➤ Compare and contrast the locational patterns associated with regional sectoral specialisation and spatial divisions of labour respectively.

3.5 New international divisions of labour

3.5.1 The process of internationalisation

The last of the three periods of industrial production outlined in this chapter is defined by globalisation which originated in the increasing internationalisation of production in the 1960s and 1970s. This reflects the efforts of TNCs to increase their rates of profit. Increases in profits are achieved through increasing the productivity of existing operations, by increasing the exploitation of labour or by developing new technology or new forms of organisation. This can continue successfully for some time, but at some point the returns to capital from remaining in its original location begin to fall, either through the growing organisation and resistance of labour, increased competition from other firms, or the achievement of productivity limits in existing technologies and methods of production. Addressing these falling profit rates requires firms to seek out cheaper locations than existing arrangements. The need to resolve this dilemma and

Table 3.3 Hymer's stereotype of the new international division of labour

Level of corporate hierarchy	Major metropolis (e.g. New York)	Type of area Regional capitals (e.g. Brussels)	Periphery (e.g. South Korea, Ireland)
1. Long-term strategic planning	A		
2. Management of divisions	D	B	
3. Production, routine work	F	E	C

Source: Sayer 1985: 37.

develop a new geography of production has been termed the **'spatial fix'** (Harvey 1982), whereby capitalists seek to set up operations elsewhere. This will often involve relocation within the existing national economy in the first instance, creating a new spatial division of labour (Massey 1984), but over time, as new limits to profitability are reached at the national level, firms will start to internationalise their operations. This is what happened in the 1970s and 1980s as TNCs responded to the crisis of the post-war Fordist regime of accumulation by relocating their operations to lower-cost locations.

Investment overseas is usually driven by the desire for access to either markets or new sources of labour and raw materials (section 3.2.4). While much of the relocation that occurred in the 1960s and 1970s was driven by labour cost considerations, market access generally accounts for a larger share of foreign direct investment (FDI). In practice, the two drivers may often be interlinked, with China, for instance, attracting large inflows of investment on account of both its huge market and abundant supplies of labour (Dicken 2015).

3.5.2 The new international division of labour

The wave of internationalisation that occurred in the 1970s and 1980s reflected a search for cheaper labour costs. This has led to a growing level of investment in the developing world, associated with a phenomenon known as the **'new international division of labour'** (**NIDL**) (Froebel *et al.* 1980), replacing or more accurately interacting with the old international division of labour inherited from the colonial era (section 3.2.4). The simple model envisaged by economist Stephen

Hymer, dating from the 1960s, was one in which TNCs developed a new geography of production by reorganising the division of labour within the firm. Higher-level decision making and research and development activities would remain concentrated in the major metropolitan regions of the advanced world (e.g. London, New York, Paris, etc.) whilst the more routine production activities would be dispersed, depending upon labour skills and levels of technology, to more peripheral locations at home (see Table 3.3), equating to a spatial division of labour at the inter-regional scale as outlined in section 3.4.2. Over time, the more basic production functions will be relocated to cheaper locations overseas with a 'race to the bottom' encouraging the search for low-cost locations in the global South. The developed countries would, however, retain the higher value-added activities such as long-term strategic planning and research and development. The NIDL represents a spatial division of labour on a global scale, reflecting the 'global shift' of manufacturing production towards lower-cost countries in the global South from the 1970s (Dicken 2015).

The NIDL has been shaped and facilitated by the interaction between three underlying factors (Wright 2002):

➤ an extended technical division of labour in production which reduced work to simple routine functions that needed minimal training, allowing them to be performed by semi-skilled workers (see section 3.3.2);

➤ the development of advanced transportation technologies such as containerisation and air freight that allow materials and finished/semi-finished goods to be moved cheaply and efficiently over large distances, making production 'footloose';

➤ the release of agricultural labour as a low-wage resource onto urban labour markets in less developed economies through the modernisation and intensification of agriculture under the 'Green Revolution'.

Wage rates vary enormously between countries (Figure 3.16), attracting investment in labour-intensive industries to countries with abundant supplies of low-cost labour. The NIDL has been particularly concentrated in high-volume, labour-intensive industries such as textiles, footwear, toys and electronics.

Relocation often occurred to geographically proximate countries in the 1960s and 1970s, for example from Western Europe to southern Europe and North Africa, from the US to Mexico and from Japan to the **newly industrialising countries (NICs)** of East Asia. For developing countries, gains in employment were

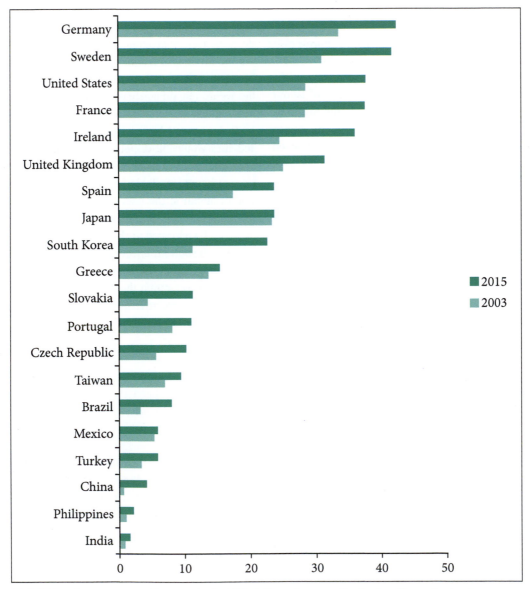

Figure 3.16 Hourly compensation costs, in US dollars, in manufacturing, for selected countries 2003, 2015

Source: www.conference-board.org. International Labor Comparisons database.

often associated with poor working conditions and low pay by global standards, though wages were often higher than for other types of work available in the host economy. Production often takes place in **export processing zones (EPZs)** which are special zones that offer additional incentives to investors, typically involving 100 per cent rebates on local taxation, the provision of infrastructure and the relaxation of rules on foreign ownership. Lower levels of regulation and poor labour standards have often prompted claims of exploitation in EPZs, particularly of young women workers. In recent years, conditions in the Asian factories of the electronics subcontractor Foxconn, which manufactures products for Apple and other global electronic brand names, have attracted criticism following a spate of worker suicides in 2010.

The *maquiladora* plants of northern Mexico are well-known examples of the NIDL pattern of investment. Originating in the 1960s and reinforced by the signing of the North American Free Trade Agreement (NAFTA) in 1994, *maquiladoras* are plants in which foreign-owned TNCs, many of them from the US, assemble products for overseas markets, particularly the US, using largely imported components. This is designed to capitalise on lower labour costs, very limited taxation and weak environmental regulation (Coe *et al.* 2013: 309). Some 5,000 *maquiladora* plants accounted for around 1.8 million jobs in Mexico in April 2011, although competition from China and the financial crisis of 2008 and its aftermath has reduced growth (ibid.).

A key trend that has been underway since the 1980s is the shift from TNCs setting up their own production facilities in developing countries to subcontracting work out to locally based firms. Such a change provides TNCs with greater locational flexibility to switch suppliers and avoid the sunk costs that are involved in committing investment to fixed assets (Clark 1994). This organisational process of outsourcing or subcontracting to another firm is often intimately tied up with the parallel process of geographical relocation or 'offshoring' to distant lower-cost countries. Outsourcing is often used as a generic shorthand for this, particularly in the business literature. 'Offshoring' is distinct from the related geographical processes of 'near-shoring', referring to the

relocation of services to (usually neighbouring) countries of a broadly similar level of development (for example, UK-Ireland, US-Canada), and 'onshoring' when services are relocated within the same country (Bryson 2006).

The clothing firm Nike is a frequently quoted example of a TNC that maintains spatial flexibility through a vast web of 'independent' suppliers. Headquartered in Beaverton, Oregon, Nike styles itself as 'a firm of marketers and designers', with no in-house production (Coffey 1996). As such, production is wholly outsourced to subcontractors' factories, most of which are located in Asia. This means that Nike indirectly employs more than 800,000 workers in its 'contract supply chains' of over 600 factories (Dicken 2015: 160). The geography of its supply chain has changed over time with continual relocation between countries in search of lower costs. Now the vast majority of its East Asian factories (167) are located in China, followed by Vietnam and Thailand (ibid.).

3.5.3 Services and the 'new' NIDL

Another key aspect of globalisation over the past couple of decades is the growing importance of FDI in services alongside manufacturing. Over the past decade or so, this has attracted enough attention to be regarded as a key process in its own right. Indeed, Bryson (2016) goes so far as to identify a 'second global shift' involving the international relocation of certain service activities to developing countries which is creating a 'new' NIDL. This process of relocation or 'offshoring' really began to gather momentum and attract attention in the early 2000s as commentators, politicians and trade unions in developed economies became increasingly concerned about the potential loss of service employment (ibid.), which had itself often served as a partial replacement for the loss of manufacturing jobs in earlier decades (Richardson *et al.* 2000). It parallels the global shift of manufacturing in the 1970s and 1980s in some respects, reflecting a combination of three main factors (Gordon *et al.* 2005: 19):

➤ technical developments in telecommunications and internet provision, ensuring that "many clerical tasks have become increasingly footloose and susceptible to spatial variations in production costs" (Warf 1995: 372);

➤ increased pressures to reduce costs across a range of service industries, as the sustained growth of the 1990s gave way to a period of slower growth and increased competition from 2000;

➤ the discovery and, to some extent, the creation, of new pools of skilled labour within certain low-wage economies.

Traditionally, the production and consumption of services has been seen as inseparable with both occurring locally. The development of ICTs, however, has allowed firms to overcome this constraint by separating them in geographically dispersed or 'stretched' value chains (Pisani and Ricart 2016). This has been termed a new NIDL, based upon outsourcing and the extension of the NIDL into the delivery of consumer services as firms take advantage of geographical variations in the costs of labour (Hudson 2016).

There is nothing particularly new about the spatial dispersal of services with 'back-office' functions – routine clerical and administrative tasks like the maintenance of office records, payroll and billing, bank checks and insurance claims – long having been subject to relocation out of city centres to surrounding suburbs where property (rent) and labour costs are typically much lower. What is new is that ICT opens up the possibility of relocation at the regional and, especially, global scales. One of the earliest examples of the international relocation of services was New York financial firms relocating administrative work to western Ireland, actively encouraged by the Irish government. These firms sent life insurance claims to a site located near Shannon airport for completion and checking before the processed data was sent back to New York via fibre-optic cable or satellite (Figure 3.17). Similarly, the Caribbean has become a favoured destination for US back-office functions. This reflects its combination of low wages and geographical proximity, giving it a similar relationship to the US in terms of back-office functions as Mexico holds for manufacturing (albeit on a smaller scale). In the example shown in Figure 3.17, American Airlines sent its tickets for processing to a subsidiary in Barbados, a process that began in 1981. The subsidiary opened a second office in the Dominican Republic, where wages are one-half as high as Barbados, in 1987.

In addition to traditional 'back-office' or business support functions, a range of other types of service

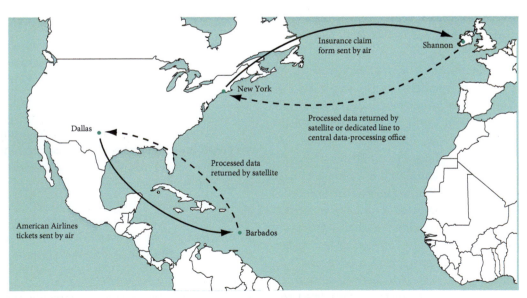

Figure 3.17 Offshore processing in airline and insurance industries
Source: Warf 1995.

activity are subject to spatial dispersal or 'offshoring' to low-wage locations, including:

➤ Call centres, focusing on marketing, routine customer enquiries and more sophisticated technical support. Recently rebranded as contact centres or customer service facilities, they have grown very rapidly since the early 1990s. Call centres were pioneered in the financial sector and have subsequently grown in areas such as travel, telecommunications, mail order and the deregulated utilities.

➤ IT functions, including data processing, codechecking, software development and modification, operations support, publishing and statistical analysis.

➤ More sophisticated back-office functions such as business services, accountancy, legal services, human relations and business and financial analysis in what is often known as 'business process outsourcing' (Bryson 2016). In particular, the big accountancy firms such as KPMG are capitalising upon the recent availability of 'big data', reflecting the latest developments in ICTs, to provide large-scale analytical services to a range of clients from a small number of integrated service hubs (Devi 2015).

Beyond the underlying emphasis on labour costs, the geography of the second global shift is determined by the educational and language abilities of workers in foreign countries (Bryson 2016). Language and cultural abilities are far more important than they were for the first global shift, generating a distinct geography, with investment focusing on countries with widely spoken 'global' languages such as English, French and Spanish. This obviously reflects the importance of customer interaction and service, meaning that potential service providers must be able to supply a pool of employees that speak the same language as customers in the home country (Box 3.9). As such, geographical distance often goes hand in hand with an underlying cultural proximity or nearness. This is typically derived from history and the experience of colonialism in the nineteenth century with colonial powers such as Britain and France having unwittingly sown the seeds of future competition through the imposition of their home language (ibid.). Time differences also play an important role through the 'follow the sun' strategy whereby firms link facilities in different time zones together to offer an extended or even 24-hour service to customers (ibid.).

Box 3.9

The growth of outsourced services in Morocco

Whilst not one of the principal global offshoring hotspots, Morocco has a substantial outsourcing industry. It grew from 300 employees in 1999 to 70,000 in 2016, reaching $780 million in revenue the previous year (Saleh 2016). This was generally based on the provision of services to French customers, reflecting Morocco's cultural and colonial heritage. For instance, Outsourica, a Moroccan firm that employs 800 people, provides a range of call-centre, e-commerce and other back-office services to major French companies such as Total, the oil major, Carrefour,

the supermarket chain and Meditel, Orange's Moroccan affiliate (ibid.). Call centre agents often adopt a French name to make themselves familiar to French clients, who do not necessarily know that they are calling from Morocco, adopting a practice widely used in Indian call centres serving the British market.

The growth of the Moroccan industry has slowed in recent years, however, reflecting two major trends. First, competition between French telecoms providers has led to cost cutting and the migration of customer care online from phone-based helplines in some

cases. Second, Moroccan back-office service providers are facing competition from lower-cost providers in other francophone African countries (ibid.). Furthermore, the outsourcing of back-office processes in France is less advanced than in the UK or US, with employment protection regulations acting as something of a barrier, in addition to the limited availability of some skills in Morocco. This has prompted the government to sign an agreement with private operators to support the sector, in the hope that it can meet the demand for more sophisticated services such as business process outsourcing.

Reflecting this colonial history, India has emerged as the leading 'offshore' destination followed by China, Malaysia and Brazil (A.T. Kearney 2016). This reflects its advantages in terms of skilled, English-speaking labour and ICT infrastructure. Firms are attracted by the large cost savings involved with starting salaries of around £2,500 in India compared to around £12,500 in the UK in the early 2000s. Despite some reports of rising costs in India prompting companies to move services back to developed economies such as the UK (Kavanagh 2011), the differential remains large with both India and the Philippines having wage cost levels below 20 per cent of the UK (Devi 2015). At the same time, call centre work is well paid in an Indian context, with the average entry-level wage equating to three times the monthly per capita income, making it highly attractive to young graduates (James and Vira 2012: 14).

The scope for firms to transfer more higher-value service work such as accountancy and data analysis to lower-cost locations is a source of recurring concern among policy-makers and trade unions in developed countries. While this trend is likely to continue, research has indicated that relocation is limited by a number of constraints such as the impossibility of standardising and digitising some service activities (often those associated with innovation and creativity which typically require face-to-face communication), the need for proximity to existing customers, the nature of information which is personal, sensitive and confidential, and national standards and regulations (Pisani and Ricart 2016; UNCTAD 2004). In a small number of cases, customer perceptions of lower quality when a service is performed in an off-shore location has prompted 're-shoring' back to developed economies. For example, the computer backup firm Carbonite moved its service centre from India to the US in 2011 in an effort to improve customer service levels (Walsh *et al.* 2012). More generally, however, the cost savings associated with offshoring remain highly attractive and developing countries such as India are likely to make further inroads into more complex and high-value tasks. Processes of automation based on robotics and the standardisation of business processes across multiple customers are set to have a further disruptive effect, requiring the development of more analytical skills in host countries as routine labour is increasingly performed by machines and built into systems (A.T. Kearney 2016).

Reflect

Compare and contrast the 'first' and 'second' global shifts, referring to the relocation of manufacturing in the 1970s and 1980s and the 'offshoring' of services over the last decade or so.

3.6 Summary

This chapter has examined the changing geography of production, supported by an account of the underlying dynamics of capitalist production which have shaped this changing geography. The basic process of production was described in terms of the circuit of capital, in which capital, labour and the means of production are combined to generate commodities that are sold in competitive markets for a profit. Firms are the main organisational form of capitalism, comprised of distinctive competencies and assets, which are expressed through specific skills, practices and knowledge. The position of labour was also considered, emphasising that, although workers are required to sell their labour to capitalists for a wage, labour should be viewed as a 'fictitious commodity' on account of its irreducible human and social qualities (Polanyi 1944). At the same time, the evolving technical division of labour in production has shaped the development of industrial capitalism, finding perhaps its fullest expression in the Fordist mass production industries that typified the middle decades of the twentieth century. The process of innovation and technological change is an inherent feature of capitalism, reflecting the pressures of profit-seeking and competition between firms. Economic historians have observed the tendency of major innovations to 'swarm' together, underpinning 'long waves' or Kondratiev cycles of economic development.

While facilitated by successive revolutions in the 'space shrinking' technologies of transport and communications, the geographical expansion of capitalism over time has been fundamentally driven by a quest for new markets, fresh sources of raw materials and unexploited reservoirs of labour. In more specific terms, the geography of production has shifted from the pattern of regional sectoral specialisation that characterised the late nineteenth and early twentieth centuries, supported by colonial markets and raw material supplies, to the increasingly elaborate international divisions of labour that characterise the contemporary economy. The former pattern involves all the main stages of production being concentrated within a particular region, whilst international divisions of labour are based on the geographical 'stretching' of production across space as different parts of the overall process are carried out in different countries. The 1970s and 1980s saw the spatial dispersal of routine manufacturing activities to lower-cost locations in the developing world (Froebel *et al.* 1980), paralleled by the second 'global shift' of selected consumer services in recent years (Bryson 2016). These successive waves of dispersal reflect the emergence of a global labour surplus in recent decades through the growth of new labour supplies in developing countries, particularly China and India (Standing 2009). This has exerted downward pressure on wages in developed economies, reflecting a serious weakening of labour's bargaining position relative to capital (see Chapter 6). At the same time, higher-value activities like research and development and financial services have become increasingly concentrated in metropolitan regions and world cities. This underlying tendency for the spatial agglomeration of high-value economic activities in large urban areas and lower-value ones to be relocated and dispersed to lower-wage locations can be seen as a key underlying pattern or 'stylised fact' (a broad simplified generalisation) in economic geography.

Exercise

Concentrate on the city, region or state in which you live. Investigate the restructuring of its economy since the early nineteenth century. Prepare an essay which addresses the questions identified below.

What have been the main industries located there? What markets did these industries serve and where were raw materials and components supplied from? In what ways has this area been affected by successive waves of industrialisation? How have the economic fortunes of the place changed over time? What key institutions have shaped the process of economic development? Can this be understood in relation to changing spatial divisions of labour and expertise? Has the area been affected by deindustrialisation since the 1970s? Have any new industries or new forms of investment grown in the same period? What are the economic prospects of the area in the early twenty-first century?

Key reading

Bryson, J. R. (2016) Service economies, spatial divisions of expertise and the second global shift. In Daniels, P., Bradshaw, M., Shaw, D., Sidaway, J. and Hall, T. (eds) *An Introduction to Human Geography*, 5th edition. Harlow: Pearson, pp. 343–64.
A useful introductory summary of the changing nature of service economies, introducing the concept of spatial divisions of expertise and assessing how it is shaping the international relocation of services.

Dicken, P. (2015) *Global Shift: Mapping the Changing Contours of the World Economy*, 7th edition. London: Sage, pp. 11–46, 114–72, 451–76, 510–38.
Quite simply the key text on the economic geography of globalisation. Chapter 2 outlines the changing contours of the global economy, particularly with reference to the shift of gravity towards East Asia, while chapter 5 assesses the role of TNCs as the primary 'movers and shakers' of the global economy. Finally, chapters 14 and 17 provide detailed case studies of the clothing industries and business services respectively.

Leyshon, A. (1995) Annihilating space? The speed-up of communications. In Allen, J. and Hamnett, C. (eds) *A Shrinking World? Global Unevenness and Inequality*. Oxford: Oxford University Press, pp. 11–54.
Worth revisiting (a Chapter 1 reading) for the historical perspective on processes of time-space compression during the nineteenth century, driven by developments such as railways and the telegraph.

Massey, D. (1984) *Spatial Divisions of Labour: Social Structures and the Geography of Production*. London: Macmillan.
An introduction to the concept of spatial divisions of labour, illustrated with reference to the changing geography of the UK.

Wright, R. (2002) Transnational corporations and global divisions of labour. In Johnston, R.J., Taylor, P. and Watts, M. (eds) *Geographies of Global Change: Remapping The World*, 2nd edition. Oxford: Blackwell, pp. 68–77.
An accessible and concise introduction to the evolution of TNCs over time and the development of the old and new international divisions of labour.

Useful websites

www.ft.com

Financial Times site containing numerous articles on international business and investment trends such as offshoring which are reshaping the geography of the world economy.

http://unctad.org.

Website of the United Nations Conference on Trade and Development which provides access to a wealth of information on trade, investment and development. This includes the annul *World Investment Report* which offers a detailed analysis of investment trends and flows.

References

A.T. Kearney (2016) *On the Eve of Disruption*. 2016 A.T. Kearney Global Services Location Index. Chicago: AT Kearney.

Baran, P. and Sweezy, P. (1966) *Monopoly Capital*. New York: Monthly Review Press.

Barnes, T.J. (1997) Introduction: theories of accumulation and regulation: bringing life back into economic geography. In Lee, R. and Wills, J. (eds) *Geographies of Economies*. London: Arnold, pp. 231–47.

Benwell Community Project (1978) *The Making of a Ruling Class*. Benwell Community Project, Final Series No. 6.

Beynon, H. (1984) *Working for Ford*, 2nd edition. Harmondsworth: Pelican.

Blackburn, R. (2008) The subprime crisis. *New Left Review* 50, March–April 2008.

Boschma, R. and Martin, R. (2007) Constructing an evolutionary economic geography. *Journal of Economic Geography* 7: 537–48.

Bryson, J.R. (2006) Offshore, onshore, nearshore and blended-shore: understanding the evolving geographies of 'new international division of service labour'. Paper presented to Association of American Geographers, 2006 Annual Meeting, Chicago, Illinois.

Bryson, J.R. (2016) Service economies, spatial divisions of expertise and the second global shift. In Daniels, P., Bradshaw, M., Shaw, D., Sidaway, J. and Hall, T. (eds) *An Introduction to Human Geography*, 5th edition. Harlow: Pearson, pp. 343–64.

Bryson, J. and Henry, N. (2005) The global production system: from Fordism to post-Fordism. In Daniels, P., Bradshaw, M., Shaw, D. and Sidaway, J. (eds) *Human Geography: Issues for the Twenty First Century*, 2nd edition. Harlow: Pearson, pp. 318–36.

Castells, M. and Hall, P. (1994) *Technopoles of the World*. London: Routledge.

Castree, N., Coe, N., Ward, K. and Samers, M. (2004) *Spaces of Work: Global Capitalism and Geographies of Labour*. London: Sage.

Chapman, K. and Walker, D. (1991) *Industrial Location*, 2nd edition. Oxford: Blackwell.

Clark, G.L. (1994) Strategy and structure: corporate restructuring and the scope and characteristics of sunk costs. *Environment & Planning A* 26: 9–32.

Coe, N., Kelly P.F. and Yeung, H.W. (2013) *Economic Geography: A Contemporary Introduction*, 2nd edition. Oxford: Wiley Blackwell.

Coffey, W. (1996) The newer international division of labour. In Daniels, P.W. and Lever, W. (eds) *The Global Economy in Transition*. Harlow: Pearson, pp. 40–61.

Cronon, W. (1991) *Nature's Metropolis: Chicago and the Great West*. New York: W.W. Norton.

Devi, S. (2015) Outsourcing moves from Poland to Asia and eventually to robots. *Financial Times*, 24 September.

Dicken, P. (2003) *Global Shift: Reshaping the Global Economic Map in the 21st Century*, 4th edition. London: Sage.

Dicken, P. (2015) *Global Shift: Mapping the Changing Contours of the World Economy*, 7th edition. London: Sage.

Doeringer, P. and Piore, M. (1971) *Internal Labour Markets and Manpower Analysis*. Lexington, MA: Heath Lexington Books.

The Economist (2004) Exit strategy. Can Silicon Valley's magic last? *The Economist*, 1 May.

Froebel, F., Heinrichs, J. and Kreye, O. (1980) *The New International Division of Labour*. Cambridge: Cambridge University Press.

Gardiner, B., Martin, R., Sunley, P. and Tyler, P. (2013) Spatially unbalanced growth in the British economy. *Journal of Economic Geography* 13: 889–928.

Gordon, I., Haslam, C., McCann, P., and Scott-Quinn, B. (2005) *Offshoring and the City of London*. London: The Corporation of London.

Gregson, N. (2000) Family, work and consumption: mapping the borderlands of economic geography. In Sheppard, E. and Barnes, T.J. (eds) *The Companion to Economic Geography*. Oxford: Blackwell, pp. 311–24.

Harvey, D. (1982) *The Limits to Capital*. Oxford: Blackwell.

Harvey, D. (1989) *The Condition of Postmodernity*. Oxford: Blackwell.

Harvey, D. (2010) *The Enigma of Capital*. London: Profile Books.

Hobsbawm, E.J. (1962) *The Age of Revolution, 1789–1848*. London: Weidenfeld and Nicolson.

Hobsbawm, E.J. (1987) *The Age of Empire, 1875–1914*. London: Weidenfeld and Nicolson.

Hobsbawm, E.J. (1999) *Industry and Empire*, 2nd edition. London: Penguin.

Hudson, R. (1989) *Wrecking a Region*. London: Pion.

Hudson, R. (2003) Geographers and the regional problem. In Johnston, R.J. and Williams, M. (eds) *A Century of British Geography*. Oxford: Oxford University Press, pp. 583–602.

Hudson, R. (2016) Rising powers and the drivers of uneven global development. *Area Development & Policy*. Online first. DOI: 10.1080/23792949.2016.1227271.

Industrial Communities Alliance (2015) *Whose Recovery? How the Upturn in Economic Growth Is Leaving Older Industrial Britain Behind*. Barnsley: Industrial Communities Alliance.

James, A. and Vira, B. (2012) Labour geographies of India's new service economy. *Journal of Economic Geography* 12: 841–75.

Kavanagh, M. (2011) Private sector backtracks on 'offshoring'. *Financial Times*, 20 June.

Knox, P., Agnew, J. and McCarthy, L. (2003) *The Geography of the World Economy*, 4th edition. London: Arnold.

Lawton, P. (1986) Textiles. In Langton, J. and Morris, R.J. (eds) *Atlas of Industrialising Britain 1780–1914*. London and New York: Methuen, pp. 106–13.

Lee, C. (1986) Regional structure and change. In Langton, J. and Morris, R.J. (eds) *Atlas of Industrialising Britain 1780–1914*. London and New York: Methuen, pp. 30–33.

Lee, N. (2016) Powerhouse of Cards? Understanding the 'Northern Powerhouse'. SERC Policy Paper 14. Spatial Economic Research Centre, London School of Economics. At www.spatialeconomics.ac.uk/textonly/SERC/publications/download/sercpp014.pdf.

Leyshon, A. (1995) Annihilating space? The speed-up of communications. In Allen, J. and Hamnett, C. (eds) *A Shrinking World? Global Unevenness and Inequality*. Oxford: Oxford University Press, pp. 11–54.

Malmberg, A. and Maskell, P. (2002) The elusive concept of localisation economies: towards a knowledge-based theory of spatial clustering. *Environment and Planning A* 34: 429–49.

Martin, R. (1988) The political economy of Britain's north-south divide. *Transactions, Institute of British Geographers* NS 13: 389–418.

Martin, R., Pike, A., Tyler, P. and Gardiner, B. (2016) Spatially rebalancing the UK economy: towards a new policy model? *Regional Studies* 50 (2): 342–57.

Marx, K. [1867] (1976) *Capital*, vol. 1. New York: International Publishers.

Marx, K. and Engels, F. [1848] (1967) *The Communist Manifesto*. London: Penguin.

Massey, D. (1984) *Spatial Divisions of Labour: Social Structures and the Geography of Production*. London: Macmillan.

Massey, D. (1988) Uneven development: social change and spatial divisions of labour. In Allen, J. and Massey, D. (eds) *Uneven Re-development: Cities and Regions in Transition*. London: Hodder and Stoughton, pp. 250–76.

Massey, D. (2001) Geography on the agenda. *Progress in Human Geography* 25: 5–17.

McCann, P. (2016) *The UK National-Regional Economic Problem*. London: Routledge.

Meegan, R (1988) A crisis of mass production? In Allen, J. and Massey, D. (eds) *The Economy in Question*. London: Sage, pp. 136–83.

Myrdal, G. (1957) *Economic Theory and the Under-developed Regions*. London: Duckworth.

Nelson, R.R. and Winter, S.G. (1982) *An Evolutionary Theory of Economic Change*. Cambridge, MA: Harvard University Press.

Peck, J. (1996) *Work-Place: The Social Regulation of Labour Markets*. New York: Guildford.

Penrose, E.T. (1959) *The Theory of the Growth of Firms*. New York: Oxford University Press.

Pisani, N. and Ricart, J.E. (2016) Offshoring of services: a review of the literature and organising framework. *Management International Review* 56: 385–424.

Polanyi, K. (1944) *The Great Transformation: The Political and Economic Origins of Our Time*. Boston, MA: Beacon Press.

Pollard, S. (1981) *Peaceful Conquest: The Industrialisation of Europe 1760–1870*. Oxford: Oxford University Press.

Rempel, G. (undated) The industrial revolution. At www1.udel.edu/fllt/faculty/aml/201files/IndRev.html.

Richardson, R., Belt, V. and Marshall, J.N. (2000) Taking calls to Newcastle: the regional implications of the growth in call centres. *Regional Studies* 34: 357–69.

Rogers, B. (2015) The social costs of Uber. *The University of Chicago Law Review Dialogue* 82: 85–102.

Saleh, H. (2016) Growth slows for Morocco's outsourcing industry. *Financial Times*, 23 March.

Saxenian, A.L. (1994) *Regional Advantage: Culture and Competition in Silicon Valley and Route 128*. Cambridge, MA: Harvard University Press.

Sayer, A. (1985) Industry and space: a sympathetic critique of radical research. *Environment and Planning D: Society and Space* 3: 3–39.

Sayer, A. (1995) *Radical Political Economy: A Critique.* Oxford: Blackwell.

Sayer, A. and Walker, R. (1992) *The New Social Economy: Reworking the Division of Labour.* Oxford: Blackwell.

Schumpeter, J.A. (1943) *Capitalism, Socialism and Democracy.* London: Allen and Unwin.

Slaven, A. (1986) Shipbuilding. In Langton, J. and Morris, R.J. (eds) *Atlas of Industrialising Britain 1780–1914.* London and New York: Methuen, pp. 132–5.

Sloman, J. (2000) *Economics*, 4th edition. Harlow: Prentice Hall.

Smith, A. (1991) *The Wealth of Nations: Inquiry into the Nature and Causes of the Wealth of Nations.* Buffalo, NY: Prometheus Books.

Smith, N. (1984) *Uneven Development: Nature, Capital and the Production of Space.* Oxford: Blackwell.

Standing, G. (2009) *Work after Globalisation: Building Occupational Citizenship.* Cheltenham: Edward Elgar.

Storper, M. and Walker, R. (1989) *The Capitalist Imperative: Territory, Technology and Industrial Growth.* Oxford: Blackwell.

Taylor, M. and Asheim, B. (2001) The concept of the firm in economic geography. *Economic Geography* 77: 315–28.

Taylor P.J. and Flint, C. (2000) *Political Geography: World Economy, Nation-State and Locality.* Harlow: Pearson.

Thompson, E.P. (1963) *The Making of the English Working Class.* London: Penguin.

Thrift, N. (1994) On the social and cultural determinants of international financial centres: the case of the City of London. In Corbridge, S., Thrift, N. and Martin, R. (eds) *Money, Power and Space.* Oxford: Blackwell, pp. 327–55.

Tickell, A. (2005) Money and finance. In Cloke, P., Crang, P. and Goodwin, M. (eds) *Introducing Human Geographies*, 2nd edition. London: Arnold, pp. 244–52.

Townsend, A. and Champion, T. (2014) The impact of recession on city regions: the British experience, 2008–2013. *Local Economy* 29: 38–51.

Turner, W. (undated) The decline of the handloom weaver. At www.cottontown.org/Health%20and%20Welfare/Working%20Conditions/Pages/Tough-Times.aspx.

United Nations Conference on Trade and Development (UNCTAD) (2004) *World Investment Report 2004: The Shift Towards Services.* New York and Geneva: United Nations.

Urry, J. (2000) *Sociology Beyond Societies: Mobilities for the Twenty-first Century.* London: Routledge.

Walker, R.A. (2006) The bomb and the bombshell: the new economy bubble and the San Francisco Bay area. In Vertova, G. (ed.) *The Changing Economic Geography of Globalisation: Reinventing Space.* London: Routledge, pp. 121–47.

Walsh, G., Gouthier, M., Gremler, D.D. and Brach, S. (2012) What the eye does not see, the mind cannot reject: can call centre location explain differences in customer evaluations? *International Business Review* 21: 957–67.

Warf, B. (1995) Telecommunications and the changing geographies of knowledge transmission in the late twentieth century. *Urban Studies* 32: 361–78.

Wright, R. (2002) Transnational corporations and global divisions of labour. In Johnston, R.J., Taylor, P. and Watts, M. (eds) *Geographies of Global Change: Remapping The World*, 2nd edition. Oxford: Blackwell, pp. 68–77.

Yueh, L. (2014) Nokia, Apple and creative destruction. At www.bbc.co.uk/news/business-27238877.

PART 2

Reshaping the economic landscape: dynamics and outcomes

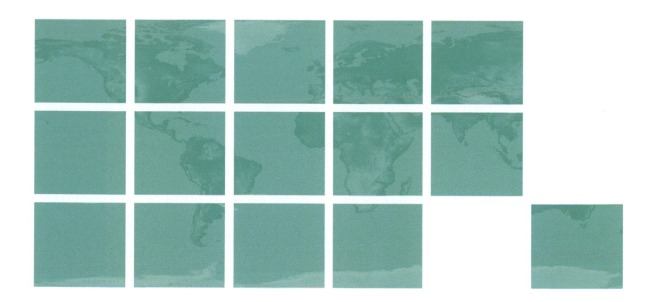

Chapter 4

Capital unbound? Spatial circuits of finance and investment

4.1 Introduction

If anyone was in any doubt about the role of money in "making the world go around", to quote the old song, the financial crisis of 2008–09 provided a timely reminder. The crisis was important in illustrating both how important money is to the basic functioning of the economy, and also how geographically entangled our global economy, with its circuits and flows of money and finance, has become. When the supply of money dries up, the danger of economic and social breakdown is very real, while the global interconnections of the economy can quickly spread the effects far and wide like an out-of-control forest fire or contagious epidemic.

The contemporary geography of money and finance is such that a housing market crash in the heart of the US can quickly ripple out to Europe, Asia and beyond.

Because of the past three decades of **globalisation** and financial **deregulation**, the reverberations of a crisis that started in the US **sub-prime housing market** sent shock waves throughout the global economy, and indeed are still with us. Only massive levels of government intervention in North America, Western Europe and Asia, particularly China, prevented the global financial system from complete meltdown. Subsequently, the wider global economy was plunged into a massive downturn from which it is still struggling to recover. The geographical effects of this have been dramatic and pronounced with many losers and few winners. The financial crisis has also spawned a major debate over effective government policy to return economies to sustainable growth with the dominance of an 'austerity' economics being questioned by many (see Chapter 5). To get to grips with these dynamic and increasingly uncertain times, this chapter is concerned with unravelling and understanding the spatial complexities of finance and how its operations shape the global economy.

4.2 Money, credit and debt

Money, finance and credit are critical to the functioning of **capitalism**. Although money dates back to the earliest large-scale human civilisations in the Middle East and Egypt more than 3,000 years ago, its pre-eminence under capitalism reflects its critical role in the provision of credit for businesses to invest in things that they expect to make a profit from. Essentially, credit and debt, and with them the role of money, are fundamental to growth and the geographical expansion of the global economy. If new technology and **innovation** provide the catalyst for change, and profit is the motor for the expansion of the capitalist economy, money and its credit form are the necessary lubricants.

The accepted wisdom in **economics** from the time of Adam Smith onwards is that money emerged historically as societies became more complex and divisions of labour developed to deal with the limitations of bartering one good against another (e.g. shoes against meat).

The problem with this account is that there is actually no empirical evidence to support it, as David Graeber (2011) has shown in his historical anthropology of money *Debt: The First 5,000 Years*. While there are other ideas on the emergence of money, Graeber and others make persuasive arguments that it was originally a means through which ambitious states and political leaders financed wars and imperial expeditions. From this perspective, money was created as a way of paying for everything, from soldiers' wages to supplies of food and clothing, to even buying off enemy forces, rather than engaging them in battle.

Although the subject of finance can appear dry and technocratic, money has always been a political construct. Contrary to the conventional wisdom of economics, it did not emerge organically between equal parties through the evolution of market exchange. Money was used by governments to secure credit to pay for essential items, but, once in existence, it took on a life of its own as a means of exchange. The earliest forms of money had an intrinsic value typically as gold and silver, followed by the printing of bank notes – originally promissory notes – that were supposed to be backed up by gold reserves.

The underlying point here is that the origins and evolution of money are bound up with underlying power relations within society. As Mann (2013: 200) states: "the political power of money resides in the fact that more than any other social relation, it appears to guarantee the continuity of most fundamental conditions of the existing order." It is typically employed to suit the interests of the most powerful in a society, a fact evident in the **Eurozone sovereign debt crisis** (see section 4.5.2 below); money is far from a neutral medium of exchange.

4.2.1 The functions and regulation of money

At a basic level, money is a measure and store of value and is anything that can be accepted as payment for goods or services. Ancient societies used durable commodities such as grains, weapons or even cattle as forms of exchange before precious metals such as gold and silver were used in the minting of coins. But money has also taken on a second and more important role as a medium of exchange and circulation. Money is crucial to the functioning of the economy, not only by allowing the trading of goods, but also through the provision of credit, whereby those individuals and groups with a surplus supply loans to firms to invest in new technologies and products. As such, credit becomes critical to economic growth and the evolution of economies (Dow: 1999). In practice, these two functions of money frequently come into conflict, where the over-extension of credit during economic boom times leads to the devaluation of financial assets during an economic crisis or downturn (see section 4.4). As economies have become more complex and globalised in recent decades, the store of value function has been displaced by money's role as a medium for transactions and circulation.

A third related function of money is as a universal equivalent enabling us to assess the relative value of commodities; for example bread versus shoes. This has geographical significance, for it allows the assessment of the value of investment in one place against that in another. As we will demonstrate, the transformation of money from material to electronic form has dramatically accelerated this **time-space compression** (Harvey 1989) as part of the broader process of globalisation.

Over time, the different functions of money have led governments to establish regulatory authorities and **central banks** to manage the contradictions inherent in the restriction and expansion of credit. In the United Kingdom, this role has been carried out traditionally by the Bank of England, a formerly independent institution that was nationalised in 1946, and given operational independence in 1997. In the United States, central bank duties are undertaken by the Federal Reserve (or 'The Fed'), while the establishment of a single currency in Europe, the euro, required the setting up of the European Central Bank.

Conventionally, central banks use interest rate policy (setting the 'price' of money which lenders effectively charge borrowers, requiring them to repay interest as well as the underlying loan) to try to influence the supply of money and credit in an economy: raising interest rates makes borrowing money more expensive and therefore is a means of restricting credit and vice versa. Ultimately, central banks are known as 'lenders of the last resort' in the sense that they have the key strategic role of ensuring that the supply of credit is maintained during periods of crisis.

4.2.2 Money and its social construction in a capitalist economy

Mainstream economics maintains that money is created by central banks and then loaned out through commercial banks, who are required to keep a particular proportion of reserves to loans. Money supply is therefore seen as 'neutral' and not a factor in the creation of booms or economic crises. The money markets are viewed as efficient at matching demand and supply under conditions of perfect competition, so that riskier investments will be avoided and the overall ratio of debt to reserves will remain stable. The mainstream view is disputed, however, by 'heterodox' economists who stress both the uncertainty and 'unknowability' of risk, the role of credit emanating from commercial banks in the creation of money, and the dangers of imbalances occurring between credit and monetary deposits. Furthermore, many Marxist, post-Keynesians and other heterodox scholars argue that, as a political and social construct, money is active in shaping the economy (Ingham 1996; Gilbert 2005) through the activities of banks and financial institutions and their decisions on what and what not to invest in – having the power to mediate the flow of funds going to businesses and individuals (Box 4.1).

Marx long ago recognised that commercial banks and investment houses control the circulation of funds and have the power to 'create capital' through the provision of credit. As the following quote demonstrates, Marx also recognised that, instead of money markets operating through perfect competition, over time the competitive dynamics of the marketplace would lead to a concentration of finance into fewer and larger organisations with immense market power.

Talk about centralisation! The credit system, which has its focus in the so-called national banks and the big money-lenders and usurers surrounding them, constitutes enormous centralisation, and gives to this class of parasites the fabulous power, not only to periodically despoil industrial capitalists, but also to interfere in actual production in a most dangerous manner – and this gang knows nothing about production and has nothing to do with it. The Acts of 1844 and 1845 are proof of the growing power of these bandits, who are augmented by financiers and stock-jobbers.

(Marx 1894: 544–5, cited in Keen 2009)

Furthermore, banking and investment activities can have damaging effects on the rest of the economy, for example in promoting speculative activity over investment in more productive capital, or in times of financial crisis restricting finance even to perfectly well-functioning firms, thus hampering normal business activities and potentially causing job losses and closures. Indeed, many argue that we have moved into a phase of financialised capitalism (see section 4.6) where the rest of the economy is becoming increasingly subservient to financial interests and speculative activity (e.g. Froud *et al*. 2006; Lapavitsas 2013).

Reflect

How are money and credit linked under capitalism? What is the relationship between money and power?

4.3 The changing geographies of money

From the outset, money and finance have been tied to the expansion of international trade. Yet there have been various stages in the development of the geography of money since the advent of the modern capitalist economy (Table 4.1). During the late eighteenth century, technological innovation was financed primarily by local and regional banks. As capitalism developed, concentration of the banking sector led to the

Box 4.1

Alternative theories of money supply: the monetarists versus the post-Keynesians

Following the work of John Maynard Keynes, a school of **post-Keynesian economics** developed in the UK and US from the 1940s onwards. Not to be confused with the neo-Keynesian and New Keynesian schools, post-Keynesian economists, who include distinguished figures such as Joan Robinson, Michal Kalecki and Nicholas Kaldor, dismiss the idea that the economy will have a tendency to return to full employment and market equilibrium without government intervention. They place a strong Keynesian emphasis upon the role of aggregate demand in the functioning of a market economy and upon the role government should play in intervening in and stabilising the economy during economic downturns. Contrary to the increasingly mainstream orientation of Keynesian

economics in the post-Second World War period, post-Keynesians emphasise the importance of uncertainty, social relations and institutions to the functioning of modern economies (Arestis 1996).

Post-Keynesians have been particularly influential in relation to understanding the workings of money in the modern capitalist economy with their emphasis upon the way banks endogenously expand the supply of money through credit. Post-Keynesian monetary circuit theory suggests that money is generated internally within the economy and therefore cannot be controlled by central banks themselves restricting the money supply, a policy associated with the monetarist theories of the neoclassical US economist Milton Friedman. Friedman's basic point was to reiterate

the mainstream perspective, known as the 'quantity theory of money', whereby the value of money is like any other commodity, and varies in accordance with the quantity in circulation. To control inflation, and the value of money, governments needed to restrict its supply. The failure of monetarist strategies to control inflation in the early 1980s led to most governments and central banks implicitly accepting the post-Keynesian critique by using the setting of interest rates as a means of attempting to control inflation and the flow of money in an economy. The financial crisis and subsequent responses to it seem to bear out the post-Keynesian view of the futility of monetarist policy and the dominant role played by commercial banks in the development of credit money.

emergence of national financial systems. In practice, this varied across different countries. For example, in countries such as the UK, the financial sector became heavily oriented around London and a national system, whereas in Germany the banking system continues to be more decentralised with regional banks (often owned by regional governments) continuing to play an important role in the financing of companies (Wojcik and MacDonald-Korth 2015).

The emergence of more nationally oriented financial systems resulted in an increased internationalisation of financial activities as banks took advantage of foreign investment opportunities. **Deindustrialisation** domestically often went alongside successful geographical diversification in foreign markets (Box 3.1). Since the 1970s, the financial system has undergone an even more profound set of changes as national banking systems have become increasingly integrated with global financial flows. As part of this, there has been an increasing separation between the financial system and what we might term the 'real' economy of productive activities. An increasing amount of financial activity is tied to speculative investment and new financial products such as **derivatives**.

4.3.1 The globalisation of the financial system

One obvious indicator of the increasing globalisation of the financial sector has been the massive growth in global financial flows; daily turnover on the major foreign exchange markets grew from between $10–$20 billion in 1973, about twice the value of trade, to an average of $1.5 trillion in 2004, which had increased again to $5.3 trillion by April 2013 (Aalbers and Pollard 2016; BIS 2013: 3), over 95 times the value of trade. This **hypermobility of capital** – made possible in part by advances in electronic technology and the ability to shift billions of dollars from one part of the globe to another at the touch of a button – is seen as having its own disciplining effect on national economies and states, encouraging policies that are pro-business and fiscally conservative (e.g. low taxes, low inflation) in order to attract mobile funds and prevent capital flight. It is creating new relational geographies (section 2.5.4) of finance as increasing flows of investment and money across national borders link places more closely together in an increasingly integrated global financial system. These global links mean that the financial crisis

Table 4.1 Geographical development of the financial system

Geography	Local/regional	National	Global
Periodisation	Eighteenth and early nineteenth centuries	Late nineteenth century to 1970s	1970s onwards
Phase of development	Industrialisation	Mature national economy, growth of services	Post-industrial, economy of flows, hyper-capitalism
Objectives of finance	Local manufacturing firms	National firms but also growing international investments	Increasing separation between financial and 'real' economy; growth of derivatives, futures, SIVs, etc.
Financial characteristics	Emergence of regional and national banks	Concentration of banking in national institutions, growth of capital markets	Growing internationalisation of banks, emergence of hedge funds
Type of finance, capital	Loans, risk capital, profit	Increasing importance of share capital	Capital and credit markets

Source: Adapted from Martin 1994: 256, Table 11.1.

of 2008–09 spread rapidly from its epicentre in the US sub-prime housing market.

A number of factors have been critical in the globalisation of the financial system. Of these, the most important are: the deregulation of financial markets; the development of advanced communication technologies; and, the emergence of new financial products such as derivatives.

➤ The deregulation of financial activities through **neoliberal** policies. Conventionally, the financial system was tightly controlled by governments through a range of detailed rules and restrictions which separated different kinds of activities – for example, banking and insurance – and limited entry of firms into the financial sector, particularly foreign ones. But an unprecedented 'crumbling of walls' has taken place since the 1970s, through major deregulation programmes. These involved the abolition of exchange and capital controls – which limited trade in foreign exchange markets and the volume of capital that could be exported – and the removal of barriers between different financial activities and limits on entry of foreign firms. A series of changes in the United States since the 1970s has both eased the entry of foreign banks into the domestic market and facilitated the expansion of US banks overseas. In the United Kingdom the so-called 'Big Bang' of October 1986 removed the barriers that previously existed between banks and securities houses and allowed the entry of foreign firms into the Stock Exchange. In France the 'Little Bang' of 1987 gradually opened up the French Stock Exchange to outsiders and to foreign and domestic banks (Dicken 2003: 448–9). In the US, a final important piece of legislation was the abolition of the Glass-Steagall (1933) Act in 1999 which removed most of the remaining barriers between retail and investment banking in the United States, allowing more speculative activities into 'everyday' banking markets such as housing and real estate.

➤ The development of advanced communications technologies. Computers have transformed payment systems, through allowing electronic money to be moved around the world at great speed. The introduction of chips (microprocessors) allows customers to pay for their purchases using plastic in the form of credit and debit cards. Technological innovations have facilitated 24-hour trading on a global basis, exploiting the overlap between the trading hours of the world's major financial centres. In essence, we live in an age of increasingly dematerialised electronic money or what some refer to as a 'cashless' society (Martin 1999). Above all, ICT has greatly increased the pace of financial activity, allowing almost instantaneous trading between distant centres (ibid.: 14). Finance is a crucial agent of time-space compression, obliterating the friction of distance in terms of the rapid movement of capital and information across space (Harvey 1989).

➤ The number of financial and monetary products, which has grown massively. Of particular significance is the rise of new financial instruments known as derivatives, which are traded and have made it easier to move money across the globe. The term encompasses a highly complex range of instruments, but these are essentially "contracts that specify rights/obligations based on (and hence derived from) the performance of some other currency, commodity or service used to manage risk and volatility in global markets" (Pollard 2005: 347). Such risk and volatility is associated with changes in prices of commodities, currencies and instruments such as interest rates.

The basic forms of derivatives include: "simple futures (where an agreement to buy a commodity at a given date and given price is made, allowing purchasers to purchase a degree of certainty" (Tickell 2000: 88); swaps (where the parties to an agreement exchange particular liabilities, for example, interest repayments, usually through an intermediary such as a financial institution); and options "which, in exchange for a premium, allow buyers to trade assets at a predetermined price at a point in the future" (ibid.). Derivatives are traded in two ways: on organised and regulated exchanges such as the London International Financial Futures Exchange or the Chicago Mercantile Exchange and 'over-the-counter' through direct agreements between the parties concerned. Since the early 1980s, derivatives have become symbolic of a highly monetised and financially driven world economy, in which ICT has rendered financial instruments increasingly mobile and separate from flows of material goods (Tickell 2003).

4.3.2 The uneven geographies of finance

Despite the growing spatial interconnections evident in these activities, it is wise not to overplay the globalising extent of financial activities (see Hirst *et al.* 2009). While the global financial system has become more complex and intertwined as a result of the processes described above, distinctive financial geographies remain. For one thing, 70 per cent of the world economy's domestic assets and global flows were accounted for by North America and Europe (including Russia) just before the financial crisis in 2007 (Thompson 2010: 130) and trading usually takes place between key financial centres within the major national economies.

In a fully globalised world, one might expect a multitude of different currencies to be traded relatively equally on foreign exchange markets. However, the US dollar continues to dominate, giving a sense of the continued geopolitical and economic significance of the United States in relation to the rest of the world (see Table 4.2). The fact that the rest of the world holds dollars as their main global trading currency gives the US considerable power and authority as the world's main superpower. Beyond the US, what is remarkable about world currencies is the relative stability in the dominance of a few currencies over the rest, despite successive **financial crises** and the growth in euro holdings since 1998. Yet the push by the Chinese state to promote the renminbi as a global currency, through a recent deal to trade its sovereign debt on the London financial markets, may challenge the financial dominance of the advanced economies of the global North in the years ahead (Moore 2016).

Although globalisation is associated with dramatic time-space compression in the financial sector, it has not resulted in the end of geography: place still matters. Whilst world financial markets may be mobile in their effects, the key actors are centred in a few key nodes in dominant world cities. The paradox of globalisation in this sense is that financial markets are dominated by the decisions of a small number of market traders and analysts, operating out of a select group of cities. London and New York have been the dominant financial centres in the global economy for over a century, a fact that continues despite various crises and downturns (Table 4.3).

Whilst groups such as financial traders, brokers, investors and executives have benefited hugely from financial globalisation, not least through the payment of massive bonuses linked to financial transactions, most of the world's population has been left behind. Groups such as the poor in developing countries and low-income groups in developed economies are experiencing growing marginalisation from the financial system. The debt crisis afflicting many developing countries can be seen as a global problem of **financial exclusion** and marginalisation with such countries only gaining access to further credit on the conditions imposed by international organisations like the IMF and World Bank. For many low-income groups in the global North, credit is often only offered on extortionate and unsustainable terms (see section 4.6).

Table 4.2 Top ranked currencies in terms of distribution of global foreign exchange market turnover (% of total and global ranking)

	1998	Rank	2007	Rank	2013	Rank
US dollar	86.8	1	85.6	1	87.0	1
Euro	. . . . .	32	37.0	2	33.4	2
Japanese Yen	21.7	2	17.2	3	23.0	3
UK pound	11.0	3	14.9	4	11.8	4
Australian dollar	3.0	6	6.6	6	8.6	5
Swiss Franc	7.1	4	6.8	5	5.2	6

Source: Bank for International Settlements.

Table 4.3 Ranking of global financial centres by Global Financial Centre Index (GFCI) September 2015

2015 Rank	City	GFCI Score*	2010 Rank
1	London	796	1
2	New York	788	2
3	Hong Kong	755	3
4	Singapore	750	4
5	Tokyo	725	5
6	Seoul	724	—
7	Zurich	715	8
8	Toronto	714	12
9	San Francisco	712	15
10	Washington	711	17
11	Chicago	710	7
12	Boston	709	13
13	Geneva	707	9
14	Frankfurt	706	11
15	Sydney	705	10
16	Dubai	704	—
17	Montreal	703	—
18	Vancouver	702	—
19	Luxembourg	700	20
20	Osaka	699	—

* index constructed using measures of competitiveness and connectivity + survey of financial sectors professionals' perceptions of centres' competitiveness.

Source: QFC: The Global Financial Centres Index 18. Available at: www.longfinance.net/images/GFCI18.

Reflect

To what extent has globalisation rendered the financial system 'placeless'? In what ways does geography continue to shape financial markets?

4.4 Financial crises and cycles

As we have already observed in earlier chapters, capitalism is prone to uneven development over time and across space. Periods of economic growth are punctuated by periods of stagnation, decline and even crisis as over-expansion of the system eventually leads to the point where firms are producing more than the market can absorb. The financial system plays an important part in this uneven dynamic, prone to the same kinds of crises and cycles that characterise the rest of the economy (Minsky 1975, 1986; Wolfson 1986). Indeed, financial crises are often constitutive of broader economic crises, because of the critical role of money. An oversupply of credit can lead to excessive speculation, increasing levels of debt and over-inflated asset prices (e.g. housing). Such circumstances create financial bubbles that eventually burst in a variety of different ways as the laws of financial gravity take effect.

Globalisation appears to have greatly amplified the crisis tendencies in the financial system with an acceleration of crises in time and space. In spatial terms, since 1980, financial crises have been "both endemic and contagious" (Harvey 2005: 94). Although much attention has recently focused on the 2008–09 financial crisis, which began in the United States and quickly spread to the rest of the world, on average 25 per cent of the world's countries suffered some form of banking crisis every year between 1986 and 2001 (Palma 2009: 849).

The increasing degree of financial mobility in the world economy and the deregulation of economies by national governments following neoliberal policies seems to be generating greater instability in the world economy with international investors now able to shift funds from one country to another in search of higher returns. Competition and the fear of incurring major losses means that such investors often act in very similar ways. As David Harvey succinctly puts it: "The 'herd mentality' of the financiers (no one wants to be the last one holding on to a currency before devaluation) could produce self-fulfilling expectations" (2005: 94–5). This can have devastating effects on individual places. While the opening up of an economy to greater foreign trade often attracts speculative investment, bringing short-term prosperity, this may be followed by a rapid out-flow of such 'hot money' if international investors lose confidence in an economy, leading to a financial crisis, the collapse of the banking system and recession.

A case in point was Mexico, which, as a major oil exporter, suffered an economic and financial crisis in 1982 following the collapse of the oil price. The gov-ernment also defaulted on the debts (largely with US financial institutions) that it had run up in the eco-nomic downturn of the 1970s. Although Mexico was

at the epicentre of the crisis, it spread to many other developing countries (Figure 4.1), which had been encouraged to borrow heavily from western finan-cial institutions in the 1970s to finance development projects, only to find themselves unable to service their debts in the early 1980s when interest rates rose sharply (Harvey 2005).

In the second half of the 1980s, and under pressure from the International Monetary Fund to address its massive debts, the Mexican Government cut its tariff barriers to overseas trade and privatised 90 per cent of its state-owned enterprises (Dicken 2003: 192). The result was a dramatic influx of foreign investment, which had the effect of driving up the value of the peso to levels that made much of Mexican industry uncom-petitive. Eventually the government was forced to devalue the currency leading to an outflow of funds and a new financial crisis in 1995. In both cases, the crises spread to other countries as financial markets became nervous about investments in developing countries.

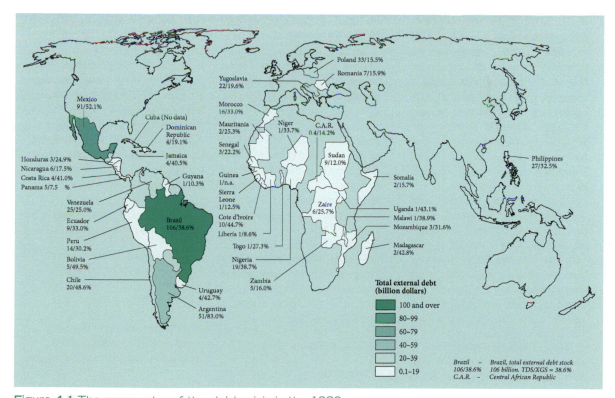

Figure 4.1 The geography of the debt crisis in the 1980s
Source: Harvey 2005: 95, Figure 4.2.

Major global financial crises have now affected every world region since 1980.

As recent events have shown, economic orthodoxy has a poor record in both understanding and predicting the tendency for financial markets to veer between boom and bust. Famously, in response to the 2008–09 financial crisis, the British Queen, Elizabeth II, who was estimated to have lost around £25 million of her personal fortune, asked academics at the London School of Economics, a centre of economic orthodoxy: "Why did no one notice it?" (Pierce 2008).

One response would have been that non-mainstream economists such as John Maynard Keynes and Hyman Minsky had long warned of the inherent instability of financial markets. Ignored for over 20 years, Minsky's theory of how unregulated financial markets create havoc in the wider economy is now being taken seriously again. His '**financial instability hypothesis**' was first developed in the 1960s. Its central message is that the accumulation of debt by financial institutions as a result of increasingly over-exuberant speculative activities eventually produces a crisis which can have knock-on effects for the rest of the economy with the potential to produce a full-blown recession (see Box 4.2). In the words of one commentator: "bankers, traders and financiers periodically [play] the role of arsonists, setting the entire economy ablaze" (Cassidy 2008).

Box 4.2

Minsky's financial instability thesis

Minsky's post-Keynesian starting point is that the capitalist economy works as an evolutionary, dynamic and unstable process in "real calendar time" (1992: 2) rather than the equilibrium model of mainstream economics. In a complex capitalist economy, banks and financial institutions are the mediators between the initial depositors of money and the firms or other organisations that borrow money in order to invest. Banks loan out money on the basis of expected profits in the future.

Minsky follows the post-Keynesian view of banks as profit-driven institutions that have an interest in developing new methods of making a profit from their activities. Financial innovations therefore are at the heart of the banking system if it is not strictly regulated: banks will find ways of expanding credit and develop new forms of credit financing in search of profits.

Minsky recognises three forms of what he terms "income-debt relations" (ibid.: 6) through which banks finance credit:

➤ Hedge units refers to borrowers who are able to pay off the initial loan and interest from their income returns from investment;
➤ Speculative units are borrowers who are able to cover the interest payments from income received, but are unable to pay off the principal debt and therefore need to either sell assets or borrow more;
➤ Ponzi units are borrowers who are unable to pay either the interest or the principal. In this situation, the borrower must rely on the appreciation of the value of the underlying asset to repay the interest and principal. This often requires selling on or further borrowing of money which, according to Minsky, "lowers the equity of a unit, even as it increases liabilities and the prior commitment of future income. A unit that Ponzi finances lowers the margin of safety that it offers the holders of its debts" (ibid.: 7).

Minsky recognised that, over time, as economic boom conditions develop, financial institutions tend to shift from hedge to the more risky speculative and Ponzi forms, meaning that the economy acts as a "deviation amplifying system" (ibid.). Furthermore, if inflation begins to get out of control and central banks raise interest rates, the extra cost of borrowing will turn speculative units into Ponzis. If asset prices fall once an investment bubble bursts, Ponzi borrowers can no longer repay their loans, resulting in a financial crisis if Ponzi-style finance has been widely used across the system.

As such, financial crises tend to be built into the system (endogenous):

The financial instability hypothesis is a model of a capitalist economy which does not rely upon exogenous shocks to generate business cycles [e.g. oil price rises, wars] of varying severity. The hypothesis holds that business cycles of history are compounded out of the internal dynamics of capitalist economies

Box 4.2 (continued)

and the system of interventions and regulations that are designed to keep the economy operating within reasonable bounds.

(Ibid.: 8)

Minsky's financial cycle model recognises five stages: displacement, boom, euphoria, profit-taking, panic. Displacement happens when investors become excited in a new sector of activity; for example, the internet, the housing market, etc. As money floods into the sector, a credit boom develops. This leads to euphoria – a key phase – in which credit is extended to more risky borrowers, including the development of new financial products (e.g. junk bonds in the 1980s, home loans linked to mortgages in the 2000s). At some point, the more intelligent traders begin to cash in their profits. As more institutions attempt to sell, the danger is that asset prices begin to fall. Confidence in the value of assets may also be affected by dramatic events such as the apparent vulnerability or collapse of banks that have overloaded on speculative or Ponzi schemes. Panic can then set in with dramatic consequences including the seizing up of credit to the 'normal' economy that had hitherto remained innocent of any dealings with the speculators.

4.5 The 2008–09 crisis, recession and faltering recovery

The financial crisis of 2008–09 triggered the worst global economic downturn since the Great Depression of the 1930s (see Figure 4.2). The US alone is estimated to have lost around $13 trillion, when lost GDP and the government funds spent to tackle the crisis are taken into account (Blyth 2013: 45). Nor is there much sense that the global economy is making anything like a sustainable recovery. As a recent UN survey of world economic prospects noted: "The world economy stumbled in 2015, amid weak aggregate demand, falling commodity prices and increasing financial market volatility in major economies" (UN 2016: 1). Nor is there much indication of where a new round of sustainable economic growth will come from, leading some commentators to question whether capitalism is facing an existential crisis (Harvey 2010).

A critical geographical political economy of the crisis needs to appreciate that "it is difficult to understand what has been going on at one or another scale of crisis crystallization without simultaneously attending to other scales, for the simple reason that they are not only interconnected but mutually constitutive" (Christophers 2015: 205–6). While, superficially, we can recognise the global flows and interconnections evident in the crisis, a deeper understanding requires appreciating the 'relational geographies' through which economic actors and institutions such as the IMF, World Bank and the European Union created the conditions for crisis over a longer time period. Global financial flows, which are orchestrated primarily in the world's major financial centres such as London and New York, have had diverse effects on different places around the globe. These flows have been facilitated by the neoliberal policies of deregulation led by major national governments such as the United States and United Kingdom in the 1980s and 1990s and 'policed' by supranational institutions (see section 5.3).

Although the epicentre of the initial crisis was the US housing market, its effects have rippled out and affected the broader global economy because of the intensified connections between countries and regions. The crisis, however, is far from over; three main phases can be identified to date: an initial phase centred around the US and the sub-prime housing market; the development of the Eurozone sovereign debt crisis from late 2009 onwards; and an emerging economic crisis in the largest developing economies of the global South which has in large part been triggered by the first two phases, through a collapse in global demand for their commodities and cheaper manufacturing products.

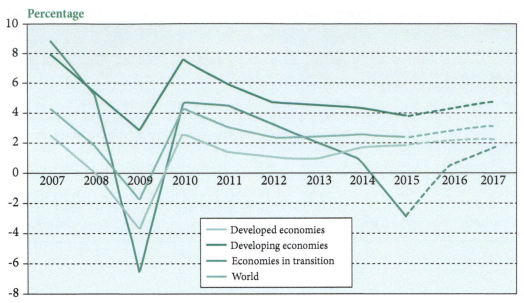

Figure 4.2 Growth of world gross product and gross domestic product by country grouping 2007–17

Note: Data for 2015 are estimated; data for 2016 and 2017 are forecast.

Source: UN 2016: 1, Figure I.1.

4.5.1 The epicentre of the crisis: the US sub-prime housing market

The trigger for the crisis, beginning in late 2006 (Table 4.4), was the growing number of households in the US's so-called 'sub-prime' housing market defaulting on their mortgages. 'Sub-prime' is a euphemistic term for that part of the housing market consisting of low-income households in relatively precarious economic conditions, who are at risk of default if circumstances change (typically if interest rates rise). In the rising and exuberant housing market of the early 2000s, banks rushed to lend to the sub-prime market with little heed to potential longer-term consequences. The situation was further exacerbated by the creation of new mortgage derivative products that aggregated individual mortgages into collateralised debt obligations (CDOs) (often packaging together sub-prime and prime mortgages). These were then sold on, thereby spreading the sub-prime risk throughout the US banking system and beyond as foreign banks also sought a share of the action.

Sub-prime borrowers began to default on their mortgages in large numbers from 2006, deflating the housing market bubble as prices fell and investors withdrew. This meant that the repackaged mortgage securities also began to lose much of their value. As a result, many major financial institutions found themselves holding assets that quickly became worthless – notably over-priced CDOs that no one wanted to buy. On 15 September 2008, the fourth-largest US investment bank, Lehman Brothers, went bankrupt because of its exposure to the sub-prime housing market and other US banks were subsequently bailed out by massive loans from the US Government. Thirty years ago, given the tightly regulated and essentially regional nature of the US banking system, and an absence of 'financial innovation', this might have been contained to a handful of states and regions. But, because of the complex ways in which the US domestic housing market had become entangled in broader financial networks, the crisis extended outwards dramatically to the rest of the global economy.

The collapse of Lehman itself provides a vivid example of the wider geographical connections through which global contagion rapidly spread. The firm was forced to file for bankruptcy when the Japanese Government intervened to force Lehman's Japanese subsidiary to retain assets in the country when it emerged that Japanese banks were Lehman's leading unsecured lenders (*Financial Times* 2008). Lehman's collapse led to over 25,000 job losses throughout its global operations but some of the major banks in a variety of countries were directly hit through their exposed trading positions with the bank (see the European banks affected in Table 4.5). The German company Deutsche Bank, for example, was one of the first firms forced to take action to safeguard two of its own property-based US funds (Blackburn 2008) while leading French bank *Société Générale* lost $7 billion in the housing crash.

As panic gripped financial markets more generally, another critical development was the collapse of wholesale or inter-bank lending markets – known as 'repo' markets – whereby banks and financial institutions lend and borrow short-term funds from each other (Blyth 2013). In what came to be known as the 'credit crunch' banks stopped lending to each other. This had devastating effects because of the number of over-leveraged institutions with diminishing or even worthless assets, unable to refinance through short-term borrowing to keep themselves afloat. The US firm Bear Stearns collapsed because of its over-exposure to mortgage securities and was taken over by JP Morgan with government funds as an incentive. In the UK,

Table 4.4 Key moments in the rolling out of the financial crisis

Date	Events
2006–07	Growing number of households defaulting on mortgage payments
2007–08	REPO market beginning to 'dry up' as banks try to sell mortgage securities and market collapses
Summer 2007	European banks' exposure to crisis becoming evident. Governments around the world beginning to intervene through cutting interest rates, direct lending to banks, swapping mortgage debt with government bonds. Northern Rock suffers bank run when it becomes clear it cannot meet its trading positions (nationalised in February 2008)
September 2008	Lehman's goes bust. REPO market freezes. Central banks and governments forced to provide liquidity to prevent collapse of banking system. Widespread nationalisation across globe follows.
October 2009	New Greek government reveals real fiscal deficit of 13% of GDP rather than 6.5%.
2010–14	Bailouts and development of TROIKA austerity regime in PIIGS across Eurozone
2014–16	Chinese growth slows, other BRIC economies slow and enter recession

Table 4.5 European banks with exposed trading positions with Lehman at time of collapse (September 2008)

Bank	Country	Outstanding funds	
		Q2 2008 (€m)	2007 (€m)
Société Générale	France	473,329	487,959
Credit Agricole	France	383,995	364,178
BNP Paribas	France	NA	597,578
Natixis	France	NA	202,928
Barclays	UK	460,423	352,133
Deutsche Bank	Germany	1,138,090	1,193,131
Credit Suisse	Switzerland	277,362	331,807
UBS	Switzerland	652,972	757,271

Source: *Financial Times* 2008.

the Newcastle-based mortgage bank Northern Rock suffered a run on its assets, following news of its over-exposure to the collapse of the repo market. The Rock's entire business model was predicated on short-term financing for its own mortgage lending. Its collapse, bailout by the UK Government, nationalisation and sub-sequent privatisation (Box 4.3) is a classic story of how an essentially solid and stable local bank can become caught up in global speculation through deregulation and high-risk management (Marshall *et al.* 2012).

As the risks of banking collapse became increasingly evident, government intervention – directly contradict-ing reified neoliberal prescriptions – intensified. The US, hitherto the bastion of free-market conservatism, led the way with the Troubled Asset Relief Program (TARP), passed by Congress in September 2008, which provided around $700 billion of public money for bank bailouts (Lapavitsas 2013: 286). Other countries followed but this was only an emergency measure. As the underlying structural weakness of the entire banking sector across the developed economies became more evident, a wave of nationalisations took place whereby the governments either took partial stakes or full ownership of banks (see Cumbers 2012, chapter 1). In the UK, the exposure of the financial system to the housing market collapse, both at home and overseas, meant that around half of the banking system was brought into public ownership at massive expense to the public purse (see Box 4.3).

Box 4.3

The UK bank nationalisations: a brief detour from 'business as usual'

The Northern Rock nationalisation in February 2008 came on the back of months of prevarication by the Labour Government, which, follow-ing its initial £25 billion financial rescue package in September of the previous year, had done everything in its power to avert public ownership, which was anathema to the neolib-eral playbook of Prime Minister Gor-don Brown and his Finance Minister Alistair Darling. The subsequent re-privatisation of the institution to Richard Branson's Virgin Group – an entity with no serious financial experience – while the state retained the more 'toxic' bad debt assets indi-cates the prevailing political econ-omy at work here.

The nationalisation of Northern Rock was followed by an even more momentous series of events, with the full or partial nationalisation of many of Britain's largest banks including the Royal Bank of Scotland and Hali-fax/Bank of Scotland (HBOS), as well as smaller ex-building societies such as the Bradford and Bingley (Table 4.6). The subsequent acquisition of HBOS by Lloyds meant that the state now held majority stakes in two out of the four leading UK banks. In total £123.93 billion were provided in the form of loans or share purchases, with a total further £332.4 million in the form of guarantees (e.g. covering pension liabilities) (NAO 2011). At its peak, once interest and fees were added, the banks owed the taxpayer £1.162 trillion in 2009 (ibid.).

Darling was quick to reassure busi-ness interests that nationalisation was a necessary evil, but a temporary measure:

> It is better for the Government to hold on to Northern Rock for a temporary period and as and when market conditions improve the value of Northern Rock will grow and therefore the taxpayer will gain. The long-term owner-ship of this bank must lie in the private sector.
>
> (Alistair Darling, BBC News, 17 February 2008)

Table 4.6 Government support (loans) and share purchases of nationalised banks during 2008–09 financial crisis (as of 31st March 2011)

Bank	£bn	Value as of 31st	Govt share (%)
Royal Bank of Scotland	45.8	36.97	84
Lloyds	20.5	16.04	43
Northern Rock	421.59	21.59	100
Bradford & Bingley	8.55	8.55	100

Source: NAO 2011.

Box 4.3 (continued)

Nationalisation was presented as a means of dealing with extraordinary events that departed from the 'normal' functioning of markets, rather than accepting, as a growing body of opinion has, that perhaps the underlying structures of the financial sector and forms of ownership and control might need a critical reassessment. As an interesting footnote to the episode, and example of the spatial and social connections between national politicians and the global financial sector, on leaving elected office both Brown and Darling have gone on to pursue lucrative careers in the financial sector; Brown with US asset management company Pimco (Parker and McLannahan 2015) and Darling with US bank Morgan Stanley, a company which was itself fined $2.6 billion by the US Justice Department for its role in the sub-prime crisis (Popper 2015).

The economic and social consequences of the financial crisis for ordinary US citizens outside the financial sector has been devastating. There was a particular geography to the sub-prime crisis in the US with the states worst affected by housing foreclosures being those that were home to the most extreme housing booms, such as Florida and western states such as California and Nevada (Figure 4.3). But other low-income areas such as the industrial Midwest and the rural south were also hit hard. Beyond housing, the knock-on effects of the 'credit crunch' were felt when banks dramatically reduced their lending to business, prompting the

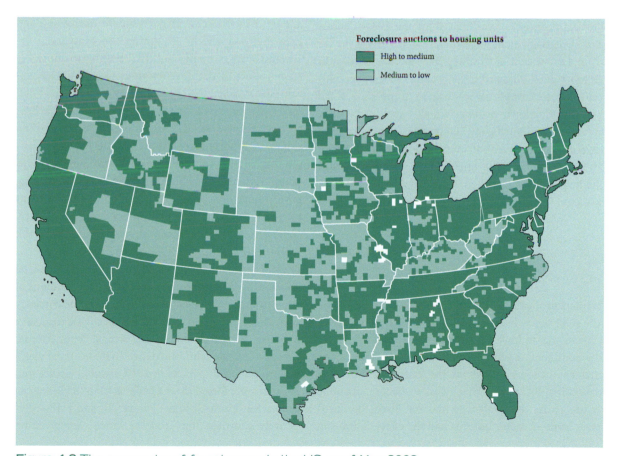

Figure 4.3 The geography of foreclosures in the US as of May 2009

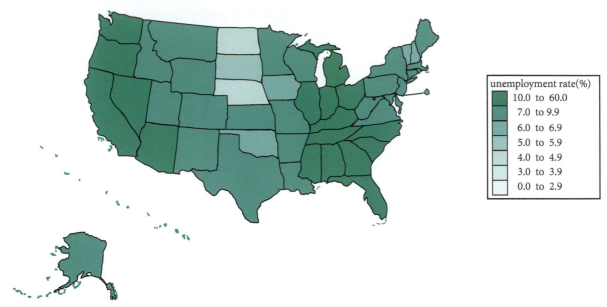

unemployment rate(%)
10.0 to 60.0
7.0 to 9.9
6.0 to 6.9
5.0 to 5.9
4.0 to 4.9
3.0 to 3.9
0.0 to 2.9

Figure 4.4 US unemployment rates by state October 2009
Source: www.bls.gov/LAU (2012)

greatest economic downturn since the 1930s depression. Unemployment more than doubled across the US from 4.6 per cent in summer 2007 to 10.1 per cent in October 2009 (Casaux and Turrini 2011). Geographically, the greatest effects were again in the areas of the housing meltdown in diverse states such as Nevada, California and Michigan (Figure 4.4). One indicator of severity was the extent of 'mass layoffs' (defined as over 50 people being made unemployed by a business) at the height of the recession, recorded as 3,059 (326,392 workers) in the month of February 2009 (ibid.).

4.5.2 The Eurozone sovereign debt crisis

While European banks were heavily involved in the US sub-prime debacle, the subsequent global credit crunch was to have much more profound effects for the continent, leading to a second phase of the crisis that has become known as the Eurozone sovereign debt crisis. By early 2009, a massive injection of public funds across the continent had stabilised the European banking sector but this had the effect of drawing attention to the scale of government borrowing and debt that had arisen, prompting concerns about the sustainability of many countries' economies. One key staging post in the developing crisis was the revelation (in October 2009) by the newly elected centre-left Greek Government that its fiscal deficit (state borrowing) relative to GDP was more than double than originally reported (see Table 4.4). This led to a downgrading of the country's credit status on its bonds (from A to BBB), which had the effect of making government borrowing, and as a consequence loans to domestic firms and consumers, more expensive, compounding the problems already facing the Greek economy from the global downturn.

Contagion set in across the continent as financial investors became concerned about the debt positions of other countries in the Eurozone – the common currency area of the European Union (see Box 4.4), especially the more peripheral countries Portugal, Italy, Ireland, Greece and Spain (christened the PIIGS). As bond yields rose on financial markets in 2010, Ireland, Greece and Portugal were forced to seek bailouts from the European Union, European Central Bank (ECB) and IMF – collectively known as the 'Troika' – which were given on condition that governments cut spending by over 20 per cent in each case and raise taxes to pay back the loans – the theory being that this would help countries to return to balanced growth. In each case, as the economist Mark Blyth has detailed in his book *Austerity: The History of a Dangerous Idea* (Blyth 2013), these neoliberal-inspired austerity measures actually worsened the depth of recession, contributing to increased unemployment rather than restoring the conditions for growth. The crisis has hit young people particularly hard, youth unemployment in Greece and Spain, for example, remained at very high levels in 2015 of 41.5 per cent and 36.1 per cent respectively.

As we set out in Box 4.4, the initial financial crisis was compounded by the design flaw in the euro which failed to recognise the very different structures and circumstances and trajectories of European economies. Creating a single currency across 19 different countries seems in retrospect a hugely ambitious project that flies in the face of the underlying realities of geographical uneven development in the European Union.

Different factors were also at work across the PIIGS in relation to the financial crisis (see Blyth 2013, chapter 3). Ireland and Spain had both experienced housing bubbles and crashes associated with the same kinds of financial recklessness of banks in the US and the UK. Ireland's property bubble was so extreme in the scale of construction over-supply outstripping demand that there were 620 'ghost estates' across the country at the start of 2010, defined as a development of 10 or more houses where 50 per cent are either vacant or under construction (Kitchen *et al.* 2012). Greece had experienced an explosion of public and private debt, financed in part by northern European banks, whereas Portugal was suffering from longer-term problems of losing market share in its main export markets of clothing and textiles from cheaper producers in the global South. Both Portugal and Italy were running long-term government deficits to cover their deteriorating trade performances, which were masked by the existence of a common interest rate within the Eurozone.

While governments had invested considerably in their more economically disadvantaged regions in an era of cheap money, none of the governments – with the possible exception of Greece – fitted the subsequent austerity narrative of reckless public overspending. As in the US, private-sector speculation was the cause of public pain. As Blyth puts it: "The fiscal crisis in all these countries was the *consequence* of the financial crisis washing up on their shores, not its *cause*" (Blyth 2013: 73). While broadly true, this statement underplays the complex geographical entanglements at work in producing the sovereign debt crisis – in particular the role of domestic European banks and economic actors such as governments and wealthy elites in being drawn into an increasingly speculative and deregulated global banking system. While the financial crisis may have been the product of Anglo-American inspired deregulation, the active role played by French and German banks (both public and privately owned), the failure of Greek governments to tax their wealthier citizens in the first place, and the reduced commitment of Europeans to the redistribution of income through traditional progressive tax policies to poorer citizens and more economically disadvantaged regions should not be underplayed.

Not only do individual countries have very different economies, but variations in economic performance within countries (e.g. between northern and southern Italy or between the more advanced economic regions of Spain such as the Basque Region and Andalucía in the south) are also important in understanding the lack of economic convergence and its implications for trying to run a single currency zone (see Fingleton *et al.* 2015). While there is some evidence of international variations in how government policy has shaped regional imbalances, with higher public debt economies helping to shelter less prosperous regions from the effects of the crisis (Crescenzi *et al.* 2016), how regions and cities are placed within broader global divisions of labour is also important. The gap between a Eurozone core – particularly Germany, France, the Benelux countries and Austria – and the geographical periphery of southern Europe and Ireland appears to have widened since the introduction of the euro, compounded by the prolonged economic crisis of recent years (Figure 4.5).

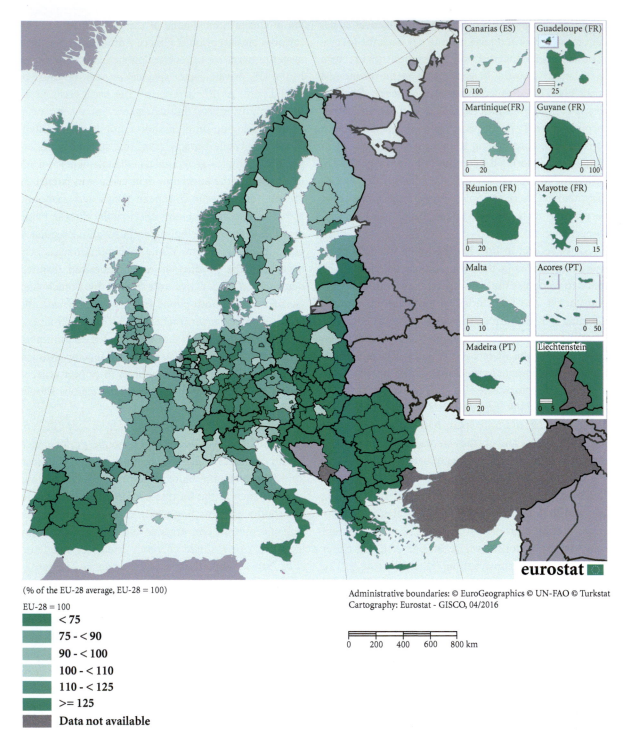

(% of the EU-28 average, EU-28 = 100)

EU-28 = 100

- < 75
- 75 - < 90
- 90 - < 100
- 100 - < 110
- 110 - < 125
- >= 125
- **Data not available**

Administrative boundaries: © EuroGeographics © UN-FAO © Turkstat
Cartography: Eurostat - GISCO, 04/2016

0   200   400   600   800 km

Figure 4.5 Gross domestic product (GDP) per inhabitant in purchasing power standard (PPS) in relation to the EU-28 average, by NUTS 2 regions, 2014

Source: http://ec.europa.eu/eurostat/cashe/RSI#?vis-nuts2.economy&lang=en.

Box 4.4

The revenge of geography? Uneven development and the Eurozone sovereign debt crisis

The euro was launched as a common currency on 1st January 1999 and currently includes 19 of the 28 EU member states (Austria, Belgium, Cyprus, Estonia, Finland, France, Germany, Greece, Ireland, Italy, Latvia, Lithuania, Luxembourg, Malta, Netherlands, Portugal, Slovakia, Slovenia and Spain). The European Union's website maintains that the euro "is the most tangible proof of European integration" (see: http://europa.eu/about-eu/basic-information/money/euro/index_en.htm, last accessed 16 June 2016) and that "benefits of the common currency are immediately obvious to anyone travelling abroad or shopping online on websites based in another EU country" (ibid.).

Yet the 2008–09 financial crisis and subsequent recession have exposed serious underlying flaws in both the management of the euro and ensuring economic convergence.

Various mainstream and more heterodox economists, including the Nobel Prize-winner Paul Krugman, foresaw problems with the whole idea of the euro because of both the wide economic disparities between European economies and the underlying 'monetarist' bias of the convergence criteria, agreed at the 1992 Maastricht Treaty. These stated that, among other things, government debt and inflation should be kept to very low levels, irrespective of economic cycles of growth and contraction. This anti-Keynesian deflationary bias has meant that the Eurozone lacks the political will and institutional capacity to embark on more expansionary economic policies during periods of crisis and downturn.

Another fundamental flaw, that became starkly evident at the time of the financial crisis, was the lack of powers given to the ECB to play the role of 'lender of the last resort'. Most national central banks have the powers to print money or allow devaluation of the currency to ease financial crises and rescue errant banks, but if you have surrendered your currency and the ability to act independently, a state's powers are limited. Because the ECB had not been given such full powers to operate independently and flexibly across the Eurozone, and governments lacked their own sovereign currencies, they were beholden to the Troika and its demands to deal with the crisis.

The more fundamental problem was the failure to recognise deep-seated patterns of geographical uneven development across Europe, particularly between the Eurozone core and southern Europe. The creation of the euro, backed by the ECB and a single interest rate, encouraged investors and financial markets to think of the rest of the EU as having the same economic conditions and

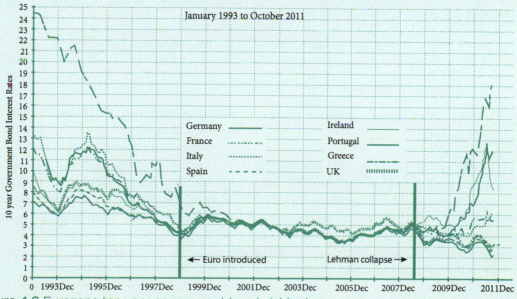

Figure 4.6 Eurozone ten-year government bond yield rates
Source: Blyth 2013: 80, Figure 3.2.

Box 4.4 (continued)

performance as Germany, despite continuing wide disparities in competitiveness and productivity. As Blyth puts it: "the introduction of the ECB and its unending quest for anti-inflationary credibility signalled to bond buyers that both foreign exchange risk and inflation risk were now things of the past. The euro was basically an expanded Deutsche Mark and everyone was now German" (Blyth 2013: 79). This meant that when the euro was created, government bond rates across the EU converged (Figure 4.6) as money and credit flowed into peripheral economies, prompting a consumption boom (of foreign imports in Greece) and housing bubbles in Ireland and Spain. But, when the financial crisis arrived, this 'cheap money' flowed out again, leading to rapid divergence of bond prices and a 'fiscal crisis' for the governments concerned.

Ultimately, as various commentaries have noted, the euro was a political project to encourage greater integration that ignored fundamental economic differences between European countries and regions which generations of economic geographers have rigorously documented (e.g. Dunford and Smith 2000; Hadjimichalis 2011). The Eurozone sovereign debt crisis has exposed the continued if not increased importance of these underlying geographical differences, undermining an integrationist project that simply assumed them away.

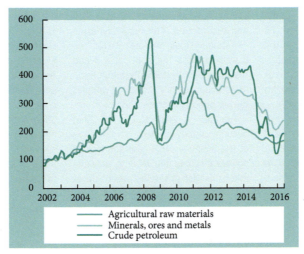

Figure 4.7 Monthly commodity price indices by commodity group, January 2002–June 2016 (Index numbers, 2002 = 100)

Source: UNCTAD 2016: 13. UNCTAD secretariat calculations, based on *UNCTADstat*.

4.5.3 The crisis reaches the global South

The third phase of the ongoing crisis involves two interrelated sets of global connections that have dragged the developing and less developed economies of the global South into a more generalised global economic downturn. The first is a slowdown in the rate of Chinese economic growth, which dropped below 7 per cent in 2014 after being almost 10 per cent for much of the 2000s (see Box 1.3). Such figures would be cause for celebration elsewhere, but have raised concerns, reflecting widespread scepticism about Chinese economic statistics and the crucial role of China in the world economy following the financial crisis in the global North. As the UN puts it: "China became the locomotive of global growth, contributing nearly one third of world output growth during 2011–2012. [It] sustained the global growth momentum during the post-crisis period, maintaining strong demand for commodities and boosting export growth in the rest of the world" (UN 2016: 3). The more sluggish growth has led many to question China's development trajectory, dependent as it is on huge levels of state investment, both in protected industrial sectors that supply the domestic market in basic goods (e.g. steel), as well as in fuelling domestic consumption through massive infrastructural spending.

The Chinese slowdown (and the continuing economic instability in the global North) has had profound implications for the other countries in the **BRICS** block (Brazil, Russia, India, China, South Africa), identified as the leading economic powers of the global South in 2001 (see Box 7.6). With the exceptions of India and China, these countries are heavily dependent on commodity (raw material) exports, resulting in economic slowdown and recession as commodity prices have fallen in recent years (Figure 4.7). While India and much of Asia continue to experience reasonably strong growth, it is the downturn in China that is the greatest cause for concern for the rest of the global economy. This is compounded by growing evidence of a speculative property bubble, fuelled partly by state action, but also by a growth in debt-fueled investment that has the potential to trigger another global economic downturn (see section 4.7).

Reflect

How does the adoption of a relational perspective help us understand the playing out of the financial crisis over time and across space?

4.6 The financialisation of the economy?

The term '**financialisation**' has become a key concept for understanding the growing influence of financial processes and actors in the global economy, first coined in the early 2000s but subsequently used across a range of disciplines including economic geography (see section 1.2.1 for a definition) (e.g. Epstein 2005; Pike and Pollard 2010; Lapavitsas 2013). There is considerable debate and a rapidly expanding literature on the subject but three key aspects can be identified for our purposes here (Stockhammer 2004; van der Zwan 2014): the growth of financial activity in relation to the rest of the economy; the relationship between finance capital and productive capital; and the growing role of finance in the everyday lives of people and communities.

4.6.1 The growth of finance

Financial activities have increased as a share of total economic activity in recent years, although, depending on the indicator used, the trends can be variable between countries. For example, Lapavitsas (2013) finds that financial assets have grown (including those held overseas) in relation to total GDP in the US, UK, Germany and Japan since 1980. The UK, however, has seen by far the strongest growth of finance in relation to the rest of the economy, which demonstrates the strength of global financial interests in relation to the domestic economy. This in turn reflects "the historical evolution of British capitalism and the continuing role of the City of London" (Lapavitsas 2013: 206). Financialisation can therefore be seen as both a more global process but also one that tends to take specific forms in different geographical contexts. The UK and US experienced greater financialisation in the 2000s than Germany or Japan, linked as we have seen to the effects of market deregulation and asset bubbles, particularly in the property market.

The other marked variation between Anglo-Saxon and more regulated economies relates to employment where the UK and US have far higher proportions employed in the sector relative to Germany and Japan. Interestingly, however, the growth in financial activities has not resulted in significant jobs growth in either country, itself the result of automation and restructuring within the banking sector during the 1990s (Lapavitsas 2013). A recent report for the UK Parliament showed that in 2014 there were 1.1 million jobs in finance and insurance sectors, around 3.4 per cent of the workforce, a figure that has remained relatively static since 1997 (Tyler 2015: 5).

Financialisation has also occurred in the world's less developed and emerging economies, though often through terms dictated by the more powerful developed economies of the global North, leading Lapavitsas to coin the phrase 'subordinate financialisation' (2013). Under the influence of the neoliberally inspired policies of the Washington Consensus, governments were encouraged to remove restrictions on the flow of capital, privatise and liberalise their banking sectors and allow the entry of foreign corporations (see section 7.3.3). The theory was that this would make their financial activities more efficient as well as encouraging the influx of finance for investment; capital would flow from rich countries to poor countries according to mainstream economic reasoning if restrictions on global finance were lifted (Lucas 1990).

While there has been some financial investment into the global South, in practice the main trend has been the opposite in the period since the mid-1990s (Figure 4.8). Private flows of capital have tended to be volatile, flowing in during periods of turbulence elsewhere – for example during the financial crisis – but just as quickly flowing out again. What such figures don't capture however is the huge amount of illicit money outflows from poorer countries as a result of deregulation and global integration. The campaigning group Global Financial Integrity estimated that at a conservative estimate around $854 billion left Africa between 1970 and 2008, largely due to political elites acting in collusion with western financial interests and offshore banking centres (Shaxson 2012).

The key capital flows, however, have been from governments. In addition to the servicing of debts to international lending organisations, states have increasingly been accumulating dollar and foreign exchange reserves, not only to safeguard themselves against increasing global uncertainty, but also to peg their currencies to the dollar as part of inflation reduction strategies advocated by the Washington Consensus (Lapavitsas 2013). From a development perspective, this means that some of the poorest countries in the world are spending scarce resources on building up dollar reserves – often by buying US government bonds and thereby helping to reduce interest rates in the US domestic economy – rather than embarking upon domestic investment programmes.

4.6.2 The hegemony of finance over production

Secondly, financialisation relates to a more qualitative change in the relationship between finance and the rest of the economy, whereby non-financial activities are becoming increasingly driven by financial imperatives and financial interests have migrated into the operation of the 'real' productive economy. This is linked to neoliberalism, the deregulation of financial markets, and a relaxation of credit and financial controls across the economy more generally (as described in earlier sections in the chapter). In addition, the spread of financial values and more aggressive profit- or rent-seeking interests is associated with a changing regime of

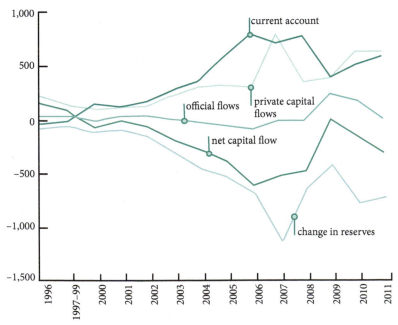

Figure 4.8 Net global capital flows for developed and developing economies 1996–2011 ($Bn)
Source: Lapavitsas 2013: 247, Figure 46.

economic governance where 'Shareholder Value' (Lazo-nick and O'Sullivan 2000; Froud *et al.* 2000) – i.e. the prioritisation of shareholders' interests in terms of real-ising greater share values, profits and dividends from companies – is increasingly dominating the interests of other stakeholders in the firm, notably employees and even consumers to an extent.

The relation between productive and financial capital has long been of interest to political economists (see the interesting review and discussion in Lapavitsas 2013, Part 1). Keynes famously called for the 'euthanasia of the rentier', because he saw financial interests as essen-tially parasitic and potentially destructive in appropri-ating profits rather than reinvesting them in productive

Table 4.7 The pathway to financialisation: a schematic depiction of the changing relations between finance, production and labour in the evolution of capitalism

Time period	Phase	Essence of business relations	Economic rationale underpinning capitalism	Geographical expression	Key power relations	Role of finance
19th century	Familial capitalism	Owner controlled	Private accumulation and enhancement	Locally based and owned but geographical market expansion into national + in some sectors international markets	Growing conflicts between industrial capitalists + workers	Provision of finance for investment in growth, expansion and facilitating concentration of industrial capitalism
20th century	Managerialism capitalism	Managerial hegemony, separation of ownership from control	Growth and market capture, securing cheaper labour supplies	Increasingly extended spatial divisions of labour; growth of dominant national + increasing multinational capital	Corporate national capital v organised labour mediated through varieties of national state regulation (i.e. Fordism)	Financial interests increasingly marginalised; formalised after 1940s in capitalist golden age – national organised capitalism + capital restrictions, Bretton Woods
21st century	Financialised or rentier capitalism?	Shareholder value	Rent seeking, short-term profit	Globally integrated financial markets and role of key centres; new forms of dependency between global North and South	Shareholder value takes precedence over productive capitalism. Capital markets versus industrial growth. Labour share of income reduced dramatically	Finance increasingly dominant across economy, growth of new corporate and institutional forms (e.g. private equity, hedge funds)

activities in a way that could regenerate and sustain capitalism (Keynes 1973), while theorists as diverse as Adam Smith and Karl Marx were aware of the conflicts between financial and industrial capitalism.

Marx tended to assume that, over time, financial interests would be marginalised by the productive forces of capitalism and capital would increasingly be absorbed into production with a growing conflict between capital and labour over the surplus produced under capitalism. In the nineteenth century, banks helped to finance the growth of industry and the growing concentration of business (Table 4.7), and the twentieth century seemed to bear out much of Marx's argument with the growth of large national and then transnational corporations. The mid-twentieth-century literature celebrated the growth of a managerial class, displacing familial capitalism, that was regarded as representing a more efficient and productive form of business organisation, with the separation of ownership from control generally seen as a positive development for capitalism (e.g. Berle and Means 1932; Chandler 1962).

More recently, financialisation seems to be bringing about another structural transformation in capitalism where new organisational forms are emerging that reinforce the position of shareholders or even create new forms of financialised ownership such as private equity firms. One of the most profound examples of this has been in the UK where the privatised utility sectors offer excellent profit-making opportunities in sectors of necessary and stable consumer demand. A report into England and Wales's local water companies, for example, found that in 2012, 13 out of 23 were either wholly or partly owned by private equity companies (Tinson and Kenway 2013). Such examples show how financialisation increases the potential conflict between the social value of services and the profit motive.

4.6.3 Financialisation and everyday life

A third feature of financialisation relates to the way that it has percolated into the lives of ordinary citizens. An important aspect of global capitalism since 1980 – and in part a result of the weakness of trade unions and the flight of capital from higher-wage economies of the global North to lower-wage countries of the south – has been the depression of real wage rates in North America and Western Europe. For example, in the United States real wages have been pretty static since the 1970s and have fallen dramatically as a proportion of the total wealth in the economy (Harvey 2010: 13).

While this might be good news for individual capitalists, keeping wage rates low and boosting their profits, it is not good for the economy as a whole because wages also represent demand for the products and services of capitalist firms. The deregulation of finance has meant that the gap has been filled by the explosion of household debt, in particular the greater availability of credit for middle- and low-income households. In real terms, the average debt for US households has risen from $40,000 to $130,000 since the late 1970s (ibid.: 18). This is not just confined to the increase in mortgage debt, but debt for all kinds of consumption goods as well as for more basic services and needs, such as higher education where high levels of student debt (to pay for university tuition and maintenance) are now the norm in many countries. While the upsurge in household debt has been most pronounced in the Anglo-American economies and less prevalent elsewhere, this has nevertheless been a global trend. For instance, a recent study by the management consultants McKinsey highlighted the growing levels of household and private corporate debt in countries such as China, Malaysia and Singapore since the 2008–09 crisis (McKinsey 2015).

There is evidence of deeper underlying changes in the economic identity of ordinary people as a consequence of financialisation (Langley 2008) where the withdrawal of states from the traditional provision and supervision of savings, pensions and welfare is associated with a gradual transformation of the relationships between money, credit and society. One key element of this transformation is "the reshaping of consumers into financial subjects, i.e. from passive savers to entrepreneurial investors" (Lai 2018: 27). States, financial actors and even the media have promoted a culture of financial entrepreneurialism where individuals are encouraged to take on increased responsibility and risk for their financial futures. The result is that unwitting consumers are drawn into dense and complex financial relations over which they have little control and often little knowledge. Predatory lending during the sub-prime crisis exposed the levels of exploitation of lower-income consumers that can result (see Box 4.5). As Lapavitsas has put it in relation to the financial crisis, "It is remarkable – and a true reflection of the content of financialisation – that the historic crisis that commenced in 2007 was triggered by the poorest layers of the US working classes defaulting on mortgage debt" (2013: 276).

4.7 Financialisation and the urbanisation of capital

One of the key manifestations of financialisation has been a growing tendency for capital to invest and speculate in the areas of land, property and housing. The heterodox economist Michael Hudson has gone so far as to suggest that the financialised economy is one in which the financial system has effectively been 'decoupled'

Box 4.5

The *Big Short* and the geographies of financialisation

In Michael Lewis's 2010 book, *The Big Short* (subsequently made into a star-studded Hollywood Oscar-winning film), a small group of financial analysts predict the financial crisis and make vast sums of money by betting against the rest of the financial markets. While the big US investment banks such as Goldman Sachs and Morgan Stanley, and many European banks are pouring money into the sub-prime housing market, this collection of individuals ignore the 'herd mentality' of the markets, and decide the housing boom is unsustainable and an inevitable bust is coming.

The first person to spot the flaws in the system is a California-based hedge fund manager called Michael Burry, who realises that he can make money by betting against a rising housing market, setting up a credit default swap deal with several major banks and investment companies who gladly take his money, thinking the rising market will continue. Burry's bet is that the housing market will begin to fail in the second quarter of 2007. Why does Burry believe this against the conventional wisdom of the rest of the global financial establishment? In the words of the film's narrator, Burry and the

others "saw the giant lie at the heart of the economy and they saw it by doing something that the rest of the suckers never thought to do. They looked." (The words are accompanied in the film by an aerial screen shot panning out from a big suburban housing estate.)

What Burry saw was the disconnection between the everyday economy of people's lives in particular places and the financial markets caught up in a speculative bubble. Early on in the film, Burry is seen poring over spreadsheets of thousands of individual mortgages from the western states of the US (one of the worst-affected regions of the subsequent foreclosure crisis as indicated in Figure 4.3). What he spots is that a large number of these mortgages are adjustable rate or variable mortgages, with people on low incomes or in precarious work struggling to meet or missing monthly payments. He realises that large numbers of people will default on their mortgage loans if interest rates rise, leading to the collapse of the whole housing market (see Box 4.2). Burry sees a way of making money but of course the social consequences are millions of ordinary people losing their homes.

In one of the most memorable scenes of the film, Mark Baum, a New York-based money manager, takes his team down to Florida (a region with a subsequent high rate of foreclosure) to test how everyday realities on the ground match their suspicions of an imminent housing market collapse. Talking to a stripper in a night club, he realises that she has absorbed the financialised culture of leveraging leaving her vulnerable to the collapse of the housing market: "You're not going to be able to refinance", the Wall Street insider tells her. "On all of my loans?" she asks. "What do you mean, 'all your loans'?" Baum responds. "I have 5 houses, and a condo", the stripper informs her client.

This scene from the film neatly captures financialisation through the spilling over of a financial 'rentier' logic into the 'real' economy of material things, people and communities. At the same time, it also illustrates the dramatic urbanisation of capital that has accompanied financialisation as property and homes become financial assets for exchange value rather than resources to meet basic human needs such as housing and shelter.

from the actual productive economy where useful goods and services are made and exchanged. Under financialisation, speculative investment in real estate is marginalising more socially useful forms of capitalism. The growing dominance of a financialised capitalism is leading to more asset bubbles, debt-driven crises and the return of the 'rentier' capitalist (who makes profits from extracting value from property rather than more socially useful forms of production and technological advancement) (Hudson 2012). Although these ideas are quite controversial, there is certainly no doubting the growing importance of property-related investment in major economies such as the US and the UK.

Spatially, this form of financialisation has become associated with what we can term the 'urbanisation of capital thesis' after David Harvey. In a series of works over two decades (e.g. Harvey 1985, Harvey 2010, Harvey 2012), Harvey has developed the French Marxist theorist Henri Lefebvre's work on the importance of urbanisation processes to capitalist development (e.g. Lefebvre 1968). A key argument from Harvey is that the history of urbanisation from a Marxist perspective can be seen as linked to the absorption of surpluses of capital and labour arising from industrialisation processes, which becomes important in resolving periodic crises in capitalism. Over time, production under capitalism produces both surplus capital which must be invested in new ventures to generate further profits (Figure 3.1), and surplus labour or unemployment as a result of restructuring and reorganisation to drive down costs. Urban infrastructure projects that can absorb such surplus labour and capital can therefore provide new spatial fixes (Harvey 1982) in securing new rounds of capital accumulation.

Harvey (2012) cites the infrastructure and construction projects of Second Empire Paris – from 1848 to 1970 – which included the reconstruction of the city's neighbourhoods from dense cluttered neighbourhoods to the architect Haussmann's grand boulevard schemes. The physical urban renewal was accompanied by a process of social cleansing in which Haussmann "deliberately engineered the removal of much of the working class and other unruly elements, along with insalubrious industries from Paris's city center, where they constituted a threat to public order, public health and of course political power" (Harvey 2012: 16). The US in the post-Second World War period went through a process of capitalist urbanisation on a nation-wide scale, whereby public and private investment flowed into new suburbs. According to Harvey, this was critical "in the stabilization of global capitalism after World War II" (Harvey 2012: 9). Yet such urbanisation processes also threaten to generate new asset bubbles and crises as capital flows in and debt-fuelled speculation gains pace (as evident in the most recent financial crisis).

Many of the financial crises that have affected the world economy since the 1970s have been at least in part produced by property bubbles or real estate speculation according to Harvey (2010). In countries as diverse as Japan, Sweden, Finland and most recently of course the UK, US, Spain and Ireland, a financialised urbanism appears to have been at the root of financial meltdown. Time and again, it is only government intervention that has prevented broader systemic collapse.

Although such tendencies have long been a feature of industrial capitalism, the period of neoliberalism has unleashed a more dramatic global phase of urban development, where city governments across the world, in tandem with finance and developer interests, have developed property-led regeneration programmes, more often than not involving the privileging of luxury developments for the rich at the expense of social housing for the poor. London has become the extreme form of this, rapidly becoming a city that attracts global investment of and for the rich at the expense of the city's working population, producing a developing housing crisis (see Box 4.6). As such, Harvey talks of a "transformation in the scale of the urban process" (Harvey 2012: 12) as cities as diverse as London, Mexico City, Mumbai, New York, Santiago, Seoul and Taipei experience debt-filled building booms. While this has spurred various forms of resistance by diverse groups of citizens now claiming their own rights to the city, such as the Occupy movements that arose in 2011 across North America and Europe (Caffentzis 2012), and social unrest in poorer neighbourhoods like Paris's *banlieues* (see Dikec 2007), there has been little success so far in reversing these trends.

China is the more recent country to have experienced a massive housing and property boom, fuelling an orgy of speculative debt-filled investment. As part of the country's opening up to the rest of the global

Box 4.6

London in the twenty-first century: a city of the global super-rich at the expense of the working poor?

London's reputation as a city of global finance is being accompanied by its growth as a magnet for investment by the super-rich. In 2014, the *Sunday Times*'s rich list showed that Britain's capital had 80 residents worth over £1 billion, more than any other world city; New York (56), San Francisco (49), Moscow (45) and Hong Kong (43) were the next ranked (Atkinson *et al.* 2016). The international elite have also been buying up London housing; at a time of faltering investments elsewhere, the property sector is seen as a safe haven for financial returns. A recent report suggested that 49 per cent of all purchases in the capital of properties over £1 million came from international investors, many of whom spend very little time in these dwellings (ibid.). The phenomenon of ghost neighbourhoods or 'lights out London' has been noted in many of the more exclusive areas such as Chelsea and Kensington and Knightsbridge where property prices have risen from £745,000 in 2002 to £3.4 million in 2015 (Cumming 2015). Across the capital there are 22,000 empty properties.

The effect of this globalised property speculation has been devastating locally, driving up prices across the city and creating a shortage of affordable housing for those living and working in London. For instance, a quarter of a million households are on the waiting list for public housing while there are over 40,000 families registered as 'homeless' and in temporary accommodation (Atkinson *et al.* 2016: 2). Not surprisingly, housing was a key battleground in the recent mayoral election with the victorious candidate, Sadiq Khan, pledging both to tackle foreign property speculation and encourage more affordable housing. The mayor told *The Guardian* newspaper: "There is no point in building homes if they are bought by investors in the Middle East and Asia . . . I don't want homes being left empty. I don't want us to be the world's capital for money laundering. I want to give first dibs to Londoners" (Booth and Bengtson 2016).

Key reading: Atkinson *et al.* (2016) International capital flows into London property, *SPERI Global Political Economy Brief No. 2*. Available at http://speri.dept.shef.ac.uk/, last accessed 7 July 2016.

economy since the 1970s, China has experienced the most dramatic period of urbanisation of any country in history, from a predominantly rural society with only 18 per cent of the population in urban areas in 1978 to over 50 per cent in the mid-2010s (Gu *et al.* 2015). The scale of this urban growth has also been vast; it now has six megacities (with populations of over 10 million) and ten cities with populations between 5 and 10 million and it will add one more megacity and six more large cities by 2030 (UN 2014). One telling statistic is the figure that in the years 2011–12 China consumed more cement than the US did in the whole of the twentieth century (Anderlini 2016). China's urbanisation has been deliberately engineered by the state in an attempt to shift the country's development trajectory away from an export-dependent one to encourage greater domestic consumption.

The signs are that the country may be fomenting its own property crash on a massive scale with housing prices beginning to decline and vast acres of ghost cities that dwarf anything seen elsewhere. Perhaps the greatest symbol of this is Sky City: a proposal by a local billionaire to create the world's tallest building in the city of Changsha in central China. The site is currently home to a "makeshift fish farming pond" among "forests of half-built or empty apartment towers" (ibid.). China now has a debt to GDP ratio higher than the US and real estate development has quickly come to dominate the Chinese economy, accounting for 60 per cent of all bank investment (ibid.). Some commentators fear that this could lead to another global crisis because of the importance of China as a market for economic activities elsewhere. As the *Financial Times* put it, "the fate of everything from Hong Kong financial institutions to German carmakers to Australian miners is now in the hands of homebuyers in places like Changsha, the city where Sky City was supposed to be built" (ibid.).

4.8 Summary

Money and credit have always been critical to the functioning of capitalism, providing the means for innovation, technical change and geographical expansion. Over time, the evolution of capitalism has seen dramatic changes in the nature and form of money and its relation to broader processes of economic development. Recent decades have seen the massive expansion of financial activities relative to the rest of the economy. This is associated with a more globally integrated financial system representing the most advanced form of time-space compression (Harvey 1989). Neoliberal deregulation and the development of ICT have together created a hypermobile form of finance capital which seems to have unleashed a period of increasing economic instability and a growing tendency for asset booms and subsequent crises to occur.

As such, the globalisation of finance is associated with an era of increasing instability and crisis tendencies. The volatility experienced since the 1990s can be seen as the latest chapter in the unstable history of capitalism, which has been characterised by cycles of growth and crisis (see section 3.2.3). For some, financialisation is leading to a new phase of capitalism where financial interests tend to dominate the economy over other interests (e.g. Lapavitsas 2013). For others (e.g. Harvey 2010), the very sustainability of capitalism as an economic system is in question, with the growing number of financial crises redolent of its increased contradictions.

In geographical terms, the globalisation of the financial system has produced a new set of geographies of connection and differentiation, accentuating existing patterns of uneven geographical development. This is evident in the growth of leading world cities as hubs that link local and national economies into global flows of money, and in other parts of the Global North and South which are either marginalised or drawn into exploitative debt relations. As the examples of the US sub-prime housing crisis or the European sovereign debt crisis show, financialisation can negatively affect the lives of citizens in more 'ordinary' cities and regions who are drawn into spatial circuits of finance through complex and often unequal relationships. Geographically, financialisation is associated with new forms of urbanisation where property and land speculation alongside gentrification are creating increasingly divided cities between economic elites and working-class groups whose 'rights' to the city may be threatened.

Exercise

Drawing upon the experience of one country from the global North and one from the global South, trace the impact of the recent financial crisis and its evolution since 2008–09 in terms of economic growth, employment and relations between cities and regions. Address the following questions:

1. What have been the relational geographies connecting each country to the financial crisis?

2. To what extent has the financial crisis accentuated processes of uneven geographical development in each country?

3. What forms does the urbanisation of capital take in the two countries?

4. What kind of government policies (at national and international levels) have been applied to stimulate new rounds of economic growth?

Key reading

Aalbers, M. and Pollard, J. (2016) Geographies of money, finance and crisis. In Daniels P., Bradshaw, M., Shaw, D., Sidaway, J. and Hall, T. (eds) *An Introduction to Human Geography*, 5th edition. Harlow: Pearson, pp. 365–78.
A good overview of the workings of the financial system and its geographies with an account of the global finance crisis.

Blyth, M. (2013) *Austerity: The History of a Dangerous Idea*. Oxford: Oxford University Press.
A good critical analysis of the impact of austerity policies in the aftermath of the financial crisis. Particularly insightful about the problems of the Eurozone.

Christophers, B. (2015) **Geographies of Finance II: crisis, space and political-economic transformation.** *Progress in Human Geography* 38 (2): 205–13.
A good overview of recent geographical perspectives on the crisis and subsequent recession.

Harvey, D. (2005) *A Brief History of Neoliberalism.* Oxford: Oxford University Press, chapter 4.
An excellent account of the uneven development of neoliberalism and financial globalisation and its geographical effects.

Kitchen, R., O'Callaghan, C., Boyle, M. and Gleeson, J. (2012) **Placing neoliberalism: the rise and fall of Ireland's Celtic Tiger.** *Environment and Planning A* 44: 1302–26.
A paper that uses the example of Ireland to make connections between the global financial crisis and its effects in terms of property speculation and the economic and social consequences.

Lapavitsas, C. (2013) **Approaching financialization: literature and theory.** In Lapavitsas, C., *Profiting without Producing: How Finance Exploits us All.* London: Verso, chapter 2, pp. 13–43.
A critical analysis of different perspectives on financialisation.

Useful websites

www.bis.org/.
The site for the Bank for International Settlements which gives good information and data relating to international financial issues and regulation.

www.debtdeflation.com/blogs/.
A website, run by University of Kingston Professor Steve Keen, a post-Keynesian economist, which gives interesting insights into the financial system and the contradictions of mainstream economic policymaking.

www.imf.org.
The website of the International Monetary Fund, the global body responsible for ensuring financial stability.

References

Aalbers, M. and Pollard, J. (2016) Geographies of money, finance and crisis. In Daniels P., Bradshaw, M., Shaw, D., Sidaway, J. and Hall, T. (eds) *An Introduction to Human Geography*, 5th edition. Harlow: Pearson, pp. 365–78.

Anderlini, J. (2016) The Chinese chronicle of a crash foretold. *Financial Times*, 24 February. At www.ft.com/content/65a584e2-da53-11e5-98fd-06d75973fe09. Last accessed 16 August 2018.

Arestis, Philip (1996) Post-Keynesian economics: towards coherence. *Cambridge Journal of Economics* 20: 111–35.

Atkinson, R., Burrows, R., Glucksberg, L., Ho, H.K., Knowles, C., Rhodes, D. and Webber, R. (2016) International capital flows into London property. *SPERI Global Political Economy Brief No. 2.* Available at http://speri.dept.shef.ac.uk/, last accessed 7 July 2016.

Berle, A. and Means, G. (1932) *The Modern Corporation and Private Property.* New York: Macmillan.

BIS (Bank for International Settlements) (2013) *Triennial Central Bank Survey.* Basel: Bank for International Settlements.

Blackburn, R. (2008) The subprime crisis. *New Left Review* 50: 63–106.

BLS (Bureau of Labor Statistics) (2012) *BLS Spotlight on Statistics: The Recession of 2007–9.* Bureau of Labor Statistics, Washington, DC, available at: www.bls.gov/spotlight.

Blyth, M. (2013) *Austerity: The History of a Dangerous Idea.* New York: Routledge.

Booth, R. and Bengtson, H. (2016) The London skyscraper that is a stark symbol of the housing crisis. *The Guardian*, 24 May.

Caffentzis, G. (2012) In the desert of cities: notes on the occupy movement in the US. *Reclamations.* Retrieved from www.reclamationsjournal.org/blog/?p=505.

Casaux, S. and Turrini, A. (2011) Post-crisis unemployment developments: US and EU approaching? *ECFIN Economic Brief, Issue 13 May.* Brussels: European Commission.

Cassidy, J. (2008) The Minsky moment. *The New Yorker*, February. At www.newyorker.com.

Chandler, A. (1962) *Strategy and Structure: The History of the American Industrial Enterprise.* Cambridge, MA: MIT Press.

Christophers, B. (2015) Geographies of finance II: crisis, space and political-economic transformation. *Progress in Human Geography* 38 (2): 205–13.

Crescenzi, R., Luca, D. and Milio, S. (2016) The geography of the economic crisis in Europe: national macroeconomic conditions, regional structural factors and short-term economic performance. *Cambridge Journal of Regions, Economy and Society* 9: 13–32.

Cumbers, A. (2012) *Reclaiming Public Ownership: Making Space for Economic Democracy.* London: Zed.

Cumming, E. (2015) It's like a ghost town: lights go out as foreign owners desert London homes. *The Guardian*, 25 January.

Dicken, P. (2003) *Global Shift*, 3rd edition. London: Sage.

Dikec, M. (2007) *Badlands of the Republic*. London: Wiley Blackwell.

Dow, S. (1999) The stages of banking development and the spatial evolution of financial systems. In Martin, R. (ed.) *Money and the Space Economy*. Chichester: Wiley, pp. 31–48.

Dunford, M. and Smith, A. (2000) Catching up or falling behind? Economic performance and regional trajectories in the 'New Europe'. *Economic Geography* 76 (2): 169–95.

Epstein, G. (ed.) (2005) *Financialization and the World Economy*. Northampton: Edward Elgar.

Financial Times (2008) Lehman Brothers file for bankruptcy, 16 September. At www.ft.com/content/52098fa2-82e3-11dd-907e-000077b07658. Last accessed 16 August 2018.

Fingleton, B., Garretsen, H. and Martin, R. (2015) Shocking aspects of monetary union: the vulnerability of regions in Euroland. *Journal of Economic Geography* 15 (5): 907–34.

Froud, J., Haslam, C., Johal, S. and Williams, K. (2000) Shareholder value and financialization: consultancy promises, management moves. *Economy and Society* 29 (1): 80–110.

Froud, J., Johal, S., Leaver, A. and Williams, K. (2006) *Financialization and Strategy: Narrative and Numbers*. London: Routledge.

Gilbert, E. (2005) Common cents: mapping money in time and space. *Economy and Society* 34: 356–87.

Graeber, D. (2011) *Debt: The First 5,000 Years*. New York: Melville.

Gu, C., Kesteloot, C. and Cook, I. (2015) Theorising Chinese urbanisation: a multi-layered perspective. *Urban Studies* 51 (14): 2564–80.

Hadjimichalis, C. (2011) Uneven geographical development and socio-spatial justice and solidarity: European regions after the 2009 financial crisis. *European Urban and Regional Studies* 18 (3): 254–74.

Harvey, D. (1982) *The Limits to Capital*. Oxford: Blackwell.

Harvey, D. (1985) *The Urbanisation of Capital*. Oxford: Blackwell.

Harvey, D. (1989) *The Condition of Post-Modernity*. Oxford: Blackwell.

Harvey, D. (2005) *A Brief History of Neoliberalism*. Oxford: Oxford University Press.

Harvey, D. (2010) *The Enigma of Capital and the Crises of Capitalism*. New York: Profile.

Harvey, D. (2012) *Rebel Cities: From the Right to the City to the Urban Revolution*. London: Verso.

Hirst, P.Q., Thompson, G.F. and Bromley, S. (2009) *Globalization in Question*, 3rd edition. Cambridge: Polity Press.

Hudson, M. (2012) *The Bubble and Beyond*. Dresden: ISLET.

Ingham, G. (1996) Some recent changes in the relationship between economics and sociology. *Cambridge Journal of Economics* 20: 243–75.

Keen, S. (2009) The roving cavaliers of credit. *Steve Keen's DebtWatch*, 31 February. At www.debtdeflation.com/blogs/2009/01/31/therovingcavaliersofcredit/. Accessed 16 August 2018.

Keynes, J.M. (1973) *The General Theory of Employment, Interest and Money*. London: Macmillan.

Kitchen, R., O'Callaghan, C., Boyle, M. and Gleeson, J. (2012) Placing neoliberalism: the rise and fall of Ireland's Celtic Tiger. *Environment and Planning A* 44: 1302–26.

Lai, K. (2018) Financialisation of everyday life. In Clark, G.L., Feldman, M., Gertler, M.S. and Wojcik, D. (eds) *The New Oxford Handbook of Economic Geography*. Oxford: Oxford University Press, pp. 611–27.

Langley, P. (2008) Financialisation and the consumer credit boom. *Competition and Change* 12: 133–47.

Lapavitsas, C. (2013) *Profiting without Producing: How Finance Exploits us All*. London: Verso.

Lazonick, W. and O'Sullivan, M. (2000) Maximising shareholder value: a new ideology for corporate governance. *Economy and Society* 29 (1): 13–35.

Lefebvre, H. (1968) *Le Droit a la Ville*. Paris: Anthropos.

Lewis, M. (2010) *The Big Short: Inside the Doomsday Machine*. New York: W.W. Norton.

Leyshon, A. and Thrift, N. (1995) Geographies of financial exclusion: financial abandonment in Britain and the United States. *Transactions of the Institute of British Geographers* NS 20: 312–41.

Lucas, R. (1990) Why doesn't capital flow from rich to poor countries? *American Economic Review* 80 (2): 92–6.

Mann, G. (2013) The monetary exception: labour, distribution and money in capitalism. *Capital and Class* 37 (3): 197–216.

Marshall, J.N., Pike, A., Pollard, J.S., Tomaney, J., Dawley, S. and Gray, J. (2012) Placing the run on Northern Rock. *Journal of Economic Geography* 12 (1): 157–81.

Martin, R. (1994) Stateless monies, global financial integration and national economic autonomy: the end of geography? In Corbridge, S., Martin, R. and Thrift, N. (eds) *Money Power and Space*. Oxford: Blackwell, pp. 253–78.

Martin, R. (ed.) (1999) *Money and the Space Economy.* Chichester: Wiley.

McKinsey (2015) *Debt and Not much Deleveraging.* London: McKinsey and Co.

Minsky, H. (1975) *John Maynard Keynes.* New York: Columbia University Press.

Minsky, H. (1986) *Stabilizing an Unstable Economy.* New Haven: Yale University Press.

Minsky, H. (1992) The financial instability hypothesis. *Working Paper* No. 74. New York: Jerome Levy Economics Institute, Bard College.

Moore, E. (2016) China issues its first renminbi sovereign debt in London. *Financial Times,* 26 May. Available at: www. ft.com/cms/s/0/f81c777a-233e-11e6-aa98-db1e01fabc0c. html#axzz4AzlTc2FV, last accessed 3 June 2016.

NAO (National Audit Office) (2011) *The Financial Stability Interventions, Extract from the Certificate and Report of the Comptroller and Auditor General on HM Treasury Annual Report and Accounts 2010–11,* HC 984, July. London: HM Treasury.

Palma, J.G. (2009) The revenge of the market on the rentiers. Why neoliberal reports on the end of history turned out to be premature. *Cambridge Journal of Economics* 33: 829–69.

Parker, G. and McLannahan, B. (2015) Alistair Darling joins board of Morgan Stanley. *Financial Times,* 8 December. At www.ft.com/content/b7c708a4-9e03-11e5-b45d-4812f209f861. Last accessed 16 August 2018.

Pierce, A. (2008) The Queen asks why no one saw the financial crisis coming. *Daily Telegraph,* 5 November. At www.telegraph.co.uk/news/uknews/theroyalfamily/3386353/The-Queen-asks-why-no-one-saw-the-credit-crunch-coming.html. Last accessed 29 October 2009.

Pike, A. and Pollard, J. (2010) Economic geographies of financialisation. *Economic Geography* 86: 29–51.

Pollard, J. (2005) The global financial system: worlds of monies. In Daniels, P., Bradshaw, M., Shaw, D. and Sidaway, J. (eds) *Human Geography: Issues for the Twenty First Century,* 2nd edition. Harlow: Pearson, pp. 358–75.

Popper, N. (2015) Morgan Stanley in $2.6 billion settlement over crisis in mortgages. *Financial Times,* 25 February. At www.nytimes.com/2015/02/26/business/dealbook/morgan-stanley-in-2-6-billion-mortgage-settlement.html. Accessed 16 August 2018.

Shaxson, N. (2012) *Treasure Islands: Tax Havens and the Men Who Stole the World.* London: Random House.

Stockhammer, E. (2004) Financialization and the slowdown of accumulation. *Cambridge Journal of Economics* 28 (5): 719–41.

Thompson, G. (2010) 'Financial globalisation' and the 'crisis': a critical assessment and 'what is to be done'? *New Political Economy* 15 (1): 127–45.

Tickell, A. (2000) Dangerous derivatives: controlling and creating risks in international money. *Geoforum* 31: 87–99.

Tickell, A. (2003) Cultures of money. In Anderson, K., Domosh, M., Pile, S. and Thrift, N. (eds) *Handbook of Cultural Geography.* London: Sage, pp. 116–30.

Tinson, A. and Kenway, P. (2013) *The Water Industry: A Case to Answer.* London: UNISON and the New Policy Institute.

Tyler, G. (2015) Financial services: contribution to the UK economy. *Commons Briefing Papers SN/EP/06193.* London: House of Commons Library.

United Nations (UN) (2014) *World Urbanisation Prospects.* New York: United Nations.

United Nations (UN) (2016) *World Economic Situation and Prospects 2016.* New York: United Nations.

United Nations Conference on Trade and Development (UNCTAD) (2016) *Trade and Development Report, 2016: Structural Transformation for Inclusive and Sustained Growth.* New York and Geneva: UNCTAD.

van der Zwan, M. (2014) Making sense of financialisation. *Socio-Economic Review* 12: 99–129.

Wojcik, D. and MacDonald-Korth, D. (2015) The British and the German financial sectors in the wake of the crisis: size, structure and spatial concentration. *Journal of Economic Geography.* Advanced access, doi:10.1093/jeg/lbu056.

Wolfson, M. (1986) *Financial Crises.* New York: M.E. Sharp.

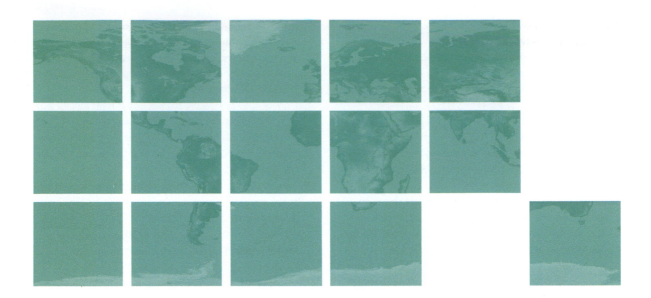

Chapter 5
Managing capitalism: states and changing forms of economic governance

5.1 Introduction

The role of **the state** in the economy is wide-ranging, but often invisible to individual consumers, shaping the provision of goods and services in ways that are not always immediately apparent. For example, whenever you go to the pub for a drink, you will find that the state decides how long it can stay open; the size of the measures of beer, wine or spirits offered; how much of the price is taken in tax; how the drinks are labelled; the standards of hygiene governing the kitchen; and the minimum wages paid to the staff (Painter 2006). A key underlying argument made in this chapter is that the state should be viewed as a dynamic process rather than as a fixed 'thing' or object (Peck 2001). Instead of focusing solely on the size of the state, expressed in terms of levels of taxation or expenditure, for example, we should examine how states intervene in economic life, the forms of economic policies that the state pursues and the effects of these on different social groups and regions (O'Neill 1997).

Our analysis of the geography of state intervention is informed by our **GPE** approach, which rejects the notion of an autonomous, self-regulating economy. The idea of a self-regulating economy is central to mainstream, neoclassical economics, emphasising the role of the market in ensuring that supply and demand are balanced through the price mechanism, consigning the state to a limited role of upholding property rights and enforcing business contracts. By contrast, our institutionalist understanding contends that the economy is regulated through various forms of social regulation, including social habits, administrative rules and cultural norms (Aglietta 1979) (see sections 2.4.3, 2.5.3). The state plays a key role in harnessing and coordinating these different mechanisms, formulating a wide range of rules and laws covering matters such as business taxation, trade policies, employment standards and financial markets. The role of the state in the economy is generally directed towards the promotion of economic growth, attempting to create the conditions that allow businesses to make profits and workers to find employment, thereby generating revenue through various forms of taxation. From a geographical perspective, states play a key role in regulating wider processes of uneven development, sometimes introducing regional policies that are focused on particular types of place (for example, depressed regions). Beyond these very general dimensions of state regulation, the specific forms and functions of the state change over time, as highlighted by the regulationist notion of **modes of**

regulation, referring to specific institutional arrangements which help to create relatively stable periods of economic growth known as **regimes of accumulation** (section 2.4.3).

5.2 Understanding the changing nature of the 'qualitative state'

The state is the basic organising unit of political life (Figure 5.1). It refers to a set of public institutions for the protection and maintenance of society (Dear 2000: 789). These institutions include parliament, the civil service, the judiciary, the police, the armed forces, the security services, local authorities, etc. As this suggests, the state is a complex entity stretching beyond what is normally referred to as government (the national executive of ministers and civil servants). States exercise legal authority over a particular territory, holding a monopoly of legitimate force and law-making ability. While they are often conflated together, it is important to distinguish the state from the nation. The latter refers to a group of people who feel themselves to be distinctive, on the basis of a shared historical experience and cultural identity, which may be expressed in terms of ethnicity, language or religion. The two come together to form nation-states in cases where the state territory contains a single nation. This is often presumed to be the norm, but there are many examples of multi-national states which contain different national groups (the UK for one is made up of English, Scots, Welsh and Northern Irish). The corollary of this is that not all national groups have their own states (for example, the Kurds or the Basques).

5.2.1 Reshaping the 'qualitative state'

As emphasised in the introduction to this chapter, it is useful to view the state as a dynamic process rather than regarding it as a fixed object or thing. This is consistent with our regulationist perspective which emphasises the role of the state in stabilising and sustaining capitalism. Rather than focusing on the size of the state and

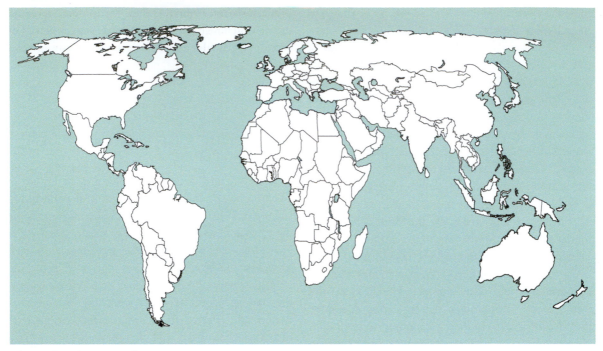

Figure 5.1 A world of states

the extent of its intervention, measured in terms of its share of national wealth, the level of taxation and the volume of welfare payments, recent work has emphasised the nature of such intervention and the social and economic goals that it is directed towards. This can be seen as a shift of emphasis from a concern with quantitative aspects of state intervention to an interest in its qualitative characteristics (Painter 2000: 363).

This new approach is best captured by the concept of the '**qualitative state**' developed by the Australian economic geographer Philip O'Neill (1997). It is based on three main points. First, O'Neill rejects the notion of the state as a single, unified actor with a highly centralised structure. Instead, it is structured by a continual process of interaction between state agencies such as the UK Treasury and non-state actors and forces, for example business organisations such as the Confederation of British Industry (CBI). Second, states always play a crucial role in the construction and operation of markets, including those operating at the international scale. Two key ways in which states have actively constructed markets in recent years are through the **privatisation** of formerly state-owned industries such as electricity or telecommunications and the fostering

of **globalisation** as a result of policies that have sought to lower trade barriers and promote competition. Third, a qualitative view of the state overcomes the politically disabling argument that the powers of the nation-state are being eroded by globalisation and **neoliberalism**, emphasising, instead, that they are being transformed in particular ways. Rather than being powerless in the face of globalising processes, governments can still regulate markets to achieve broader social goals such as full employment or universal health care.

The notion of the 'qualitative state' can be used to inform analyses of contemporary processes of state restructuring. It helps to focus attention on the changing forms and functions of the state in relation to the interrelated processes of globalisation and neoliberal reform. The qualitative perspective highlights the importance of the state itself as an actor in processes of globalisation and economic restructuring, rather than assuming that the state is always acted upon as a victim of globalisation by other more powerful forces, such as TNCs and financial markets. States have contributed to the process of globalisation through measures such as the reduction of trade barriers and the abolition of controls on the movement of capital.

5.2.2 Rescaling the state

The dominant view of the relationship between globalisation and the state has been the neoliberal or free market perspective on globalisation as an inevitable and irreversible process (section 1.2.1). This claims that globalisation has eroded the capacity of nation-states to regulate their economies, leaving them unable to intervene meaningfully in markets, with some commentators even claiming that this equated to the 'death of the nation state' (see Anderson 1995), but the overstated nature of this claim has been increasingly recognised in recent years (Peck 2001). Rather than offering an objective analysis of state restructuring, it is informed by neoliberal ideas, representing a form of wishful thinking that views the reduction of state powers as a 'good thing'. Instead, the more nuanced 'transformationalist' understanding that we adopt stresses that states are subject to multifaceted and ongoing processes of qualitative adjustment rather than a simple quantitative erosion or diminution of their powers (ibid.). This multifaceted and ongoing process of reorganisation can, in part, be viewed as a response to globalisation pressures.

The geographical structure of the state has been profoundly reshaped by these processes of reorganisation as powers have been transferred between the national, supranational and sub-national scales of state activity (section 1.2). In the post-Second Word War period, the national level enjoyed a taken-for-granted primacy, reflecting its role as the locus of government power in exerting a coordinating influence over economic activity. Since the 1970s, however, state powers have been subject to ongoing processes of '**rescaling**', referring to changes in the relationships between geographical scales. These have involved shifts of power and responsibilities in two principal directions: from the national to the supranational level; and from the national to the sub-national level of cities and regions. This has been termed the 'denationalisation' of the state by the sociologist Bob Jessop (2002). It must be understood as a relative shift, meaning that the national scale continues to play a central role in economic **governance**, with the supranational and sub-national levels becoming more prominent compared to the post-Second World War period. While states may have lost powers in some areas

(trade, monetary policy), they have largely retained them in others (foreign and defence policy) and may have even extended their reach in certain areas (security, welfare).

The emergence of a more prominent supranational tier of government involving bodies such as the WTO and EU is partly the result of state action in that states have come together to create such bodies. Membership provides a forum for states to assert and extend their power, requiring regular contact and discussion with other states. It is state representatives who negotiate over trade in the WTO whilst states have sought to exert some supervision over international financial transactions through bodies such as the Bank for International Settlements. Regional economic integration offers states access to larger markets and protection against competition from outside the regional bloc. The best example is the EU which has evolved from a customs union between six countries into a monetary union containing 28 states (Box 5.1). Other regional trade blocs include the North American Free Trade Association (NAFTA), incorporating Canada, Mexico and the US and AFTA in South-east Asia (Table 5.1). States have certainly lost some powers to the EU, particularly in the economic sphere, such as those over trade, competition and monetary policy (for those within the Eurozone), as well as in sectors such as agriculture and fisheries. Even in these areas, though, state representatives develop policy and national governments continue to exercise some authority. In other spheres there is increased cooperation between national states in face of threats such as terrorism, drug trafficking, organised crime, etc. The operation of such organisations, then, seems to reinforce and reconstruct state power, rather than usurping it.

Since the 1970s, many different governments around the world have sought to transfer power to sub-state governments, meaning that political decentralisation or **devolution** has become a key 'global trend' of recent decades (Rodríguez-Pose and Gill 2003). Devolution can be defined as "transfer of power downwards to political authorities at immediate or local levels" (Agranoff 2004: 26). It typically involves the establishment of elected political institutions (governments and/or assemblies) at the regional scale of government. These institutions vary considerably between countries,

Table 5.1 Major regional economic blocs

Regional group	Membership	Date	Type
EU (European Union)	Austria, Belgium, Bulgaria, Croatia, Czech Republic, Cyprus, Denmark, Estonia, France, Finland, Germany, Greece, Hungary, Ireland, Italy, Latvia, Lithuania, Luxembourg, Malta, Netherlands, Poland, Portugal, Romania, Slovakia, Slovenia, Spain, Sweden, United Kingdom	1957 (European Communities) 1967 (European Communities) 1992 (European Union)	Economic union
NAFTA (North American Free Trade Agreement)	Canada, Mexico, United States	1994	Free trade area
EFTA (European Free Trade Association)	Iceland, Norway, Liechtenstein, Switzerland	1960	Free trade area
MERCOSUR (Southern Common Market)	Argentina, Brazil, Paraguay, Uruguay, Venezuela (2006)	1991	Common market
ANCON (Andean Common Market)	Bolivia, Colombia, Ecuador, Peru	1969 (revived 1990)	Customs union
AFTA (ASEAN Free Trade Agreement)	Brunei Darussalam, Cambodia, Indonesia, Laos, Malaysia, Myanmar, Philippines, Singapore, Thailand, Vietnam	1967 (ASEAN), 1992 (AFTA)	Free trade area
China-ASEAN Free Trade Agreement	Brunei Darussalam, Cambodia, China, Indonesia, Laos, Malaysia, Myanmar, Philippines, Singapore, Thailand, Vietnam	2010	Free trade area

Source: Dicken 2015: 211.

Box 5.1

The EU and state reorganisation

The institutional structure of the EU is unique. It is neither a federation nor a simple inter-governmental organisation. Decision making is centred upon the 'institutional triangle' formed by the European Commission, Council and Parliament. The Council of Ministers is the main decision-making body, representing the member states. The heads of government and foreign ministers meet at least twice a year in the European Council, high-profile summits that attract widespread media coverage. The main roles of the Commission are to initiate legislation and proposals, to implement European legislation, budgets and programmes and to represent the Union on the international stage. The Parliament, consisting of directly elected MEPs, has powers over legislation and the budget which are shared with the Council. It also exercises democratic supervision over other EU institutions, particularly the Commission.

The EU operates on the basis of agreement and negotiation between independent states, although the member states have delegated some of their functions to the central institutions. Member states continue to shape how the powers of the EU are exercised and implemented, often seeking to further their national interests within the supranational space it provides. This is evident even in areas like trade where the EU is the decision-making body. Whilst the EU is represented as one body in trade negotiations, different member states have adopted different stances. The UK's liberal free trade views, for example, have often been countered by the more cautious position of countries like France, although this tension may be eased by the former's departure from the EU following the 'Brexit' vote of 23 June 2016.

according to the nature of the national political system and the specific sets of powers and responsibilities devolved to the sub-national level. Devolution is common to both federal and unitary states, defined by the 'vertical' division of sovereignty between the national and sub-national levels in the former and its concentration at the national level in the latter. Whilst federal states have devolved more power to established sub-national institutions, these have generally been established by acts of devolution in unitary states.

Devolution has been introduced in response to pressures exerted on established states from both 'below' and 'above', in relation to demands from groups within the state for more say over their own affairs and the effects of processes of globalisation and supranational integration respectively. In response to these conflicting pressures, governments have sought to modernise and reinvent the state, making it more flexible and responsive to the challenges of securing prosperity, security and solidarity in an increasingly turbulent world. Devolution has been an important part of this project, promising to empower citizens to address the key issues affecting their regions and to help renew the state more broadly.

In addition to 'denationalisation', Jessop (2002) identifies two other underlying trends reshaping the operation of the qualitative state. First, the 'destatisation' of the political system refers to an 'outwards' movement of responsibilities from the state to various arms-length agencies, private interests and voluntary bodies. Second, the internationalisation of policy regimes refers to a growing volume of linkages between national, regional and local institutions and personnel in different countries. This enables policy mobility and transfer as initiatives (for example privatisation or welfare reform) introduced in one country are taken up and adapted by officials in another. Processes of denationalisation, destatisation and internationalisation have come together to shape a new system of urban and regional governance in the 1990s. Governance is a much broader term than the traditional notion of government, emphasising the increased involvement of various special-purpose and semi-autonomous agencies (quangos), business and the voluntary sector in the delivery and management of services. This has created a complex and fragmented organisational landscape comprised of a number of actors and organisations at the local and regional scales, generating real problems of policy coordination and political accountability.

Reflect

How does the qualitative concept of the state as a process differ from traditional understandings of the state as a fixed object with a widely accepted function and a highly centralised structure?

5.3 States as managers of national and regional economies

States play a key role in managing the economy, establishing and enforcing the 'rules of the game', referring to the laws, norms and regulations that govern economic activity (Gertler 2010) (section 2.5.2). These include the maintenance of private property rights, the regulation of money, the upholding of labour standards and the provision of transport infrastructure. Within their geographical boundaries, states have sought to define and promote national economies, building integrated national markets through the creation of common legal standards and financial rules, the expansion of transport and communication systems and the regulation of flows of goods, money and people across their borders. As economies have become more complex and integrated over time, the role of the state has tended to expand, moving, for example, from the fairly minimalist liberal state of the nineteenth century to the highly interventionist welfare states of the 1960s and 1970s. National states contain considerable internal diversity in terms of economic conditions, cultural values and political allegiances. The problems of territorial management created by such internal diversity have generally been recognised through the creation of a tier of local government which administers state programmes and represents local interests (Duncan and Goodwin 1988).

States have never existed in isolation from the wider international economy with the management and regulation of trade representing one of their key functions (Dicken 2015). Here, we can identify a long-standing tension between the opening up of national economies to international trade and investment flows and pressures for the greater protection of domestic firms and workers from international competition through increased tariffs and restrictions on imports. The balance between these conflicting pressures has changed over time with, for instance, the international integration of the 'gold standard' era of the late nineteenth century giving way to renewed protectionism in the depressed economic conditions of the 1920s and 1930s. More recently, of course, economic globalisation means that flows of goods, money and information across national borders have grown rapidly (section 1.2), but this is arguably creating renewed protectionist pressures, particularly in the wake of the financial crisis of 2008–09 and the Great Recession (Johnson 2016). At the same time, states remain important actors within globalisation (Dicken 2015), not least in prosecuting the interests of their own multinationals in overseas markets and in international trade forums such as the WTO.

The wide range of economic activities undertaken by the state can best be understood in relation to a smaller number of overarching economic roles that it plays in managing its national economy. Our approach here is based on the framework set out by Coe *et al.* (2013: 88–99), identifying five key functions. Before outlining these functions, it is important to recognise that the state should not be seen as a wholly separate entity from the economy, seeking to somehow intervene in the latter from outside; instead, its long-standing and wide-ranging economic interests mean that the state is itself part of the national economy that it seeks to manage and regulate.

5.3.1 The state as ultimate guarantor

Here, the state plays a critical role in supporting and underpinning the operation of basic economic instruments and institutions, including the currency, financial system and certain firms that are deemed to be of strategic importance to the economy. This essentially involves the state stepping in and guaranteeing the continued operation of these instruments and institutions in the event of serious disruption or failure, acting as the institution of last resort (ibid.: 88). Particular circumstances in which states tend to step in include financial crises and natural disasters which may involve them taking over responsibilities for the debts and financial liabilities of private institutions such as banks. This basic economic function consists of three central elements:

➤ *Maintenance of property rights and the rule of law.* This aspect of the state's role is fundamental to the operation of the capitalist economy through the provision and protection of private property rights as its most basic institution, allowing individuals and firms to generate income and rent from the ownership of various economic assets. Here, property includes land, buildings, equipment, ideas and brands or trademarks, with the last two often being defined as intellectual property. The state establishes rules and laws for the acquisition, disposal and transfer of property rights between individuals and firms, enabling them to enter into economic contracts with other actors. Its existence as a kind of neutral referee or arbiter of property rights and claims that stands apart from private economic interests is widely recognised and accepted, even by ardent neoliberals such as the highly influential economist and philosopher Friedrich Hayek. For Hayek and his supporters, however, the actions of the state should be confined to this rather minimal 'night watchman' role of maintaining private property rights.

➤ *Guaranteeing national economic instruments.* This part of the state's role is of central importance to the operation of the economy, particularly in relation to trade and the functioning of the financial system. Here, states seek to uphold the value of their currency and other financial instruments that they issue, particularly government bonds, generating confidence among firms, traders and investors. Central banks (section 4.2.1) often play a key role in the management of the currency, acting as a kind of purchaser of last resort in the event of the currency coming under pressure from speculative activities in the foreign exchange markets. This involves the central bank spending its reserves to buy its own currency in the markets, aiming to uphold its price and thereby maintain the confidence of investors. Processes of supranational integration and globalisation have tended to weaken the power of states in this respect, as was starkly demonstrated by the failure of the Bank of England's effort to protect the pound on 'Black Wednesday', 16 September 1992, against currency speculators, forcing it out of the EU's Exchange Rate Mechanism (ERM), a precursor of the euro. As outlined in section 4.5.2, the introduction of the euro was itself underpinned by the establishment of the ECB to manage and defend the new currency.

➤ *Responding to financial crises.* As we emphasised in Chapter 4, financial crises are a recurrent feature of the evolution of capitalism, requiring the state to step in to contain the damage and prevent a wider economic collapse. Acting as 'lenders of last resort' to maintain the flow of credit is a key task of central banks (section 4.2.1). It reflects the importance of finance and money to the operation of the economy more widely, meaning that governments often see financial institutions as 'too big to fail'. With the exception of the US investment bank Lehman Brothers, this was evident in the financial crisis of 2008–09. Led by the US and UK, a number of governments decided to bail out their largest banks through a range of direct support packages, including loans, and share purchases, alongside guarantees, that amounted to a substantial share of GDP (Table 5.2).

5.3.2 The state as regulator

As emphasised by the concept of the qualitative state, states are closely involved in the operation and regulation of markets and economic activities. This encompasses overarching macroeconomic policies in addition to interventions in particular markets and efforts to govern economic flows across national borders. Such regulation is often designed to ensure that markets operate efficiently and fairly by subjecting them to the influence of prevalent social norms, values and rules, reflecting the fact that the state ultimately derives its legitimacy from its capacity to defend the interests of its citizens (Coe *et al.* 2013: 90). The precise nature of social regulation is, of course, subject to variation over time and across space, that is between different periods and between different countries and regions (section 5.4). From the 1980s, for instance, many states have adopted more market-friendly forms of regulation informed by the individualist neoliberal values of private property,

Table 5.2 The costs of bailing out the banks, selected countries as of March 2011, percent of 2010 GDP

	Direct support	Recovery	Net direct cost
Ireland	30.0	1.3	28.7
Netherlands	14.4	8.4	6.0
Germany	10.8	0.1	10.7
UK	7.1	1.1	6.0
USA	5.2	1.8	3.4
Greece	5.1	0.1	5.0
Belgium	4.3	0.2	4.1
Spain	2.9	0.9	2.0
Average	6.4	1.6	4.8
In billions ($)	1,528	379	1,149

Source: Kitson *et al.* 2011: 293. Data from IMF 2011: 8.

competition and enterprise compared to the more collectivist social democratic ethos of the 1960s and 1970s (Jessop 2002). Alternative values continue to be held by some sections of society, however, exerting recurring pressures for the re-regulation of markets and underlining the contested nature of state regulation.

Two types of macroeconomic policy are adopted by the state to manage and regulate its national economy (Dicken 2015: 185–6):

➤ **Fiscal policies** based on the raising or lowering of taxes and the level of public expenditure. Such taxes are sources of revenue for state activities, making states financially dependent on the health of the national economy. Tax policy influences the level of economic activity with increased taxes leading to a reduction in domestic demand (reducing people's disposable income) and possible economic contraction, whilst reducing taxes stimulates domestic demand. This is not an automatic process, however, ultimately depending upon individual actors' responses to tax changes (ibid.). Raising or lowering public expenditure has broadly similar effects, with increases stimulating demand and economic activity and vice-versa. Fiscal measures play a particularly prominent role in Keynesian policies of demand management, based on the need for adjustment across the economic cycle, reducing demand through increased taxes and/or reduced expenditure in the growth phase, whilst maintaining demand through reduced taxes and/or increased expenditure in the contraction phase.

➤ **Monetary policies** concerned with managing the money supply and its rate of circulation in the form of credit. The control of inflation through interest rate policy is of central importance here. This function is often undertaken by central banks. Lowering interest rates should stimulate economic activity by making it cheaper for firms and consumers to borrow money, while increases in interest rates have the opposite effect. In a similar fashion to fiscal policies, monetary policies have been used to manage the economy across the economic cycle – raising interest rates in growth periods and reducing them when activity slows. They have often been privileged by policy-makers since the 1970s, reflecting the influence of monetarist policies and the commitment to low inflation, alongside the neoliberal aversion to increased taxes and public expenditure. Yet a period of near zero interest rates since the depths of the economic crisis in 2009 means that there is little scope to further reduce them in response to any further downturn, prompting policy-makers to turn to other forms of monetary policy such as quantitative easing (expanding the volume of money in the economy).

In addition to these overarching policies, states also regulate market and economic flows in other ways. In the regulation of markets, for instance, states often act to ensure that competition is fair and open to new entrants and that large existing firms do not exert monopoly powers. In many cases, governments have to approve mergers between firms, giving them the power to promote competition. The US has historically been very proactive through 'anti-trust' policies designed to prevent the formation of large conglomerates or trusts, resulting in the break-up of numerous large corporations during the twentieth century, such as Rockefeller's Standard Oil in 1911 and the telecoms giant AT&T in 1984.

The state is also concerned with the regulation of economic flows of goods, services, capital and labour across its borders. Traditionally, states have levied tariffs on imported goods and services and established capital and exchange controls to manage movements of capital and foreign exchange (currency), while labour has been subject to immigration policies. In recent decades, however, economic globalisation is associated with the reduction or abolition of controls on the flows of capital. At the same time, restrictions on the movement of labour have been maintained or even reinforced in response to populist concerns about immigration.

5.3.3 The state as a strategic economic actor

States are strategic actors in the economy, not only reacting to economic events such as crises, but also operating in a more proactive fashion through deliberate policies that are designed to promote particular objectives. In general, such policies will be designed to increase national prosperity and create a competitive advantage over rival countries. They are more specific in nature than the fiscal and monetary policies outlined above, which are concerned with the overall management of the economy. They target particular activities and sectors. Here, we follow Coe *et al.* (2013: 91–2) in focusing on four areas: trade, foreign direct investment, industry and labour markets.

➤ *Trade.* States are generally responsible for the negotiation of trade deals, working either through the WTO in multilateral discussions or directly with other governments in bilateral arrangements. Here, states generally seek to manage trade in the interests of domestic producers, particularly in terms of negotiating access to overseas markets, although this is balanced by the pressure to offer other countries access to their own domestic markets (Box 5.2). This reflects the long-standing tension between openness to trade and protectionism. In general, the post-Second World War period saw successive reductions in tariffs, particularly in manufactured goods, although this has stalled in recent decades. Many developing countries adopted a fundamental change of strategy in the 1960s and 1970s, moving from **import-substitution industrialisation (ISI)**, based upon the domestic production of goods that were formerly imported, to **export-oriented industrialisation (EOI)**, based on producing goods for external markets. EOI was often combined with the protection of strategic '**infant industries**' from outside competition until they were strong enough to compete in global markets. Certain developed countries such as Australia also sought to open their economies up to international competition in the 1980s and 1990s (Box 5.2).

➤ *Foreign direct investment* (FDI). Here again, the underlying trend over recent decades has been towards increased openness to FDI, which has been widely viewed as a source of jobs, investment and technology, having grown faster than trade since 1985 (Dicken 2015: 20). States seek to attract FDI through a range of measures such as the provision of land and industrial premises, direct financial incentives, assistance with workforce training and the development of links to local suppliers (Phelps *et al.* 2003). As outlined in section 3.5.2, many developing countries have established Export Processing Zones to attract FDI on the basis of reduced taxes and exemptions from certain national regulations.

➤ *Industry strategies.* These usually involve the strategic support of particular sectors of the economy on the basis of their economic significance through various forms of financial and non-financial assistance (Chang *et al.* 2013). Industrial policy has become controversial in some quarters, with neoliberal critics in particular arguing that it is tantamount to the state 'picking winners' which is something that should be left to the market. Accordingly, it became discredited in more liberal states such as the UK and US whilst persisting in more interventionist ones like France and Japan. In the wake of the Great Recession, governments have expressed renewed interest in industrial strategy as a means of promoting economic recovery and stimulating key sectors (Bailey *et al.* 2012). Between 2011 and 2013, for instance, the UK Government published industrial strategies for 11 key sectors of the economy.

➤ *Labour market strategies.* Governments generally seek to promote employment and skills training as a way of making the economy both more economically competitive and socially inclusive. An important element of this is measures to address unemployment, particularly long-term unemployment, which usually requires direct government intervention. As well as the encouragement of employers to provide additional training for those in work, many developed countries have sought to promote more flexible labour markets in response to processes of globalisation, requiring workers to become more responsive to the requirements of employers. This is attractive to business, but has generated greater insecurity and precarity for some groups of workers, particularly the lower paid, as we examine in greater depth in section 6.4.

5.3.4 The state as an owner

Despite widespread policies of privatisation in many countries over recent decades, the state continues to directly own firms and even industries. State ownership is often related to the pursuit of broader political and

Box 5.2

Reorienting a national economy: Australian trade policy in the 1980s and 1990s

Australia embarked upon a series of major economic reforms in the 1980s and 1990s, influenced by the neoliberal (free market) economic orthodoxy then sweeping the English-speaking world and a domestic re-appraisal of Australia's economic malaise in the 1960s and 1970s (Kelly 1992: 15). Australia had long been one of the most protected developed countries, reflecting the nature of its political-economic settlement after it became an independent state in 1901. This was based upon a rural export base of wool, wheat and minerals, combined with the protection of manufacturing, seen as essential to the construction of a stable industrialised nation and the promotion of social objectives of full employment and fair wages. By the 1960s, the manufacturing sector employed over 25 per cent of the workforce, though much of this was in small, externally owned branch plants (Weller and O'Neill 2014: 513). This established economic arrangement became unsustainable by the 1970s, with Australia experiencing a serious decline in its **terms of trade** as the value of agricultural and commodity exports decreased relative to industrial goods and services, while

the protected manufacturing sector was failing to compete with emerging producers in neighbouring Asia. This convinced some policy-makers of the need for radical reform.

These reforms were carried out by the Labour Government following its victory in the General Election of 1983, led by Prime Minister Bob Hawke, a former trade union leader, and Paul Keating as Treasurer. They represent the most radical shift in Australia's political economy since the establishment of the system of protection in the early decades of the twentieth century, deploying the strategic powers of the state to change the direction of national economic development. Their first key move was to float the Australian dollar in response to the growing difficulties of trying to manage the exchange rate in the face of increased international capital flows. This was linked to the **deregulation** of the financial system and the abolition of exchange controls. It was followed by the reduction of tariffs in 1988, which were cut from over 20 per cent to either 15 per cent or 10 per cent by 1992, succeeded by a further across-the-board cut to 5 per cent, announced

in 1991 (Leigh 2002). These tariff reductions were seen as essential to the restoration of Australia's international competitiveness.

Over time, the reforms have had decidedly mixed effects. While they benefited mining, which experienced an unprecedented boom in the 2000s, agriculture and finance, they are associated with the continued, even accelerated decline of manufacturing, echoing the effects of trade **liberalisation** on protected sectors across much of the global South (section 7.5). Manufacturing's share of employment fell from 25 per cent in the 1960s to 8 per cent in 2013, reflecting the inability of domestic firms to compete internationally (Weller and O'Neill 2014). The uneven effects of reform have also been expressed geographically with a divide opening up between the states of western Australia and Queensland, where employment in mining boomed in the 2000s and Sydney, which has experienced growth in financial services, on the one hand, and the manufacturing regions of Victoria, South Australia and New South Wales which have suffered from serious deindustrialisation, on the other.

social objectives, including national security and the maximisation of benefits from the extraction of natural resources. Accordingly, state ownership is more common in resource and infrastructure sectors such as energy, electricity and transport in which the state plays a particularly important regulatory role and which lend themselves to a natural monopoly. A common distinction is made between State Owned Enterprises (SOEs) and Government Linked Corporations (GLCs). While the former are directly owned and managed by the state, it only owns a share in the latter and does

not exercise direct managerial control (Coe *et al.* 2013: 94–5). In addition, states have also developed their own national funds in recent decades, termed sovereign wealth funds, which they use to invest, more indirectly, in a range of firms and projects, many of them overseas.

China is renowned for its high levels of state ownership, including some 275 known SOEs such as Petrochina which is the country's largest oil producer. Many other developing countries also have prominent SOEs in the sectors identified above, which often provide a vehicle for the state pursuing wider development objectives.

Many European countries also have relatively high levels of state ownership, particularly in more coordinated welfare states. In Norway, for instance, the state owned 33 per cent of the market value of companies in 2003, compared to 30 per cent owned by private investors and 28 per cent by foreign capital (Figure 5.2) (Norwegian Ministry of Trade and Industry 2011: 8–9). Prominent SOEs include the oil company Statoil and the energy utility Statkraft, which are 67 per cent and 100 per cent owned by the state respectively.

Rather than being solely national in their operation, state-owned firms are often active participants in globalisation, particularly through FDI in foreign markets. For instance, Chinese SOEs have been investing heavily in energy and resource projects in the global South, particularly sub-Saharan Africa, reflecting China's need for access to natural resources to fund its growth and prompting talk of a 'new scramble for Africa' between emerging powers, particularly the **BRICS** countries (Carmody 2011). At the same time, Norwegian SOEs have also internationalised their activities with Statoil, for example, having operations in some 36 different countries including stakes in several UK offshore wind farms.

5.3.5 The state as a service provider

The State typically assumes responsibility for the delivery of a range of public services, funded by taxation, including education, health, infrastructure and social welfare. This reflects the broader social importance of these services and their role in sustaining or 'reproducing' the workforce as well as the limited incentives for private provision. Despite the growth of privatisation and deregulation policies in recent decades, the state continues to play a leading role in the provision of public services in most countries, although some have made greater efforts to increase private involvement in service delivery.

This role of the state is hugely significant in terms of employment, meaning that the state is often the largest employer in a country across its range of activities, including organisations such as hospitals, schools, the national civil service, local authorities, public transport companies, prisons and many others. As Figure 5.3 shows, public sector employment averaged over 20 per cent across the members of the Organisation for Economic Co-operation and Development (developed

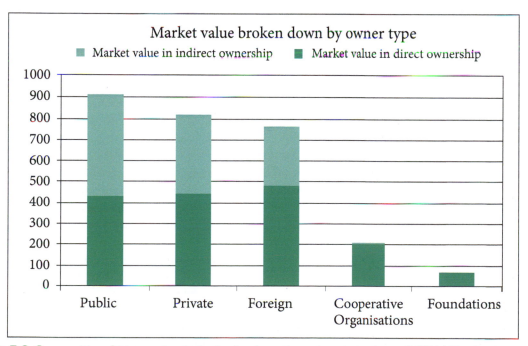

Figure 5.2 Ownership of Norwegian industry broken down by owner type (NOK billion)
Source: Norwegian Ministry of Trade and Industry 2011: 8–9 (Jakobsen and Grünfeld 2003).

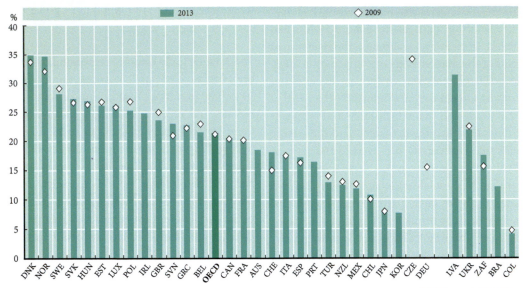

Figure 5.3 Public sector employment as a percentage of total employment, 2009 and 2013
Source: International Labour Organization (ILO), *ILOSTAT* (database).

and industrialised economies), although there is considerable variation between individual countries, with continental European welfare states tending to have the highest levels of public employment (section 5.4).

State employment also varies regionally within countries, often being relatively higher in poorer regions in which the private sector is less developed. This is evident in the poorer regions of the UK, for instance in northern England, Northern Ireland and Wales (Table 5.3). State bodies are often the largest employers in such regions, which can be regarded as 'state anchored', along with other regions that may depend upon large state facilities such as defence plants (Markusen 1996). While often viewed in negative terms as inhibiting or 'crowding out' private sector growth and investment, public sector employment is very important as a source of employment, income and economic stabilisation for such regions. This makes them less vulnerable to unforeseen economic shocks and downturns, although these may still affect state employment through associated reductions in public spending, often termed **austerity** policies (section 5.5).

In overall terms, the state remains a hugely important economic actor across these five distinct roles. As the economic crisis of 2008–09 revealed, it remains the guarantor of last resort, coming to the rescue of

insolvent banks through large-scale 'bailouts'. Despite ongoing processes of globalisation, it continues to be the key regulator of national economies and the flows of goods, services, capital and labour across their borders. The state is also a key strategic actor through its efforts to influence and manage the future direction of economic activity across the key domains of trade, FDI, industry policy and labour markets. Furthermore, states play an important role as the owners of a significant number of firms, many of which have become active investors in overseas markets. Finally, states are the main providers and organisers of public services in most countries, making them key employers, particularly in poorer regions. As such, the state is unique in the collective scale and scope of its economic activities, operating as the key coordinator and manager of its national economy in the interests of broader social and political goals such as prosperity, security and inclusion.

Reflect

In what ways does the financial crisis of 2008–09 and subsequent Great Recession underline the enduring economic role of the state?

Table 5.3 Regional dependence on the public sector: share of public sector in regional employment and output, UK, 2010

	Percent of total employment	Percent of total output (GVA)
Greater London	18.0	14.2
South East	21.0	15.6
Eastern	18.9	16.4
South West	22.0	19.8
East Midlands	19.9	17.9
West Midlands	22.3	20.1
Yorkshire–Humberside	22.4	19.9
North West	23.1	19.8
North East	28.6	25.3
Wales	25.2	23.7
Scotland	21.9	20.1
Northern Ireland	26.0	25.5

Source: Gardiner *et al.* 2013: 920.

5.4 Varieties of capitalism

While the overarching roles discussed in the previous section reflect the basic nature of the state as an economic actor, the particular ways in which these functions are exercised varies considerably between different types of state. This point has been emphasised by leading political economists and comparative sociologists who have identified the existence of different '**varieties of capitalism**'. This reflects an underlying sense that 'institutions matter', referring particularly to the role of the state in establishing and upholding the 'rules of the game' that govern economic life (section 2.5.2). In a particularly influential contribution, Hall and Soskice (2001) identify two main varieties of capitalism, focusing heavily on the national scale: coordinated-market economies which are associated with countries such as Germany, Japan and Sweden; and liberal-market economies, exemplified by the likes of the UK, US and Australia. As many critics have pointed out, however, capitalism is actually far more diverse than the reliance upon these two archetypes would suggest. The economic geographers Jamie Peck and Nik Theodore (2007) argue that this diversity should be seen in terms of the ongoing 'variegation' of state types over time as they interact with broader global forces, rather that the simpler idea of fixed national 'varieties'. Our approach is based upon the work of Amable (2003) which offers a richer account of institutional diversity than Hall and Soskice's twofold classification (Table 5.4).

Neoliberal states are characterised by a reliance upon market mechanisms with the state playing a key role in overseeing the rules that govern the operation of markets. As the label suggests, such states have been in the vanguard of neoliberal reform from the early 1980s, embracing policies of deregulation, liberalisation and privatisation (section 5.5). Accordingly, they have actively promoted labour market flexibility as a way of adapting to increased competition, favouring local wage bargaining between companies and trade unions at the firm or plant level, while employment protection has been weakened and the role of trade unions reduced (Table 5.4). The financial system plays an important role in the economy, based upon the maximisation of shareholder value and leaving firms dependent on accessing credit from banks and other lenders. Participation in higher education is strong, but vocational training is weak with limited company involvement.

Table 5.4 Characterising five types of state

Types of state	Labour markets	Finance	Welfare	Education & training	Examples
Neoliberal states	Flexibility, company-based wage bargaining, limited role for trade unions, limited employment protection	Market-based financial system; emphasis on shareholder value; orientation towards higher risk capital markets; strong venture capital sector	Liberal model of welfare state with residual provision in some countries with more universalist model in others	Stronger higher education system; weak vocational training with limited company involvement	US, UK, Canada, Australia, New Zealand
Developmental states	Regulated with employment protection	Bank-based financial system; key co-ordinating role of the state; close links between the state and industry	Low level of social protection; low public social expenditure; limited health expenditure	Private tertiary education system	Japan, Singapore, South Korea, Taiwan
Continental European states	Co-ordinated labour markets with variations in the degree of employment protection; co-ordinated wage bargaining between firms and trade unions	Bank-based financial system with loose links between banks and firms; importance of insurance companies	Employment-based benefits and reliance on insurance arrangements to cover unemployment, sickness and old age	Strong vocational training system with close company involvement; emphasis on secondary education	Germany, France, Austria, Switzerland
Social democratic states	Regulated labour markets; active labour market policies; high rates of trade union membership	Bank-based system	Universalist model; importance of family services	High public expenditure on tertiary education	Sweden, Denmark, Finland, Norway
Mediterranean states	Regulated labour markets; conflictual industrial relations; limitations to temporary work	Bank-based system; weak corporate governance; ownership concentration	Limited welfare state; importance of old age expenditures	Weak education system; low expenditure on education; weakness in science and technology education	Greece, Italy, Portugal, Spain

Source: Adapted from Amable 2003: 174–5.

Welfare provision tends to be more residual in nature; that is, confined to the provision of services to disadvantaged groups such as the unemployed, although countries such as the UK and Australia have maintained a stronger universalist tradition, despite decades of neoliberal reform, reflecting their social democratic pasts.

Developmental states are defined by the active role of the state in economic policy, based upon the pursuit of developmental goals aimed at enhancing the productive powers of the nation and closing the economic gap with the industrialised countries. It is typified by the newly industrialised countries of East Asia (Box 5.3). The state plays a key coordinating role in the economy, particularly through the channelling of savings and financial surpluses into investment through the banking system and the establishment of a strong government department to coordinate and promote industrial development. The prototype of the powerful coordinating agency is the Japanese Ministry of International Trade and Industry (MITI), created to promote export-driven growth. The state also develops close cooperative ties with business, allowing the development of new technologies and methods through joint projects. In South Korea, for instance, the state facilitated the development of the *chaebol* – large family-based corporations such as Samsung, Hyundai, Daewoo and Lucky-Goldstar – that

have dominated the economy. Labour markets are tightly regulated with limited rights for workers and trade unions, reflecting a strong authoritarian element. This is associated with relatively low levels of social protection, linked to the imperative of channelling the surpluses from growth into productive investment.

Continental European states are highly coordinated in nature, based upon a strong social partnership between employers and trade unions, although there are considerable variations between countries in the extent of employment protection. They are characterised by close links between finance and industry with regional banks often playing a key role in channelling finance towards small firms in particular. The welfare state is highly corporatist in nature with guaranteed social rights funded through social insurance schemes. Vocational training is a particularly strong feature of this regime, and is closely linked to the needs of firms.

Box 5.3

Development states in action: the case of Singapore

By far the smallest of the East Asian **newly industrialising countries (NICs)**, Singapore has been described as the world's most successful economy (Lim, quoted in Huff 1995: 1421). In common with the other NICs, the role of the state has been central in driving and shaping development. Although it is a parliamentary democracy, Singapore has been governed by one party, the People's Action Party (PAP) since the early 1960s. Indeed, one powerful politician, Lee Kuan Yew, was in charge until his retirement in 1990, becoming the longest-serving Prime Minister in the world. Singapore occupied an important position within the British Empire, becoming a major trading centre and port, based upon its strategic geographical position astride the major East–West global trade route and its fine natural harbour.

The break from Malaysia in 1965 greatly reduced the size of the domestic market, militating against the pursuit of an ISI-based strategy. In response, the PAP regime pursued an aggressive policy of export-oriented, labour-intensive manufacturing development based on attracting FDI. This was seen as a question of national survival for a newly independent state without any natural resources. The policy was highly successful with manufacturing employment increasing fourfold between 1967 and 1979 and manufacturing exports growing from 13.6 to 47.1 per cent of GDP over the same period (Huff 1995: 1423). Investment was focused in sectors such as electronics, petroleum and shipbuilding. Control of labour was central to the export-based strategy with the state achieving a rare degree of acquiescence and cooperation from the labour movement. This was based on the incorporation of labour into corporatist bodies such as the National Wages Council and the

redistribution of some of the proceeds of growth in the form of employment, housing, education and healthcare programmes (Coe and Kelly 2002).

In response to fears about the erosion of its labour-costs advantage by low-wage competitors, the government changed policy direction in the 1980s, focusing on the expansion of the financial and business services sector. This is closely linked to a strategy of 'regionalising' the economy by establishing Singapore as the control centre of a regional division of labour, hosting high-level functions such as corporate headquarters, business services and research and development whilst labour-intensive manufacturing and assembly is carried out in lower-wage neighbours (Dicken 2015: 203–4). The Singapore–Johor Bahru (Malaysia)–Batam/Bintan (Indonesia) growth triangle is the best-known manifestation of this policy (Figure 5.4).

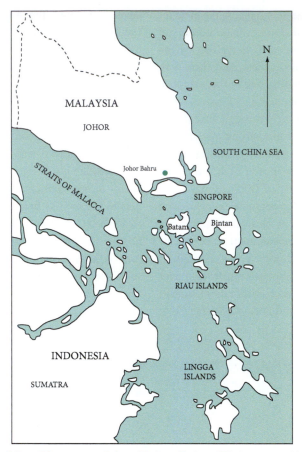

Figure 5.4 The location of the Singapore-Johor Bahru-Batam/Bintan growth triangle in South-east Asia
Source: Sparke *et al*. 2004: 487.

Germany is usually seen as the archetype of this model, which also includes other European countries such as France, Austria and Switzerland.

Social democratic states are characterised by a high degree of universality in service provision, based on an underlying consensus between business, unions and the state. They have regulated labour markets, with a strong commitment to active labour market policies designed to support full employment and the rapid reintegration of the unemployed back into the workforce. The welfare state has been designed to promote equality between citizens through universality, supported by high levels of taxation, in contrast to the residual approach adopted by neoliberal states. Similarly, tertiary education benefits from high levels of public expenditure.

This model is often described as Scandinavian, being closely associated with Sweden, Denmark, Finland and Norway.

Mediterranean states are also associated with regulated labour markets, but relations between business and trade unions tend to be more conflictual in nature. Corporate governance is often weak and the ownership of the financial and industrial sectors is highly concentred in the control of a small number of key actors. State welfare provision is limited with civil institutions such as the church, family and private charity assuming an important role. These states have relatively weak education systems with a low level of expenditure and serious weakness in science and technology which reduce their competitiveness. This model is rooted in southern Europe.

This typology does not offer a full picture, reflecting its roots in analyses of European welfare states (see Epsing-Andersen 1990). Some accounts also include *authoritarian* states, which combine a highly centralised political system with active participation in capitalist markets (Coe *et al.* 2013: 105). This model is typically found in some former socialist countries such as China and Russia that have embarked upon fundamental economic reform whilst maintaining tight political control of society. Elements of the developmental state are found in other parts of the global South beyond Asia, particularly newly industrialised Latin American states such as Brazil and Mexico. The classification also does not readily accommodate the group of Latin America countries, including Bolivia and Venezuela, which elected radical left-wing governments in the late 1990s and 2000s (section 7.6). There are also cases of *failed* states which lack the institutional capacity, economic strength and political authority to manage national economies effectively, often as a result of wars, natural disasters or external pressure (for example, Haiti or Somalia).

There are important differences between states within each of the five types outlined above, for example between the US's residual welfare system and the UK's universalist tradition in the neoliberal category or between German corporatism and the French state in the continental European regime. The typology should be seen in dynamic rather than static terms, particularly according to how these types of state are engaging with the broader pressures of neoliberalism, the economic crisis and austerity policies, which are covered in the next section. The fact that some states may combine elements of more than one model in practice (see Box 5.4) often reflects how they have adapted to these broader pressures over time.

Reflect

Consider how the five types of states identified here have been affected by the financial crisis of 2008–09 and subsequent 'Great Recession'.

5.5 Neoliberalism, crisis and austerity

Established state structures have undergone considerable restructuring and change since the late 1970s, driven by neoliberal reform programmes and the globalisation of the economy. In Europe and North America, these structures were based on strong interventionist states focused on managing their economies according to Keynesian theories of demand management and dispensing universal welfare services. Most attention has focused on the abandonment of Keynesianism and the reform of welfare states in the developed world, but states in developing countries have faced similar pressures, not least through the activities of international organisations like the IMF and World Bank. Neoliberalism can be seen as providing the basis of the new mode of regulation that has emerged since the early 1980s, informing the introduction of a range of institutional experiments and reforms. Whether it offers the stability and order required for the consolidation of a coherent 'post-Fordist' regime of accumulation is, however, highly questionable (Peck and Tickell 2002).

The financial crisis of 2008–09 was interpreted by many as representing a profound challenge to neoliberal policies, given that these were deeply implicated in the crisis, particularly through the deregulation of the financial system in the 1980s and 1990s. This was reflected in the initial response to the crisis, which saw western governments return to Keynesian policies of stimulating demand through the injection of additional public expenditure into the economy. Over time, however, it has become clear that neoliberal ideas have survived the crisis largely unscathed – what the political economist Colin Crouch (2011) refers to as the 'strange non-death of neoliberalism'. By early 2010, the Keynesian moment was coming to an end as governments sought to reduce public expenditure through austerity policies in response to the budget deficits that had arisen from the crisis in many countries. The effects of the crisis and subsequent austerity policies in squeezing living standards have triggered various populist backlashes in recent years, from leftist protests against inequality and austerity to the more conservative resentments against immigration and the assumptions of established elites that have been mobilised by Donald Trump in the US, the 'leave' campaign in the UK's EU referendum of 2016 and the National Front in France.

5.5.1 The evolution of neoliberalism

The late 1970s and early 1980s represent a key turning point in the recent history of capitalism (Harvey 2005: 1). Following the 1973 military coup, Chile introduced neoliberal economic policies, whilst the Chinese communist leader, Deng Xiaoping, launched a far-reaching economic reform programme in 1978. A year later, Margaret Thatcher came to power in the UK, espousing a radical new brand of free market liberalism. This was followed by the election of a former Hollywood actor named Ronald Reagan as US President in 1980. "From these several epicentres . . . revolutionary impulses spread out and reverberated to remake the world around us in a totally different image" as the doctrine of neoliberalism was "plucked from the shadows of relative obscurity . . . and transformed . . . into the central guiding principle of economic thought and management" (Harvey 2005: 1–2).

As a political and economic ideology, neoliberalism is based on a belief in the virtues of individual liberty, markets and private enterprise. Neoliberals are hostile towards the state, believing that its role in the economy should be minimised to that of enforcing private property rights, free markets and free trade (ibid.: 2). Furthermore, in areas where markets do not exist, because of excessive state intervention and regulation, the state should create them through policies of privatisation (transferring state-owned enterprises into private ownership), liberalisation (opening up protected sectors to competition) and deregulation (relaxing the rules and laws under which business operates). Beyond this, however, the state should not venture, since it cannot possess enough information to second-guess market signals (prices) based on the preferences of millions of individuals (ibid.: 2).

Since the 1970s, neoliberalism has evolved considerably. Three distinct phases can be identified (Peck and Tickell 2002). The first, proto-liberalism, refers to its early development in the 1970s when ideas that were deeply unfashionable for most of the twentieth century were developed and promoted by a New Right group of important thinkers and politicians in the UK and US (including the economists Milton Friedman and Friedrich von Hayek) in think tanks, universities and the media. Their views became increasingly influential, appearing to offer radical solutions to the economic crisis of the 1970s in terms of reducing inflation, cutting welfare spending and restricting trade union power whilst restoring individual liberties through the promotion of free markets. The New Right essentially sought to reassert traditional nineteenth-century liberal principles in the circumstances of the 1970s (hence the term neoliberalism).

After the election victories of Thatcher and Reagan, a second phase of 'roll back' neoliberalism ensued. Neoliberal ideas involving the reduction of state intervention in the economy and the curbing of trade union rights were put into practice. Inflation was tackled by applying the monetarist theory of Friedman, based on reducing the supply of money in the economy. This form of 'shock therapy' succeeded in lowering inflation, in the short term, but at the expense of deepening the recession of the early 1980s and increasing unemployment. State intervention in the economy was reduced through policies of privatisation, liberalisation and deregulation. Many conservative politicians and commentators criticised the welfare state in the US and UK during the 1980s and 1990s, attacking it for encouraging individuals to become dependent on the state, undermining work incentives and imposing a high tax burden. Whilst successive 'reform' programmes have been launched amid considerable fanfare, it has proved more difficult to achieve significant reductions in welfare expenditure.

In the early 1990s, a new form of 'roll-out' neoliberalism emerged. By this stage, neoliberalism had become normal, regarded as simple economic 'common sense'. As such, it could be implemented in a more technocratic and low-key fashion by governments and agencies such as the World Bank and IMF. Following the conversion of key figures in the early 1980s, the latter two agencies in particular played a crucial role in spreading neoliberal doctrines across the globe, acting as "the new missionary institutions through which these ideas were pushed on the reluctant poor countries that often badly needed their grants and loans" (Stiglitz 2002: 13). In the early 1990s, neoliberal 'shock therapy' in the form of privatisation and liberalisation was rapidly implemented in the former communist countries of Central and Eastern Europe.

As such, neoliberal policy prescriptions became consolidated into the so-called **Washington Consensus**, reflecting how this agenda has been embraced and enforced by the US Treasury, World Bank and IMF, all headquartered in the city. The Washington Consensus consists of the following key elements (Peet and Hartwick 1999: 52):

➤ Fiscal discipline: minimising government budget deficits.

➤ Public expenditure priorities: promoting economic competitiveness not the provision of welfare or redistribution of income.

➤ Tax reform: lowering of tax rates and strengthening of incentives.

➤ Financial liberalisation: determination of interest rates and capital flows by the market.

➤ Trade liberalisation: eliminating restrictions on imports.

➤ Foreign direct investment: removing barriers to the entry of foreign firms.

➤ Privatisation: selling off of state enterprises.

➤ Deregulation: abolishing rules which restrict competition.

Discipline over national governments is exercised by the power of the IMF and World Bank to refuse debt rescheduling for developing countries and declare them uncreditworthy and by the prospect of capital flight (investors withdrawing their money) if alternative policies stressing full employment or the redistribution of wealth are adopted. In response to growing criticism, the IMF and World Bank modified their approach in the 2000s to place more emphasis on governance and poverty reduction rather than structural adjustment and privatisation as part of the so-called post-Washington Consensus (Sheppard and Leitner 2010), although they retained an underlying commitment to neoliberal models of development (section 7.3.3).

The implementation of neoliberal policies has been a highly uneven process, as individual states have tended to adopt particular aspects of the neoliberal package whilst ignoring others. Elements of neoliberalism have, moreover, interacted with pre-existing institutional arrangements and practices in complex ways, generating a wide range of distinctive local outcomes (Box 5.4).

Box 5.4

Neoliberalism 'with Chinese characteristics' (Harvey 2005: 120)

Following the momentous decision of the Communist leadership to open up to foreign trade and investment in 1979, China has experienced a rate of economic growth almost unsurpassed in recent history, averaging 10 per cent a year between the early 1980s and 2012 (Box 1.3). Reform was initially couched in terms of the 'four modernisations' (referring to agriculture, industry, education and science and defence), before a period of retrenchment after the Tiananmen Square massacre of 1989. The pace of reform accelerated again in the early 1990s, after an ageing Deng Xiaoping declared, in a tour of the southern regions in 1992, that "to get rich is glorious" and "it does not matter if it is a ginger cat or a black cat as long as it catches mice" (quoted in Harvey 2005: 125). The implementation of neoliberal reforms in China has created a curious hybrid of communism and capitalism, creating real tensions between economic

Box 5.1 (continued)

liberalisation and political authoritarianism, in the form of the continuing control of the Communist Party.

The Party leadership initially sanctioned the establishment of four economic zones to attract foreign investment as local experiments which would have little effect on the rest of the economy. Three of these zones were located in the southern Guangdong province, adjacent to Hong Kong, and the other was situated in Fujian province, across the straits from Taiwan (Figure 5.5). Such experiments proved hugely successful, with the Guangdong province in particular acting as a magnet for foreign capital. Two-thirds of foreign investment in China was being channelled through Hong Kong in the mid-1990s (Harvey 2005: 136). Other externally oriented areas such as coastal cities and export processing zones were created in the 1990s with the growth of Shanghai, China's largest city, proving particularly explosive. China's main comparative advantage lay in labour-intensive goods such as textiles, footwear and toys with hourly wages in textiles standing at around $2 an hour compared to $5 in Slovakia, $7 in South Korea and around $18 in the US (Dicken 2015: 458).

Rapid economic growth has created a number of social and political contradictions and tensions in China. One of the major difficulties that China has faced is how to absorb its huge labour surplus, fuelled by massive migration from the rural areas to the coastal cities, officially estimated at 114 million workers in the reform period (Harvey 2005: 127). The principal response in recent years has been the development of massive infrastructure projects, financed by borrowing, such as the Three Gorges Project to divert water from the Yangtze to the Yellow River. Regional inequalities have deepened as the southern and eastern coastal zones have surged ahead of the interior and north-eastern 'rustbelt' region. Rapid economic growth and the spread of a modern consumer culture have helped to contain demands for political liberalisation, but indications of an economic slowdown in recent years may require a change of approach from the ruling Communist Party (see Box 1.3).

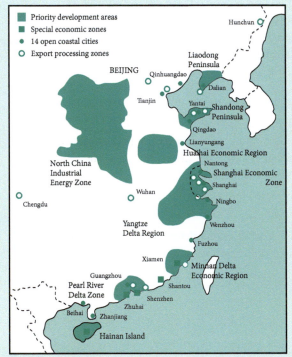

Figure 5.5 The geography of China's 'open door' trading policy
Source: Dicken 2003: 190.

5.5.2 Neoliberalism and the crisis

The 2008–09 financial crisis and subsequent Great Recession represented a profound challenge to neoliberal ideas which denied that such a crisis was possible, based on the assumption that markets are essentially self-correcting and immune to the kind of catastrophic failure that occurred in 2008–09. This was particularly the case for deregulated financial markets which were viewed as operating in a particularly pure fashion according to the 'efficient market hypothesis'. This stated that markets always clear or determine the optimum price through the interaction of supply and demand, reflecting the rational self-interested behaviour of individual actors. Such ideas, however, were confounded by the financial crisis which showed that the self-interested behaviour of individual actors in seeking to maximise profits did not have optimal aggregate outcomes, but had generated large-scale failure, with the supposed spreading of risk among investors through new mortgage derivative products actually resulting in the massive amplification of risk across the system (section 4.5.1). As such, the financial crisis was "also a crisis of the ideas that had made these instruments and institutions possible" (Blyth 2013: 43).

As outlined above, this was reflected in the initial response to the crisis which saw policy-makers return to Keynesian policies of increasing public expenditure to stimulate demand. These had some success in ensuring that the Great Recession of 2008–09 did not turn into a catastrophic economic collapse like the Depression of the 1930s. Even the likes of the IMF accepted the need for fiscal stimulus, prompting Keynes's biographer, Lord Skidelsky (2009), to proclaim the 'Return of the Master'. But this Keynesian moment was to prove short-lived. By early 2010, a new response was taking shape, led by the ECB and German Government, who had not supported stimulus policies in 2008–09. Prompted by concerns about budget deficits, particularly in the European PIIGS countries, and the spectre of future inflation, this neoliberal response was concerned to end the stimulus and control spending, pitting conservative commentators and neoclassical economists against Keynesians. The former won the day at the June 2010 meeting of the G20 in Toronto, heralding a new era of austerity politics (Blyth 2013). The political success of this neoliberal counter-offensive is evident in the continuing sense that it is the political left that was most bewildered and wrong-footed in the aftermath of the crisis (Mirowski 2014), meaning that there was little real alternative on offer in the economies of Europe and North America.

The turn to austerity reflects a rapid return to "the pre-crisis mindset" (Elliott 2011), based upon the idea that competitiveness is best restored by the reduction of prices, wages and public spending through the cutting of the state's budget and debts (Blyth 2013: 2). In this sense, neoliberal ideas have been used to solve the very crisis they created (Mirowski 2014). This reflects a deep attachment to such ideas among economic and political elites, particularly in supranational agencies such as the IMF and World Bank, national government

departments, conservative think tanks and the financial and economic press.

While neoliberal theory emphasises the need for reduced government intervention, in practice neoliberals have proved adept at using the powers of the state to advance their project. Indeed, neoliberalism is best viewed as a redirection of the state away from social welfare and towards the promotion of economic competiveness and business interests. As such, measures such as the 'bailout' of insolvent banks can be seen as reinforcing the class project of neoliberalism, as state power is used to support financial interests (Harvey 2005). The subsequent introduction of austerity packages indicates how "neoliberal policy prescriptions" are "deeply embedded in the dominant circuits of corporate, financial and political power" (Peck 2013: 138). This is particularly evident in the entrenched structural power of financial institutions and investors and international regulatory agencies such as the IMF and European Commission.

5.5.3 Austerity politics

The introduction of austerity measures is evident across a number of different countries since 2010, particularly neoliberal and Mediterranean states, although also continental European and social democratic states. Examples include the EU's imposition of cuts in exchange for bailout programmes in the PIIGS, deficit reduction in the UK and fiscal tightening alongside some element of stimulus in the US and France. This reflects rising debt levels in the wake of the financial and economic crisis, with the average gross debt relative to GDP rising sharply after 2007 (Figure 5.6). It is crucial to stress here that these rising levels of debt were more an effect than a cause of the crisis. This is a common side effect of economic downturns in modern capitalist economies, reflecting the dependence of the state upon economic growth for revenue. In a recession, tax revenues fall substantially, while public expenditure generally goes up through increased demand for unemployment and welfare payments. These are often referred to as the automatic stabilisers, reflecting how state expenditure works to support the economy in a recession, maintaining a level of consumption and providing an income floor for unemployed people. For the advanced G20 economies, the IMF (2010) estimated that half of the 39 per cent increase in debt as a proportion of GDP from 2008 to 2015 was due to reduced tax revenues. In the 2008–09 financial crisis, the fiscal strain on states was exacerbated by the rescue of the banks (see Table 5.2), which accounted for 35 per cent of the increase in debt in the UK, for instance (Blyth 2013: 46). This prompts Blyth to describe the neoliberal re-christening of public debt as the result of excessive expenditure by states rather than the financial crisis as the biggest bait-and-switch operation in modern history, advancing a deliberately misleading narrative of the crisis.

Europe's sovereign debt crisis was triggered in 2009–10 by market panic about the large public debts incurred by the government of the 'PIIGS' economies in particular (section 4.5.2). This led to sharp increases in the bond yields or interest rates demanded by investors from these countries (Figure 4.6), leaving them unable to service their debts. The policy response to this crisis reflects the rescaling of the state outlined in section 5.2.2, controlled by the supranational institutions of the EU, ECB and IMF, the so-called 'Troika'. Each of the crisis economies was granted a bailout loan from the Troika in exchange for large-scale reductions in public expenditure and the structural reform of their economies, embodying underlying neoliberal assumptions about how economies should operate. Thus, Greece received a loan of 110 billion euros in exchange for a 20 per cent cut in public sector pay and tax increases in May 2010, while Ireland received 85 billion euros in exchange for a 26 per cent cut in public expenditure in November 2010 (Box 5.5) (Blyth 2013: 71). This echoes the structural adjustment programmes implemented by the World Bank and IMF across much of the global South in the 1980s and 1990s (section 7.3.3).

In the case of the Eurozone, the commitment to austerity reflects a lack of other adjustment options for underperforming economies which cannot devalue their currencies or inflate their economies by printing more money since they no longer have an independent currency. Beyond this, while US and UK banks were 'too big to fail', Blyth argues that Eurozone banks have "become too big to bail" with the assets of the three biggest French banks, for example, amounting to 316 per cent of French GDP in 2008 (ibid: 83). This reflects the speculative activities of Eurozone banks in acquiring cheap debt financing on international markets, particularly in US mortgage-backed products and bonds issued by southern European countries. If these banks were to become insolvent as a result of the peripheral Eurozone countries defaulting on their debts, neither their governments nor the EU/ECB would be able to bail them out.

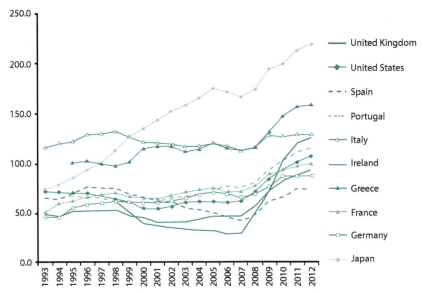

Figure 5.6 Trends in general government debt, as a percentage of GDP, selected countries 1993–2012
Source: OECD (http://stats.oecd.org/OECDStat–Metadata/).

Box 5.5

Ireland: from growth to crisis

Traditionally regarded as one of Western Europe's most impoverished countries, Ireland experienced an unprecedented economic boom in the 1990s and 2000s, followed by a dramatic bust associated with the financial crisis. During the growth years, Ireland was held up as a model of economic openness by economists, politicians and think tanks. It was labelled the most globalised country in the world by *Forbes Magazine* in 2004, reflecting its small size and highly international nature. These make it more akin to a regional economy in some respects, particularly its dependence upon FDI and high levels of labour migration (see Figure 5.7). Ireland had abandoned its post-independence ISI policy as long ago as the late 1950s, but prosperity remained elusive into the 1980s, despite high levels of FDI from US firms in particular.

After experiencing a severe economic crisis in 1986, the government,

business and trade unions came together to agree a plan for national recovery. This new approach to economic policy was associated with a dramatic improvement in economic performance, with GDP growing by an average of 8.1 per cent between 1993 and 2000 (Drudy and Collins 2011: 341), while unemployment fell from 19 per cent in 1991 to 4.3 per cent in 2002 (House and McGrath 2004: 37), reversing Ireland's historical pattern of labour emigration (Figure 5.7). Growth was underpinned by membership of the EU, providing access to the single market and investment from the structural funds. In the late 1990s, it represented a form of export-driven catch-up with the rest of Europe, but this increasingly gave way to domestic consumption in the early 2000s, driven by an intense construction boom. This was reflected in a house-building frenzy with Ireland, along with Spain,

producing more than twice as many units per head of population than the rest of Europe (Kitchin *et al.* 2012: 1308–9). The construction boom was funded by large-scale bank lending to developers and encouraged by the government's philosophy of light-touch regulation, based upon reduced taxes on property and a range of tax incentive schemes for developers.

The economic boom burst in 2007 when the credit crunch meant that the banks could no longer borrow cheaply in international capital markets to fund their loans to developers. Ireland experienced a particularly severe economic contraction with GDP declining by an annual average of 2.4 per cent between 2007 and 2011. Unemployment rose to a peak of around 15 per cent in 2010–11, leading to a restoration of the traditional pattern of net emigration (Figure 5.7). The major Irish banks, the Allied Irish

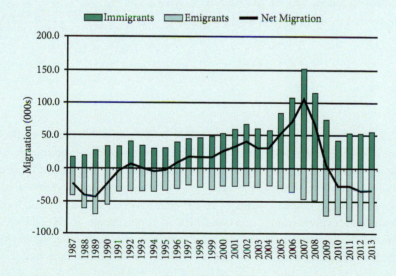

Total emigration from Ireland in the year to April 2013 is estimated to have increased to 89,000. The number of immigrants also increased to 55,900, resulting in total net outward migration remaining broadly constant with the previous twelve month period. Amongst Irish nationals, net outward migration is estimated to have increased significantly, rising from 25,900 to 35,200. This represents a significant loss of skills.
Ranking: n/a

Figure 5.7 Net migration, Ireland (000s), 1987–2013
Source: Forfas 2014. Figures from OECD.

Box 5.5 (continued)

Bank, Anglo Irish Bank, Irish Nation-wide and the Educational Building Society, were effectively insolvent as a result of bad loans to developers, prompting the government to first guarantee all bank deposits in September 2008. This was followed by the establishment of the National Asset Management Agency (NAMA) to acquire 88 billion euros of toxic assets from the banks. The costs of this bailout (see Table 5.2), combined with a one-third decline in tax revenues between 2007 and 2010 and a growing welfare bill, led to a fiscal crisis of the state, triggering the EU bailout of November 2010. The economy has recovered since 2013, experiencing strong growth of over 5 per cent in 2014 to again become the fastest-growing economy in the EU, leading to increased job creation and some reduction in government debt (European Commission 2016).

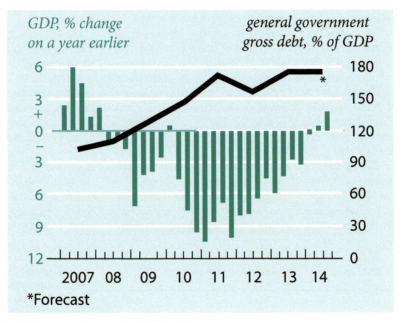

Figure 5.8 Greek GDP and debt levels 2007–14
Source: Eurostat; European Commission; Haver Analytics.

Austerity is often justified as a necessary step to restore the competiveness of an indebted economy. In practice, however, the reduction of expenditure, prices and wages entails a contraction of the economy, which requires a source of stronger countervailing demand from elsewhere – specifically, demand for exports from larger economies which are growing. This has not been the case in recent years, when much of the global economy has been characterised by low growth or stagnation. Thus, while austerity can work for a single economy in specific circumstances as outlined above, economies cannot all cut their way to growth simultaneously. This will result only in an overall shrinking of the economy. The effects of austerity in undermining economic growth are starkly evident in the case of Greece, with the economy suffering a particularly severe contraction of 26 per cent between 2009 and 2014, leading to increased debt as a percentage of GDP (Figure 5.8). Economic contraction has been reflected in the need for further bailouts with Greece receiving a second loan of 100 billion euros from the Troika in 2012 and a third one of 85 billion euros in 2015. At the same time, austerity tends to have regressive social consequences, affecting lower-income groups more than the affluent as they are generally more dependent upon state services and welfare payments. As such, it can be seen as a means of deciding who bears the costs of the economic adjustment associated with a crisis. This sense of the poor paying for the actions of the rich in creating the financial crisis and associated recession in the first place is seen as highly unjust by many critics and protestors (Fishwick 2016).

Blyth (2013) draws an interesting parallel between the operation of the euro and the gold standard exchange rate system of the late nineteenth and early twentieth centuries which pegged the value of national currencies to fixed rates of gold, providing a mechanism of adjustment for international trade. As well as relying upon the convertibility of the value of the currency to gold, the gold standard was based upon the flexibility of domestic wages and prices to adjust to trade imbalances, with governments unable to devalue their currencies or inflate their economies. As we have seen, this reliance on domestic deflation (austerity) is echoed, even magnified, in the Eurozone, as there is no scope for devaluation by coming out of the euro – as Greece discovered in 2015. By contrast, countries could come off the gold standard, which represented a lower level of monetary integration than a common currency. Crucially, the operation of the gold standard pre-dated the introduction of modern mass democracy after the First World War. Accordingly, Blyth argues that the emphasis on adjustment through deflation means that the euro is incompatible with democracy as people will not continue to support a system that reduces their living standards. This tension has been evident in the **Eurozone sovereign debt crisis** whereby the Troika have imposed austerity policies and at times influenced the appointment of government ministers and officials, for example the Italian Prime Minister of 2011–13, Mario Monti (a former EU Commissioner and Goldman Sachs adviser) and his cabinet of unelected technocrats. In early July 2015, the EU effectively rejected the result of the Greek referendum which had voted against further austerity measures, requiring the Greek Government to accept these. This has fuelled growing public resentment against established economic and political elites, particularly supranational institutions and national governments.

5.5.4 The populist backlash against globalisation and neoliberalism

A distinct populist backlash has emerged in recent years against established political institutions and parties, based upon diverse protests against the mainstream political agenda of globalisation, economic integration and austerity. The established political parties of the centre-right and centre-left have broadly supported globalisation since the 1990s in many developed countries, introducing many of the key reforms that led to increased capital mobility, including financial deregulation and the abolition of capital controls, as well as accepting the increased mobility of labour. As the Harvard economist Dani Rodrik (2016) argues, the "popular revolt that seems to be underway is taking diverse, overlapping forms: reassertion of local and national identities, demand for greater democratic control and accountability, rejection of centrist political parties, and distrust of elites and experts".

As some economists have long recognised, the benefits of globalisation tend to be unevenly distributed between social groups, favouring the most affluent and skilled, while increased competition reduces wages for lower-skilled workers in the developed economies of Europe and North America (Standing 2009). Economic insecurity has been magnified by the effects of the economic crisis and austerity policies with 20 of the 34 OECD countries still to regain their pre-crisis employment rates (the proportion of working-age people in employment) and the likes of Ireland, Spain and Greece being among the worst affected countries (Figure 5.9).

The populist backlash against globalisation has taken both a left-wing and a right-wing form. Sections of the political left have long mobilised against globalisation, stretching back to at least the large-scale protests of the late 1990s and early 2000s (section 1.2.1). In Latin America, globalisation was generally experienced as a trade and foreign-investment shock (Rodrik 2016), particularly in terms of liberalisation reducing employment in traditional industries. This fuelled the rise of leftist social movements and political parties which were elected to government in several countries. More recently, the experience of austerity in Europe and of

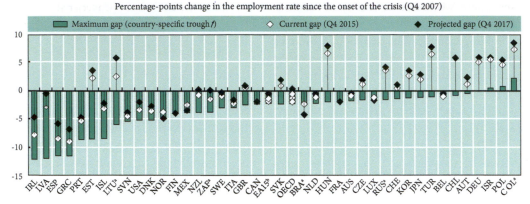

a) Annual values calculated using employment data from the *OECD Economic Outlook Database* and UN population projections.
b) Aggregate of 15 OECD countries of the euro area.

Figure 5.9 The employment gap in OECD countries
Source: OECD calculations based on *OECD Economic Outlook Database*; and United Nations, *World Population Prospects: The 2015 Revision*.

growing insecurity in the US has prompted new waves of political mobilisation that have both created new political parties and infiltrated existing ones. Examples of the former include the rise of Syriza to win the Greek general election of January 2015 and the electoral success of Podemos in Spain and the Five Star Movement in Italy, while the latter incorporates the wave of support that propelled Jeremy Corbyn to the UK Labour Party leadership in 2015 and the high-profile campaign of Bernie Sanders for the Democratic Party Presidential nomination in the US. Support for these movements has been particularly strong among young people who have emerged as key losers from the crisis as represented by very high levels of youth unemployment, peaking at over 50 per cent in Greece and Spain (OECD 2016: 45).

A neo-conservative backlash against globalisation and its discontents (Stiglitz 2002) is also evident, becoming particularly prominent in 2016 with Donald Trump's successful campaign for the US Presidency and the UK's 'Brexit' vote to leave the EU. This reflects growing opposition to immigration in particular, alongside the effects of competition on wage and living standards and an upwelling of hostility to established elites. It is underpinned by a growing sense of economic insecurity among lower-skilled workers who have seen their wages and living standards decline or stagnate and who feel threatened by immigration, free trade and technological change. As such, it is feeding growing

protectionist pressures in the wake of the crisis. As the economist Nouriel Roubini (2016) observes, the divisions in the UK's Brexit vote were clear: "rich versus poor, gainers versus losers from trade/globalisation, skilled versus unskilled, educated versus less educated, urban versus rural and educated versus less educated communities". This was expressed geographically in the strong 'leave' vote in disadvantaged working-class areas of the north and Midlands (Figure 5.10), surprising many commentators, where the traditional dominance of the Labour Party was challenged by the growth of the anti-immigration United Kingdom Independence Party (UKIP) (Ford 2016).

Reflect

To what extent does the reversion to austerity policies since 2010 reflect a qualitatively distinctive phase of neoliberalism?

5.6 Governing urban and regional development

One of the contradictory effects of globalisation is to have increased rather than diminished the significance of geographical differentiation and place. A key

Key: ■ Majority leave ■ Majority remain ■ Tie ■ Undeclared

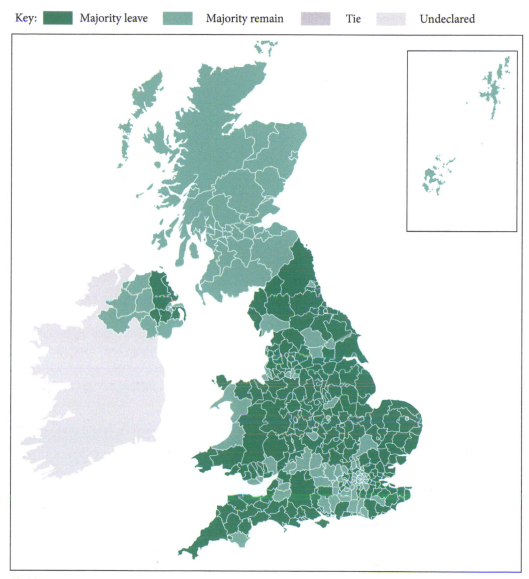

Figure 5.10 The overall result of the UK referendum on EU membership, 23 June 2016

element of this has been the resurgence of local and regional levels of government as part of the broader rescaling of the state (section 5.2.2), particularly through the devolution of powers from the central state. As states' control of flows across their national borders has diminished in the face of globalisation and neoliberalism, cities and regions have become increasingly exposed to the effects of global competition. Devolution is important in granting them additional powers and responsibilities to manage their own economies rather than relying upon the central state for resources. It is closely bound up with the shift away from the old 'top-down' model of **regional policy** towards a new 'bottom-up' emphasis on economic competitiveness and growth (section 5.6.2). Yet recent years have also seen the devolution of austerity programmes to the urban and regional scales, particularly in Europe and North America, reducing the resources available to support economic development initiatives (section 5.6.3).

5.6.1 Devolution and economic governance

The political rationale for devolution has evolved in recent decades, moving from an emphasis on cultural, ethnic and linguistic factors to a growing emphasis on it as a means of promoting economic competitiveness (Rodríguez-Pose and Sandall 2008). In particular, devolution is seen as a necessary step to allow cities and regions to adapt to the challenges of globalisation and fulfil their economic potential. This is bound up with the growth of the '**new regionalism**', based upon the identification of regions and city-regions as key units of economic organisation in an increasingly global economy. The 'new regionalism' focuses attention on the importance of cities and regions developing their own economic development strategies and initiatives to enhance their competitiveness and support growth, particularly though the promotion of innovation, learning and entrepreneurship. Such arguments reflect the underlying economic rationale for devolution, which is essentially threefold (Rodríguez-Pose

and Gill 2005). First, decentralisation takes account of regional preferences for different levels and forms of public sector provision and taxes. Second, devolution can encourage policy innovation based on closer local knowledge of regional conditions and needs, particularly when devolved governments are responsible for raising their own revenues. Third, devolution can improve political accountability and transparency by reducing the distance between politicians and their electorates. At the same time, devolution also creates economic risks, including: inefficiencies through over-spending by sub-national governments; increased inequalities between regions; and increased institutional burdens through the replication of the same administrative functions across the different devolved governments (ibid.).

Devolution has generally been associated with a period of increased regional disparities following a period of regional convergence under 'spatial Keynesianism' in the post-war era (Table 5.5). The magnitude of regional inequalities is generally fairly modest for developed countries, but greater for developing

Table 5.5 Regional inequalities. Variance of the log of regional GDP per capita

Country	Percentage change		
	1980–90	1990–00	1980–00
Developing countries			
China	–16.31	20.21	0.61
India	7.11	16.96	25.27
Mexico	–1.29	13.57	12.11
Brazil	–17.16	1.33	–16.06
Developed countries			
USA	11.75	–2.69	8.74
Germany	2.18	–0.96	1.2
Italy	1.55	3.01	4.6
Spain	–3.92	10.47	6.14
France	8.63	–0.31	8.3
Greece	1.22	0.13	1.35
Portugal		1.82	
European Union		11.25	

Source: Rodríguez-Pose and Gill 2004: 2098.

countries with the exception of Brazil where regional inequalities fell. At the same time, devolution takes different forms in different countries, reflecting the specific timing of the key reforms and their particular rationales. In the US, for instance, it has been closely associated with neoliberal efforts to increase states' rights and reduce the redistributive role of the federal government. Increased regional inequalities reflect the adverse effect of federal cutbacks on the poorest states and the greater capacity of richer states to benefit from increased powers and resources. Decentralisation in China is closely associated with economic reform and marketisation, involving substantial fiscal decentralisation to states, leading to a rise in regional (inter-provincial) inequalities from 1990 to the mid-2000s before this gave way to a slight decline (Candelaria *et al.* 2013). Devolution in Spain was bound up with the transition to democracy in the late 1970s and early 1980s, and further devolution in the 1990s was associated with a rise in regional inequalities (Rodríguez-Pose and Gill 2004: 207).

5.6.2 Changing regional policy paradigms

Regional policy can be defined as those policies that are designed to address regional disparities and achieve a more geographically balanced pattern of economic and social development (Garretsen *et al.* 2013). This involves various publicly funded initiatives and programmes to support employment and wealth creation. Regional policy was established in many developed countries in a period of sustained economic growth, fiscal expansion and low unemployment in the 1950s and 1960s. It has changed considerably since the 1980s:

> regional policy, originally a top-down, subsidy-based group of interventions designed to reduce regional disparities, now typically involves much broader policies designed to improve regional competitiveness. Today, national governments increasingly favour regional growth over redistribution, in pursuit of national or regional competitiveness and balanced national development. Territorial development instruments have become broader in scope . . . and have

adapted to the requirements of individual regions. This policy approach has been accompanied by a growing trend of decentralisation to the regional levels. Regional strategic programmes and programming have grown in prominence, reflecting a general policy shift towards support for endogenous development and the business environment, building on regional potential and capabilities, and aiming to foster innovation-orientated initiatives.

> (OECD 2014: 67)

As this suggests, two broad paradigms of regional policy can be identified: the old paradigm which was prevalent in the 1960s and 1970s, and a new paradigm which has become influential since the early 1990s (Table 5.6).

The old regional policy paradigm represented a response to the problem of regional disparities in income, infrastructure and employment. It aimed to foster equity through balanced regional development, closing the gap in income and wealth between rich and poor regions. It was targeted at lagging regions, typically adopting a sectoral approach that was reactive and short term in nature. This was a highly top-down approach, with the central government offering grants and financial incentives to companies to locate factories or offices in lagging regions. At the same time, development in core regions such as south-east England and Paris was restricted. Old-style regional policy reached its peak in the 1960s and 1970s, helping to reduce the income gap between rich and poor regions in Europe (Dunford and Perrons 1994).

In contrast to traditional regional policy, the new paradigm has sought to facilitate growth and enhance the competitiveness of the regional economy. This reflects a relaxing of the objective of spatial redistribution and associated interventionist measures in the face of uncertain economic conditions and neoliberal ideology. The underlying problem here is identified as one of a lack of regional competitiveness and under-used regional potential for economic growth. As such, the objective is that of competitiveness alongside the traditional focus on equity. Rather than focusing solely on lagging regions, the new paradigm of regional policy is concerned with all regions, adopting an integrated and comprehensive rather than narrowly sectoral approach. It is more

Table 5.6 Regional policy paradigms

	Old paradigm	New paradigm
Problem recognition	Regional disparities in income, infrastructure stock, and employment	Lack of regional competitiveness, underused regional potential
Objectives	Equity through balanced regional development	Competitiveness and equity
General policy framework	Compensating temporally for location disadvantages of lagging regions, responding to shocks (*e.g.* industrial decline) (*Reactive to problems*)	Tapping underutilised regional potential through regional programming (*Proactive for potential*)
– theme coverage	Sectoral approach with a limited set of sectors	Integrated and comprehensive development projects with wider policy area coverage
– spatial orientation	Targeted at lagging regions	All-region focus
– unit for policy intervention	Administrative areas	Functional areas
– time dimension	Short term	Long term
– approach	One-size-fits-all approach	Context-specific approach (place-based approach)
– focus	Exogenous investments and transfers	Endogenous local assets and knowledge
Instruments	Subsidies and state aid (often to individual firms)	Mixed investment for soft and hard capital (business environment, labour market, infrastructure)
Actors	Central government	Different levels of government, various stakeholders (public, private, NGOs)

Source: OECD 2010: 13.

'bottom up' and long term in nature, involving different levels of government, particularly local and regional agencies, and favours a context-specific or place-based approach. The new paradigm focuses on endogenous local assets and knowledge within regions, emphasising the need to develop local skills and stimulate enterprise. In this sense, "locally-orchestrated regional development has replaced nationally-orchestrated regional policy" (Amin *et al.* 2003: 22). This includes initiatives to stimulate innovation and learning within firms, measures to try and increase entrepreneurship in terms of the number of new firms that are being created and efforts to develop and upgrade the skills of the workforce through a range of training and education programmes.

These measures are focused on the **supply side** of the local or regional economy, defined in terms of the quality of the main factors of production such as labour (training, skills), capital (enterprise, innovation and finance) and land (sites and infrastructure for investors). Improving these supply-side factors is seen as vital to the competitiveness of the regional economy. This can be contrasted with the **demand-side** emphasis of the old Keynesian paradigm, which assumed that the government could alter demand conditions in the lagging regions through financial transfers from the central government and large-scale public investments.

At the same time, however, there are strong continuities and overlap between the two paradigms with elements of each tending to coexist in many countries, reflecting how new elements have been grafted onto more established commitments (OECD 2010). Thus, the new objective of competitiveness is often combined with the longer-standing one of regional equity in many cases. While regional policy has been extended to cover most if not all regions, a degree of spatial concentration remains in place in many countries, particularly through the designation of regional aid areas, larger funding allocations to lagging regions and specific regional targeting (ibid.: 19). In addition, the central state often remains the key funder and coordinator of local and regional development programmes. The continuities between the 'old' and 'new' models of regional development are evident in initiatives such as South Korea's proposals for the relocation of the national capital which combines an underlying commitment to equity and balanced national development with a growth-oriented approach (Box 5.6).

5.6.3 Austerity urbanism

Urban and regional development has been shaped by the experience of austerity since 2010, particularly in Europe and North America. This new round of what Peck (2012) calls 'austerity urbanism' is based upon a devolution or 'downloading' of the financial and budgetary pressures outlined in section 5.5.3 to the urban scale as the costs of austerity measures fall disproportionately on sub-national governments. The effects of these reductions in public spending are highly geographically uneven both between and within

---

Box 5.6

Balanced national development in South Korea

As one of the four East Asian NICs or 'tiger' economies (see Box 5.3), South Korea has experienced rapid development since the early 1970s, successfully transforming itself from an agricultural economy to a leading world industrial power. This has been the result of a deliberate national development plan which fostered export-oriented industrialisation in the heavy and chemical industries (OECD 2012). The speed and scale of economic transformation has fostered a high degree of regional imbalance, with development concentrated in the Seoul metropolitan region and, to a lesser extent, the south-east coastal region around Busan in line with the government's export-oriented strategy. Approximately 50 per cent of national GDP is generated in Seoul and the surrounding Gyeonggi province (ibid.). Such regional imbalance is common in developing countries experiencing rapid growth as the regions in which economic activity is concentrated pull ahead of the remainder (World Bank 2009).

This problem of regional imbalance has been addressed through successive regional policies. The government adopted the 'old-style' approach in the 1970s and 1980s, principally measures to control growth in Seoul and efforts to disperse heavy industries to other regions (Lee 2009). These were abandoned in the 1990s, encouraging further concentration in Seoul. In response, the government of President Roh Moo-Hyun (2003–08) pursued an explicit strategy of 'balanced national' development, embracing a 'new regionalist' approach. This included the establishment of a Presidential Commission on Balanced National Development and the Special Act on Decentralisation, alongside a project to develop a new capital region outside Seoul.

The New Capital Regional Development Plan focused on building a new capital city in Sejong in South Chungcheong province. While this was passed by the National Assembly, the project was halted when the constitutional court ruled it unconstitutional in October 2004. This resulted in the project being downgraded from the relocation of the capital itself to a process of administrative decentralisation. The plan is to relocate 12 of 18 government ministries and 30 other state agencies to Sejong City by 2030, with a target population of 500,000 and a budget of 46 trillion won. This is still a highly ambitious initiative in an international context: while the dispersal of government offices and functions is a long-standing tool of regional policy, it is rarely contemplated on this scale. The initiative is motivated by traditional concerns about equity and regional balance, though it is part of a broader 'new regionalist' package of measures, pointing to the continuities between paradigms (Table 5.6).

countries. They are most pronounced in more decentralised systems such as the US where cities and local governments are most reliant on raising much of their own revenues. On the intra-national scale, budget cuts have tended to be most severe in many of the poorest cities and localities in both the US and the UK. Particularly in the decentralised context of the US, cities' exposure to austerity reflects the specific interaction between the downloading of budget cuts from the national and state level and the state of the local economy, with cities experiencing economic decline the most vulnerable due to falling local tax revenues. This has seen several US cities going 'bust' by declaring bankruptcy in the wake of the financial crisis and Great Recession, most notably Detroit. In the UK, local authorities have also borne a disproportionate share of spending cuts with the poorest urban areas often experiencing the greatest reductions (Beatty and Fothergill 2014). As such, not only has a financial crisis been translated into a state crisis through austerity, but that state crisis has also been translated into an urban crisis (Peck 2012: 651).

The city of Detroit has become emblematic of this urban crisis, following its filing for bankruptcy in 2013. This reflects both falling tax revenues as a result of the severe decline of the Detroit economy and the loss of its tax base through suburbanisation, and the retrenchment of federal and state finances with the state of Michigan withholding over $700 million from Detroit in the decade to 2014 (Peck and Whiteside 2016: 257). As Peck and Whiteside (2016) demonstrate, this prompted the city to become increasingly active in seeking new forms of credit to sustain spending, involving a range of instruments such as bonds, swaps, bond insurance and private financing, allowing it to refinance its debts through deals with Wall Street creditors. This kind of activity is not confined to Detroit, but is representative of the growing **financialisation** of urban governance in the US under conditions of low growth and reduced federal and state expenditure. Rather than the growth machines of the 1970s and 1980s, Peck and Whiteside suggest that cities are increasingly becoming 'debt machines' that are dependent on financial markets to access credit. As the example of Detroit illustrates, this entails a movement of power from local political and business elites to more distant financial interests. This

is apparent in the nature of Detroit's post-bankruptcy adjustment plan which has tended to privilege the interests of creditors over retirees and citizens, with cuts in pension obligations and service reductions.

> **Reflect**
>
> Is devolution to the regional scale a necessary condition for the transition to the new paradigm of regional policy?

5.7 Summary

The state refers to a set of institutions which holds sovereignty over a designated territory, exercising a monopoly of legitimate force and law-making ability. The key concept of the 'qualitative state' has been used to frame and inform this chapter, emphasising that we should focus on the "nature, purpose and consequences" of state intervention in the economy rather than its extent or magnitude (O'Neill 1997: 290). States have experienced a process of rescaling since the 1980s through the transfer of powers 'upwards' to supranational organisations and 'downwards' to urban and regional authorities through various forms of devolution. The state plays a key role in managing national economies, establishing and enforcing the 'rules of the game' that govern the behaviour of other economic actors according to broader social and political interests such as prosperity, security and inclusion. The role of the state in, for instance, acting as ultimate guarantor was starkly illustrated by the financial crisis of 2008–09 when many governments stepped in to rescue their insolvent banks. As this chapter has demonstrated, the specific ways in which states exercise their functions vary considerably in character, however, with the chapter identifying five types of state: neoliberal, developmental, continental European, social democratic and Mediterranean.

The shift away from the Keynesian welfare state originates in the political and economic crisis of the 1970s, triggering the political response of the New

Right. Their neoliberal agenda has subsequently spread across the globe, underpinning the Washington Consensus, implemented through bodies like the WTO, World Bank and IMF. Neoliberalism was challenged by the financial crisis of 2008–09 and associated Great Recession, undermining its assumptions about the efficiency of unfettered financial markets and the benefits of deregulation. Paradoxically, however, despite an initial reversion to Keynesian stimulus measures, neoliberal ideas have been reasserted in the wake of the crisis through austerity policies. At the supranational scale, austerity programmes have curtailed growth in southern Europe, whilst ushering in a new era of austerity urbanism at the sub-national scale, particularly in the US. Yet, despite recurring attempts to shrink its activities, the state remains central to the management of contemporary capitalism. The overriding conclusion of this chapter is that the national state remains a crucial actor in the regulation of the economy, not least through its role in coordinating activities across different geographical scales.

Exercise

Select a particular city or region, referring to basic economic statistics (GDP, income, growth, employment and unemployment) available from the appropriate government publications or website to get a basic sense of economic conditions within it. Examine and review the current economic strategy for that city or region, identifying its strengths and weaknesses. What are the key agencies and organisations involved in the formulation and implementation of the strategy? What are the key elements of the strategy? How realistic or appropriate is it in relation to regional economic conditions and needs? What assumptions is the strategy based on? To what extent is it framed by the new paradigm of regional policy identified in section 5.6.2? Are there any major omissions from the strategy? Are there any potential tensions or conflicts between different objectives? What alternative objectives or priorities would you like to see included?

Based on your analysis, sketch your own economic strategy for the city or region. How would you characterise this strategy (e.g. Keynesian, neoliberal, alternative)?

Key reading

Blyth, M. (2013) *Austerity: The History of a Dangerous Idea.* **Oxford: Oxford University Press.**
An authoritative and accessible history of austerity as an economic policy concept in the context of its adoption by governments since 2010. Argues powerfully that austerity doesn't work, invariably resulting in low growth alongside increases in inequality. Contains an excellent analysis of the Eurozone crisis.

Dicken, P. (2015) *Global Shift: Mapping the Changing Contours of the World Economy,* **7th edition. London: Sage, pp. 173–225.**
A thorough review of the role of the state in managing and regulating the economy, with a particular emphasis on policies to promote globalisation. Covers a number of policy areas, including trade, FDI and industry as well as economic strategy and the proliferation of regional trade agreements in recent decades.

Harvey, D. (2005) *A Brief History of Neoliberalism.* **Oxford: Oxford University Press.**
A stimulating account of the impact of neoliberalism across the world from the leading Marxist geographer. Harvey assesses the growth of neoliberal theories, their impact on state policies and the effects on economic growth and development. He views neoliberalism as a project to restore upper-class power and wealth which was developed in response to the economic crisis of the 1970s.

O'Neill, P. (1997) Bringing the qualitative state into economic geography. in Lee, R. and Wills, J. (eds) *Geographies of Economies.* **London: Arnold, pp. 290–301.**
The key article in which the concept of the qualitative state is developed. O'Neill criticises the focus on the extent of state intervention, focusing attention on the "nature, purpose and consequences" of state action (p. 290). Instead of heralding a decline in the importance of the state, the period since the 1970s has seen the nature and purpose of its intervention in the economy change significantly.

Peck, J. (2012) Austerity urbanism: American cities under extreme economy. *Cities* **16: 626–55.**
The key article which coined the term 'austerity urbanism', demonstrating how responsibility for cutting public expenditure is being devolved to the urban scale in the US particularly. Views this as the latest mutation of neoliberalism, with budget deficits in the wake of the crisis triggering further efforts to shrink the state.

Rodríguez-Pose, A. and Gill, N. (2005) On the 'economic dividend' of devolution. *Regional Studies* **39: 405–20.**
An informative discussion of the so-called 'economic dividend' of devolution, discussing the main economic arguments for and against the devolution of power and resources to urban and regional authorities. Argues that devolution can have negative as well as positive economic effects and that these are contingent upon the actors who are driving devolutionary policies.

Useful websites

http://europa.eu/index_en.htm
The official site of the European Union. Contains a wealth of information on the EU's activities, divided into specific topic areas. See especially the sections on 'economic and monetary affairs', 'enterprise', 'external trade' and 'regional policy'.

www.oecd.org/regional/regional-policy/multi-level governance.htm
The official site of the Organisation for Economic Cooperation and Development (OECD) which contains a range of publications and statistics on regional policy and multi-level governance, including the role and finances of sub-national governments.

http://web.inter.nl.net/users/Paul.Treanor/neoliberalism. html
Offers a useful introduction to neoliberalism, covering its origins, theoretical background and definition.

References

Aglietta, M. (1979) *A Theory of Capitalist Regulation: The US Experience*. London: New Left Books.

Agranoff, R. (2004) Autonomy, devolution and intergovernmental relations. *Regional & Federal Studies* 14: 26–65.

Amable, B. (2003) *The Diversity of Modern Capitalism*. Oxford: Oxford University Press.

Amin, A., Massey, D. and Thrift, N. (2003) *Decentring the Nation: A Radical Approach to Regional Inequality*. London: Catalyst.

Anderson, J. (1995) The exaggerated death of the nation state. In Anderson, J., Cochrane, A. and Brooks, C. (eds) *A Global World: Re-ordering Political Space*. Oxford: Open University and Oxford University Press, pp. 65–112.

Bailey, D., Lenihan, H., Arauzo-Carod, J.M. (2012) *Industrial Policy Beyond the Crisis: Regional, National and International Perspectives*. London: Routledge.

Beatty, C. and Fothergill, S. (2014) The local and regional impact of the UK's welfare reforms. *Cambridge Journal of Regions, Economies and Societies* 7: 63–80.

Blyth, M. (2013) *Austerity: The History of a Dangerous Idea*. Oxford: Oxford University Press.

Candelaria, C., Daly, M. and Hale, G. (2013) Persistence of regional inequality in China. *Federal Reserve Bank Of San Francisco Working Paper* 2013–06. At www.frbsf.org/ publications/economics/papers/2013/wp2013-06.pdf, last accessed 7 August 2015.

Carmody, P. (2011) *The New Scramble for Africa*. Cambridge: Polity.

Chang, H.-J., Andreoni, A. and Ming, L.K. (2013) International industrial policy experiences and the lessons for the UK. *Future of Manufacturing Project: Evidence Paper* 4. London: Foresight, Government Office for Science.

Coe, N. and Kelly, P. (2002) Languages of labour: representational strategies in Singapore's labour control regime. *Political Geography* 21: 341–71.

Coe, N.M., Kelly P.F. and Yeung, H.W. (2013) *Economic Geography: A Contemporary Introduction*, 2nd edition. Oxford: Wiley Blackwell.

Crouch, C. (2011) *The Strange Non-death of Neoliberalism*. Cambridge: Polity.

Dear, M. (2000) State. In Johnston, R.J., Gregory, D., Pratt, G. and Watts, M. (eds) *The Dictionary of Human Geography*, 4th edition. Oxford: Blackwell, pp. 788–90.

Dicken, P. (2003) *Global Shift: Reshaping the Global Economic Map in the 21st Century*, 4th edition. London: Sage.

Dicken, P. (2015) *Global Shift: Mapping the Changing Contours of the World Economy*, 7th edition. London: Sage.

Drudy, P. and Collins, M. (2011) Ireland: from boom to austerity. *Cambridge Journal of Regions, Economy and Society* 4: 339–54.

Duncan, S.S. and Goodwin, M. (1988) *The Local State and Uneven Development*. Cambridge: Polity.

Dunford, M. and Perrons, D. (1994) Regional inequality, regimes of accumulation and economic development in contemporary Europe. *Transactions, Institute of British Geographers* NS 19: 163–82.

Elliott, L. (2011) The strategy of stagnation. *The Guardian*, 30 May.

Epsing-Andersen, G. (1990) *The Three Worlds of Welfare Capitalism*. Princeton, NJ: Princeton University Press.

European Commission (2016) Country report Ireland 2016. *Commission Staff Working Document*. Brussels: European Commission.

Fishwick, C. (2016) Anti-austerity protestors: why we want David Cameron to resign. *The Guardian*, 16 April.

Ford, R. (2016) The 'left-behind', white, older, socially conservative voters turned against a political class with values opposed to theirs on identity, EU and immigration. Commentary. *The Observer*, 26 June.

Forfas (2014) *Ireland's Competitiveness Scorecard 2014*. Dublin: National Competitiveness Council.

Gardiner, B., Martin, R., Sunley, P. and Tyler, P. (2013) Spatially unbalanced growth in the British economy. *Journal of Economic Geography* 13: 889–928.

Garretsen, H., McCann, P., Martin, R. and Tyler, P. (2013) The future of regional policy. *Cambridge Journal of Regions, Economy and Society* 6: 179–86.

Gertler, M.S. (2010) Rules of the game: the place of institutions in regional economic change. *Regional Studies* 41: 1–15.

Hall, P. and Soskice, D. (eds) (2001) *Varieties of Capitalism: The Institutional Foundations of Comparative Advantage*. Oxford: Oxford University Press.

Harvey, D. (2005) *A Brief History of Neoliberalism*. Oxford: Oxford University Press.

House, J.D. and McGrath, K. (2004) Innovative governance and development in the new Ireland: social partnership and the integrated approach. *Governance* 17: 29–57.

Huff, W.G. (1995) The developmental state, government and Singapore's economic development since 1960. *World Development* 23: 1421–38.

International Monetary Fund (IMF) (2010) Navigating the fiscal challenges ahead. *Fiscal Monitor*, May 14. Washington, DC: IMF.

International Monetary Fund (IMF) (2011) Shifting gears. *Fiscal Monitor*, April. Washington, DC: IMF.

Jakobsen, E.W. and Grünfeld, L. (2003) *Hvem eier Norge? Eierskap og verdiskaping i et grenseløst næringsliv*. Oslo: Universitetsforlaget.

Jessop, B. (2002) *The Future of the Capitalist State*. Cambridge: Polity.

Johnson, C. (2016) Rising tide of protectionism imperils global trade. *Financial Times*, 18 February.

Kelly, P. (1992) *The End of Certainty*. Crow's Nest, New South Wales: Allen and Unwin.

Kitchin, R., O'Callaghan, C., Boyle, M., Gleeson, J. and Keaveney, K. (2012) Placing neoliberalism: the rise and fall of Ireland's Celtic Tiger. *Environment and Planning A* 44: 1302–26.

Kitson, M., Martin, R. and Tyler, P. (2011) The geographies of austerity. *Cambridge Journal of Regions, Economy and Society* 4: 289–302.

Lee, Y.-S. (2009) Balanced development in globalizing regional development? Unpacking the new regional policy of South Korea. *Regional Studies* 43: 353–67.

Leigh, A. (2002) Trade liberalisation and the Australian Labour Party. *Australian Journal of Politics & History* 48: 487–508.

Markusen, A. (1996) Sticky places in slippery space: a typology of industrial districts. *Economic Geography* 72: 293–313.

Mirowski, P. (2014) *Never Let a Serious Crisis Go to Waste: How Neoliberalism Survived the Financial Meltdown*. London: Verso.

Norwegian Ministry of Trade and Industry (2011) *Active Ownership – Norwegian State Ownership in a Global Economy*. Oslo: Norwegian Ministry of Trade and Industry.

OECD (Organisation for Economic Co-operation and Development) (2010) Regional development policy trends in OECD member countries. In *Regional Development Policies in OECD Countries*. Paris: OECD Publishing.

OECD (2012) *Industrial Policy and Territorial Development: Lessons from Korea*. Development Centre Studies. Paris: OECD Publishing.

OECD (2014) *OECD Regional Outlook 2014*. Paris: OECD Publishing.

OECD (2016) *Employment Outlook 2016*. Paris: OECD Publishing.

O'Neill, P. (1997) Bringing the qualitative state into economic geography. In Lee, R. and Wills, J. (eds) *Geographies of Economies*. London: Arnold, pp. 290–301.

Painter, J. (2000) States and governance. In Sheppard, E. and Barnes, T.J. (eds) *A Companion to Economic Geography*. Oxford: Blackwell, pp. 359–76.

Painter, J. (2006) Prosaic geographies of stateness. *Political Geography* 25: 752–74.

Peck, J. (2001) Neoliberalising states: thin policies/hard outcomes. *Progress in Human Geography* 25: 445–55.

Peck, J. (2012) Austerity urbanism: American cities under extreme economy. *Cities* 16: 626–55.

Peck, J. (2013) Explaining (with) neoliberalism. *Territory, Politics, Governance* 1: 132–57.

Peck, J. and Theodore, N. (2007) Variegated capitalism. *Progress in Human Geography* 31: 731–72.

Peck, J. and Tickell, A. (2002) Neoliberalising space. *Antipode* 34: 380–404.

Peck, J. and Whiteside, H. (2016) Financialising Detroit. *Economic Geography* 92: 235–68.

Peet, R. and Hartwick, E. (1999) *Theories of Development*. New York: Guildford.

Phelps, N.A., MacKinnon, D., Stone, I. and Braidford, P. (2003) Embedding the multinationals? Institutions and the development of overseas manufacturing affiliates in Wales and North East England. *Regional Studies* 37: 27–40.

Rodríguez-Pose, A. and Gill, N. (2003) The global trend towards devolution and its implications. *Environment and Planning C, Government and Policy* 21: 333–51.

Rodríguez-Pose, A. and Gill, N. (2004) Is there a global link between regional disparities and devolution? *Environment and Planning A* 36: 2097–117.

Rodríguez-Pose, A. and Gill, N. (2005) On the 'economic dividend' of devolution. *Regional Studies* 39: 405–20.

Rodríguez-Pose, A. and Sandall, R. (2008) From identity to the economy: analysing the evolution of decentralisation discourse. *Environment and Planning C, Government and Policy* 21: 54–72.

Rodrik, D. (2016) The surprising thing about the backlash against globalization. At www.weforum.org/agenda/2016/07/the-surprising-thing-about-the-backlash-against-globalization?utm_source=feedburner&utm_medium=feed&utm_campaign=Feed%3A+inside-the-world-economic-forum+(Inside+The+World+Economic+Forum), 15 July. Last accessed 9 August 2016.

Roubini, N. (2016) Globalization's political fault lines. *Project Syndicate*. At www. project-syndicate.org/commentary/globalization-political-fault-lines-by-nouriel-roubin—2016–07, 4 July. Last accessed 27 July 2018.

Sheppard, E. and Leitner, H., 2010: *Quo vadis* neoliberalism? The remaking of global capitalist governance after the Washington Consensus. *Geoforum* 45: 185–94.

Skidelsky, R. (2009) *The Return of the Master*. New York: Penguin.

Sparke, M., Sidaway, J.D., Bunnell, T. and Grundy-Warr, C.V. (2004) Triangulating the borderless world: geographies of power in the Indonesia–Malaysia–Singapore growth triangle. *Transactions of the Institute of British Geographers* NS 29: 485–98.

Standing, G. (2009) *Work after Globalisation: Building Occupational Citizenship*. Cheltenham: Edward Elgar Publishing Ltd.

Stiglitz, J. (2002) *Globalisation and its Discontents*. London: Penguin.

Weiss, L. (2000) Developmental states in transition: adapting, dismantling, innovating, not 'normalising'. *Pacific Review* 13: 21–55.

Weller, S.A. and O'Neill, P.M. (2014) De-industrialisation, financialisation and Australia's macro-economic trap. *Cambridge Journal of Regions, Economy and Society* 7 (3): 509–26.

World Bank (2009) *World Development Report 2009: Reshaping Economic Geography*. Washington, DC: The World Bank Group.

Chapter 6
Restructuring work and employment

6.1 Introduction

For many people regular interaction with the 'economy' takes place through work in its different forms, and our chances of making a decent living are shaped by our relationship to prevailing forms of work and employment. In the developed capitalist economies, the majority of us make a living through forms of paid employment, where we exchange our human labour for a wage. In developing world countries, much of the population is still engaged in non-capitalist forms of work linked to subsistence agriculture. The increasing integration of the world into a single global capitalist economy often threatens such traditional lifestyles, although it also opens up the possibility for some groups to escape feudal or more traditional forms of oppression to work as waged labour. As such, the processes of geographical **uneven development** and economic restructuring that we have considered in earlier chapters take on particular significance in terms of how they affect our conditions of employment and livelihoods.

In this chapter, our purpose is to examine the nature of employment change in the contemporary economy, exploring in particular the transformation of work that has occurred since the 1970s and the role played by geography in shaping change. We pay particular attention to the emergence of a global workforce and the implications for different cities and regions through new forms of global connection. We also emphasise that, unlike other factors of production, labour is not passive to processes of economic restructuring but plays a more active role in shaping the global economy. There are important **divisions of labour**, however, in terms of the bargaining power that different groups and individuals have to shape their work prospects, based on class, gender, ethnicity, nationality and other forms of social identity.

6.2 Conceptualising labour, work and social reproduction

6.2.1 Definitions

Human beings have to perform basic work tasks (e.g. hunting and gathering food, finding shelter, making clothes, looking after and raising children, etc.) to reproduce daily life. In this sense, work is essential to all societies, however primitive or advanced. How work is organised has varied and changed dramatically over time as societies have developed from early and rudimentary nomadic peoples to the more advanced global capitalist society of today. In section 3.2.2, we examined the labour process under **capitalism** and developed an understanding of how a more complex technical and social division of labour emerged with the growth of industrial capitalism. The concept of a social division of labour allows us to differentiate between different forms of work, some of which are paid and others which receive no financial reward.

The emergence of capitalism and its expansion to become a global system has fundamentally changed the way that we think about work, its relationship to the economy and how divisions of labour work between different social groups and different places. In a capitalist economy, premised on the generation of profit, labour is treated primarily as a commodity, however 'fictitious' this is in reality (see section 3.2.2). This means that work is essentially reduced to its economic value. Thus there tends to be a distinction between formally paid employment – where workers sell their labour to an employer in the labour market – and other forms of work that do not carry any commodity value under capitalism. Typically, most of the labour that goes into reproducing workers themselves – through household work – and predominantly carried out by women in most countries is unpaid. Thus, there is a key divide around work between employment that goes into capitalist production and that concerned with **social reproduction**. Such a division of labour is socially and politically constructed as part of capitalism's value system, rather than naturally occurring and has important social and spatial consequences.

The organisation of work, the valuing of different forms of labour and the distribution of this between social groups gives rise to the concept of the social division of labour (section 3.2.2). As Pahl observes:

> Someone arriving from another planet might be surprised and puzzled by the way that we distinguish between work and employment and the differential rewards that are paid to employees based on the kind of work they do and the kind of person they are. Interesting, creative and varied employment is highly rewarded; dull, repetitive and routine work is poorly rewarded. Men receive more than women, and this is related to social attitudes and conventions more than the actual amount or quantity of work that the individual or the gender category does.
>
> (Pahl 1988: 1)

The broader point is that work is highly differentiated and these differences reflect facets of social identity such as class, gender, ethnicity and age. In the first instance, divisions of labour in any society are intimately related to class (Wright 2015) and how work is organised between different social groups.

6.2.2 Social divisions of labour

For Marx, class is the fundamental concept in his analysis of power relations at work where a capitalist class (bourgeoisie) dominates a class of workers (proletariat) because of its ownership of the means of production (section 3.2.2). Sociologists of work have subsequently developed complex and multi-layered analyses of how class position shapes people's employment and income-generating prospects and subsequent life chances (see for example Box 6.1). This emphasises the role of social and cultural identities as well as economic power relations (Bourdieu 1984; Skeggs 1997; Savage *et al.* 2013). Getting on and being successful in life is a complex mix of the conditions you were born into and the opportunities available to you from this class position as well as the social and cultural networks and resources that you utilise in forging a livelihood. Clearly, there are major inequalities between different individuals and social groups in this respect.

Box 6.1

What class are you? Work and lifestyles in the twenty-first century

A group of UK sociologists recently carried out a large online survey of class for the BBC with 161,400 respondents (Savage *et al.* 2013). The survey wanted both to account for recent changes in employment structure and the weakening of traditional class boundaries in British society, and to develop a more sophisticated understanding of class that went beyond existing approaches centred on occupational

Table 6.1 Summary of social classes

	% GfK	% GBCS	Description
Elite	6	22	Very high economic capital (especially savings), high social capital, very high highbrow cultural capital
Established middle class	25	43	High economic capital, high status of mean contacts, high highbrow and emerging cultural capital
Technical middle class	6	10	High economic capital, very high mean social contacts, but relatively few contacts reported, moderate cultural capital
New affluent workers	15	6	Moderately good economic capital, moderately poor mean score of social contacts, though high range, moderate highbrow but good emerging cultural capital
Traditional working class	14	2	Moderately poor economic capital, though with reasonable house price, few social contacts, low highbrow and emerging cultural capital
Emergent service workers	19	17	Moderately poor economic capital, though with reasonable household income, moderate social contacts, high emerging (but low highbrow) cultural capital
Precariat	15	<1	Poor economic capital, and the lowest scores on every other criterion

Source: Savage *et al.* 2013.

Box 6.1 (continued)

status. This more rounded approach recognised the empirical reality of growing income inequalities in society but also the way that class inequality is also about "forms of social reproduction and cultural distinction" (ibid.: 223). The approach draws upon the French social anthropologist Pierre Bourdieu's recognition of the importance of social capital (i.e. people's ability to draw upon social networks), and cultural capital (i.e. having the right social and educational background to act and operate successfully within elite and higher social-class networks), alongside 'economic capital' (income and material wealth).

Using this more nuanced approach, the researchers identified a sevenfold classification of class operating in the UK (see Table 6.1). A key finding was the identification of a narrow elite, whose position at the apex

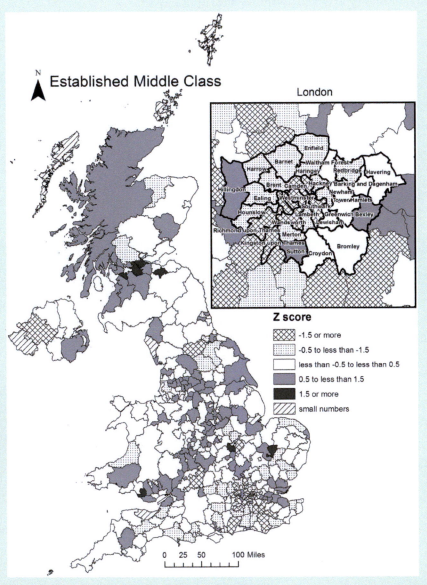

Figure 6.1 Mapping the elite in twenty-first-century Britain
Source: Savage *et al.* 2013: 235.

Box 6.1 (continued)

of power in British society was due both to their overall economic wealth (in terms of income, savings, family inheritance, property) and their backgrounds, particularly attendance at elite universities such as Oxford, Cambridge, LSE, Imperial and Kings College, London. Their results complicate conventional UK understandings of class as being divided into a predominantly 'white-collar middle class' of professional groups versus a 'blue-collar working class' of manual labour (with a very small 'upper-class

aristocracy') to identify multiple differences in relation to power, status and economic resources.

The research reaffirms Massey's earlier analysis of the **spatial divisions of labour** that exist in the UK (Massey 1984), with a core of higher-level managerial and decision-making jobs in London and the south-east of England compared to the rest of the country. An emphasis upon social and cultural capital also draws attention to the role of elite institutions (e.g. universities, elite schools, media

institutions such as the BBC, headquarters of large corporations, public and third-sector organisations) whose spatial concentration in London and other large cities and absence from more traditional working-class and rural areas reinforces a sense of alienation and exclusion for many (Figure 6.1). Such geographical divisions were evident in the 2016 'Brexit' vote to leave the European Union.

See the BBC's The Great British Class Survey at: www.bbc.co.uk/news/magazine-22000973.

For Erik Olin Wright (Wright 2015) advanced capitalist societies such as the US now contain at least five identifiable classes according to their relationships to work: a rich capitalist or managerial class who exercise considerable economic and political power; a large middle class who have traditionally benefited from high levels of education and training and occupy well-paid and secure jobs; a manual working class that has traditionally been unionised and enjoyed some form of employment protection, regular and decent wages; a lower working-class group of people who are in poorly paid and insecure work; and an 'underclass' without the skills, cultural resources or social networks to access decent employment and income-generating opportunities.

Intertwined within such 'economic' divisions are gender, race and other facets of social identity that shape a person's life chances and lead to wider patterns of discrimination, especially in cities (see Derickson 2016). Race is particularly prevalent in the US but also many European countries where "the working poor and the marginalized population are disproportionately made up of racial minorities" (Wright 2015: 17). Recent analysis by the Pew Research Centre (Patten 2016) found that black (73 per cent) and Hispanic (69 per cent) men earned substantially less than their white counterparts in the US whilst the corresponding figures for women (of white men's earnings) were 65 per cent (black) and

58 per cent (Hispanic). White women still only earn 82 per cent of white male hourly wages. The same report found that whilst Asian men earn more than white men per hour (117 per cent), Asian women earn 87 per cent less (ibid.).

Gender is therefore another critical fault-line for the division of work and employment. Until relatively recently, women were discouraged and actively discriminated against in the labour market, through government legislation, employer attitudes and wider social norms such as exclusion from university and higher education. In the UK a 'male breadwinner' model persisted into the 1960s, with women largely expected to give up paid employment once married to concentrate on the unpaid labour of bringing up children and looking after the household. Although the Sex Disqualification Removal Act was passed in 1918 by the government, theoretically allowing married women the same rights to work as men, local governments used local 'marriage bars' to ban married women from certain professions. Society at large also 'frowned upon' women taking what were regarded as male jobs. The rationale was usually based on the high levels of unemployment and reintegrating men into the labour market from the armed forces following the end of the First World War. But, it illustrates how a combination of social norms and labour regulations work to segment labour markets, in this case reinforcing patriarchy by enshrining employment rights for men at the expense of women. One of our maternal grandmothers – who had been a primary school teacher in South Wales before getting married in the 1920s – used to describe how she would cry every morning as she cleaned the floor of the marital home on hearing children passing the front door outside on their way to school.

Although most advanced capitalist societies now have laws promoting equal employment opportunities between different social groups, pay differentials persist. With regard to gender, the pay gap has diminished over time (Figure 6.2) though still at a slow pace, although there is considerable geographical variation between countries. In northern European countries such as Finland, the gender wage gap is lower than elsewhere because of active labour market policies such as rights to maternity and paternity leave, paid childcare and quotas for women on company boards.

6.2.3 Geography and work

From the discussion above, it should be obvious that geography is critical in understanding how work and employment are organised. This is true in two senses. First, following on from our discussion about social divisions around work above, distinctive forms of national and local labour market regulation mean that place is critical in mediating the work experiences of different groups. An important area of labour research in geography has drawn attention to the gendered power relations of **local labour markets** in emphasising both the differential positioning of women in paid employment and the unequal power relations that women face in the workplace, sometimes suffering abuse and violence (Hanson and Pratt 1995; McDowell 1997; Pratt 2004; Wright 2006). As the example above from South Wales in the inter-war period showed, national legislation worked in favour of men and against women in creating various forms of labour market exclusion around who was able to take up particular forms of employment in particular places. In this case, a highly patriarchal **labour control regime** was enforced spatially by a combination of 'hard' institutions (national laws) and the soft institutions of established societal practice and customs around what was regarded as men's and women's work.

More generally, women are often faced with a politics of spatial confinement, where laws and social norms not only prevent them from entering the workplace but also restrict the conditions under which they inhabit public spaces. Such conditions are harshest in more traditional societies and cultures in Asia and Africa, but can also persist through familial networks of patriarchy in ethnic minority communities in more liberal capitalist societies. Recent work by Linda McDowell and colleagues on Asian women workers engaging in

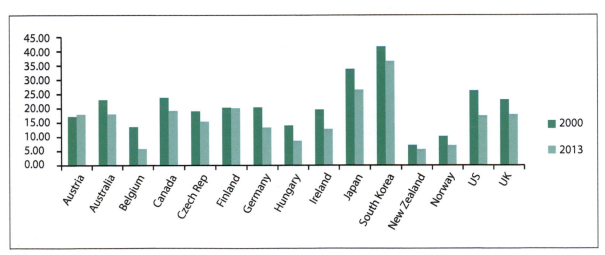

Figure 6.2 Gender pay gap in selected OECD countries (% difference between male and female median wages)

Source: OECD statistics database: http://stats.oecd.org/.

strike action in west London notes how such women face multiple axes of power, relating to class, gender and race in negotiating their rights to decent pay and conditions.

As McDowell *et al.* succinctly put it:

> women's disadvantage in the labour market is a consequence of their construction as 'women' and their evident embodied difference from a masculine norm that constructs them as 'out of place' in both occupations and workplaces. For minority women, this embodied disadvantage is exaggerated by their construction as Other: a location that carries with it various associations including tradition, the family, fragility, patriarchal dominance. . . .
>
> (McDowell *et al.* 2012: 148–9)

The right to be recognised as an employee engaged in industrial action – as well as a mother, daughter or migrant 'other' – is a continuing struggle for many women in local labour markets. Women are often discriminated against as much by other (male) workers as they are by employers and patriarchal family structures.

A second way in which geography shapes work is – as we have noted in past chapters – through the way that the economy under capitalism is characterised by the spatial differentiation of work between places. Through the development of large multi-plant corporations and the construction of a spatial division of labour (section 3.4.2), a highly differentiated landscape of employment emerges reflecting variations in the nature and type of work between different places. Through **globalisation** and the activities of TNCs this has been translated into a **'new international division of labour' (NIDL)** (section 3.5). The important point to note here is that these spatial variations in labour (at national and international levels) shape future rounds of capital investment, allowing firms to make location decisions based upon the different types of labour (e.g. rates of pay, levels of skill) available across the economic landscape.

Spatial divisions of labour are reflected in variations in employment conditions between places, most obviously in the wide discrepancies in wages between developed and developing countries (see Figure 3.16), encouraging the relocation of low-wage production by TNCs. These spatial relations are dynamic, with, for example, formerly low-wage economies in Eastern Europe and Asia (e.g. Slovakia, China, South Korea) becoming more costly in recent years, as labour becomes scarcer and is able to demand a higher share of income, sometimes through successful collective worker organisation and action. The ability for places to sustain jobs and economic prosperity as wages rise depends on the nature of these jobs, the extent of trans-local competition for TNC investment, and the ability of countries and regions to move up the global division of labour. In general, higher-value activities are less susceptible to competition from low-wage regions.

Wide spatial disparities in the opportunities for paid work exist within countries as well as between them. Data from the US Bureau of Labor Statistics indicates considerable local variation in the incidence of unemployment, for example as a measure of the availability of jobs (see for example Figure 4.4). Areas hit by **deindustrialisation** in the Midwest exhibit high levels of unemployment, as do many rural areas as a result of the decline of agricultural employment and the absence of jobs in growth sectors. The 2008 financial crisis and subsequent recession, however, has also hit formerly prosperous states such as California and Nevada (section 4.5.1). Spatial variations in employment not only relate to wage or unemployment differentials, but also reflect different cultures of labour associated with particular industries (see Peck 1996). These include differences in working practices, levels of unionisation and the way local labour markets operate (e.g. through different kinds of recruitment and training strategies).

Reflecting the dynamic nature of capitalism, work and employment relations within capitalism are rarely static but subject to considerable change with different phases of capitalism. Indeed 'creative destruction' is intimately bound up with **labour geography** in the way that the fusion of the search for new forms of profit and technological dynamism more often than not entail a process of destroying jobs, livelihoods and even communities in one place only to recreate employment opportunities and new industrial spaces elsewhere. The shift from a craft-based form of production to large-scale factory-based production was critical to the reorganisation of work and the introduction of more complex social and spatial divisions of labour in the nineteenth century (sections 3.2 and 3.3).

Similarly, the advent of mass production and Fordism once again shifted the geographies of employment to new manufacturing regions in no small part at the expense of traditional older industrial regions in the twentieth century (section 3.4). The period since the 1970s has once again been one of dramatic change linked to the consequences of globalisation, deindustrialisation, technological change and the growth of high-technology sectors, all of which are reshaping the employment landscape.

6.3 Globalisation and the restructuring of work

6.3.1 The changing global workforce

Although paid employment has become the dominant category of work in the global economy, it is complemented by a range of other forms of work (Box 6.2).

At the same time, its geographical incidence remains highly uneven. At the end of the twentieth century it was estimated that three-quarters of world employment (waged labour) was located in just 22 countries with "almost half of the world's labour force located in four countries: China, India, the United States and Indonesia" (Castree *et al.* 2004: 11). Within the global South, the incidence of wage labour varies considerably, reflecting differences in levels of integration within the global economy. In many parts of Africa, Asia and Latin America, forms of subsistence agriculture persist although attempts to modernise economies have often led to the forced destruction of such traditional ways of 'making a living', without always replacing them with sustainable alternatives. For much of sub-Saharan Africa in particular, integration into the global economy continues to be defined by the production of primary commodities such as minerals, and agricultural products such as coffee and cotton (Chapter 7), with the higher value-added activities such as manufacturing and R&D taking place in Europe, North America or South-east Asia.

Box 6.2

Categories of 'other work' in the global economy

Although it has spread with the development of globalisation, it is important to emphasise that over half of the world's adult population at any one time are not in paid employment. As well as housework, there are also other forms of work that are unpaid, including voluntary work and the work of many carers (particularly family members), looking after older people, children and the disabled. Attempts to reduce state welfare provision in many countries since the 1980s have meant that unpaid care work has become increasingly critical both to supporting the economy – through the supervision of children and their 'socialisation' for future employment – and in providing for the more disadvantaged sections of society. Unpaid housework is still the dominant form of domestic labour in both advanced and less developed economies, although the situation is slightly complicated by the existence of paid work for some household tasks (e.g. nannying, cleaning), particularly where both members of a household are engaged in full-time paid employment.

Alongside paid work, other categories include the self-employed, the definition of which often varies in time and space according to differences in employment laws and regulation. Additionally, a growing number of the global labour force is subject to unemployment or underemployment. The extent of this again varies over time, dependent upon the pendulum of uneven development. A growing number of people in Western Europe have been exposed to unemployment in various forms since the 1970s, whilst after the collapse of Communism unemployment rocketed in Eastern Europe during the early 1990s.

Another important category of work is child labour – the employment of children under the age of 16 – which is also commonplace in the global South, more often than not under extremely exploitative and badly paid working conditions. In China, for example, up to 80 per cent of the workforce in one electronics factory that supplies Samsung were children, employed for 11 hours per day, but paid only 70 per cent of the wages of other workers (China Labor Watch 2012). There have been major international campaigns against the practice of children working, based

Box 6.2 (continued)

upon not only the negative individual effects, but also the detrimental development consequences of low education. As a result, the use of child labour fell by one-third, from 246 million in 2000 to 168 million by 2012. Despite this, what remains shocking is the proportion of working children (70 per cent) engaged in hazardous work in mining, chemical industries or working with harmful pesticides in agriculture or dangerous machinery (ILO 2013).

A final category of work to comment on here is slave or **forced labour** which sadly remains a feature of the global economy despite its near universal rejections by states in the modern economy (e.g. McGrath 2013). The International Labour Office's 1998 Declaration on Fundamental Principles and Rights at Work has drawn attention to the continuing use by employers of what it terms 'forced labour' in the global economy. One of the more pernicious outcomes of globalisation has been an increase in the trafficking of people in conditions of forced labour, particularly for the sex trade. Illegal immigrants who lack the citizenship rights and status of domestic residents in their destination countries are particularly vulnerable to highly exploitative employers with an estimated 2.5 million people engaged in forced labour as a result of trafficking (ILO 2005: 14). Forced labour is defined as "all work or service which is exacted from any person under the menace of any penalty and for which the said person has not offered themselves voluntarily" (ibid.: 5).

Although its illegality makes it difficult to accurately measure, a conservative estimate is that over 12 million people are in some form of forced labour globally, with Asia and the Pacific Region dominating the trade (ibid.: 12). Two million people are in 'state or military imposed' forced labour, such as prisons, whilst another growing category is sexual exploitation with over 1 million 'workers' forced to sell their bodies for sex. Forced labour is a highly gendered affair and often involves children: women and girls account for 56 per cent of total forced labour and 98 per cent of the sex trade.

The twenty-first century is likely to see major changes to this situation as more countries and workers seek to become part of the global capitalist workforce. The last quarter of the twentieth century was already characterised by massive structural changes to global employment with the emergence of China, India, Mexico and other selective locations in the global South as competitors for jobs and economic opportunities with established firms, workers and places in the global North. This competitive process is likely to intensify further in the next couple of decades. To give an idea of the scale of competitive pressures at work, the global workforce is estimated to have risen from 960 million people in 1980 to 1,460 million by 2000 (Freeman 2010).

To put it bluntly, there has been a virtual doubling of the global labour supply in a 20-year period, in turn putting massive downward pressure on wages, leading to the flight of millions of jobs from the advanced industrial economies of North America and Western Europe to emergent capitalist economies in the South. One estimate is that the global workforce has actually trebled when one takes into account the entry into the market economy of former socialist economies such as the countries of the former Soviet Union, Cambodia and Vietnam – the latter alone having a population of over 80 million with wage levels at US$100 per month that have been static in real terms for 20 years (Standing 2011: 28). With the increased focus in China, India and other countries in the global South on raising student numbers at university, allied to massive levels of investment in research and development activities, the pressure in the future will be on higher-level jobs as well as for more routine low-wage activities (ibid.) (see also section 3.5).

An increasing supply of workers competing for jobs globally is linked to a number of other longer-term processes causing a dramatic transformation in contemporary employment conditions. The period since the mid-1970s in particular has been accompanied by three interrelated shifts in the global workplace: deindustrialisation; a shift towards services and more post-industrial forms of work; and the increased **automation** of work. These trends have also had pronounced spatial effects.

6.3.2 Deindustrialisation and its spatial consequences

In the most advanced industrial economies of the global North, there has been a generalised shift of employment out of manufacturing and industrial activities into services (see Figure 6.3). The extent of deindustrialisation has however varied spatially across the developed economies; manufacturing remains more important as a source of value added in countries such as Germany and Italy – which have even seen a modest growth relative to the rest of the economy since the financial crisis – compared to countries like France, the UK and US (Figure 6.3).

Spatially, the economic and social effects of deindustrialisation, to date, have been particularly concentrated

EMPLOYMENT

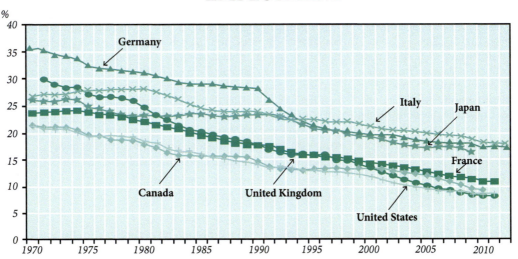

VALUE ADDED, CURRENT PRICES

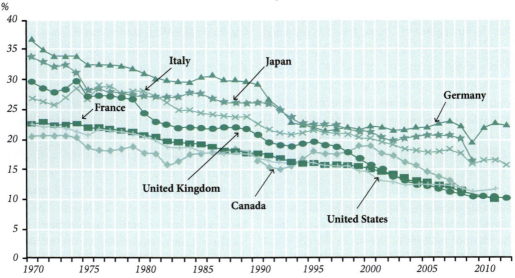

Figure 6.3 Share of manufacturing in employment and value added (current prices) for G7 countries 1970–2012

Source: De Backer *et al.* 2015: 8.

in old industrial cities and regions in North America (see Box 6.3) and Western Europe. In the latter, deindustrialisation has affected areas of traditional and heavy industry (e.g. coal mining, steelmaking and shipbuilding, textiles, auto production) particularly hard. However, experiences here have also been variable between countries. Research comparing the fate of UK old industrial regions with German, French and Spanish counterparts suggests that the other European regions fare better in both retaining manufacturing jobs and creating new activities (Birch *et al.* 2010), highlighting the importance of national and regional policy in mitigating processes of global economic integration.

At the same time, there is a blurring of the boundaries between traditional categories so that it is harder to differentiate between services and manufacturing work. As the OECD put it in a recent report: "A number of firms that are classified as service firms are in reality manufacturing firms that have re-organised their activities on an international scale within GVCs [global value chains]. The competitiveness of manufacturing firms in OECD countries is increasingly linked to 'intangible' services activities like design, R&D, sales, logistics, etc." (OECD 2015: 5).

Much of the world's computer industry has been transformed in this way. IBM is the most notable example, selling its entire PC manufacturing capacity to the Chinese firm Lenovo in the early 2000s to focus on software, consultancy and logistics. Dell and Apple outsource much of the lower-skilled and basic manufacturing activity to suppliers. Costs are reduced not only by accessing cheaper overseas labour supplies and not having to invest in costly infrastructure, but also by outsourcing the problem of labour control to other parties. As long ago as 1982 Apple Computer President Mike Scott remarked that "Our business was

Box 6.3

Detroit's collapse: an extreme case of deindustrialisation

Nowhere is more synonymous with the effects of deindustrialisation and the longer-term process of spatially uneven development than the US city of Detroit, which has seen its economic base and population collapse dramatically as a result of the contraction of the automobile industry in a long period of decline since its Fordist heyday in the mid-twentieth century (see section 3.4). Increasing competition from overseas and the relocation of work to cheaper non-union locations in the US south and west have had a devastating effect on Detroit and the surrounding Great Lakes industrial belt.

Detroit was home to Ford and the assembly line revolution, at the centre of the global auto industry in the twentieth century with the growth of a mass industrial workforce that yielded secure and well-paid work, strong levels of unionisation and collective worker rights. The city had the largest concentration of manufacturing jobs in the US during its heyday in the 1950s with almost half of total employment in manufacturing and of this about 13 per cent at the massive Ford plant at Dearborn on the outskirts of the metropolitan region (McDonald 2014).

The city had been experiencing a decline in manufacturing jobs already in the 1950s and 1960s from a peak of 349,000 in 1950 to 201,000 in 1970 but in common with many US cities, this reflected more a movement of population and jobs to the outer metropolitan region as part of postwar suburbanisation trends rather than urban decline per se. However, between 1970 and 1990, the city experienced massive job losses in its manufacturing sector, falling by 65.7 per cent from 201,000 to 69,000 (ibid.: 3320). The situation stabilised somewhat during the relative boom years of the 1990s with new service jobs created, including in the casino sector, but a further decline in the city's manufacturing base occurred in the early 2000s with manufacturing employment falling to around 20,000 in the city in 2010 (ibid.).

Employment decline has led to the effective abandonment of large parts of the city with population falling cataclysmically from a peak of 1.85 million in 1950 to around 714,000 by 2010. A sharply deteriorating tax base forced the city to file for bankruptcy in 2013 (section 5.6.3). Like all cities, some of Detroit's problems were historically and geographically specific: racial tensions caused by its pattern of large in-migration from the south of the US being a key one (Galster 2012). But the scale of its collapse marks it out as an extreme case of deindustrialisation.

designing, educating and marketing. I thought that Apple should do the least amount of work that it could and . . . let the subcontractors have the problems" (Chan *et al.* 2013: 104).

The geographical consequences of these organisational shifts can be profound and seem likely to accentuate existing spatial divisions of labour within and between countries. On the one hand they create new job opportunities in high-tech hotspots such as Silicon Valley and in creative and design-intensive clusters in metropolitan centres such as London, New York, Los Angeles, Paris and even Shanghai, whilst creating low-value and unstable jobs in peripheral regions that are in competition for the more routine production of components and assembly work.

Deindustrialisation is no longer confined to the developed economies of the global North but is a developing trend in the global South as the nexus of global competition for lower-skilled manufacturing jobs is increasingly a South versus South struggle. China and other East Asian economies have been accused of undermining and undercutting (with suggestions of unfair state subsidies and dumping on foreign markets) other countries' manufacturing bases in sectors as diverse as clothing and consumer electronics (in Mexico) or steel (India and Brazil). Nor has China itself been immune. Overcapacity in the global steel industry led to an announcement in February 2016 that the Chinese government was cutting almost 2 million jobs in its domestic industry (Yang 2016).

6.3.3 Service growth and narratives of post-industrial knowledge work

The idea of the decline of employment in manufacturing, and the growth of service-based work, signifying a post-industrial future is long established (e.g. Bell 1973). Certainly it is hard to deny that in the advanced industrial economies, most people's working lives and nature of their jobs has changed dramatically in the space of 40 years. Fewer and fewer people – in countries like the US and UK now less than 10 per cent of the employed workforce – are involved in the manufacturing sector (OECD 2015).

An important aspect of the post-industrial narrative is the claim that we live in an increasingly **knowledge-based economy** where advanced ICTs are providing new forms of work that require a more educated and skilled workforce. The idea that the post-industrial economy is knowledge intensive and provides plenty of jobs as long as the workforce has the appropriate skills is a powerful myth which has shaped domestic economic policy in both Western Europe and North America since the early 1990s. Robert Reich, President Clinton's Secretary for Labour during his first term of office (1992–96), famously claimed:

> The most rapidly growing job categories are knowledge-intensive; I've called them 'symbolic analysts'. Why are they growing so quickly? Why are they paying so well? Because technology is generating all sorts of new possibilities . . . The problem is that too many people don't have the right skills.
>
> (Quoted in Henwood 1998: 17)

The implications of this are clear: that advanced economies in the global North should prioritise knowledge-based activities for job growth and increase the number of people going through higher education. Such thinking has become deeply embedded within policy discourse at national government level and supranational institutions such as the EU and OECD. It tacitly accepts as fact that more routine activities in both services and manufacturing will be increasingly exposed to low-wage competition from the global South, and the futility of policy directed at attracting and retaining this kind of work.

However, the evidence does not support this prognosis. During the period of economic growth in the late 1990s and early 2000s, the fastest-growing job categories have been in more menial work that does not require high skill or education levels (Thompson 2004), while there has been a continuing decline in "middle level, craft and skilled manual employment" (ibid.: 30). In this respect, the discourse of the knowledge-based economy is arguably more useful for national politicians in the context of the disappearance of relatively well-paid and secure jobs in the manufacturing sector and their replacement by lower-paid, lower-skilled work in service activities. It places the responsibility on workers to upgrade themselves through education and training, rather than on employers or the state to create decent work. Yet, rather than the creation of high-quality jobs, there has been a further growth in less well-paid and more insecure jobs, particularly since the financial crisis (Kalleberg 2013; Standing 2011) (see section 6.4). The shift towards services seems predominantly associated with job degradation than upgrading.

6.3.4 A workless future? Automation and its social implications

A recent special feature in *The Observer* newspaper welcomed us "to a world without work" (Avent 2016: 37). The central message of the article was that in a world of increased globalisation, with more countries and workers competing for work, and the increased use of automation and robots replacing human labour, society faces a future of an "excess of labour and a shortage or work". Given the connections under capitalism between making a living, paid employment and work, this vision, if it comes to pass, presents massive problems for global society.

As we have already shown, the replacement of jobs by technology has been a recurring feature with the evolution of capitalism (section 3.2), and automation is already a reality in much of the global economy. The loss of jobs in manufacturing in recent years is as much down to automation and what is euphemistically referred to by mainstream economists as increases in labour productivity (e.g. OECD 2015). This is why successful advanced manufacturing economies such as Germany can realise significant wealth (in the form of measured value added) from production despite having a greatly reduced domestic labour force. Automation and computerised processing technologies are also replacing labour in many routine service sectors such as banking, retailing and leisure. But a growing number of commentators are forecasting that a new technology revolution is underway that will push automation even further, involving the robotisation of everything from driverless cars and other modes of transport to the increased use of drones for transportation and delivery of goods and services.

While a healthy scepticism is always required with such predictions, there are some worrying labour market trends that highlight the problems facing both the more developed and less developed economies in trying to find meaningful and gainful employment for particular sections of the population. For example, in the US, male workforce participation rates have fallen steadily in the last 25 years, from 75 per cent in 1990 to 69 per cent in 2015, a gap of around 9 million jobs (Avent 2016: 38). This is an economy that performs better than most in job creation terms. Yet, the World Bank (2016) estimated that an additional 600 million jobs were needed globally to address current population growth by 2030; this seems to be a rather tall order given recent trends and the lack of secure employment growth in most countries.

The impact of current trends is being most heavily felt by young people, with a growing global crisis of youth unemployment (ILO 2015) and a dramatic fall globally in rates of youth (15–24) labour force participation since the early 1990s (see Table 6.2). The recent upsurge in global protest in the period 2010–12 through moments such as the Arab Spring in North Africa and the Middle East and the Occupy movement in North America and Western Europe has in large part been linked to the worsening economic prospects facing young people (Mason 2012). Those in work are disproportionately more likely to be experiencing poverty, temporary and more precarious forms of work than their elders (ILO 2015). While some of the trends are undoubtedly shorter-term effects of the recent financial crisis, ongoing recession and austerity policies (Chapter 4), a concern remains around whether a future of jobs-related sustainable growth can be achieved. More automation of work without some more fundamental social and political changes to the prevailing economic order will only make this more difficult.

Reflect

What are the key social and spatial dimensions of recent employment restructuring?

Table 6.2 Youth labour force participation by region and sex 1991 and 2014

Region	1991			2014		
	Total	Male	Female	Total	Male	Female
World	59.0	67.0	50.6	47.3	55.2	38.9
Developed Economies and European Union	55.6	58.7	52.4	47.4	49.1	45.5
Central and South-Eastern Europe (non-EU)	50.2	56.3	44.0	40.6	47.9	33.0
East Asia	75.7	74.9	76.6	55.0	57.0	52.9
South-East Asia and the Pacific	59.3	65.8	52.7	52.4	59.4	45.2
South Asia	52.2	70.4	32.5	39.5	55.2	22.6
Latin America and the Caribbean	55.5	71.3	39.6	52.5	62.1	42.6
Middle East	35.6	57.3	12.6	31.3	47.2	13.8
North Africa	37.0	51.8	21.5	33.7	47.2	19.7
Sub-Saharan Africa	54.3	58.6	50.1	54.3	56.6	52.1

Source: ILO 2015: 9, Table 2.1.

6.4 Polarisation, flexibilisation and precarity in the labour market

6.4.1 Polarisation of income

Globalisation, deindustrialisation and the increased supply of labour globally have greatly weakened the bargaining power of workers and consequently their employment conditions. At the same time, the wealth accruing to those at the top of the class system – the elite of chief executives – has increased exponentially over this period. Data from the United States suggests that the gulf between the average worker's pay and that of the top chief executives of leading stock market corporations has increased from a ratio of 1:42 in 1980 to 1:373 in 2014 (www.aflcio.org/Corporate-Watch/Paywatch-2015, last accessed October 2016). Across the social spectrum there has been a massive increase in inequality and social polarisation between rich and poor. In the United States, the period since the late 1970s has seen a marked increase in the household

income of the richest 5 per cent compared to the rest, which is itself dwarfed by the riches accruing to the very wealthiest individuals.

Although the US is at the extreme end of the spectrum, growing income inequalities between rich and poor are prevalent elsewhere. Data for some of the world's richest countries suggests a generalised divergence between the income of the richest 10 per cent and the bottom 10 per cent, though with variations between countries (Dicken 2015: 314). Levels of income inequality in the developing countries tend to be even higher although there are variations here too; notably between East Asian and Latin American countries such as Brazil, Chile and Mexico (Table 6.3).

6.4.2 The pursuit of labour flexibility

Linked to the growing polarisation of income, the period since 1980 has also seen a growth in **labour market flexibility** (Rodgers 2007; Standing 2009). While this was initially a response by employers in the advanced industrial economies in the 1970s to a

Table 6.3 Distribution of income within selected developing countries

Country (year)	Lowest 10%	Lowest 20%	Highest 20%	Highest 10%
Brazil (2009)	1	3	59	43
Chile (2009)	2	4	58	43
Mexico (2000)	2	5	53	38
Malaysia (2009)	2	5	51	35
Philippines (2009)	3	3	50	34
Singapore (1998)	2	5	49	33
China (2009)	2	5	47	30
India (2010)	4	9	43	29
Korea (1998)	3	8	37	22

Source: Dicken 2015: 319, Table 10.1.

slowdown in growth in product markets and growing competition, strategies have evolved to manage and organise labour in more flexible ways to reduce costs and give employers more adaptability in the face of fluctuating market conditions. Globalisation and the threat of job relocation have strengthened employers by disciplining workforces to agree to more flexible forms of employment, supporting neoliberal-inspired government policies (see section 5.5) of labour market deregulation and freeing employers to manage labour according to their own requirements (Standing 2011).

From an employers' perspective, labour market flexibility can be divided into four key elements (Rodgers 2007; Standing 2009):

➤ *numerical*, which relates essentially to the ability of employers to 'hire and fire' workers, or reduce the amount of labour more broadly according to their requirements;

➤ *functional*, which refers to the ability to move workers flexibly between job tasks, to redeploy labour within the firm, or to reorganise the division of labour more fundamentally within a firm or even broader supply chain;

➤ *wage* flexibility, meaning the ability to lower wages and non-wage compensation given to workers, but also to change conditions of payment to the firm's advantage (e.g. changing the contract status of an

employee to a third party or self-employed to avoid welfare and pension costs);

➤ *time* flexibility, partly related to numerical flexibility but referring more broadly to the ability of firms to control the hours of work of employees.

A 2015 report by the OECD estimated that half of all jobs created in the advanced industrial economies since 1995 were in non-standard forms of employment; part time, temporary and self-employment (OECD 2015). The rise of part-time work as a proportion of total employment in most advanced economies has been evident since the 1980s, linked to the decline of manufacturing and the shift to services (Figure 6.4). More broadly, the ILO recently estimated that only one-quarter of the world's workforce were employed on permanent contracts (ILO 2015).

Flexibility has become a key plank of neoliberal labour market reform, alongside welfare reforms; to put it simply and brutally, policies that make it easier to hire and fire workers, pay them less and vary their working hours according to business needs, have run alongside policies that penalise those out of work and coerce them back into work (Standing 2011). It became an 'article of faith' for mainstream labour market policy-makers from the mid-1990s when the OECD proposed that the US's better record at creating jobs than the EU's was due to its less rigid and more

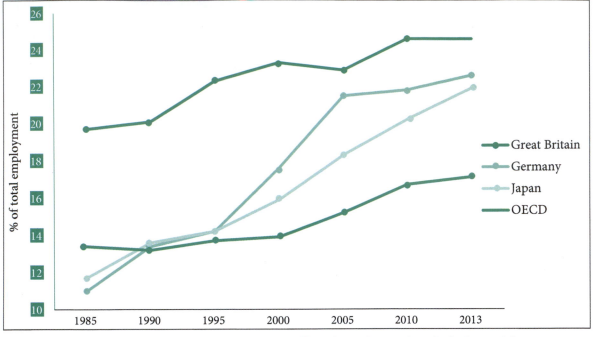

Figure 6.4 Part-time employment as a proportion of total employment: selected countries
Source: OECD statistical database.

flexible labour market regime (OECD 1994). More recently, in the wake of the financial crisis and austerity (see section 4.5.2), flexible labour market policies – particularly new legislation to make it easier to fire workers and reduce employees' compensation – were part of the austerity measures inflicted upon southern European economies and their workers (Moreira *et al.* 2015). Such measures reflected the continuing grip of neoliberal reasoning that 'labour market reform', code for reducing the regulations and constraints on business and undermining employment protection and workers' rights, is necessary to allow the smooth functioning of markets, a return to economic growth and the creation of jobs.

Flexibility may, of course, be beneficial to workers: the ability to agree flexible working time arrangements can be particularly useful for those who have to balance paid employment commitments against family time and caring work. Also, the ability to move relatively easily between jobs might suit certain types of workers, particularly those who are younger and more footloose. There is little doubt, however, that it is employers who have benefited most from increased flexibility (Kalleberg 2013; Standing 2011). In a remarkable volte face, given its previous advocacy of labour market deregulation and flexibility, the OECD recently noted that increased inequality since the 1990s was strongly linked to the growth of such flexible jobs, which were lower in both quality and wages, and involved greater insecurity and stress than more permanent forms of employment (ibid.). The period since the financial crisis has also seen an acceleration of these trends: "In the six years since the global economic crisis, standard jobs were destroyed while part-time employment continued to increase" (OECD 2015: 189).

Beyond the labour market flexibility imposed on southern European countries in the wake of the Eurozone crisis, there are persistent differences in flexibility between countries. Of particular importance here is the continuing contrast between the more regulated approaches to employment in continental European economies compared to the US and UK, evident, for example, in the legal protections afforded to workers against dismissal (Figure 6.5).

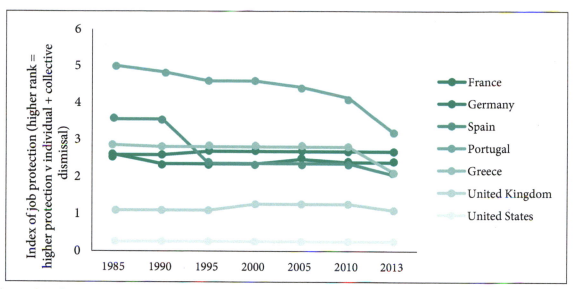

Figure 6.5 Comparisons in levels of legal job protection in selected countries 1985–2013
Source: OECD Employment protection statistics.

Box 6.4

Zero hours contracts as the ultimate in labour flexibility

The lack of legal rights for workers in the UK has resulted in the proliferation of non-standard forms of work, which have accounted for all of the country's net jobs growth since 1995 (OECD 2015). This has included the growth of 'zero hours' contracts, which represent the ultimate source of employment flexibility, whereby

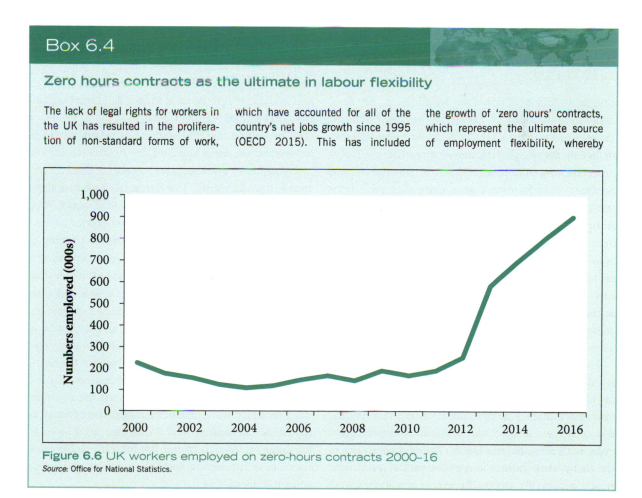

Figure 6.6 UK workers employed on zero-hours contracts 2000–16
Source: Office for National Statistics.

Box 6.4 (continued)

employers take on workers with no guarantees of weekly working hours. The number employed on such contracts has risen dramatically to nearly 1 million people by 2016 (see Figure 6.6), around 2.5 per cent of the total UK workforce (Farrell and Press Association 2016)

Although employers have defended the practice for providing valued flexibility for both firms and workers, **trade unions** point to them as being highly precarious and marginalised forms of work where flexibility ultimately works in one direction. As Frances Grady, General Secretary of the TUC notes:

The so-called flexibility these contracts offer is far too one-sided. Staff without guaranteed pay have much less power to stand up for their rights and often feel afraid to turn down shifts in case they fall out of favour with their boss.

(Ibid.).

Average wages for zero hours workers are £188 per week compared to £479 for permanent workers. Zero hours contracts also reinforce existing patterns of labour market segmentation: women, young people and students are disproportionately represented in their number, as are low-paid sectors

such as health and social care, hotels and food processing. The subject has risen to public prominence in recent years through the activities of a number of notorious employers, notably the firm Sports Direct, owned by billionaire football club chairman Mike Ashley. The firm developed a business model whereby it was using zero hours contracts for 90 per cent of its 20,000 employees and pursuing harsh managerial practices likened to 'Victorian workhouse conditions'. Through the creative use of zero hours contracts, many of its workers were being paid less than the national legal minimum wage of £7.20 per hour (Farrell and Butler 2016).

6.4.3 The rise of the global precariat

Commentators have noted an overall rise in job insecurity and instability as a result of the trends outlined above (Kalleberg 2013; OECD 2015). An influential thesis is Guy Standing's depiction of a new class: a global **precariat**, to signify how processes of globalisation and flexibility have fostered a generalised shift away from stable and permanent occupations to a rise in insecure and precarious forms of work (Standing 2011). Standing suggests that contemporary labour market segmentation, driven by the dismantling of established social protections, the decline of trade unions and collective bargaining cultures, is shrinking the number of people in stable and permanent forms of work and creating a growing number in more temporary and contingent forms of employment.

He outlines a sixfold classification of work that is emerging in the twenty-first century: a tiny 'elite' of "absurdly rich global citizens" (ibid.: 7), who have immense influence on governments and policy makers; a 'salariat' consisting of those in stable full-time employment with pensions and regular holidays who work in the major corporations and public sector; 'proficians', which is a smaller group (the term is a combination of

professional and technicians) that has specialist skills that they can trade in the labour market for high wages. This group can take advantage of flexible labour markets and do not feel the need to seek employment security. Next is what Standing refers to as the 'old working class' or "a shrinking core of manual workers" (ibid.: 8), battered by the fierce winds of deindustrialisation and globalisation so that they have lost the old sense of collective class solidarity. The 'precariat' lie below these groups and below it is an underclass of the unemployed and incapacitated individuals unable to work.

Critically for Standing, the precariat is growing as a proportion of total employment; he estimates that up to a quarter of all workers are now in the precariat, supported by the kinds of trends in flexible and insecure work identified earlier. What marks this class out in the contemporary period is its insecure status in relation to work and occupational identity, compared to the post-war period of full employment in the developed economies where the dominant class socially and politically was the industrial proletariat. He contrasts the precariat's work conditions with the security of industrial citizenship, which included high levels of job security, decent incomes, opportunities for promotion and skills training and enhancement, and strong collective bargaining.

The growth of the precariat is a global trend, a recognisable class in the less developed economies in Sub-Saharan Africa as much as in the developed economies of Western Europe, North America and the rapidly developing economies of South and East Asia. Even Japan, whose post-war social contract was based upon a system of lifetime guaranteed work for the majority of workers, has seen a massive growth in temporary work, now accounting for one-third of the workforce (ibid.: 15). Nevertheless, the precariat is a diverse group representing a range of experiences and social groups who enjoy different levels of status – compare teenagers who move between jobs in cafes, call centres and restaurants with illegal migrant workers who are forced to work in the black economy always fearing police and security services. The single mother working in a variety of low-paid and insecure part-time jobs represents another type of precarious worker.

Despite such variations, what unifies these workers is the lack of basic employment rights and security that marks the class out as "denizens", rather than "citizens" (ibid.: 14) in the sense of a lack of recognition compared to employees on more standard regular contracts. As a result of its rootlessness and lack of a stake in the existing social and political order, the precariat is in some ways a "dangerous class" because its material conditions of hyper-commodification, atomisation and alienation mean that it lacks a sense of community and "empathy" (ibid.: 23).

Reflect

How does the pursuit of labour market flexibility create new forms of labour market segmentation and income polarisation?

6.5 Labour agency in the global economy

6.5.1 The perspective of labour geography

Standing's work, in common with many macro-level analyses of economic globalisation, is often criticised for treating labour as a passive victim of broader restructuring processes, shorn of the capacity to act in its own right (Cumbers *et al.* 2010). While it is necessary to point out that such approaches are focused on charting broader processes of change rather than upon questions of agency per se, it remains important to address the possibilities and constraints for workers in attempting to improve their working and living conditions in the global economy. Geographers have made an important contribution to this task in recent years with the emergence of the sub-discipline of labour geography (Herod 2001; Coe and Jordhus-Lier 2010; Bergene *et al.* 2010). A critical insight has been to emphasise the way that labour – both individually and collectively through trade unions and other forms of coordinated struggle – also shapes the emerging landscapes of the global economy alongside capital:

> the production of the geography of capitalism is not the sole prerogative of capital. Understanding only how capital is structured and operates is not sufficient to understand the making of the geography of capitalism. For sure, this does not mean that labor is free to construct landscapes as it pleases, for its agency is restricted just as is capital's – by history, geography, by structures that it cannot control, and by the actions of its opponents. But it does mean that a more active conception of workers' geographical agency must be incorporated into explanations of how economic landscapes come to look and function the way they do.
>
> (Herod 2001: 34).

Labour as a human actor rather than a 'fictitious commodity' retains both the power to collectively organise for higher wages and better conditions, and the ability, as individuals, households and communities, to take decisions of its own volition, which may involve everything from quitting an unpleasant job to migrating away from a region either permanently in the hope of a better future, or temporarily to earn a better income elsewhere that can support the household and community at home.

A salient point to remember, despite deindustrialisation, automation and the increased flexibilisation of employment outlined above, is that ultimately capital requires labour for surplus accumulation to happen. The changes wrought by globalisation may have shifted the balance of power from labour to capital, through the sheer weight of numbers of new workers in the global economy and the competition for jobs this entails (section 6.3.2). Yet, in the same way that David Harvey reminds us that capital's spatial fix always comes up against a limit because of the tendency for profits to begin to fall at some point, capital always comes up against another spatial limit in its relations with labour. In other words, it cannot forever drive down labour costs globally through spatial switching from one location to another and setting one group of workers against another. Capital can never completely escape its 'labour problem' (Cumbers *et al.* 2008). At some point, it always needs to settle (at least for a fixed period of time) in particular places to produce things that make a profit, which in turn allows space for labour resistance and agency (see Box 6.5 below). As Marx long ago put it: "capital is constantly compelled to wrestle with the insubordination of the workmen . . . with the refractory hand of labour" (Marx, 1965, cited in Holloway 2005: 161, 191).

One way that capital tries to overcome this problem, in order to construct sustainable conditions for profit accumulation, is through local labour control regimes (Burawoy 1983; Jonas 1996; Kelly 2001) where local and even national government actors act in concert with businesses to create local environments for the effective production and reproduction of a labour force. This goes beyond active labour market and welfare policies and includes creating what are effectively spaces of labour control where the provision of housing and even leisure facilities is part of the overall mix. Local labour control regimes are particularly notorious in **EPZs** in developing economies such as China, Mexico and the Philippines where anti-union strategies by corporations and local governments often run alongside harsh disciplinary regimes for workers (e.g. Kelly 2013).

6.5.2 Globalisation and trade unions

In general terms, globalisation has undoubtedly had a negative effect on trade unions in the more developed economies in the global North, with job losses in highly unionised manufacturing industries accompanied by

a shift into services where unions have not, thus far, been able to develop a significant presence. While the rate of decline has slowed, and at the same time some countries – notably the Nordic countries – continue to have a strong trade union presence, these trends coupled with flexible labour markets, and insecure and precarious work, pose important questions about the very future of the movement (Figure 6.7). Trade unions are also struggling to organise in emergent and developing economies in recent years. In some of the larger developing economies such as Korea, South Africa and Brazil, the successful growth of trade unions in the 1980s, linked to broader projects for democracy, has been reversed in the recent period in the context of increased outsourcing and the use of informal and contract labour.

The more perceptive commentators have suggested that contemporary patterns of decline represent the crisis of a particular form of trade unionism in time and space, which derived its strength from the full employment and longer-term growth of the post-1945 period, centred around large manufacturing-based workforces, nationally regulated economies and systems of collective bargaining, and the social model of a male breadwinner in permanent employment (Munck 1999). Trade unions have been caught out by the shifting nature of capitalism itself. An amusing parable is offered up by Peter Waterman (2014: 40):

> The Capitalists and the Unions meet in the traditional World Labour Cup. The Unions arrive, all kitted up, from shirts to boots. But they find, to their horror, that the customary green pitch has been replaced by a shiny white skating rink. They protest loudly but the Capitalists say, 'This is New Football, it's faster, it's more profitable, so get your skates on or go away'. The Unions complain to the Referee but he hoists his shoulders and says, 'What can I do? If I make it an issue, they'll simply move the match somewhere else'.

Trade unions remain rooted in the past and incapable of keeping up with the fast-changing dynamic of global capitalism. They remain western and even Euro-centric, over-concentrated in declining sectors and shrinking forms of employment (ibid.) and need to make themselves more relevant to the dispersed and fragmented economy of services by appealing to non-traditional

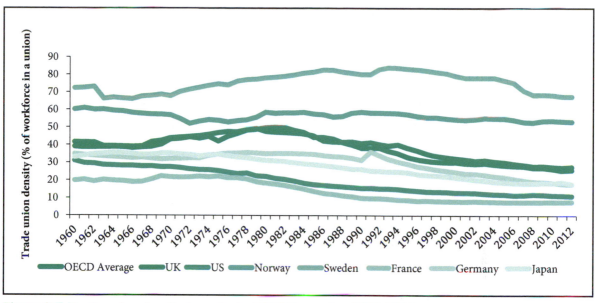

Figure 6.7 Trade union density by selected OECD country 1960–2012
Source: OECD statistical database.

constituencies such as women and those in non-standard work, who have little social protection. Some have also called for unions to become less narrowly oriented to workplace issues, but instead to work for broader social and political goals, forging alliances with other social movements within local communities on issues such as housing and combating poverty (Wills 2003; Jordhus-Lier 2013).

Perhaps the most serious issue facing trade unions worldwide is the growth of a massive non-unionised and highly exploited labour force in the global South, stemming from the increasingly complex divisions of labour. The fastest-growing area of employment in Latin America and Africa is the urban informal economy which includes a growing number of 'unregistered' workers engaged in outsourced work linked into global production networks, but with minimal social protection and employment regulation (ILO 2015). Not only are such workers subject to considerable exploitation by managers and key customers, but their presence also drives down the wages and conditions of workers in formal employment. Nor is this a problem solely of the global South with the growth of precarious, informal and insecure employment in much of the global North too. A recent report by the ILO found that working

poverty in the EU (those earning less than 60 per cent of the national median wage) has increased as a proportion of total employment from 11.9 per cent in 2005 to 13.3 per cent in 2012 (ILO 2015: 12). Tackling the unorganised workforce is now a common problem facing unions in both the global North and South.

In responding to these developments, unions and workers face three critical problems, which they are attempting to address.

➤ Developing more effective transnational networks to hold global corporations to account. While the labour movement already has its own transnational structures, such as the International Trade Union Confederation (ITUC) and Global Union Federations (GUFs), the union movement remains predominantly nationally centred (Cumbers *et al.* 2008). Many in the trade union hierarchy have prioritised the importance of developing a set of minimal international labour standards (with regard to pay and conditions, gender equality, the right to join unions and the end to child and forced labour) by lobbying governments and key global institutions such as the World Trade Organisations but, as the trends identified above suggest, these have borne little fruit. A more interesting development has been the establishment of Global Frame Agreements (GFAs), where GUFs sign agreements with key TNCs to provide decent work and minimum labour standards in their global supply chains. Fifty-four GFAs have been signed between 2009 and 2015 (Hadwiger 2015) and although they are an important step forward, they remain voluntary codes that have no legal basis, leaving them prone to abuse.

➤ Developing effective and independent transnational labour networks (Waterman 2014). This approach tends to be more militant, viewing attempts by union leaders to forge social partnerships such as GFAs at an international level with governments and employers as inevitably compromised. Indeed, as our China example shows below in section 6.5.3, many official trade union associations have often done more to suppress workers' rights in the pursuit of broader national development goals rather than being independent organisations for worker representation. However, whilst grassroots initiatives between workers of different countries may result in short-term victories they can often prove unsustainable without effective

legislative rights that can enforce and entrench such gains (Cumbers *et al.* 2016).

➤ Organising workers in the dispersed and fragmented landscape of the post-industrial service-based economy (Wills 2005; Tufts 2007; Jordhus-Lier and Underthun 2014). As labour geographers have been quick to point out, not all capital is mobile. Much service work in particular is tied to place; restaurants and hotels, cleaning and security work, and employment in public services all have to locate where the customers are and, in particular, within large urban areas. Nevertheless, as we have already documented, such work is often low-paid and non-unionised. The geographical dilemmas facing unions in this context are about developing more effective strategies for organising across a city in multiple workplaces, rather than focusing upon the single large mass workplaces typical of the industrial economy. An important example is Living Wage Campaigns: coalitions of trade unions, church groups and other social activists have achieved considerable success in recent years, in an otherwise hostile industrial relations climate in winning 'living wage campaigns' at the local level (Box 6.5).

Box 6.5

Organising city-wide for low-wage workers: the 'Living Wage Movement'

One of the most significant anti-poverty campaigns in recent years has been the movement for Living Wages that started in the United States but has now spread to the UK, Ireland, New Zealand and Canada. Living wages guarantee decent wage levels with the aim to provide an income level that will take workers out of poverty. The campaign in the US started because of the inadequacies of official minimum wage policy, which at the current rate of $7.25 per hour is below the poverty threshold. If the wage had kept pace with productivity gains since the 1960s the rate would be $21 per hour (Cooper 2015).

The Living Wage campaign in the US has cleverly linked pay to public spending, suggesting that the latter should not be used to subsidise 'poverty wages', and its success can be measured by the fact that Republican as well as Democrat city authorities have signed living wage ordinances. In some cities, a 'living wage' ordinance has been used as a local marketing strategy, in effect using a 'progressive localism' to attract more socially concerned investors. The first living wage coalition was launched in Baltimore in 1994 when, in the context of deindustrialisation and a process of gentrified waterfront regeneration

that had done little to deal with the problems of poverty and alienation affecting the traditional workforce, an alliance between trade unionists and religious leaders forced the city government to pass a law requiring a 'local living wage' for public sector workers.

Subsequently 140 cities across the US followed suit and passed some form of living wage ordinance. Although the financial crisis stalled the momentum of the campaign amidst city government fears about rising costs, the campaign has gained new momentum in the wake of the Occupy protests of 2011 and a series of strikes by fast

Box 6.5 (continued)

food workers in a number of US cities. Focus has shifted scales to encourage state governments to substantially raise minimum wage levels with some success; a new campaign, 'Fight for $15', is aimed at encouraging federal states and counties to raise the wages of the poorest workers. Two big recent victories are California's vote to raise the minimum wage from $10 per hour to $15 by 2022 and New York City's commitment that all employers should pay $15 per hour by 2019 (Luce 2016). It is estimated that the rises will benefit 9 million workers (ibid.).

In the UK, living wage campaigns have been pursued by a number of unions, activist groups and civil society organisations, including London Citizens. Although such campaigns have been successful in raising the profile of low pay, they have thus far been reliant on voluntary agreements with local and regional government and private employers. Living wage employers now include a diverse range of organisations including multinational banks, universities and the Scottish Government. In addition to paying the living wage, employers typically provide employees with at least 20 days paid holiday plus bank holidays, eligibility for 10 sick days per year and access to join a trade union. The effects of the living wage remain more symbolic than real in the UK; the lack of statutory legislation means that it is restricted to only 14,000 of the UK's 6 million low-wage workers (McBride and Muirhead 2016). However, as in the US, the living wage has become a live national political issue in the UK with public concerns around inequality, low pay and the casualisation of work, even prompting the Conservative Government to introduce its own Living Wage of £7.20 per hour from April 2016. This is regarded by many as a political ruse rather than a serious commitment to tackling low pay; although above the previous minimum wage of £6.70, the figure is still well below the £8.45 per hour recommended by the Living Wage Foundation (see www.livingwage.org.uk/what-living-wage, last accessed November 2016).

A comparison of the UK and US suggests that the greater level of decentralisation of employment policy in the latter has been an important catalyst for low-wage and poverty campaigns to upscale discourses of decent pay. In the UK the ability to legislate on wage levels is a monopoly of the UK Government in London whereas individual states in the US have the ability to set their own rates, and even city and county administrations have the power to set their own pay ordinances at the local level.

6.5.3 New forms of labour resistance

Beyond formal union organisation, collective labour agency takes other forms. Some of the most unexpected and largest labour resistance in recent years has come in China where there has been a dramatic upsurge in strikes and collective action by workers in the period since 2000, challenging the external and often stereotypical perception of a passive and compliant workforce. Growing labour shortages (partly the result of the country's draconian one-child population policy) in the country's booming export-led manufacturing sectors have both driven up wages and strengthened the bargaining power of workers (Chan et al. 2013). Independent of this, workers have begun to organise themselves collectively to protest against harsh working conditions, low wages and employers' infringement of legal employment rights. The Hong Kong-based China Labour Bulletin has estimated that there have been at least 4,000 labour-related collective disputes since 2010.

The situation is all the more remarkable because of the harsh political environment facing workers who have no legal right to form independent trade unions. The one legally sanctioned trade union, the All China Federation of Trade Unions (ACFTU) is state controlled and acts to maintain and manage employment relations, usually on behalf of employers, rather than representing workers. Migrant workers are at the forefront of such disputes. The relaxation of the country's strict internal migration controls in the late 1970s paved the way for a mass influx of rural migrants into China's booming cities in the export-led industrialisation zones of the south and east; over 200 million people have migrated in the period since 2008 (Chan and Selden 2016: 1). An upsurge of labour unrest struck the country's export-led industrialisation heartland of southern Guangdong in the Pearl River Delta in 2010 with a landmark strike at the Honda car plant in

Nanhai District when 1,800 workers demanded better wages and independent union representatives (ibid.). The strike brought Honda's entire south China supply chain to a standstill and led to workers receiving an 800 yuan per month pay rise and an agreement that they could directly elect some worker representatives to the enterprise trade union. However, disillusion has set in among the workforce with the outcome that the company union chair remained in post and the 'elected' members turned out to be higher-level management officials (ibid: 7).

The Chinese case represents an interesting example of the spatially entangled politics of labour. In theory, workers in China have quite significant national employment rights compared to many countries, including maximum 40-hour working weeks, social insurance and health care, maternity pay and severance pay. Yet, given that the implementation and monitoring of these is by local governments, many of which are understaffed, it is easy for local workers to be exploited by unscrupulous company managers in tandem with local bureaucrats. Many disputes – as well as being about wage levels – are often attempts to secure these basic rights (see Box 6.6). While workers are often successful at raising wages and agreeing concessions with employers, with the Chinese state keen to restore 'industrial peace', this has stopped short of granting much local autonomy or democratic agency to workers (Chan and Selden 2016).

The 'labour question' is not easily contained, however: the Hong Kong-based China Labour Bulletin notes a spread of grassroots collective action to the Chinese service sector with a growing number of strikes in retail units, including those owned by foreign corporations such as Walmart and McDonalds (see: www.clb.org.hk/ last accessed 25 October 2016). The labour question raises more fundamental issues for the Chinese Communist Party, as the dominant state actor, about its ability to sustain its growth model and retain social harmony. It also shows how different spatial scales interact in both the political economy of state regulation and trade union representation.

Box 6.6

The re-emergence of the 'labour problem' in China's export-led industrial complexes: the case of Foxconn

One of the most celebrated cases of industrial action in China has been at Foxconn's massive industrial complex in the city of Shenzhen, which employs approximately 400,000 people. Foxconn is a Taiwanese company that has become the leading contractor for manufacturing consumer electronics in the world, supplying most of the leading computer and mobile phone brands, including Apple, Samsung and Dell. Benefiting from China's entry into the WTO in 2001 and from considerable state subsidies, Foxconn has constructed a vast network of manufacturing bases in the country with over 30 plants and 1 million workers. It has become a major contributor to Chinese exports and the country's global competitive advantage.

Firms such as Apple put enormous pressure on Foxconn to lower costs and provide extreme flexibility in responding rapidly to its fast-changing product requirements. This in turn leads to massive pressure on workers and often a deterioration in employment standards which the Chinese state has, to date, turned a blind eye to. Consequently, Foxconn's workforce of predominantly young migrant workers face appalling working and living conditions; it was common for people to be working for 10 hours per day on just $1 per hour. Harsh disciplinary conditions in the workplace also include strict supervision and Taylorist-style speeding up of work tasks and denouncement of underperforming workers in front of colleagues (Chan *et al.* 2013).

A good illustration of the spatial connections between the exacting demands of Apple management in the US, the employment consequences and the human costs for the Foxconn workforce is provided by this quote from a local HR manager:

When Apple CEO Steve Jobs decided to revamp the screen to strengthen the glass on iPhone four weeks before it was scheduled to shelf in stores in June 2007, it required an assembly overhaul and production speedup in the Longhua facility in Shenzhen. Naturally, Apple's supplier code on worker safety and workplace standards and China's labour laws are all

Box 6.6 (continued)

put aside. In July 2009, this produced a suicide. When Sun Danyong, 25 years old, was held responsible for losing one of the iPhone 4 prototypes, he jumped from the 12th floor to his death.

(Quoted in Chan *et al.* 2013: 107)

Foxconn and Apple came under increasing pressure following 18 such suicide attempts and 14 deaths. Foxconn's first response to young workers throwing themselves off the upper floors of company buildings was to install safety nets to catch them (Moore 2012), which led to a further global outcry against both firms. Following an all-out strike by workers, Foxconn was forced to double the wages of staff.

Surveillance also extends outside the factory to the dormitories where many of the workers live. In

September 2012, just as Apple was about to launch its new iPhone 5, two workers were beaten up by security guards for not showing their ID cards. In response, tens of thousands of workers attacked dormitories, production facilities and even police vehicles as part of spontaneous protests (Chan *et al.* 2013). Production was shut down for a day and state officials, police and security officers violently suppressed the dispute with many arrests. The use of cell phones and social media by workers to publicise the dispute subverted attempts to contain the situation. Continuing pressure from the company to adhere to strict quality control and delivery guidelines brought a series of further factory disputes and work stoppages, which, despite further threats and intimidation from management and state officials, led to a partial victory in February 2013 when Foxconn

agreed to directly elected union representatives. As such, the workers at Foxconn are beginning to assert themselves as a collective force with interesting implications for workers in other parts of the global South. As one set of commentators puts it:

Apple and Foxconn now find themselves in a limelight that challenges their corporate images and symbolic capital, hence requiring at least lip service in support of progressive labour policy reforms. If the new generation of Chinese workers succeeds in building autonomous unions and worker organisations, their struggles will shape the future of labour and democracy not only in China but throughout the world.

(Ibid.: 112)

6.5.4 Labour's agency beyond the workplace

A growing number of geographers are emphasising the importance of considering labour's agency beyond the workplace (e.g. Rogaly 2009; Carswell and De Neve 2013). As we noted earlier, paid employment in the capitalist economy is only one element of the total work effort required in the production and reproduction of societies and economies. One important point relates to the way that low pay in the formal sector is effectively subsidised by households. Philip Kelly's work on the Cavite region of the Philippines, for example, shows how firms are able to both keep wages low and casualise work for resident employees who have family networks to cover the costs of social reproduction and periods out of work, thus helping to facilitate a more flexibilised labour regime (Kelly 2013). However, Kelly also points out that many households choose other

more fruitful strategies, notably migration for temporary work with over 20,000, predominantly males, working overseas in 2007.

An alternative perspective to the focus on the workplace- and enterprise-driven global dynamics is to explore work and economic activities from the perspectives of household strategies 'to get by'. How do individuals, families and extended family networks develop sustainable social reproduction strategies? From this vantage point, engaging in waged work is only one alternative among many others in harnessing resources and materials to sustain livelihoods. Cindi Katz's work is useful here in highlighting the "creative strategies that people [use] to stay afloat and reformulate the conditions and possibilities of their everyday lives" (2004: x). Her work in Harlem, New York, drew attention to the self-help networks through which women in more deprived communities offered reciprocal support with childcare and other basic tasks. Household strategies are not always progressive, but may include working in high-risk and potentially harmful activities such as prostitution and drug trafficking. Nevertheless, from the perspective of the individual, these can often be preferable to the kinds of work available in the formal economy. One study in a poorer community in Glasgow found that some teenagers could earn over twice as much working for a drug gang (£500 per week) than in a minimum-wage job paying the youth rate of £3.30 per hour (Cumbers *et al.* 2010).

6.5.5 Spaces of labour mobility

A focus on household or individual labour strategies beyond the workplace draws attention to the role of migrant labour in the global economy. A recent report by the World Bank estimated that one in seven people are either internal or international migrants (World Bank 2014). China's internal patterns of labour migration alone have accounted for over 200 million in the period since 1980 (Chan and Selden 2016). In major global cities such as London and New York, low-paid migrant work – often by 'illegal' workers – has become fundamental to the functioning of the city economy itself (Wills *et al.* 2010). Although international labour migration can be taken as evidence of worker agency in moving to find work, requiring considerable resourcefulness, ingenuity and courage, migrant workers continue to be among the most exploited groups in the global economy (Box 6.2) and there is a growing amount of work in labour geography focused on the topic (e.g. Rogaly 2009; Buckley *et al.* 2016).

Different kinds of labour migration are apparent, which reflect the nature of underlying power relations within the capitalist economy and the divisions that exist within the global workforce. On the one hand, we can identify a small group of highly skilled international migrants who form a transnational capitalist elite, working in the areas of global finance and management. This privileged group of 'workers' operate in a global space of flows (Castells 2000), located in and moving freely between the headquarters of the world's most powerful TNCs in global cities such as London, New York, Tokyo and Paris. A second group of skilled

professional workers (e.g. academics, scientists, consultants, lawyers, etc.) also enjoy a similarly privileged and mobile status in negotiating the global economy, equating to Standing's 'salariat' (Standing 2011).

Most migrants have a far less exalted status to these two groups, and are differentiated variously by the extent to which they are temporary or permanent, skilled or unskilled, and voluntary as opposed to 'forced' (see Box 6.2), and legal or illegal. Undocumented or illegal workers are routinely abused by unscrupulous employers, sometimes having their lives endangered as well as suffering appalling working conditions. One of the worst incidents in recent times was the 2004 drowning of 23 Chinese migrant workers employed as cockle pickers in the Morecambe Bay area of Lancashire in the UK (see http://news.bbc.co.uk/1/hi/england/lancashire/3827623.stm). But abuse of migrant workers is also apparent in legally sanctioned work. In preparations for the 2022 football World Cup in Qatar, the ITUC documented 1,200 deaths among the Nepali and Indian migrant workforce by the end of 2013. Appalling working conditions include long hours, severe heat and squalid dormitory conditions (ITUC 2014).

As Castree *et al.* (2004: 191) note, despite the rhetoric of globalisation and a borderless world, "national governments actively regulate the international migration of workers", filtering, in the same way that firms do, workers on the basis of their 'desirability'. Not only are migrants increasingly assessed in terms of education and skill levels, but racial and ethnic characteristics typically still inform immigration policy. Racist discourses and nationalistic stances towards immigration often lead to policies that are contradictory from the point of view of capital – for example, tight immigration controls operating during periods of low unemployment and economic boom, when there is a high demand for foreign labour. A hardening of attitudes towards migrants in many developed economies in recent years has led to growing discourses on tightening border policing and security. The difficulties of migrating legally are forcing many people – fleeing areas of poverty or political turmoil – to make increasingly dangerous and often fatal journeys in search of better lives, as attested to by the tragic deaths of many thousands of men, women and children who since 2010 have tried to cross the Mediterranean – often at the mercy of unscrupulous people-smuggling gangs and flimsy vessels – from the Middle East and North Africa to Western Europe (Collyer and King 2016).

6.6 Summary

The nature of employment has changed dramatically in the period since the 1970s through processes of globalisation, deindustrialisation, changing strategies of employers and the quest for greater labour flexibility. In this chapter, we have stressed the importance of understanding these changes in terms of social and spatial divisions of labour: your experience of employment is shaped profoundly by who you are and where you live. Global employment restructuring overall has precipitated a shift to greater flexibility and insecurity, and increased divisions in income and opportunity between different places and groups, typified by the spectre of the 'precariat'.

However, geography matters in understanding the spatial unevenness of these trends. Continuing variations in the political and institutional framework within which work is embedded between countries and regions continues to account for spatial variation in the adoption of new forms of work. Another key theme was the continuing agency of labour in shaping the economic landscape. Far from being a passive victim of broader processes of restructuring, workers, both through their agency in the workplace and through their household strategies to make a decent living, are a critical component in forging the global economy through their own networks and flows.

Exercise

Using the three different conceptualisations of class discussed in this chapter (Savage, Wright and Standing),

assess which has the most resonance in encapsulating the changes to work and employment in the early twenty-first century.

1. What are the strengths and weaknesses of each approach?
2. How applicable are they to different geographical contexts?
3. What does each have to say about labour's own spatial fix?
4. What might a more spatially sensitive class assessment look like?
5. How might future trends in global employment restructuring and technological change complicate such analyses?

Key reading

Chan, J., Pun, N. and Selden, M. (2013) The politics of global production: Apple, Foxconn and China's new working class. *New Technology, Work and Employment* 28 (2): 100–115.
Important article charting the conditions of labour in China and the emergence of an organised working class.

Cumbers, A., Featherstone, D., MacKinnon, D., Ince, A. and Strauss, K. (2016) Intervening in globalisation: the spatial possibilities and institutional barriers to labour's collective agency. *Journal of Economic Geography* 16 (1): 93–108.
A critical and sober assessment of the challenges faced by workers and trade unions in responding to processes of globalisation, corporate restructuring and state rescaling.

McDowell, L., Anitha, S. and Pearson, R. (2012) Striking similarities: representing South Asian women's industrial action in Britain. *Gender, Place and Culture* 19 (2): 133–52.
An insightful paper that develops a spatially relational analysis of labour agency sensitive to place, class, gender and ethnicity and how they complicate workers' collective struggles.

Peck, J. (2013) Making space for labour. In Featherstone, D. and Painter, J. (eds) *Spatial Politics: Essays for Doreen Massey*. Chichester: Wiley-Blackwell, pp. 99–114.
A good critical overview of the state of play of labour geography, assessing the strengths and limitations of the sub-field.

Standing, G. (2011) *The Precariat: The New Dangerous Class*. London: Bloomsbury.
A seminal thesis on the conditions of flexibility and security in the world of work under globalisation in the twenty-first century.

Useful websites

www.ilo.org
International Labour Organisation's website: the UN body that carries out research into global labour issues. Produces data and research papers on a range of labour issues from trade unions membership to trends in labour flexibility.

www.ituc-csi.org/
International Trade Union Confederation. The main website for the umbrella body that represents the international trade union movement.

www.aflcio.org
The website for the main US trade union federation.

www.tuc.org.uk
British trade union confederation site.

www.bls.gov/
US Bureau of Labor Statistics

www.nomisweb.co.uk
UK Official Labour Market Statistics

www.clb.org.hk/
Hong Kong Labour Bulletin – a website with useful information about labour organising in China.

References

Avent, R. (2016) Welcome to a world without work. *The Observer*, 9 October, pp. 36–8.

Bell, D. (1973) *The Coming of Post-industrial Society*. New York: Basic Books.

Bergene, A.-C., Endresen, S.B. and Knutsen, H.M. (eds) (2010) *Missing Links in Labour Geography*. Aldershot: Ashgate.

Birch, K., MacKinnon, D. and Cumbers, A. (2010) Old industrial regions in Europe: a comparative assessment of economic performance. *Regional Studies* 44 (1): 35–53.

Bourdieu, P. (1984) *Distinction*, London: Routledge.

Buckley, M., McPhee, S. and Rogaly, B. (2016) Labour geographies on the move: migration, migrant status, and work in the 21st century. *Geoforum*. Available online at: http://dx.doi.org/10.1016/j.geoforum.2016.09.012.

Burawoy, M. (1983) Between the labor process and the state: the changing face of factory regimes under advanced capitalism. *American Sociological Review* 48: 587–605.

Carswell, G. and De Neve, G. (2013) Labouring for global markets: conceptualising labour agency in global production networks. *Geoforum* 44: 62–70.

Castells, M. (2000) *The Information Age, Volume 3: End of Millennium*. Oxford: Blackwell.

Castree, N., Coe, N., Ward, K. and Samers, M. (2004) *Spaces of Work: Global Capitalism and Geographies of Labour*. London: Sage.

Chan, J. and Selden, M. (2016) The labour politics of China's rural migrant workers. *Globalizations*. At http://dx.doi.org/10.1080/14747731.2016.1200263.

Chan, J., Pun, N. and Selden, M. (2013) The politics of global production: Apple, Foxconn and China's new working class. *New Technology, Work and Employment* 28 (2): 100–115.

China Labor Watch (2012) *Samsung's Supplier Factory Exploiting Child Labor*. China Labor Watch Report 63. Available at: www.chinalaborwatch.org/report/63.

Coe, N. and Jordhus-Lier, D. (2010) Constrained agency? Re-evaluating the geographies of labour. *Progress in Human Geography* 35: 211–33.

Collyer, M. and King, R. (2016) Narrating Europe's migration and refugee crisis. *Human Geography* 2: 1–12.

Cooper, D. (2015) Raising the federal minimum wage to $12 by 2020 would lift wages for 35 million American workers. *Economic Policy Institute Briefing Paper #405*, Washington, DC: Economic Policy Institute.

Cumbers, A., Featherstone, D., MacKinnon, D., Ince, A. and Strauss, K. (2016) Intervening in globalisation: the spatial possibilities and institutional barriers to labour's collective agency. *Journal of Economic Geography* 16 (1): 93–108.

Cumbers, A., Helms, G. and Swanson, K. (2010) Class, agency and resistance in the old industrial city. *Antipode* 42 (1): pp. 46–73.

Cumbers, A., Nativel, C. and Routledge, P. (2008) Labour agency and union positionalities in global production networks. *Journal of Economic Geography* 8 (2): 369–87.

De Backer, K., Desnoyers-James, I. and Moussiegt, L. (2015) 'Manufacturing or services – that is (not) the question': the role of manufacturing and services in OECD economies. *OECD Science, Technology and Industry Policy Papers* No. 19. Paris: OECD Publishing.

Derickson, K. (2016) The racial state and resistance in Ferguson and beyond. *Urban Studies* 53: 2223–37.

Dicken, P. (2015) *Global Shift: Mapping the Changing Contours of the World Economy*, 7th edition. London: Sage.

Farrell, S. and Butler, S. (2016) Sports Direct ditches zero-hours jobs and ups worker representation. *The Guardian*, 6 September. At www.theguardian.com/business/2016/sep/06/sports-direct-to-ditch-zero-hours-contracts.

Farrell, S. and Press Association (2016) UK workers on zero-hours contracts rise above 800,000. *The Guardian*, 9 March. Available online at: www.theguardian.com.

Freeman, R. (2010) What really ails Europe (and America): the doubling of the global workforce. *The Globalist*, 5 March. Available online at: www.theglobalist.com/what-really-ails-europe-and-america-the-doubling-of-the-global-workforce/, last accessed October 2016.

Galster, G. (2012) *Driving Detroit*. Philadelphia, PA: University of Pennsylvania Press.

Hadwiger, F. (2015) *Global Frame Agreements: Achieving Decent Work in Global Supply Chains*. Background Paper. Geneva: International Labour Office.

Hanson, S. and Pratt, G. (1995) *Gender, Work and Space*. Chichester: Wiley.

Henwood, D. (1998) Talking about work. In Meiskins, E., Wood, P. and Yates, M. (eds) *Rising from the Ashes? Labor in the Age of Global Capitalism*. New York: Monthly Review Press.

Herod, A. (2001) *Labor Geographies: Workers and the Landscapes of Capitalism*. New York: Guilford.

Holloway, J. (2005) *Changing the World without Taking Power: The Meaning of Revolution Today*, 2nd edition. London: Pluto.

ILO (2005) *A Global Alliance against Forced Labour*. Geneva: International Labour Organisation.

ILO (2013) *Marking Progress against Child Labour: Global Estimates and Trends 2000–2012*. Geneva: International Labour Office.

ILO (2015) *World Employment and Social Outlook*. Geneva: International Labour Office.

ITUC (2014) *The Case against Qatar: Host of the 2022 World Cup*. Brussels: International Trade Union Confederation.

Jonas, A. (1996) Local labour control regimes: uneven development and the social regulation of production. *Regional Studies* 30: 323–38.

Jordhus-Lier, D. (2013) The geographies of community-oriented unionism: scales, targets, sites and domains of union renewal in South Africa and beyond. *Transactions of the Institute of British Geographers* 38: 36–49.

Jordhus-Lier, D. and Underthun, A. (2014) *A Hospitable World? Organising Work and Workers in Hotels and Leisure Resorts*. London: Routledge.

Kalleberg, A. (2013) *Good Jobs, Bad Jobs: The Rise of Polarized and Precarious Employment Systems in the United States, 1970s to 2000s*. New York: Russell Sage Foundation.

Katz, C. (2004) *Growing up Global: Economic Restructuring and Children's Everyday Lives*. Minneapolis: University of Minnesota Press.

Kelly, P. (2001) The local political economy of labour control in the Philippines. *Economic Geography* 77 (1): 1–22.

Kelly, P. (2013) Production networks, place and development: thinking through Global Production Networks in Cavite, Philippines. *Geoforum* 44: 83–92.

Luce, S. (2016) And a Union. *Jacobin*, November. Available at: www.jacobinmag.com, last accessed November 2016.

Mason, P. (2012) *Why It's Kicking Off Everywhere: The New Global Revolutions*. London: Verso.

Massey, D. (1984) *Spatial Divisions of Labour: Social Structures and the Geography of Production*. London: Macmillan.

McBride, S. and Muirhead, J. (2016) Challenging the low wage economy: living and other wages. *Alternate Routes: A Journal of Critical Social Research* 27: 55–86.

McDonald, J.F. (2014) What happened to and in Detroit? *Urban Studies* 51 (16): 3309–29.

McDowell, L. (1997) *Capital Culture: Gender at Work in the City*. Oxford: Blackwell.

McDowell, L., Anitha, S. and Pearson, R. (2012) Striking similarities: representing South Asian women's industrial action in Britain. *Gender, Place and Culture* 19 (2): 133–52.

McGrath, S. (2013) Fuelling global production networks with slave labour: migrant sugar cane workers in the Brazilian ethanol GPN. *Geoforum* 44: 32–43.

Moore, M. (2012) Mass suicide protest at Apple manufacturer Foxconn factory. *The Daily Telegraph*, 11 January. At www.telegraph.co.uk/news/worldnews/asia/china/9006988/Mass-suicide-protest-at-Apple-manufacturer-Foxconn-factory.html.

Moreira, A., Dominguez, A.A., Antunes, C., Karamessini, M., Raitano, M. and Glatzer, M. (2015) Austerity driven labour market reforms in Southern Europe: eroding the security of labour market insiders. *European Journal of Social Security* 17 (2): 202–25.

Munck, R. (1999) Labour dilemmas and labour futures. In Munck, R. and Waterman, P. (eds) *Labour Worldwide in the Era of Globalization*. Basingstoke: Macmillan.

OECD (1994) *Jobs Study: Evidence and Explanations*. Paris: Organization for Economic Co-operation and Development.

OECD (2015) *In It Together: Why Less Inequality Benefits All*. Paris: Organization for Economic Co-operation and Development.

Pahl, R. (1988) Historical aspects of work, employment, unemployment and the sexual division of labour. In Paul, R. (ed.) *On Work: Historical, Comparative and Theoretical Approaches*. Oxford: Blackwell, pp. 1–7.

Patten, E. (2016) Racial, gender wage gaps persist in U.S. despite some progress. *Pew Research Center Briefing note*. Available at: www.pewresearch.org/fact-tank/2016/07/01/racial-gender-wage-gaps-persist-in-u-s-despite-some-progress/. Last accessed September 2016.

Peck, J. (1996) *Work-Place: The Social Regulation of Labour Markets*. London: Guilford.

Pratt, G. (2004) *Working Feminism*. Philadelphia: Temple University Press.

Rodgers, J. (2007) Labour market flexibility and decent work. *DESA Working Paper* No. 47. New York: UN DESA.

Rogaly, B. (2009) Spaces of work and everyday life: labour geographies and the agency of unorganised temporary migrant workers. *Geography Compass* 3 (6): 1975–87.

Savage, M., Devine, F., Cunningham, N., Taylor, M., Li, Y., Hjellbrekke, J., Le Roux, B., Friedman, S. and Miles, A. (2013) A new model of social class? Findings from the BBC's Great British Class Survey experiment. *Sociology* 47 (2): 219–50.

Skeggs, B. (1997) *Formations of Class and Gender*. London: Sage.

Standing, G. (2009) *Work after Globalisation: Building Occupational Citizenship*. Cheltenham: Edward Elgar.

Standing, G. (2011) *The Precariat: The New Dangerous Class*. London: Bloomsbury.

Thompson, P. (2004) *Skating on Thin Ice: The Knowledge Economy Myth*. Glasgow: Big Thinking.

Tufts, S. (2007) Emergent labour strategies in Toronto's hotel sector: towards a spatial circuit of union renewal. *Environment and Planning A* 39: 2383–404.

Waterman, P. (2014) The international labour movement in, against and beyond, the globalized and informatized cage of capitalism and bureaucracy. *Interface* 6 (2): 35–58.

Wills, J. (2003) Community unionism and trade union renewal in the UK: moving beyond the fragments at last. *Transactions of the Institute of British Geographers* 26: 465–83.

Wills, J. (2005) The geography of union organising in low paid service industries in the UK: lessons from the Dorchester Hotel, London. *Antipode* 37: 139–59.

Wills, J., Datta, K., Evans, J. and Herbert, J. (2010) *Global Cities at Work: New Migrant Divisions of Labour*. London: Pluto.

World Bank (2014) Migration and remittances: development and outlook. *Migration and Development Brief 22*. Washington, DC: World Bank.

World Bank (2016) *World Development Indicators*. Washington, DC: World Bank.

Wright, E.O. (2015) *Understanding Class*. London: Verso.

Wright, M. (2006) *Disposable Women and Other Myths of Global Capitalism*. London: Routledge.

Yang, Y. (2016) China to shed 1.8 million coal and steel jobs. *Financial Times*, 29 February. At www.ft.com/content/3a8dd2e0-deb4-11e5-b072-006d8d362ba3.

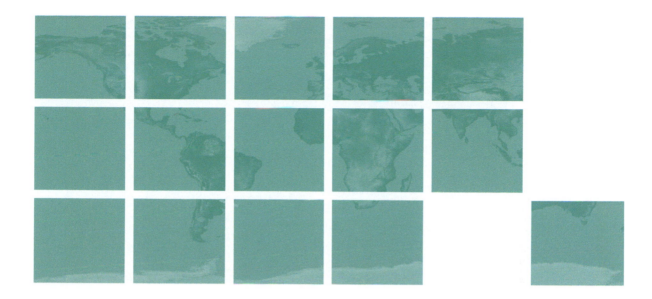

Chapter 7
Geographies of development

7.1 Introduction

Development is defined by the *Concise Oxford English Dictionary* as a "Gradual unfolding, fuller working out; growth; evolution . . .; a well-grown state, stage of advancement; product; more elaborate form . . ." (quoted in Potter *et al.* 2008: 4). In the context of economic policy, the term conveys a sense of positive change over time, applied to a particular country or region. Such change involves growth or progress as countries become more prosperous and advanced. As an economic and social policy, **development** has been directed at those 'underdeveloped areas' of the world that require economic growth and modernisation. The 'underdeveloped world', which we are concerned with in this chapter, is traditionally comprised of Africa, Latin America and Asia, representing the main focus of international development policy since the 1950s. The division between this periphery and the core of developed countries in Europe, North America, Japan and

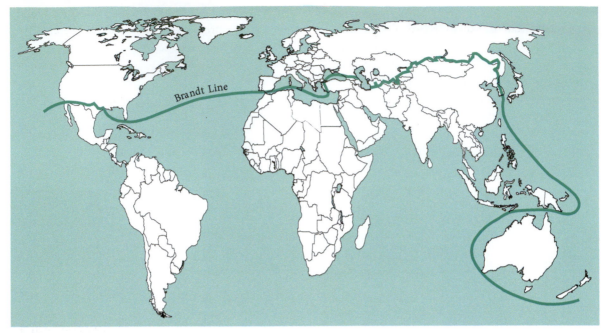

Figure 7.1 The global North and South
Source: Adapted from Knox *et al.* 2003: 24.

Australasia can be expressed as a geographical divide between the global North and South (Figure 7.1).

The problems of developing countries in the global South have attracted much attention in recent years with a number of campaign groups and social movements raising concerns about global inequalities. At the same time, and partly in response, development organisations and political leaders have focused on issues of poverty reduction and climate change, setting high-profile targets such as the Sustainable Development Goals. The promotion of development as the solution to the problem of large-scale poverty in Africa, Latin America and Asia is not new, however. During the **Cold War** (late 1940s to late 1980s), western capitalist countries, led by the US, established large-scale development programmes in the newly liberated colonies to 'save' them from the 'evils' of communism, stressing the need for economic progress and modernisation. While post-war development was driven by Cold War geopolitics, the recent interest in global poverty and development is shaped by opposing perspectives on **globalisation** (section 1.2.1). For pro-globalists, globalisation is a long-term process which will eventually lift all people out of poverty so long as their governments pursue market-oriented policies. For 'counter-globalisation' activists, however, neoliberal globalisation is the problem, not the solution, and the global economic system requires major reform if not outright revolution.

7.2 The project of development

The countries of Africa, Latin America and Asia were designated as the 'underdeveloped world' in the 1950s in a context of Cold War geopolitics and decolonisation. As part of its efforts to break the old imperial trading blocs and create a world open for international trade and investment, the US sought to encourage economic growth and modernisation in the newly liberated colonies. In his inaugural address in 1949, President Truman emphasised the need to tackle poverty through the promotion of development:

> We must embark on a bold new program for making the benefits of our scientific advances and industrial progress available for the improvement and growth of underdeveloped areas . . . For

the first time in history, humanity possesses the knowledge and the skill to relieve the suffering of these people . . . I believe that we should make available to peace-loving peoples the benefits of our store of technical knowledge in order to help them realize their aspirations for a better life . . . What we envisage is a program of development based on the concepts of democratic fair-dealing . . . Greater production is the key to prosperity and peace. And the key to greater production is a wider and more vigorous application of modern scientific and technical knowledge. Only by helping the least fortunate of its members to help themselves can the human family achieve the decent, satisfying life that is the right of all people.

(www.bartleby.com/124/pres53.html)

This 'program of development' was not only an inherent good, overcoming the 'ancient enemies' of hunger, misery, and despair; it would also provide a necessary bulwark against communism, viewed as a disease to which those living in poverty were particularly prone. As such, the former colonies in Africa, Latin America and Asia became key sites of the Cold War struggle as the US and Soviet blocs competed for influence and power.

The notion of the 'Third World' was invented as a label for the 'underdeveloped areas', in contrast to the 'First World' of western democracies and the 'Second World' of communist states in the USSR and Eastern Europe. The term was applied to vast areas of the globe, reflecting a kind of negative labelling of them as backward and in need of assistance from outside. It is a view that has been perpetuated over the decades, supported by powerful images of poverty and underdevelopment. As the development geographer Morag Bell remarks, however, this dramatic image masks a less clear-cut reality:

What is the geography of the Third World? Certain common features come to mind: poverty, famine, environmental disaster and degradation, political instabilities, regional inequalities and so on. A powerful and negative image is created that has coherence, resolution and definition. But behind this tragic stereotype there is an alternative geography, one which demonstrates that the introduction of

development into the countries of the Third World has been a protracted, painstaking and fiercely contested process.

(Bell 1994: 175)

The presentation of the 'Third World' as a monolithic bloc is highly simplistic and misleading, ignoring a complex and diverse geography. Development is never straightforward, involving a range of policies and programmes devised by leading development organisations and received or 'consumed' by the nations, communities and households which they are aiming to assist. It has resulted in different outcomes in different places, shaped by a range of locally specific factors such as social attitudes, environmental conditions, labour skills and farming practices. It is important to recognise that economic development policy is not confined to the so-called 'Third World' as underdeveloped areas in developed countries have also been the subject of state-sponsored programmes and initiatives (section 5.6).

Following Truman's crucial speech and the hardening of Cold War divisions, a development 'industry' emerged in the 1950s and 1960s, funded by the US and other western capitalist countries. This was defined by a shared belief in economic planning, modern technology and outside investment as the key drivers of change. Academic disciplines, particularly economics, played a significant role in providing expert knowledge about the process of development and the conditions shaping it. International organisations such as the World Bank, International Monetary Fund (IMF) and United Nations (UN), established as part of the new post-war order, were charged with the responsibility of promoting development in poor countries. Driven by an ideology of modernisation and development, such organisations sought to introduce modern knowledge and investment in order to overcome the 'ancient enemies' which held back progress in the underdeveloped lands. The UN, for instance, designated the 1960s as the 'decade of development'. Other key players include the governments of developing countries and **non-government organisations** (**NGOs**) working in the development field. NGOs are organisations, often of a voluntary or charitable nature, which make up the so-called 'third sector' belonging to neither the private nor public sectors. Oxfam is a good example of an NGO focused on development (Box 7.1). Others include Christian Aid, Save the Children and the Red Cross.

Economic aspects of development were heavily stressed in the 1960s and 1970s as policy focused on the need to foster investment and growth at the national level,

Box 7.1

Oxfam

Oxfam has grown to become one of the most important NGOs operating in the field of development. It is an independent British organisation, registered as a charity, working with partners, volunteers, supporters and staff of many nationalities. It had a total income of £401.4 million in 2014–15 and employs 3,198 staff working overseas and 2,292 staff in Great Britain as well as over 22,000 volunteers working in almost 700 shops (Oxfam 2015: 30, 52).

Oxfam's origins lie in the Second World War when the Oxford Committee for Famine Relief was set up in response to famine in Greece in 1942. The committee remained in existence after the war, focusing on 'the relief of suffering in consequence of the war' in Europe in the late 1940s. After 1949, the scope of its operations expanded to encompass the world. In the 1960s, Oxfam's income trebled, reflecting growing concern for the world's poor. Support for self-help schemes whereby communities improved their own farming practices, water supplies and health

provision became the major focus of activity. Community involvement and control remain key principles of Oxfam's work.

As well as famine relief and self-help schemes, Oxfam campaigns to raise awareness of global poverty and its structural causes in terms of the debt burden, unfair **terms of trade** and inappropriate agricultural policies. Its current activities focus on three main areas: emergency response to disasters; development work which addresses poverty; and campaigning for changes in policy.

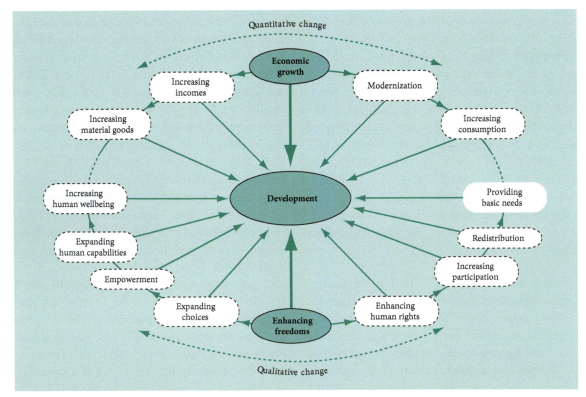

Figure 7.2 Changing conceptions of development
Source: Potter *et al.* 2004: 16, Figure 1.2.

assuming that this would generate increased incomes and employment opportunities for individuals. Development indicators concentrated on the material well-being of a country or region, with Gross Domestic Product (GDP) and Gross National Income (GNI) by far the most commonly used measures of development. This corresponds to the quantitative change section of Figure 7.2.

Since the late 1970s, however, the notion of development has broadened to include more qualitative aspects of change, encompassing broader social and political goals such as quality of life, choice, empowerment and human rights (Potter *et al.* 2008: 16–17). The work of the Nobel Prize-winning, Indian-born economist, Amartya Sen, has been particularly important in advancing the idea of 'development as freedom'. Through factors such as better education, increased political participation and free speech, working alongside the process of economic growth, people are liberated from 'unfreedoms' such as starvation, undernourishment, oppression, disease and illiteracy. This broader conception of development is reflected in the

Box 7.2

Measuring development

As indicated above, the main economic measures of development are GDP and GNI, usually expressed on a per capita basis. GDP is a measure of the total value of goods and services produced within a country whilst GNI is the sum of the incomes of residents of a country. Unlike GDP it includes income generated from investments abroad and excludes profits repatriated by foreign TNCs to their home countries. GDP and GNI were the key measures employed by international development agencies in the post-war era, but began to attract criticism from development activists and analysts in the late 1960s and 1970s for neglecting social aspects of development. They remain important, however, providing a useful summary measure of development, emphasising the divide between the global North and South, and growing

divergence between the regions of the South (section 7.4).

A wide range of social indicators of development were published in the 1970s and 1980s, focusing on issues such as poverty, education, health and gender. The proliferation of these social indicators threatened to generate considerable confusion, however, as different measures could be used to show different things, and it was almost always possible to find some statistics to 'prove' a particular argument (Potter *et al.* 2008: 9). What seemed to be required was some kind of summary measure constructed out of key economic and social indicators.

The **Human Development Index (HDI)** – published annually since 1990 by the United Nations Development Programme (UNDP) in the *Human Development Report* – has met this need, becoming widely used and adopted.

The HDI measures the overall achievement of a country in three basic dimensions of human development – health, education and standard of living. The specific measures used are life expectancy, mean and expected years of schooling and Gross National Income (GNI) per capita in US dollars. In recent years, the HDI has been supplemented by the introduction of other related indexes, including the Inequality-Adjusted HDI, the Gender Development Index and the Gender Inequality Index and the Multidimensional Poverty Index. Countries with HDI scores above 0.800 are classified as very high human development; ones with scores between 0.700 and 0.799 as high human development; ones between 0.550 and 0.699 as medium human development; and ones with scores below 0.550 as low human development (Table 7.1).

Table 7.1 The UNDP Human Development Index

HDI rank	HDI value 2014	Life expectancy at birth (years) 2014	Expected years of schooling 2014	Mean years of schooling 2014	GNI per capita (PPP $)
VERY HIGH HUMAN DEVELOPMENT					
1 Norway	0.944	81.6	17.5	12.6	64,992
25 Slovenia	0.880	80.4	16.8	11.9	27,852
48 Kuwait	0.816	74.4	14.7	7.2	83,961

Box 7.1 (continued)

HDI rank	HDI value 2014	Life expectancy at birth (years) 2014	Expected years of schooling 2014	Mean years of schooling 2014	GNI per capita (PPP $)
HIGH HUMAN DEVELOPMENT					
50 Russian Federation	0.798	70.1	14.7	12.0	22,352
77 Saint Kitts & Nevis	0.752	73.8	12.9	8.4	20,805
105 Samoa	0.702	73.4	12/9	10.3	5,327
MEDIUM HUMAN DEVELOPMENT					
106 Botswana	0.698	64.5	12.5	8.9	16,646
124 Guyana	0.636	66.4	10.3	8.5	6,522
143 Cambodia	0.555	68.4	10.9	4.4	2,949
LOW HUMAN DEVELOPMENT					
145 Kenya	0.548	61.6	11.0	6.3	2,762
167 Sudan	0.479	63.5	7.0	3.1	3,809
188 Niger	0.348	61.4	5.4	1.5	908

Source: UNDP 2015: 208–10.

Table 7.2 Progress in achieving the Millennium Development Goals

Goal	Progress
Eradicate extreme poverty and hunger	The extreme poverty rate in developing countries fell from 47 per cent in 1990 to 14 per cent in 2015, while the proportion of undernourished people in developing regions fell by almost half since 1990
Achieve universal primary education	The primary school net enrolment rate rose from 83 per cent in 2000 to 91 per cent in 2015
Promote gender equality and empower women	Not fully achieved, although the majority of developing regions did achieve parity in primary education
Reduce child mortality by two-thirds	The global under-five mortality rate declined by more than half from 1990 to 2015
Reduce maternal mortality by three-quarters	Since 1990 the maternal mortality rate declined by 45 per cent worldwide
Combat HIV/AIDS, malaria and other diseases	The spread of HIV/AIDS has been reversed, while the global malaria incidence rate fell by an estimated 37 per cent 2000–2015
Ensure environmental sustainability	The target of halving the number of people without access to safe water sources was met in 2010, but greenhouse gas emissions continue to rise, although deforestation has slowed somewhat
Develop a global partnership for development	Official development assistance from developed countries increased significantly in the early 2000s but has plateaued in recent years

Source: United Nations 2015a.

Box 7.3

The Sustainable Development Goals

Goal 1. End poverty in all its forms everywhere

Goal 2. End hunger, achieve food security and improved nutrition and promote sustainable agriculture

Goal 3. Ensure healthy lives and promote well-being for all at all ages

Goal 4. Ensure inclusive and equitable quality education and promote lifelong learning opportunities for all

Goal 5. Achieve gender equality and empower all women and girls

Goal 6. Ensure availability and sustainable management of water and sanitation for all

Goal 7. Ensure access to affordable, reliable, sustainable and modern energy for all

Goal 8. Promote sustained, inclusive and sustainable economic growth, full and productive employment and decent work for all

Goal 9. Build resilient infrastructure, promote inclusive and sustainable industrialization and foster innovation

Goal 10. Reduce inequality within and among countries

Goal 11. Make cities and human settlements inclusive, safe, resilient and sustainable

Goal 12. Ensure sustainable consumption and production patterns

Goal 13. Take urgent action to combat climate change and its impacts

Goal 14. Conserve and sustainably use the oceans, seas and marine resources for sustainable development

Goal 15. Protect, restore and promote sustainable use of terrestrial ecosystems, sustainably manage forests, combat desertification, and halt and reverse land degradation and halt biodiversity loss

Goal 16. Promote peaceful and inclusive societies for sustainable development, provide access to justice for all and build effective, accountable and inclusive institutions at all levels

Goal 17. Strengthen the means of implementation and revitalize the Global Partnership for Sustainable Development.

(UN 2015b: 14)

evolution of development indicators with more emphasis now focused on assessing social and political aspects of development alongside the economic dimension (Box 7.2).

Since the establishment of the **Millennium Development Goals** (**MDGs**), progress has been assessed against a set of specific objectives and targets. Adopted by the UN in 2000, the MDGs enabled outcomes to be monitored on an annual basis. The MDGs were only partly achieved by 2015. Although significant progress was made in most areas, this generally tended to fall short of the specific MDG targets (Table 7.2). Progress in eradicating extreme poverty has proved particularly controversial, underlining the contested nature of development and the complexities of measuring progress. Critics argue that the reduced poverty rate reflects population growth in developing countries (increasing the denominator for calculating the proportion in poverty), the movement of the starting point back to 1990 (which allowed the MDGs to take credit for rapid growth and poverty reduction in China during the 1990s) and the raising of the extreme poverty line from $1 per day to $1.25 per day (which increased the absolute number but makes the trend from 1990 look better) (Hickel 2016).

In 2015 the UN agreed a new set of goals to replace the MDGs, the **Sustainable Development Goals** (**SDGs**) (Box 7.3). These 17 goals and 169 associated targets for 2030 encompass a far broader agenda than the MDGs, which were strongly focused around poverty reduction (Fukuda-Parr 2016). Instead, the SDGs reflect a global agenda for sustainable development, covering all countries not just poor ones. They incorporate agendas such as inequality, **governance** and human rights and economic dimensions of inclusion and sustainability alongside poverty reduction. Their achievement is likely to prove highly challenging, however, in the face of pressures to simplify and adapt them to national agendas.

Reflect

To what extent are broader understandings of development reflected in the establishment of development goals such as the Millennium Development Goals and Sustainable Development Goals?

7.3 Theories of development

The process of development has changed considerably over time, guided by influential theories which have served to frame and direct development practice. Ideas about development policy and practice are often highly contested, reflecting the influence of political ideologies and moral and ethical judgments over what development should consist of (Desai and Potter 2014: 79). Development is managed and promoted by a range of actors, including international agencies, national governments and various NGOs, who often become committed to particular approaches. Concentrating on the post-Second World War period, the key theories are summarised in Table 7.3, outlining their key arguments and recommendations and the problems that they encountered, and further elaborated below.

7.3.1 Modernisation theory

This approach was dominant in the 1950s and 1960s, shaping and informing the efforts of development planners and agencies (Table 7.3). It emerged from the writings of theorists and experts based in the global North, notably the US, which was playing a key role in funding and supporting the project of development as the leading economic power in the western capitalist bloc. The work of the US economist Walt Rostow was particularly

Table 7.3 Theories of development

Theory	Key theorists	Time frame	Main argument	Recommendations	Problems
Modernization theory	Rostow	1950s and 1960s	Development occurs in distinct stages; developing countries undergo linear process of modernization, akin to developed countries in the nineteenth century	Need for external funds and expertise, along with modern planning and investment methods, to generate growth	Eurocentric; ignores structure of world economy; growth fails to alleviate poverty
Structuralism and dependency theory	Prebisch, Frank	1960s and 1970s	Metropolitan core of world economy exploits 'satellites'. Development and underdevelopment are opposite sides of the same coin	Import substitution or withdrawal from world economy	Crude and static view of global relationships. Undermined by experience of E. Asian NICs
Neoliberalism	Lal and Balassa in development circles	1980s and 1990s	Developing countries should reduce state intervention in the economy and embrace the free market	Economic reform through SAPs, free trade, promotion of exports	Focus on growth failed to alleviate poverty; SAPs and privatization led to cuts in public services, and the introduction of user charges
Grassroots approach	Emerged from activities of development NGOs and activists	1970s–	Development agencies should focus on meeting the everyday needs of the poor	Local participation and self-help should be encouraged	Limited funds. Tends to alleviate the symptoms of poverty rather than addressing the causes

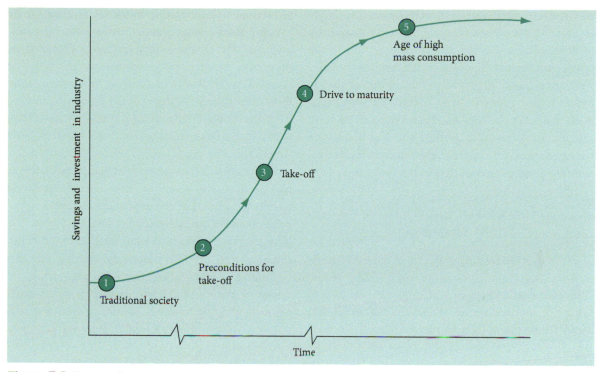

Figure 7.3 Rostow's stages of economic development
Source: Potter *et al.* 2004: 91.

influential, setting out a model of the stages of economic growth that was applied to the situation of developing countries in the 1960s (Figure 7.3). The process of 'take off' is triggered by a combination of external and internal factors such as technological innovation, increased rates of investment and saving, the development of modern banks and manufacturing industry. The metaphor invoked here is that of an aeroplane accelerating along a runway before gathering power to become airborne (Power 2016).

The basic idea is that developing countries undergo a linear process of transformation (modernisation), analogous to the changes experienced by developed countries in the nineteenth century following the industrial revolution. The assumption was that developing countries would simply follow this existing model, resulting in the consolidation of western norms of economic and social organisation – for example, a large manufacturing sector, commercialised agriculture, the importance of class groups rather than family or tribal structures and governments based on democratic election rather than tribal or religious loyalties (Willis 2014a: 299). The process of modernisation

would be driven by the injection of external funds and expertise, coupled with national government intervention and planning to mobilise resources and stimulate investment. Economic growth was paramount, generating increased income and employment opportunities which, it was assumed, would 'trickle down' to the poorest groups in society.

From this perspective, the problems of underdevelopment are internal to the countries concerned, reflecting the attachment to traditional values and the absence of modern technology and scientific knowledge. As such, the solution was to import these factors from outside, giving western experts and agencies a key role in assisting development. When linked to initiatives and reforms carried out by developing country governments, this would generate the momentum required for 'take off'. The Eurocentric assumption that western values and methods are always superior to those found in developing countries has been criticised as arrogant and condescending.

The focus on internal factors means that the structure of the world economy, particularly in terms of the relationships between richer and poorer countries, was neglected by modernisation theorists. It is simply assumed that it is just as easy for developing countries to develop as it was for developed countries in the nineteenth century. This ignores the fact that the latter were not required to compete with a group of already industrialised countries who dominated world production of manufactured goods and services. As such, the policies followed by Europe and North America were not necessarily appropriate to the problems confronting developing countries in the 1950s and 1960s.

7.3.2 Structuralism and dependency theory

The idea that the existing structure of the world economy was impeding the development of 'Third World' countries was a key point of departure for radical critics of **modernisation theory** in the 1950s and 1960s. These critics focused on the political economy

of economic development. In particular, they sought to apply Marx's insights about the historical and geographical unevenness of capitalist development to the experience of developing countries in the post-war period (Power 2016). In contrast to modernisation theory, structuralist or **dependency theories** emanated from the global South, being particularly associated with a group of theorists and activists based in Latin America.

This approach can be termed structuralist for its characteristic focus on the structure of the world economy, particularly the relationship between developed countries and developing countries (Table 7.3). It focused on the mode of incorporation of individual countries into the world economy, viewing this as a key source of exploitation. From this perspective, the causes of poverty and underdevelopment are external to developing countries, stemming from the relationship between them and the wider world economy.

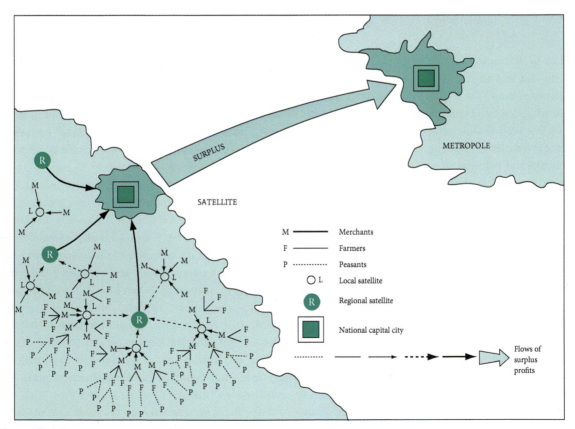

Figure 7.4 Dependency theory
Source: Potter *et al.* 2004: 111.

According to Andre Gunder Frank, the most influential of the dependency theorists in the 1960s, the metropolitan core exploits its 'satellites', extracting profits (surplus) for investment elsewhere (Figure 7.4). **Colonialism** was a key force here, creating unequal economic relations which were then perpetuated by the more informal imperialism characteristic of the post-Second World War period.

Many developing countries have found it difficult to overcome the legacy of colonialism, confining them to the export of primary commodities (agricultural goods, minerals, fuels), while the European powers export manufactured goods. Over time, the price of the former have tended to fall relative to the latter, reducing the export earnings of developing countries and making them less able to pay for imports. At the same time, **TNCs** based in Europe and North America have been able to repatriate the profits from their plantations and factories in developing countries to their home countries, paying low wages and having very few links with the host economy in which they are operating.

The implications of dependency theory are that developing countries should protect themselves against external forces or even withdraw from the world economy altogether. Development and underdevelopment are opposite sides of the same coin: closer links between core and periphery merely widen the economic gap between them. The milder form of structuralism associated with Raul Prebisch and the UN Commission for Latin America in the 1950s advocated that countries should follow protectionist policies rather than simply embracing external assistance (Willis 2014a: 302–3). In particular, they should focus on **import-substitution industrialisation** (ISI) by developing domestic industries to produce goods that are currently imported, using import tariffs to protect them against competition from the established industries of Europe and North America (section 5.3.3). For the radical version of dependency developed by Frank, however, the solution was withdrawal from the global economy and the creation of alternative forms of society based on socialism.

Dependency theory has been heavily criticised, relying, again, on a simple divide between 'core' and 'satellite' economies. Their view of global relationships is static and crude, assuming that the patterns established under colonialism will inevitably persist. This assumption was undermined by the experience of rapid economic growth in the **NICs** of East Asia such as South Korea, Taiwan and Singapore in the late 1970s and 1980s (Box 5.3). The success of these countries was based on policies of **export-oriented industrialisation** (EOI) where industrial development is based on serving international rather than domestic markets. As such it seemed to undermine some of the key foundations of dependency theory, indicating that countries could overcome the legacy of colonialism. By contrast, the notion of withdrawal from global markets seemed impractical, promising only further economic stagnation and decay.

7.3.3 Neoliberalism

Since the late 1980s, development policy has been shaped by neoliberal (free market) theories (sections 1.2.1, 5.5). These emphasise the need to reduce government intervention in the economy, encouraging the growth of private enterprise and competition. The neoliberal principles developed by key theorists such as Friedman and Hayek in North America and Western Europe were translated into development thinking in the early 1980s, following the counter-revolution in development economics when neoliberal ideas overturned the Keynesian orthodoxies of the 1960s and 1970s. Key figures included Depak Lal and Bela Balassa, both based at leading US universities, who argued in favour of free trade and the application of standard economic (neoclassical) principles to the developing world (Peet and Hardwick 1999: 49–50).

Neoliberal principles have underpinned the so-called **Washington Consensus** which has structured economic development policy since the early 1990s (section 5.5.1). Key strands include a focus on low inflation, the reduction of barriers to trade, openness towards foreign direct investment (FDI), the **liberalisation** of the financial sector and the **privatisation** of state enterprises. The World Bank and IMF have played a crucial role in requiring developing countries to sign up to this policy agenda. In the context of the **debt crisis**, developing countries were often in desperate need of further financial assistance from those organisations (Box 7.4). This allowed the World Bank and IMF to set conditions requiring developing countries to reform their economies. These reforms have become known as **structural adjustment programmes (SAPs)**, encompassing a range of measures requiring countries to open up to trade and investment and to reduce public expenditure (Box 7.5).

Developing countries should seek to compete in the global market through the development of competitive export sectors, with the experience of the East Asian NICs often cited in support of this argument. In a similar fashion to modernisation theory, **neoliberalism** is based on the imposition of a set of externally derived solutions from the global North through mechanisms such as SAPs (Willis 2014a: 305).

Box 7.4

The debt crisis

The debt crisis has been a key factor shaping North–South relations over the past 35 years, with many developing countries struggling to service loans originally taken out in the 1970s. It first came to attention when Mexico defaulted on its loans in August 1982 with other Latin American countries such as Brazil and Argentina also experiencing major problems. Much of sub-Saharan Africa and parts of Asia have also been severely affected. The debt crisis has not only seriously undermined development efforts, requiring developing countries to spend their limited export earnings on servicing debts; it has also threatened the viability of the world financial system, forcing northern governments and bodies such as the IMF and World Bank to intervene.

The origins of the debt crisis lie in the interactions between three sets of factors:

➤ The borrowing of large sums by developing countries from northern banks and institutions in the 1970s. Following the sharp rise in oil prices that occurred in 1973, the world economy was awash with funds invested by the oil-exporting companies. These so-called 'petrodollars' were lent by leading banks and government

agencies to developing countries to pay for oil imports and to finance large-scale industrial development programmes.

➤ Most of the loans were made for a period of 5–7 years, denominated in US dollars and subject to floating interest rates (Corbridge 2008: 508). In the late 1970s and early 1980s, interest rates rose markedly, following the introduction of monetarist policies in the US and UK particularly. For example, the London Inter-Bank Offered Rate (LIBOR), the main index of the price of an international loan, rose from an average of 9.2 per cent in 1978 to 16.63 in 1981 (ibid.). As a result, developing countries were faced with hugely increased debt repayments, while the world economy was in recession.

➤ The collapse of commodity prices in the early 1980s was "such that in 1993 prices were 32 per cent lower than in 1980; and in relation to the price of manufactured goods, they were 55 per cent lower than in 1960" (Potter *et al.* 2008: 369). This amounted to a serious deterioration in the terms of trade for developing countries, reducing their export revenues relative to the price of imported

goods. As a result, they were faced with a 'scissors' crisis of declining export revenues and mounting debt repayments in a strengthening dollar (Corbridge 2008: 508). This is what led to a number of countries defaulting, threatening the viability of the international financial system.

While Latin American and Asian countries tended to have the largest absolute debts, it is African countries that have been worst affected in terms of the relationship between debt and GDP. Africa's indebtedness is estimated to have increased from approximately 28 per cent of its GDP to 72 per cent between 1980 and 1999, compared with 40 per cent for Latin America (Potter *et al.* 2008: 369). In some cases, annual repayments to creditors – around $15 billion every year – outweighed expenditure on education and health. From 2000, the overall debt burden for developing countries declined rapidly, from 12 per cent in 2000 to 3.2 per cent in 2013, reflecting debt relief, trade expansion and better debt management (UN 2015a: 66). Increased borrowing since 2010 and falling commodity prices since 2013, however, means that the debt burden of developing countries is likely to rise again.

Box 7.5

Structural adjustment programmes (SAPs)

The central aim of SAPs is debt reduction through reform packages designed to enable countries to pay their debts whilst maintaining economic growth and stability. Agreement to introduce a SAP became an essential precondition for developing countries to obtain finance from the World Bank, IMF and other private donors, a precondition that few countries were in a position to refuse. As a result, they rapidly spread across the global South. The first SAP was introduced in Turkey in 1980 and 187 had been negotiated for some 64 developing countries by the end of the 1980s (Potter *et al.* 2008: 299).

SAPs were based on four main objectives (Simon 2008: 87):

➤ The mobilisation of local resources to foster development.
➤ Policy reform to increase economic efficiency.
➤ The generation of foreign revenue through exports, involving diversification into new products as well as expansion of established ones.
➤ Reducing the economic role of the state and ensuring low inflation.

The measures required to achieve these objectives are generally divided into two types (ibid.):

1. Stabilisation measures which were immediate steps designed to address the economic difficulties facing developing countries in the short term, providing a foundation for longer-term measures.

➤ A public sector wage freeze – to reduce wage inflation and the government's salary bill.
➤ Reduced subsidies on basic foods and other commodities, and on health and education – to lower government expenditure.
➤ Devaluation of the currency – to make exports cheaper and more competitive, and to deter imports.

2. Adjustment measures which were to be implemented as a second phase, having a longer-term impact. Their objective was to ensure the structural adjustment of the economy, creating a platform for future growth.

➤ Export promotion through incentives for enterprise (including increased revenues and access to foreign currency) and diversification.
➤ Downsizing the civil service through the retrenchments following a programme of rationalisation to reduce 'overstaffing', duplication, inefficiency and cronyism.
➤ Economic liberalisation – relaxing and removing regulations and restrictions on economic activities. Examples include import tariffs and quotas, import licences, state monopolies, price fixing, subsidies and restrictions on the repatriation of profits by overseas firms.
➤ Privatisation through the selling off of state enterprises and corporations.
➤ Tax reductions to create stronger incentives for individuals and firms to save and invest.

SAPs proved highly controversial. Their impacts have often been harsh with ordinary people rather than elites bearing the brunt of the adjustment costs (ibid.: 88). In general, large traders, merchants and rural agricultural producers have benefited from increased export opportunities, often at the expense of the urban poor, who have suffered from the abolition of food subsidies and reductions in public expenditure. The effects of SAPs have been felt through cuts in public services, privatisation and the introduction of user charges for services. According to many critics, SAPs have reinforced inequality and poverty by reducing access to crucial social services and opening up developing economies to outside interests such as TNCs, although some local traders and farmers have benefited (Simon 2008).

SAPs were refined in the late 1980s and 1990s, taking better account of local needs and circumstances with the longer-term adjustment element rebranded as Economic Recovery Plans (ERPs) (ibid.: 90). From 1999, SAPs were replaced by Poverty Reduction Strategies (PRSs), which required national governments to produce a comprehensive plan for reducing poverty, involving consultation with the World Bank, IMF, NGOs and local communities. Key elements of the neoliberal Washington Consensus model remained in place, however, having essentially been augmented by the new focus on poverty reduction and 'good governance' (Sheppard and Leitner 2010).

7.3.4 Grassroots development

This approach is rather different in nature and orientation from the other theories discussed above. Instead of advancing a set of overarching prescriptions about development policy, based on abstract economic analyses, it is directly concerned with the practical problems and needs of poor people in developing countries, the ultimate targets of development aid (Table 7.3). Whilst the previous models are designed and implemented in a top-down fashion, focusing on the national scale, grassroots development, as the term implies, is focused on the local level. In addressing local needs in a 'bottom-up' manner, this approach is based on the observation that increased economic growth does not necessarily reduce poverty; in some cases, growth may widen inequalities within society, benefiting some groups at the expense of others. NGOs are closely associated with grassroots strategies, often working in partnership with local agencies and groups (Townsend *et al.* 2004).

This approach to development has evolved from the **'basic needs'** strategies of the 1960s and 1970s. These focused attention on the everyday lives of the poor, reflecting concerns that these were being neglected by orthodox modernisation policies. Recent work shares the same concern for identifying and meeting such needs in relation to food, shelter, employment, education, health, etc. The ethos is one of 'helping people to help themselves', involving small-scale projects which directly benefit individuals, families and households, supporting local services and livelihoods. In urban

areas, these are often based on the large informal sector of the economy, supporting the efforts of small-scale producers and traders to survive. In rural areas, farming is usually the focus of attention, with the establishment of schemes offering targeted assistance and support. An example is Oxfam's cow loan scheme whereby families are given cows from the loan scheme, fertilising crops and providing milk which not only provides sustenance for the family, but can be sold to generate an income, allowing them to purchase food and clothes (Figure 7.5) (Willis 2014a: 308). Calves are returned to the scheme to be loaned out to other families or sold.

The main constraint on grassroots development is the limited resources available to NGOs and other bodies. Whilst the number of NGOs has increased significantly since the 1970s, their resources are still outweighed by the scale of problems confronting them. Furthermore, as much of their money is derived from donations by individuals and governments, projects often reflect the concerns of donors rather than the priorities of local people. At the same time, the proliferation of small-scale local projects can sometimes be associated with a lack of collaboration and integration between agencies and initiatives. Grassroots

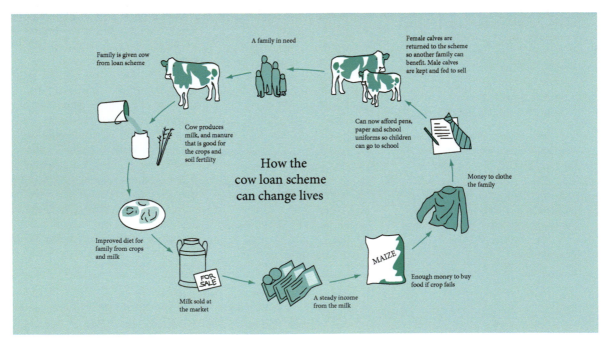

Figure 7.5 Oxfam cow loan scheme
Source: Willis 2005: 196.

development could be said to focus on alleviating the symptoms of poverty rather than addressing the underlying causes. This helps to explain some of the World Bank's and IMF's reluctance to embrace such approaches, concentrating instead on economic reform programmes. Such a view is rather harsh, however, with NGOs and others doing much valuable work in meeting local needs and highlighting the failure of mainstream initiatives to reach those most in need. The adoption of explicit poverty reduction strategies by the World Bank indicates that such concerns have influenced the mainstream agenda.

Reflect

Which of the theories offers the most appropriate model for developing countries? Do elements of different theories need to be combined? If so, which ones? Justify your answers.

7.4 Divergent trajectories of development

Average annual growth in GDP per capita in developing countries was actually lower in the 1980s and 1990s than in the 1960s and 1970s (UNCTAD 2016: 37). By contrast, the first decade of the 2000s was a period of rapid growth across all developing regions, although this has been eroded by a marked economic slowdown in recent years. Since 1980, there has been increased divergence between the major regions of the developing world, with only the NICs of East Asia demonstrating real convergence with the US, followed by China and South Asia from 2000 (Figure 7.6). Despite persistent economic growth, the gap with other developing regions remains as wide as ever.

In the 2000–2014 period, GDP growth rates in developing regions have outpaced sluggish levels of growth in Europe and North America with South Asia and sub-Saharan Africa performing particularly well, although rapid growth has done little to close the overall gap between developed and developing countries (Figure 7.6). Such growth was underpinned by a boom in commodity prices, driven particularly by rising demand for raw material imports from China. This generated increased export revenues for the producers of such commodities, greatly improving their terms of trade as they could purchase more imported goods with these revenues (Figure 7.7). The rapid growth of leading developing economies, particularly the **BRICS** group of Brazil, Russia, India, China and South Africa, attracted great attention in the 2000s, prompting claims about a fundamental shift in global economic geography (Box 7.6) (Hudson 2016).

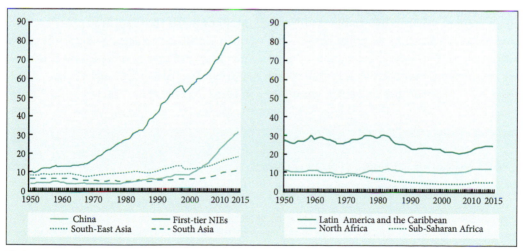

Figure 7.6 Ratio of GDP per capita of selected countries and country groups to GDP per capita of the United States, 1950–2015 (per cent)

Source: UNCTAD 2016: 39. UNCTAD secretariat calculations, based on The Conference Board, *Total Economy Database*, May 2015.

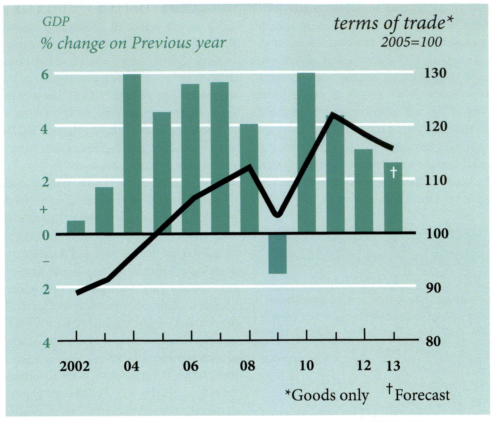

Figure 7.7 GDP growth and changing terms of trade in Latin America, 2002–13
Source: *The Economist* 2014.

Since 2014, however, development has been undermined by falling commodity prices, as the effects of the global economic crisis have spread further into the global South (section 4.5.3). While Europe and North America were worst affected in the initial phase of the crisis in 2008–09, GDP growth has subsequently slowed across the developing world. Sharp falls in commodity prices since 2013–14 mean that commodity-exporting countries in the global South are experiencing a decline in their terms of trade, reducing their real incomes and contributing to recessions in major economies such as Brazil and Russia (Box 7.6). Economic slowdown in Latin America has been particularly sharp, with growth falling from an annual average of 5 per cent between 2003 and 2010 to 0.2 per cent in 2015 and a projected -0.2 per cent in 2016 (*The Economist* 2014; UNCTAD 2016: 5).

Box 7.6

The rise and fall of the BRICS group

The term BRICS was originally coined in 2001 by the economist Jim O'Neill of Goldman Sachs Asset Management as an acronym for the rapidly emerging economic powers of Brazil, China, India and Russia with South Africa added in 2010. O'Neill envisaged them as the future engines of global economic growth,

Box 7.6 (continued)

predicting that they would eventually overtake the six largest developed economies in economic size and power by 2041 (Tett 2010). The term quickly became "a brand, a near ubiquitous financial term, shaping how a generation of investors, financiers and policy-makers viewed the emerging markets" (Pant 2013, quoted in Degaut 2015: 92). It also promoted growing political and diplomatic cooperation between the countries from 2006, reflecting their shared aspirations to superpower status.

Despite their rather disparate economies, the BRICS do share some basic underlying characteristics (see Degaut 2015). First is their size in both area and population terms which provides them with large domestic markets and labour forces. Second, they all recorded impressive economic growth rates in the 2000s and early 2010s (Table 7.4), far outstripping the G7 countries. Third, they are all the most powerful countries in their respective regions. Fourth, they were also committed to advancing their strategic interests, involving various forms of military modernisation

and improvement (ibid.). To varying degrees, they also share many of the underlying weaknesses of developing economies, including high levels of poverty, overdependence on commodity exports, entrenched regional inequalities, dependence on FDI, underdeveloped financial sectors, poor regulation and exposure to corruption (ibid.).

As Hudson (2016) argues, the rise of the BRICS is closely bound up with changing patterns of **uneven development** in the global economy and shifts towards new international divisions of labour (section 3.5). In particular, the relocation of production to lower-cost economies as part of the **NIDL** underpinned the rapid industrialisation of these countries, especially China. Furthermore, rapid growth in China and India particularly created strong demand for raw materials, greatly benefiting Brazil, Russia and, to a lesser extent, South Africa, alongside other commodity exporters, reflecting the emergence of distinct economic linkages between BRICS economies. Crucially, the BRICS also had strong states, reflecting their size and regional power status,

enabling them to take a proactive role in managing national economic development.

As economic growth has faltered in recent years, however, the BRICS 'brand' has lost some of its lustre. The economies of all five countries have experienced economic slowdown or recession with China and India maintaining relatively high rates of growth while Brazil and Russia are the worst performers (Table 7.4). The effects of recession in Brazil have been compounded by a corruption scandal and political instability that saw the overthrow of the President in August 2016. At the same time, the diplomatic alliance between the BRICS countries has failed to make any real progress, reflecting a lack of common ground between them (Degaut 2015). While the term does reflect underlying long-term processes in the world economy that are shifting power away from developed countries in Europe and North America towards emerging ones in East Asia and potentially other parts of the global South, this is a far more complex and uneven trend than the BRICS acronym indicated.

Table 7.4 Actual and projected annual economic growth figures (GDP% change) in BRICS economies 2003–15

	2003	2006	2009	2011	2013	2015
Brazil	1.1	4.0	-0.3	2.7	2.3	-3.8
China	10	12.7	9.2	9.3	7.7	6.9
India	7.9	9.3	8.5	6.3	4.4	7.3
Russia	7.3	8.2	-7.8	4.3	1.5	-3.7
South Africa			-1.5	3.5	1.8	1.3

Source: World Bank, http://databank.worldbank.org/data/reports.aspx?source=2&series=NY.GDP.PCAP.CD&country=#.

7.5 Trade, liberalisation and livelihoods

In general, developing countries have become more export-oriented in recent decades, accounting for a growing share of world trade. The global South's share of world exports rose from 29.6 per cent in 1980 to 44.7 per cent in 2012 (Horner 2016: 406). Furthermore, South–South trade grew by 677 per cent between 1995 and 2012, with southern exports to the rest of the world growing by 312 per cent (ibid.: 407). This has prompted discussion of the 'rise of the South' (UNDP 2013), although, as we saw in the previous section, this is a highly geographically uneven process with much South–South trade taking place in East Asia as the most dynamic region of the world economy (Dicken 2015). Moreover, developing countries have experienced a sharp fall in both imports and exports since 2014, reflecting the slowdown in economic growth, reduced demand, particularly from China, and falling commodity prices (UNCTAD 2016).

International trade has generally been subject to considerable liberalisation as tariff barriers have declined since 1995 to historically low levels, with developing countries particularly active in reducing tariffs in recent years. From an orthodox economic and neoliberal perspective, trade liberalisation should act as a key driver of economic development, with access to global markets generating rapid growth, increased employment and higher wages (section 7.3.3.). In practice, however, the developmental effects of trade liberalisation are complex and often socially unequal and geographically uneven (Smith 2015). Interestingly, UNCTAD (the United Nations Conference on Trade and Development) (2016) finds that, for a selected group of industrialising countries, the growth of manufactured exports, as measured by their share of global manufacturing exports, has typically been associated with a relative deterioration of national wage income compared to the world level

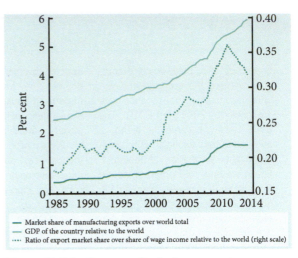

Figure 7.8 India: manufacturing exports, wage earnings and GDP relative to the rest of the world, 1985–2014
Source: UNCTAD 2016: 22. UN Global Policy Model using transitional data compiled from UN Comrade, UNCTADstat and UNSD.

since the mid-1980s (see, for example, Figure 7.8). Thus, while export-oriented manufacturing is associated with increased economic growth and employment creation, it has not led to a relative improvement of wages, reflecting the downward pressure on wages exerted by the capacity of TNCs to relocate production to other, often lower-cost, developing countries (ibid.: 21).

As we indicated in section 3.5.2, the relocation or **offshoring** of employment in industries such as textiles and electronics to developing countries often involves low wages and poor working conditions (Dicken 2015: 459). In the case of Tunisia, for example, whilst textiles and clothing firms employ 45 per cent of manufacturing workers, and account for almost one-third of manufacturing exports to the EU, they pay the lowest wages in the Tunisian manufacturing sector, conforming to the classic NIDL pattern of low-skill, female, labour-intensive employment (Smith 2015). The textiles industry is also characterised by a distinct pattern of geographically uneven development through its concentration in the eastern and northern coastal regions (Figure 7.9).

At the same time, trade liberalisation has resulted in the decline of non-competitive sectors, which have been undercut by more efficient producers elsewhere, causing substantial job loss and social dislocation. The development economist Dani Rodrik (2016) terms this process premature **deindustrialisation**, whereby many

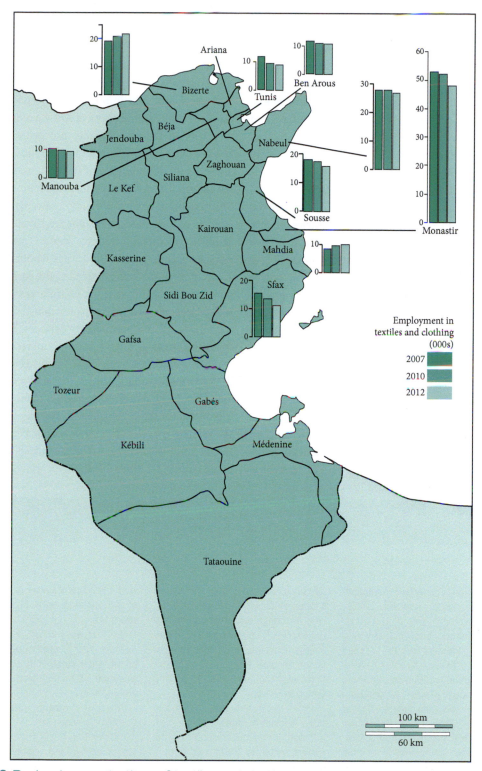

Figure 7.9 Regional concentrations of textiles and clothing industry employment in the main production locations, Tunisia
Source: Smith 2015: 447.

developing countries are undergoing deindustrialisation at a much earlier stage of development than was the case for developed countries from the 1960s and 1970s. This is likely to have seriously detrimental effects on development, removing the principal mechanism through which growth occurred in developed economies and East Asia. Such premature deindustrialisation has been evident in Latin America and sub-Saharan Africa, particularly since the 1980s, in contrast to East Asia where manufacturing has grown. Essentially, Rodrik argues, deindustrialisation was imported from abroad through trade liberalisation as lower-cost production elsewhere, particularly China, undercut existing industries in Latin America and sub-Saharan Africa, with employment loss particularly concentrated in the former region.

Rodrik's analysis raises the question of how people adapt to the employment loss associated with premature deindustrialisation. This focuses attention on the agency of individuals and households in the global South to develop other income-generating and employment strategies often as a means of survival. The **livelihoods approach** in development studies has provided a particularly influential framework for understanding this process of economic adaptation and diversification. A livelihood is defined as comprising the "capabilities, assets (including both material and social resources) and activities required for a means of living" (Carney 1998: 7). The concept of assets is an important building block of the livelihoods approach, encompassing a range of material and social resources such as land and other natural materials, labour, skills, knowledge, equipment, food, livestock, money and social relations and supports (De Haan 2012).

A key theme is the importance of livelihood diversification in enabling members of low-income households to 'get by' in the face of wider processes of change. It is often based upon forms of 'portfolio employment' involving the combination of several jobs, sometimes in the informal as well as the formal economy (Box 7.7). Most attention has been directed towards farmers and members of farm households engaged in other economic activities including trading, crafts, manufacturing, working on other farms, and wage labour in construction and manufacturing, reflecting the rural orientation of much of this literature. In urban areas, the informal sector has played a key role in absorbing labour through the growth of low-productivity services. Furthermore, livelihood strategies have become increasingly multi-local in nature through various forms of economic migration, with remittances, money and goods sent back 'home' by migrants working abroad often providing an important source of household income. According to the World Bank, global remittances reached US$483 billion in 2011, with US$351 billion of this going to developing countries (Willis 2014b: 214).

Box 7.7

Adapting to deindustrialisation in the Zambian Copperbelt

The Zambian Copperbelt developed as a mining region from the 1920s, rapidly becoming "one of the greatest concentrations of industry and urban development on the African continent" (Fraser 2010: 4). After independence in 1964, the Zambian mining industry was nationalised, forming Zambia Consolidated Copper Mines (ZCCM) in 1982. The plummeting price of copper from 1974, however, saw mining employment fall from 66,000 in 1976 to 22,280 in 2000 (Simutanyi 2008: 7). Given Zambia's economic dependence on copper which generated over 80 per cent of foreign exchange earnings in the early 1970s, this fuelled a prolonged economic crisis that saw per capita income fall by 50 per cent between 1974 and 1994 (Fraser and Lungu 2007: 8).

Prompted by the World Bank and IMF, economic liberalisation culminated in the privatisation of the mining sector as ZCCM was 'unbundled' and sold off in seven packages between 1997 and 2000. Job losses were particularly severe in the period of privatisation with an estimated two-thirds of the ZCCM workforce laid off (Mususa 2012). From 2002, however, rising copper prices as part of the wider commodity boom led to a revival of mining production and employment.

Given the broader crisis of the Copperbelt economy in the late 1990s and early 2000s, the regional labour market was unable to absorb the redundant workers. Accordingly, the principal form of livelihood

Box 7.7 (continued)

diversification was widespread informalisation, particularly before the payment of redundancy benefits to ex-miners. This entailed an important shift in gender relations as women and children moved into the informal sector as a means of ensuring household survival, in stark contrast to the past dominance of formal, unionised male employment in the mines.

Housing represented an important socioeconomic asset since many ex-miners had been offered their houses at subsidised prices, enabling them to generate income from renting out rooms and growing food in their backyards (ibid.). Many cultivators were simultaneously involved in other economic activities, particularly trading and the operation of small hair salons. In addition, household members, particularly women and children, became increasingly involved in small-scale and illegal mining activity as prices rose. At the same time, a limited number of ex-miners engaged in more profitable forms of diversification. Examples included the claiming of small items from mines when the new private owners were selling off assets and using these to supply the new owners, or deploying capital from redundancy packages to start construction or retail businesses. The contrast between a minority of former miners and residents who have prospered through the establishment of new enterprises and the struggles of the remainder to 'get by' has led to greater socio-economic differentiation in Copperbelt towns (Mususa 2010).

Reflect

To what extent do you think that trade policy should take into account the social effects of liberalisation on people's livelihoods?

7.6 Contesting development: new forms of social and political mobilisation

Orthodox models of development based on the promotion of economic growth and open markets have often failed to reach those groups most in need of assistance. Indeed, neoliberal policies of economic liberalisation and privatisation have tended to undermine livelihoods and increase externally owned TNCs' control over local resources and services. As part of this project of development, western values and expertise have been privileged over local knowledge and culture (Routledge 2014). Since the 1990s, the limitations and unequal effects of mainstream, top-down approaches to development have spurred new forms of mobilisation in developing countries. This provides the context for assessing the activities of local social movements in resisting particular aspects of neoliberal globalisation and the development projects undertaken by individual states. In a small number of Latin American countries such as Bolivia, such movements have been central to the election of radical '**post-neoliberal**' governments (Box 7.8).

Social movements in developing countries can be seen as expressions of conflicts between different groups in society over the control and use of space. Whilst local groups wish to retain control of local economic resources such as land, forests and water, utilising them to meet their material needs on a day-to-day basis, states and private interests often want to exploit them for economic gain, threatening the basis of local livelihoods (Bebbington and Bebbington 2010). Deforestation schemes, mineral extraction projects and the construction of dams for hydro-electric power and irrigation are often the focus for conflicts which can be seen as examples of 'resource wars' between the different parties (Box 7.8). Many of the local resistance movements are place-based, asserting local values and identity in the face of the knowledge and power of states and private corporations. One of the effects of policies of privatisation and liberalisation has been to make it easier for external corporations and interests to gain control over local resources, resulting in increased conflict with local groups. Whilst most movements are locally based, their struggles have often assumed a wider dimension, involving imaginative use of the media and internet to gain international support.

While social movements have been formed in a range of developing countries and regions in developing countries, including South and South-east Asia and sub-Saharan Africa, some of the most interesting examples are found in Latin America, highlighting the unequal social relations between poor and indigenous peoples, on the one hand, and the state and private capital, often in the shape of externally owned TNCs, on the other. One of the best known is the Zapatista guerrilla movement in the Chiapas region of Mexico which represents the interests of the indigenous Mayan people (Routledge 2014). The Zapatistas' protests are against poverty and the exploitation of local resources, problems which were compounded by the North American Free Trade Agreement (NAFTA) – between the US, Canada and Mexico in 1994. This fuelled the growth of intensive agriculture for international markets, leading to the emergence of a small group of wealthy farmers and a large class of landless Indian labourers.

Another example is the Movimento Sem Terra (MST) in Brazil, a mass national social movement of some 220,000 members founded in 1984 (ibid.). It is made up of mainly landless labourers and peasants from rural areas, reflecting the highly uneven distribution of land in Brazil. While land reform remains the main goal of the MST, it has come to recognise that the struggle is not only directed against the Brazilian *latifundio* (system of large estates with a single owner), but also the neoliberal economic model, campaigning vociferously against the proposed Free Trade Area of the Americas (FTAA). The strategy of the Movement involves targeting large unused private estates and illegally squatting and occupying the land. Over 600,000 people have been resettled since 1991, leading to considerable violence as large landowners and their private armies have attacked and killed squatters (ibid.). The MST has organised several marches and congresses in the capital city of Brasilia to publicise its agenda for agricultural reform (Figure 7.10).

Latin America experienced a distinct 'turn to the left' in the late 1990s and 2000s as radical left-wing and centre-left governments were elected in several countries, including Venezuela, Bolivia, Ecuador, Argentina, Brazil and Paraguay (Sader 2011). In some cases, this has been directly linked to processes of popular mobilisation by the types of social and political movements

Figure 7.10 An MST protest march
Source: Luciney Martins.

outlined above (Box 7.8). More broadly, it reflects the crisis of the neoliberal model which the region had embraced in the 1990s, resulting in greater inequality, higher levels of unemployment and informality and increased environmental degradation (Escobar 2010). This crisis prompted a new wave of mobilisation as social movements, activists and political leaders sought to resist and reverse neoliberal reforms such as privatisation, liberalisation and **deregulation**.

The New Left governments of Latin America are termed 'post-neoliberal' by Yates and Bakker (2014), viewing them as engaged in both a utopian-ideological project of transcending neoliberalism and a set of practical policies and initiatives to implement this project. More specifically, the Latin American form of post-neoliberalism revolves around the dual aims of redirecting a market economy towards social ends, and reviving citizenship through a new politics of participation

Table 7.5 Principles and practices associated with post-neoliberalism in Latin America

Principles		Practices
Re- socialization	Re-founding the state (around the social sphere)	➤ Re-regulation of the social sector and social services (reforms to welfare: public provision of basic services, particularly in relation to public goods such as water) ➤ Nationalization ➤ Regulation of big business ➤ Domestic market stimulation and the regulation of capital
	(Re-)Socialization of the market economy	➤ Building a solidarity economy (cooperatives, associations, community organizations) ➤ Strengthened labour relations ➤ Decommodification ➤ Re-establishing common property rights (territorial and collective governance) ➤ Participatory budgeting
Deepened democracy	Re-politicization of civil society (autogestión)	➤ Spaces of consensus building (place-based: issue- or resource-based; identity-based), which may challenge dominant, hierarchical scales of decision making ➤ Institutionalization of participatory decision-making mechanisms ➤ Pluri-nationalism and pluri-culturalism ➤ Social mobilization as 'politics-as-usual' (incorporation of movements into referendum politics for stability: national identity as hegemonic struggle)
	Regional integration (new regional political economy)	➤ Regional co-operation (economic trade; knowledge exchange) ➤ Financial autonomy (from international financial institutions) ➤ (Regional) Political autonomy (anti-imperialism)

Source: Yates and Bakker 2014: 71.

and alliances across socio-cultural sectors and groups (ibid.: 64). These aims are reflected in the principles and practices set out in Table 7.5. The term is not meant to imply a complete break with neoliberalism which has continued to structure the broader economic and political environment in which the New Left governments operate.

Substantial differences have been apparent between post-neoliberal regimes in Latin America. Yates and Bakker adopt Calderon's distinction between four political-economic ideologies: practical reformism, popular nationalism, indigenous neo-developmentalism, and conservative modernisation. Aspects of practical reformism, for instance, were clearly evident in Brazil, Chile and Peru where more conventional social democratic governments have emphasised the importance of economic growth and stability alongside poverty allevi-ation and democratic participation. Popular national-ism has characterised the process adopted in Venezuela and Argentina in particular, focusing on the breaking of neoliberal hegemony. Whilst reflecting elements of pop-ular nationalism, Bolivia and Ecuador are more symp-tomatic of indigenous neo-developmentalism, based upon efforts to develop pluri-national and pluri-cultural states supported by the harnessing of natural resources (Box 7.8). Finally, the conservative modernisation category points to the limitations of post-neoliberal politics in the region, emphasising the continued exis-tence of conventional neoliberalism pursued by the state, TNCs, and international economic organisations such as the World Bank and IMF. While post-neoliberal governments have met with some success in supporting

Box 7.8

Indigenous development and resource nationalisation in Bolivia

Whilst rich in natural resources, Bolivia is one of the poorest countries in South America, leaving Bolivians to commonly refer to themselves as "beggars sitting on a throne of gold" (Kohl and Farthing 2012: 227). In response to problems of hyperinflation and debt, neoliberal economic reforms were implemented from 1985, involving austerity and the liberalisation and privatisation of key sectors. These led to the collapse of formerly protected industries with over 20,000 miners left unemployed and some 35,000 manufacturing jobs lost (Perreault 2006: 155).

A major wave of social and political protest erupted in the early 2000s, including the Cochabamba 'Water War' of 2000 against the privatisation of the municipal water service, and the 'Gas War' of 2003 which opposed neoliberalism and the control of gas by foreign interests, leading to the resignation of the President. After a two-year hiatus, the Movimento al Socialismo (MAS) came to power in 2005, rooted in the social movements that had organised

the preceding resource wars and based on a campaign that promised to reclaim Bolivia's natural resource wealth 'for the people' (Kohl and Farthing 2012: 229). MAS is headed by Evo Morales from the indigenous Aymara group, and represents coca growers, miners, rural labourers and some indigenous people, whose livelihoods had been undermined by liberalisation and structural adjustment. Morales was re-elected in 2009 and 2014, although he narrowly lost a 2015 referendum which would have allowed him to stand for a fifth term.

Reflecting the model of 'indigenous neo-developmentalism', the new development plan established by MAS in 2006 is based on the indigenous concept of 'living well', based upon a harmonious balance between people and the natural environment. The passing of Supreme Decree 28701 on 1 May 2006 is the key to its alternative developmental model, involving a redistribution of the economic surplus from resource extraction through social programmes. This involved

the nationalisation of the gas sector, formerly controlled by the Brazilian partly state-owned enterprise Petrobras and the Spanish firm Repsol, by transferring a 51 per cent stake to the Bolivian State Oilfields Company (YPFB). At the same time, combined royalties and taxes were increased from 18 to 50 per cent (Kaup 2010: 129). Yet a more cautious approach was adopted towards the mining sector through partnership with foreign companies rather than nationalisation, alongside increased taxes.

Like other Latin American countries, the commodity price boom of the 2000s greatly improved Bolivia's terms of trade, but reinforced the economy's dependence on exports (Simarro and Antolin 2012). The tax increases outlined above meant that the value of hydrocarbon (oil and gas) taxes soared as the sector came to account for half of total tax revenue (Figure 7.11). Yet the government's approach has been constrained by past policies and the material constraints that surround natural gas extraction, transport and use. In

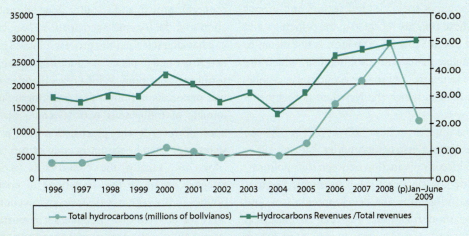

Figure 7.11 Hydrocarbon taxes in Bolivia 1996–2009
Source: Simarro and Antolin 2012: 547.

Box 7.8 (continued)

particular, a 20-year supply agreement requires Bolivia to send the majority of its natural gas to Brazil until 2019, enabling it to only renegotiate price and tax rates, not quantities (Kaup 2010). In addition, the use of revenues for social programmes has resulted in little investment back into the hydrocarbon sector with YPFB starved of funds and expertise.

The key social programmes introduced by MAS include an 87.7 per cent increase in the minimum wage between 2005 and 2014 (O'Hagan 2014), the introduction of a 'dignity pension', which covers 687,000 people, and investment in health and education. These have been successful, with the proportion of the population living beneath the national

poverty line falling from 66.4 per cent in 2000 to 39.1 per cent in 2014 (http://data.worldbank.org/country/bolivia). While the rate of economic growth fell from 6.8 per cent in 2013 to 4.8 per cent in 2015 as a result of falling commodity prices (ibid.), this remains much higher than most other countries in the region.

economic growth and reducing poverty (Box 7.8), their efforts have been undermined by the economic crisis since 2014, spurring renewed social unrest and political change in countries such as Argentina, Brazil and Venezuela (Trinkunas and Davis 2016).

Reflect

To what extent was the rise of social movements and post-neoliberal regimes in Latin America a reflection of a failure of conventional development policies in the region?

7.7 Summary

The development of the former colonies in Africa, Asia and Latin America can be seen as a key political and economic project of the post-war era, launched by the US and its allies in the late 1940s against a backdrop of Cold War tensions. The so-called Third World was constructed as a realm of universal poverty and backwardness with development requiring an injection of external knowledge and resources. The establishment of SAPs assumed particular importance in the context of the debt crisis in the 1980s and 1990s. Since the mid-1990s, the emphasis on economic reform has given way to a focus on poverty reduction, generating a new consensus amongst development agencies, governments and NGOs. In reality, however, the World Bank and IMF continue to dictate the terms of assistance, maintaining a regime of

'conditionality' where developing countries have to meet specific requirements before qualifying for assistance. A commitment to economic liberalisation, openness and adjustment remains central alongside the new focus on poverty reduction. The responsibility for alleviating poverty is allocated to poor countries themselves, while broader structural constraints such as the global rules underpinning trade and investment are largely ignored.

Despite some improvements over the last 30 years, global inequalities remain massive. Within the global South itself, processes of uneven development have resulted in a growing divergence between regions. Sustained growth in East Asia and parts of South Asia contrasted with economic decline in sub-Saharan Africa and stagnation in Latin America in the 1980s and 1990s. Rising commodity prices generated high levels of growth across most of the global South in the 2000s, but this has been insufficient to close the gap with developed economies. Since 2013–14, falling commodity prices have reduced growth, with Latin America the worst-affected region. While the relocation of manufacturing production as part of a NIDL has supported growth and job creation in selected countries, others have been affected by premature deindustrialisation as economic liberalisation has undermined the competitiveness of key industries (Rodrik 2016). This is associated with widespread livelihood diversification and the growth of the informal sector. Economic liberalisation and privatisation spurred a new wave of social and political mobilisation in some developing regions, particularly Latin America in the 2000s, leading to the election of New Left governments in several countries.

These governments have experienced mixed fortunes with the relative success of Bolivia contrasting with severe economic crisis in Brazil and Venezuela. This experience underlines the contested nature of development and the efforts of key actors, including national governments, international economic organisations, TNCs, NGOs and local communities, to reshape development models and practices in particular historical and geographical contexts.

Exercise

Select a developing country. Review its experience of development since the 1980s, using the websites listed below as starting points for your research.

Assess the country's economic performance over time (referring to figures on growth, employment, income, education, health, etc.). What have been the key forces shaping development? What development strategies has the country adopted? Who has set the development agenda – the government, the World Bank/IMF, foreign TNCs, NGOs, domestic interests (e.g. landowners, industrialists, traders, workers, farmers)? Have development strategies been informed by any of the theories discussed in section 7.3? Are there any examples of local social movements contesting particular development projects? How would you describe the country's development prospects at the present time?

Key reading

Desai, V. and Potter, R. (eds) (2014) *The Companion to Development Studies*, **3rd edition. London: Arnold.**
Contains a number of short overviews of various development themes and issues written by leading authorities. Extremely comprehensive, dealing with a wide array of topics including the meaning of development, the main theories, rural development, urbanisation, industrialisation, the environment, gender and population, health and education, violence and instability and key agents of development.

Hudson, R. (2016) Rising powers and the drivers of uneven global development. *Area Development & Policy.* **Online first, DOI: 10.1080/23792949.2016.1227271.**
An authoritative account of the emergence of the rising powers in the global economy, principally the BRICS countries. Links their rise to evolving patterns of uneven development in the world economy, particularly changing global divisions of labour. Emphasises the differences between the rising powers and comments on their future economic prospects.

Potter, R., Binns, T., Elliott, J.A., Nel, E. and Smith, D. (2017) *Geographies of Development: An Introduction to Development Studies*, **4th edition. London: Routledge.**
Probably the best contemporary textbook on development issues in geography. Provides a comprehensive and integrated treatment of development which covers the main theories and the historical legacy of colonialism; assesses the role of population, resources and key institutions; and examines spaces of development within developing countries.

Power, M. (2016) Worlds apart: the changing geographies of global development. In Daniels, P., Bradshaw, M., Shaw, D., Sidaway, J. and Hall, T. (eds) *Human Geography: Issues for the Twenty First Century*, **5th edition. Harlow: Pearson, pp. 170–85.**
An engaging discussion of development issues, emphasising the scale of global differences and inequalities. Covers the key theories, institutions and history of development, highlighting its unequal impact on households and individuals within developing countries.

Routledge, P. (2014) Survival and resistance. In Cloke, P., Crang, P. and Goodwin, M. (eds) *Introducing Human Geographies*, **3rd edition. London: Routledge, pp. 325–38.**
A useful account of the growth of local social movements against particular projects. Covers two of the three examples outlined in section 7.6, stressing both the place-based origins of the movements and the increasing global links between them.

Willis, K. (2014) Theories of development. In Cloke, P., Crang, P. and Goodwin, M. (eds) *Introducing Human Geographies*, **3rd edition. London: Routledge, pp. 297–311.**
A very clear summary of the main theories of development, covering the modernisation school, dependency theory, neoliberalism and grassroots development. Contains useful bullet point summaries of each approach.

Useful websites

www.oxfam.org.uk/index.htm
The website of one of the leading British development NGOs, providing details of its history and current strategies and

highlighting some of the key projects and campaigns it is undertaking. Contains information on how you can support Oxfam's activities through volunteering and fund-raising.

http://hdr.undp.org/

The UNDP site provides access to information about human development in the form of current statistics, the annual Human Development Report and details of how these measures such as the Human Development Index are compiled. Contains interactive maps and tools illustrating human trends and outcomes.

www.worldbank.org/

The official World Bank site contains a wide range of information, reports and projects. Key resources include the annual World Development Report, World Development Indicators and the speeches of the President and other senior figures, giving an indication of current thinking. Data profiles of countries and country groupings are also available.

www.mstbrazil.org

The website of Brazil's Landless Workers Movement, the Movimento Sem Terra (MST). Contains a range of information about the movement's history, objectives and campaigns.

References

Bebbington, A. and Bebbington, D.H. (2010) An Andean Avatar: post-neoliberal and neoliberal strategies for promoting extractive industries. *BWPI Working Paper* 117. Manchester: Brooks World Poverty Institute, The University of Manchester.

Bell, M. (1994) Images, myths and alternative geographies of the Third World. In Gregory, D., Martin, R. and Smith, G. (eds) *Human Geography: Society, Space and Social Science*. Basingstoke: Macmillan, pp. 174–99.

Carney, D. (ed.) (1998) *Sustainable Rural Livelihoods: What Contribution Can We Make?* Nottingham: Department of International Development, Russell Press Ltd.

Corbridge, S. (2008) Third world debt. In Desai, V. and Potter, R. (eds) *The Companion to Development Studies*, 2nd edition. London: Arnold, pp. 508–11.

Degaut, M. (2015) *Do the BRICS Still Matter?* Washington, DC: Center for Strategic and International Studies.

De Haan, L.J. (2012) The livelihood approach: a critical exploration. *Erdkunde* 66: 345–57.

Desai, V. and Potter, R. (eds) (2014) *The Companion to Development Studies*, 3rd edition. London: Arnold.

Dicken, P. (2015) *Global Shift: Mapping the Changing Contours of the World Economy*, 7th edition. London: Sage.

The Economist (2014) The great deceleration, 22 November.

Escobar, A. (2010) Latin America at a crossroads. *Cultural Studies* 24: 1–65.

Fraser, A. (2010) Introduction – boom and bust on the Zambian Copperbelt. In Fraser, A. and Larmer, M. (eds) *Zambia, Mining and Neoliberalism: Boom and Bust on the Globalised Copperbelt*. Basingstoke: Palgrave Macmillan, pp. 1–30.

Fraser, A. and Lungu, J. (2007) *For Whom the Windfalls: Winner and Losers in the Privatisation of Zambia's Copper Mines*. Lusaka: Civil Society Trade Networks of Zambia.

Fukuda-Parr, S. (2016) From the Millennium Development Goals to the Sustainable Development Goals: shifts in purpose, concept, and politics of global goal setting for development. *Gender & Development* 24: 43–52.

Hickel, J. (2016) The true extent of global poverty and hunger: questioning the good news narrative of the Millennium Development Goals. *Third World Quarterly* 37: 749–67.

Horner, R. (2016) A new economic geography of trade and development? Governing south-south trade, value chains and production networks. *Territory, Politics, Governance* 4: 400–420.

Hudson, R. (2016) Rising powers and the drivers of uneven global development. *Area Development & Policy*. Online first, DOI: 10.1080/23792949.2016.1227271.

Kaup, B.Z. (2010) A neoliberal nationalisation? The constraints on natural gas-led development in Bolivia. *Latin American Perspectives* 37: 123–38.

Knox, P., Agnew, J. and McCarthy, L. (2003) *The Geography of the World Economy*, 4th edition. London: Arnold.

Kohl, B. and Farthing, L. (2012) Material constraints to popular imaginaries: the extractive economy and resource nationalism in Bolivia. *Political Geography* 31: 225–35.

Mususa, P. (2010) 'Getting by': life on the Copperbelt after the privatisation of Zambian Consolidated Copper Mines. *Social Dynamics* 36: 380–84.

Mususa, P. (2012) Mining, welfare and urbanisation: the wavering urban character of Zambia's Copperbelt, *Journal of Contemporary African Studies* 30: 571–88.

O'Hagan, E.M. (2014) Evo Morales has proved that socialism doesn't damage economies. *The Guardian*, 14 October.

Oxfam (2015) Annual Report and Accounts 2014–2015. Oxford: Oxfam.

Peet, R. and Hardwick, A. (1999) *Theories of Development*. New York: Guildford.

Perreault, T. (2006) From the *Guerra del Agua* to the *Guerra del Gas*: resource governance, neoliberalism and popular protest in Bolivia. *Antipode* 31: 150–72.

Potter, R., Binns, T., Elliott, J.A. and Smith, D. (2004) *Geographies of Development*, 2nd edition. Harlow: Pearson.

Potter, R., Binns, T., Elliott, J.A. and Smith, D. (2008) *Geographies of Development*, 3rd edition. Harlow: Pearson.

Power, M. (2016) Worlds apart: the changing geographies of global development. In Daniels, P., Bradshaw, M., Shaw, D., Sidaway, J. and Hall, T. (eds) *Human Geography: Issues for the Twenty First Century*, 5th edition. Harlow: Pearson, pp. 170–85.

Rodrik, D. (2016) Premature deindustrialisation. *Journal of Economic Growth* 21: 1–33.

Routledge, P. (2014) Survival and resistance. In Cloke, P., Crang, P. and Goodwin, M. (eds) *Introducing Human Geographies*, 3rd edition. London: Routledge, pp. 325–38.

Sader, E. (2011) *The New Mole: Paths of the Latin American Left*. London: Verso.

Sheppard, E. and Leitner, H. (2010) Quo vadis neoliberalism? The remaking of global capitalist governance after the Washington Consensus. *Geoforum* 45: 185–94.

Simarro, R.M. and Antolin, M.J.P. (2012) Development strategy of the MAS in Bolivia: characterization and an early assessment. *Development and Change* 43 (2): 531–56.

Simon, D. (2008) Neoliberalism, structural adjustment and poverty reduction strategies. In Desai, V. and Potter, R. (eds) *The Companion to Development Studies*, 2nd edition. London: Arnold, pp. 86–92.

Simutanyi, N. (2008) Copper mining in Zambia: the developmental impact of privatisation. Institute for Security Studies (ISS) Paper 165, July. Pretoria, South Africa: ISS.

Smith, A. (2015) Economic (in)security and global value chains: the dynamics of industrial and trade integration in the EU-Mediterranean region. *Cambridge Journal of Regions, Economy and Society* 8: 439–58.

Tett, G. (2010) The story of the BRICS. *Financial Times*, January 15.

Townsend, J., Porter, G. and Mawdsley, E. (2004) Creating spaces of resistance: development NGOs and their clients in Ghana, India and Mexico. *Antipode* 36: 781–889.

Trinkunas, H. and Davis, C. (2016) Why Latin America's 2016 elections might produce big changes. At www.brookings.edu/blog/order-from-chaos/2016/01/06/why-latin-americas-2016-elections-might-produce-big-changes/. Last accessed 25 October 2016.

United Nations (2015a) *The Millennium Development Goals Report 2015*. New York and Geneva: United Nations.

United Nations (2015b) *Transforming our World: The 2030 Agenda for Sustainable Development*. Resolution 70/15. New York and Geneva: United Nations.

United Nations Conference on Trade and Development (UNCTAD) (2016) *Trade and Development Report, 2016: Structural transformation for inclusive and sustained growth*. New York and Geneva: UNCTAD.

United Nations Development Programme (UNDP) (2013) *Human Development Report 2015. The Rise of the South: Human Progress in a Diverse World*. New York: UNDP.

United Nations Development Programme (UNDP) (2015) *Human Development Report 2015. Work for Human Development*. New York: UNDP.

Willis, K. (2005) Theories of development. In Cloke, P., Crang, P. and Goodwin, M. (eds) *Introducing Human Geographies*, 2nd edition. London: Arnold, pp. 187–99.

Willis, K. (2014a) Theories of development. In Cloke, P., Crang, P. and Goodwin, M. (eds) *Introducing Human Geographies*, 3rd edition. London: Routledge, pp. 297–311.

Willis, K. (2014b) Migration and transnationalism. In Desai, V. and Potter, R. (eds) *The Companion to Development Studies*, 3rd edition. London: Arnold, pp. 212–16.

Yates, J.S. and Bakker, K. (2014) Debating the post-neoliberal turn in Latin America. *Progress in Human Geography* 38: 62–90.

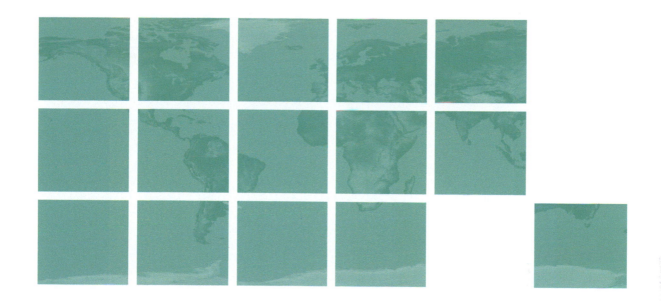

PART 3

Reworking urban and
regional economies

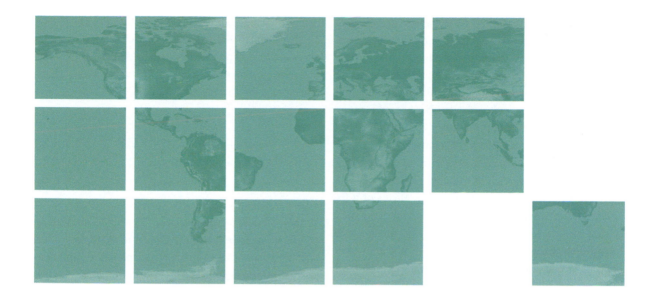

Chapter 8

Connecting cities: transport, communications and the digital economy

8.1 Introduction

The growth of transport infrastructures and ICTs is bound up with the changing spatial organisation of the economy, enabling goods, materials, information and money to be moved at lower costs across greater distances. As such, the 'space-shrinking technologies' of transport and ICT have underpinned the geographical expansion of the capitalist economy over time, facilitating the contemporary process of **globalisation** (section 1.2.1). As this suggests, the role of these technologies is essentially enabling or facilitating: they make new organisational and geographical arrangements possible rather than determining particular outcomes (Dicken 2015: 75). While business and media accounts of technological change often adopt a deterministic perspective, asserting that advances in technology will drive social, economic, and geographical change, this simplistic approach is eschewed here. Instead, technological change is treated as a socially and culturally embedded process shaped by peoples' choices between the range of options provided by technology (ibid.). While developments in ICTs have been particularly rapid in recent years through digital technologies, smart phones and social media, this basic point still applies.

A central theme of this chapter is the effects of ICTs and transport investment on the geographical organisation and structure of the economy, particularly in terms of the balance between the forces of spatial concentration (**agglomeration**) and dispersal. Prompted by the spread of modern ICTs linked to the personal computer (PC) from the 1980s, alongside the growth of low-cost air travel and high-speed trains, a number of commentators have proclaimed the 'end of geography' (O'Brien 1992) or the 'death of distance' (Cairncross 1997). This claim is based on the prospect to being able to distribute goods and services instantaneously and almost without cost over electronic networks (Gillespie *et al.* 2001). The direct implication is that spatial concentration will give way to wholesale dispersal as businesses are able to operate from any location with internet access, enabling them to stay in touch with key markets. This would undermine the economic rationale of cities based on the clustering of information and knowledge. Much subsequent research has, however, criticised this rather simplistic argument, pointing to the continued evidence of the clustering of higher-value economic activities in cities, particularly large metropolitan regions, which have the most advanced transport and communications infrastructures (section 8.3). Such concentration coincides with the widespread dispersal of other, often lower-value activities, typically to lower-cost cities and regions in the global South. This pattern may, however, be challenged by further technological improvement and by the explosive growth of large metropolitan regions in emerging economies such as China, fuelled by large-scale investments in transport and communications networks. Cities remain the main focus of activities in the '**sharing economy**' as "it is the very scale, proximity, amenities and specialisation that mark city life that

enable sharing economy firms to flourish" (Davidson and Infranca 2016: 218), while efforts to develop **smart cities** are also dependent on urban density to generate the necessary critical mass of infrastructures, users, objects and devices (section 8.5).

8.2 Transport infrastructure and urban and regional development

In the simplest terms, transport is crucial in "linking people to jobs; delivering products to markets; underpinning supply chains and logistics networks; and supporting international trade" (Eddington 2006: 3). This underpins the seminal concept of **time-space convergence** which emphasises how "places approach each other in time-space . . . as a result of transport innovation[s]" that reduce the travel time between them (Janelle 1969: 351). As indicated above, it overlaps considerably with the associated concept of **time-space compression** (section 1.2.1).

Investment in transport infrastructure has been widely used by policy-makers as an instrument to promote urban and regional growth and reduce regional disparities. This is predicated upon a belief in the positive economic effects of such investment in fostering connectivity and underpinning development. This belief is exemplified by regional development policies in the EU, which have been heavily based on investment in transport infrastructure to reduce regional disparities and promote economic development (Crescenzi and Rodríguez-Pose 2012).

8.2.1 Transport and regional development in the global North

From a theoretical perspective, the expectation that transport infrastructure investment will reduce regional disparities has been questioned by the '**new economic geography**' (**NEG**) (section 2.3) which views the relationship between transport costs and regional concentration as exhibiting an inverted U-shaped curve (Figure 8.1). Thus, high transport costs in the early stages of development discourage the transportation of goods between city-regions and production remains dispersed based on proximity to local markets. Falling transport costs and increasing returns, however, encourage the inter-regional movement of goods, leading to the increasing concentration of production in a small number of core regions that benefit from agglomeration economies and market size (Ding 2013). As this process continues, however, and dis-economies of agglomeration such as congestion emerge, alongside increasing competition between firms, production will be dispersed to peripheral locations where costs are lower, accompanied by the re-agglomeration of activities in these new locations (Fujita 2011).

This model has substantial policy implications, implying that core regions are better placed than lagging ones to take advantage of transport improvements. According to the economist Diego Puga (2008: 117), "cross-border infrastructure projects connecting lagging regions with key markets make it easier for firms in lagging regions to reach new customers, but also expose them to fiercer competition from firms in more developed areas". The results from studies that have

investigated the regional effects of transport infrastructure investment are rather mixed, underlining the fact that this is ultimately "an empirical matter that depends upon existing industrial structures and market size, the importance of transport costs and the magnitude of any change" (Vickerman 2012: 25).

In a review of surveys on transport infrastructure and regional economic development in Europe, Linneker (1997) distinguishes between the spheres of consumption and production. Improved accessibility relating to consumption definitely leads to an improvement in welfare for the population, with increased competition resulting in lower prices, although this is an individual-level outcome that has little bearing on questions of regional concentration and convergence. For the sphere of production, after making allowances for regional disparities, the question remains open, allowing very different answers to be put forward. Here, an economic study of 31 road link schemes in Great Britain between 1980 and 2007 found that these schemes did generally increase local employment and the number of industrial plants in the surrounding area (Gibbons et al. 2017). Interestingly, these transport improvements fostered

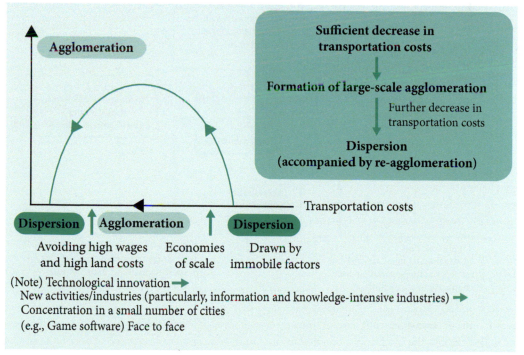

Figure 8.1 Inverted-U-shaped impact of declining transportation costs on agglomeration
Source: Fujita, 2011.

employment growth through growth in the number of plants based upon the establishment of new plants and the associated employment in these plants, while incumbent plants that were present in the local area prior to the road improvements actually cut back on employment. The authors' interpretation of this is that transport improvements attract new firms to the area which forces up wages, prompting incumbent firms to respond by shedding labour and substituting this with purchases of goods and services.

Reflecting the essentially enabling role of transport, Linneker (1997: 60) concludes that "whether further development towards higher or lower levels of economic development potential is realised . . . is determined by a large number of other factors outside the transport sector." This point is developed by Banister and Berechman (2001) who identify a series of necessary conditions that must be in place for transport investment to stimulate regional economic development. The three key conditions are: (a) positive economic externalities, basically meaning a well-functioning local economy, particularly in terms of the links between firms

and suppliers and the operation of the labour market; (b) investment factors, referring to the availability of funds, the quality of the overall network and the timing of the investment; and (c) a favourable political environment, in terms of other supporting policies and a generally enabling policy framework (Figure 8.2). All three factors must be in place for transport investment to have a positive impact on the regional economy. If only one or two of these factors are present at the time of investment, certain effects such as an improvement of accessibility may occur – but no regional economic development.

The relationship between transport infrastructure investment and urban and regional economic development has attracted renewed attention in the context of the growth of high-speed rail (HSR), defined in terms of speeds of at least 250 kilometers per hour (km/h) on specially built lines and 200 km/h on upgraded lines (EU 1996). This invests the economist Colin Clark's view of transport as 'the maker and breaker of cities' with new meaning as cities have lobbied for HSR links as a means of enhancing their competitiveness

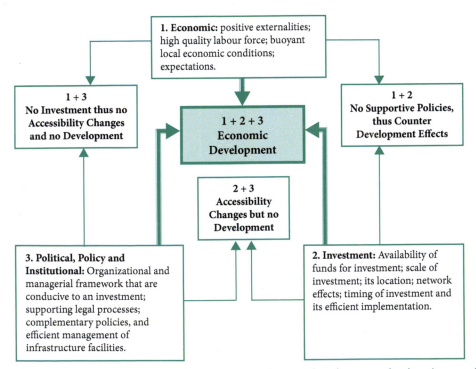

Figure 8.2 Necessary conditions for transport to stimulate regional economic development
Source: Banister and Berechman 2001: 210.

and growth (Chen and Hall 2011). HSR has grown considerably in recent decades, ushering in a new round of time-space convergence. It began in Japan in 1964 before being introduced in a number of other countries, including France's TGV system from 1981, Spain, Germany, Italy, South Korea and China (Table 8.1). China's network has expanded particularly rapidly since 2012, now accounting for more than half the high-speed traffic in the world (Table 8.1).

Studies of the urban and regional development impact of HSR are far from conclusive, but do indicate differential impacts on different types of city regions (Chen and Hall 2012). Echoing the predictions of the NEG, there is some evidence that HSR has increased the relative imbalance between core and lagging regions in both Europe and Asia (Albalate and Bel 2010). Here, a distinct 'hub' effect appears to operate, favouring the large central nodes rather than more peripheral areas (see Box 8.1). In France, Paris seems to have gained the most from faster travel and the growth of key regional hubs such as Lyon and Lille has not spread to their regional hinterlands (Box 8.1). While research on Spain showed that the peripheral cities enjoyed the biggest increases in accessibility from HSR extensions, major cities such as Madrid and Barcelona retained their privileged position (Monzón et al. 2013). Despite the objective of regional rebalancing, the greatest increases in accessibility in South Korea were in the central corridor between Seoul and Daejon (Kim and Sultana 2015). Despite the lack of evidence of HSR contributing to reductions in regional inequality, governments continue to justify its introduction on this basis as demonstrated by the UK government's claim that the new HS2 line between London and northern England will support the geographical 'rebalancing' of the economy (Tomaney and Marques 2013).

Table 8.1 High-speed rail lines in the world, selected countries

Country	Time period	In operation (km)	Under construction (km)	Planned (km)	Total (km)
Japan	1964–2035	3,041	402	179	3,622
France	1981–2018	2,142	634	1,786	4,562
Germany	1988–2025	1,475	368	324	2,167
Italy	1981–2020	923	125	221	1,269
Spain	1992–2018	2,871	1,262	1,327	5,460
China	2003–2020	21,688	10,201	1,945	33,834
South Korea	2004–2017	598	61	49	708
Turkey	2009–2016	688	469	1,134	2,291
United Kingdom	2003–2032	113	-	543	656

Source: UIC High Speed Department (2016) *High Speed lines in the World*. Updated 1 November 2016. Available at www.uic.org/IMG/pdf/20161101_high_speed_lines_in_the_world.pdf. Last accessed 17 January 2017.

Box 8.1

The uneven urban and regional impacts of the Channel Tunnel rail link

The Channel Tunnel opened in 1994, enabling the introduction of the Eurostar HSR service between London, Paris and Brussels. Following the completion of the final section of the rail link (now known as HS1) to St Pancras International in 2007, the travel time is 2 hours between London and Brussels and 2 hours 15 minutes between

Box 8.1 (continued)

London and Paris. The Eurostar service has competed very successfully with airlines on these routes with its market share exceeding 80 per cent by 2011 (Thomas and O'Donoghue 2013). Passenger numbers have, however, failed to reach the rather optimistic forecasts made before its opening in the context of lower-cost airlines offering faster and cheaper connections to a range of destinations in mainland Europe (ibid.).

While the Eurostar connects the major metropolitan centres of London, Paris and Brussels, it also passes through the economically lagging regions of Kent (and east Kent particularly) and Nord-Pas-de-Calais (Figure 8.3) with GDP per capita levels well below the west European average by the mid-1980s, providing an important test of the impact of HSR on intermediate regions (Vickerman 1994).

Most of the benefits of the new rail links have accrued to the major cities of London, Brussels, Paris and Lille. Connectivity between the capital cities has been increased, boosting their visitor numbers and tourism sector. In conjunction with its role as a hub for the domestic TGV service, Lille has become a major transport hub, supporting a number of flagship urban regeneration initiatives including the new Eurostar station and associated Euralille renewal area. Yet the growth of Lille has not had a wider multiplier effect on the wider surrounding region, reflecting the limited functional links between it and its hinterland (Thomas and O'Donoghue 2013: 108). On the UK side, there is no Kentish counterpart to Lille with the Kent region

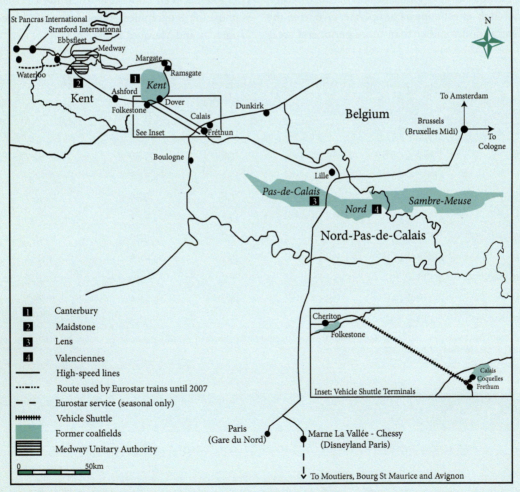

Figure 8.3 The Channel Tunnel and the Eurostar Rail network
Source: Thomas and O'Donoghue 2013: 105.

deriving little direct economic bene-fit beyond the town of Ashford which became a stop on the line with direct services to London St Pancras in 37 minutes (ibid.: 109). This underlines the tendency of HSR to benefit the major nodes in the networks rather than the intermediate spaces in between which typically incur many of the costs of construction without realising any of the potential benefits (Vickerman 2012).

8.2.2 Transport and urban development in the global South

The problems of urban transport in the developing world are huge, not least in the burgeoning mega cities such as Bangkok, Mexico City and Cairo where the basic infrastructure is simply overwhelmed by population growth (Gwilliam 2003). The streets of many such cities are among the most congested, least regulated and most polluted on the planet. Road congestion is endemic, reflecting the inadequacy of the infrastructure – road space typically takes up 10–12 per cent of land in Asian mega cities, compared to 20–30 per cent in US cities (ibid.: 202) – and the rapid growth of private vehicle traffic from a low base. Public transport remains inaccessible and unaffordable to the majority of the poor, particularly those living in the sprawling shanty towns around the urban fringe. In the developing world, then, the 'fundamental paradox of urban transport', namely the coexistence of excess demand for road space and the underfunding of other modes (ibid.: 212), is compounded by population growth, inadequate infrastructure and mass poverty.

Access to transport remains highly unequal in the cities of the global South where only a small minority travel by private car and cycling and walking are the main means of mobility for the urban poor. These inequalities are being compounded by transport policies and urban development strategies based upon the private car and prestigious megaprojects designed to promote economic competitiveness rather than social need (Lucas and Porter 2016). Motorised transport is highly expensive for many poor urban residents, and public transport tends to be unreliable and of poor quality. In the Colombian capital of Bogota, for instance, low-income households spend more than 20 per cent of their income on motorised transport to reach centres of employment, in addition to long walks to reach public transport, especially in peripheral neighbourhoods (Hernandez and Davila 2016: 184). Nonetheless, the use of informal motorised taxis in many sub-Saharan African and some Latin American cities has grown rapidly in recent years, reflecting their comparative affordability (Lucas and Porter 2016). They highlight ordinary people's effort to address the acute transport difficulties that afflict their daily lives in the absence of supportive transport policies, often providing the only way of accessing the city apart from walking.

In general, there is clearer evidence of the local economic benefits of transport investment in developing countries where transport investment was in much poorer initial condition than in developed countries with more advanced transport systems (Gibbons 2017: 6). Some cities have made improvements with Bogota, for instance, introducing over 300 km of cycle lanes since 1998 (Sietchiping et al. 2012). Other affordable and achievable transport improvements such as street realignment and widening, street traffic intersection control, and segregation of pedestrians and vehicles can also enhance mobility and economic productivity in cities. The problems of the residents of urban slums in terms of accessing employment opportunities and basic services remain particularly intractable, requiring a range of measures to make transport more affordable and accessible as well as 'non-transport interventions' which locate public facilities and services in more accessible areas (Olvera et al. 2003). One general lesson is that relatively minor cost-effective projects that better target the poor such as cycle lanes and pedestrian facilities can make a considerable difference to mobility and access to public health clinics, schools and employment. Transport schemes are unlikely to be sufficient to generate sustained economic development on their own, however, reflecting the essentially enabling role of transport and requiring other measures to stimulate urban and regional economies, generate large-scale investment and foster a more supportive political environment (Banister and Berechman 2001).

Reflect

Consider, in the light of the discussion in this section, the extent to which transport infrastructure investment can reduce disparities between core and peripheral regions.

8.3 Changing geographies of the digital economy

While modern transport systems underpin the movement of goods, physical materials and people between different places in the global economy, ICTs play a crucial role in enabling the movement of information,

ideas, money and organisational practices. The current generation of ICTs is based upon the convergence of two initially distinct technologies: communication technologies concerned with the transmission of information across distance, and computer technologies concerned with the processing of information (Dicken 2015, p. 80). Crucially, both now operate in the form of digital rather than analogue technology involving the storing of information in numerical form as electronic 'digits'. The combination of digitisation and the internet is of great significance, allowing information to be collected, packaged and distributed more rapidly and efficiently than before, alongside the spread of digital devices and applications. As a general purpose technology that underpins a huge number of more specific applications, information technology (IT) and the internet had permeated virtually all sectors of the economy by the 2000s (Figure 8.4). This means that the **digital economy** can be defined as "the pervasive use of IT (hardware, software, applications and telecommunications) in all aspects of the economy" (Moriset and Malecki 2008: 259). At the same time, access to ICT and the internet remains highly unequal between social groups, generating **digital divides** between those who enjoy rapid, convenient access and those who do not.

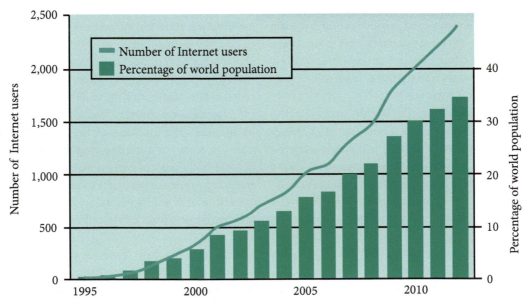

Figure 8.4 Exponential growth of the internet
Source: Dicken 2015: 91.

8.3.1 Processes of agglomeration and dispersal in the digital economy

Despite the influential 'death of distance' thesis (Cairn-cross 1997), the tendency for ICT-enabled digital and creative industries to be characterised by geographical clustering is well established. While the global reach of the internet creates the theoretical possibility for business activity to become widely dispersed across locations with internet access, its physical infrastructure is not equally distributed across space. Instead, as many studies have emphasised, it is particularly concentrated in large cities and metropolitan regions which act as the critical nodes in global communications networks. This reflects the economic logic of infrastructure provision which is driven by the geographically uneven demand for services associated with the clustering of firms and internet users in major cities (Gillespie *et al.* 2001). These cities benefit from not only their status as key nodes in telecommunications networks, but also from direct end-to-end links with the main world cities in which TNCs are based, providing secure, fast and reliable connections (Tranos 2012).

More broadly, major cities benefit from what Castells (2010: 2741) calls "a multi-dimensional infrastructure of connectivity: multimodal transport on air, land and sea; telecommunication networks; computer networks; advanced information systems; and the whole infrastructure of ancillary services (from accounting and security to hotels and entertainment) required for the functioning of the node". This provides them with distinct competitive advantages, providing better access to the digital economy and leading to reduced costs and increased revenues for firms. In the US, Mack (2014) found that broadband provision was associated with specialisation in knowledge-intensive clusters in core metropolitan counties, alongside the outer fringes of knowledge-intensive metro areas (enabling firms to reduce costs whilst still benefiting from proximity to knowledge networks) and former manufacturing centres which have lower business costs and good transport links.

The re-agglomeration of ICT-enabled informational and creative industries in cities reflects a set of wider factors operating in conjunction with the concentration of ICT infrastructure. Geographical proximity between firms remains crucial in many high-value activities and sectors, allowing for the creation of trust and the exchange of tacit knowledge and meaning through face-to-face communication (Box 8.2). This remains vital to the operation of business in sectors such as finance, business services and creative industries which exhibit high levels of clustering in world cities. Here, Castells (2010: 2741) draws a useful distinction between what he terms the "micro network of the high-level decision-making process, based on face-to-face relationships" and a linked "macro network of decision implementation, which is based on electronic communication networks". As such, face-to-face meetings are still central to the agreement of financial or political deals, "particularly where there is a need for absolute discretion in the case of decisions that provide a competitive edge" (ibid.).

The availability of skilled labour is also central to the clustering of creative and digital economy activities in cities (Box 8.2) (section 10.2.1). As well as substituting for routine forms of labour, computers also complement and reinforce the capacities of non-routine labour (Scott 2011). This is reinforcing the segmentation of the labour force into a group of highly skilled 'symbolic workers' engaged in various forms of sophisticated analysis, judgement, creative problem-solving, communication and social interaction, and a low-wage tier of workers involved in routine service provision (section 6.3.4). The former group of workers are highly, although not exclusively, concentrated in cities, providing a large pool of well-educated and relatively young workers who are open and receptive to new ideas and fashions. Their employment generally takes the form of time-limited, project-oriented work, with close interaction leading to mutual learning and joint discovery. The clustering of these activities provides the volume of demand that allows creative workers to move from one job to another and build continuous employment and regular face-to-face communication (Pratt 2013).

At the same time, the digital economy is also characterised by patterns of spatial dispersal. This has largely been confined to routine, lower-value activities thus far (Warf 2013), although there are signs of this extending into higher value-added activities through processes

Box 8.2

The video games industry in Montreal

The video games sector is an important and rapidly growing part of the digital economy. Born digital, the sector's growth has been based on rapid software development, enabling it to provide user-friendly and increasingly intuitive services on a very large scale (De Prato *et al.* 2010: 14). The production of a video game requires collaboration between a range of creative specialists: writers, game designers, graphic artists and sound engineers. It is characterised by geographical concentration in major cities around the world.

The video games industry in Montreal emerged in the early 1990s to become the largest cluster in Canada, growing from approximately 400 employees in 1996 to over 8,000 by 2012 (Darchen and Tremblay 2015:

321). It is comprised of major international developers as key anchor investors, a number of start-up firms and hardware, middleware and software companies, alongside a range of service organisations, mainly clustered in central districts of Montreal (Figure 8.5). Whilst the cluster was affected by the post-2008 economic crisis, experiencing some job losses, it is regarded as a mature cluster with important underlying strengths that are likely to sustain it into the future.

The growth of the industry in Montreal reflects its position as a kind of cultural bridge between Europe and North America, based upon its promotion of French language and culture, as well as a strong digital animation culture and sustained policy support.

The head office of the National Film Board (NFB) is located in Montreal, which created a basis for the growth of digital animation firms. The provincial government of Quebec provided financial support for job creation by firms locating in its Multimedia City zone in central Montreal, paying 25 per cent of their wage bill. This support was crucial to the attraction of the French firm Ubisoft, which established a video game studio in 1997. Cross-fertilisation with related sectors such as film, TV and animation has increased over time as incoming firms and actors have become embedded in the city, resulting in increased knowledge exchange. Montreal has gained a reputation as a 'low-cost high creativity' cluster, reflecting a culture of innovation and entrepreneurship coupled

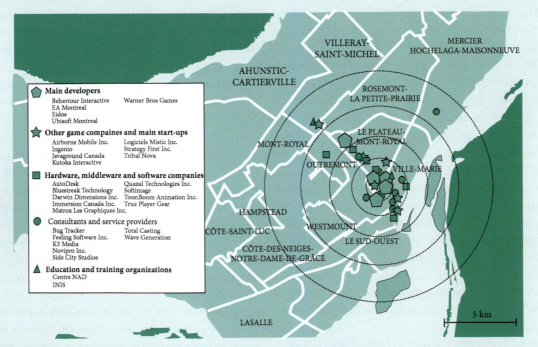

Figure 8.5 The video games industry in Montreal, Quebec, Canada
Source: Grandadam *et al.* 2013: 1707.

Box 8.2 (continued)

with low wages and rents (Darchen and Tremblay 2015).

The cluster is supported by a number of collective associations and events. These include the Society for Arts and Technology (SAT), founded in 1996, functioning as both a creative laboratory and multipurpose venue which presents a large number of live events (Grandadam *et al.* 2013). Another key group is the Alliance Numerique, founded in 2001, which acts as the business network for Quebec's new media and interactive digital content (ibid.). Since 2004 it has organised the annual Montreal International Game Summit which fosters interaction between local and international firms, generating new ideas and capabilities. These events and activities are strongly supported by the key video games companies in Montreal who actively encourage their employees to participate.

of automation. Here, ICT enables the vertical disintegration or 'unbundling' of value chains into a number of discrete elements or modules, referring to not only material inputs such as electronics components, but also various non-material goods and services like processor testing, customer relations and sales support (Moriset and Malecki 2008). Furthermore, the rapid communication and scope for remote coordination offered by ICT allows TNCs to engage in policies of global sourcing, based upon the **outsourcing** of modules to specialist suppliers and the '**offshoring**' of this work to different countries (section 3.5).

Examples of this process of dispersed production include the production of Boeing's 787 Dreamliner aircraft which involved the combination of Boeing's own internal design and management expertise with the outsourcing of parts production and elements of final assembly to a total of around 50 Tier 1 suppliers or partners who, in turn, sourced inputs and parts from Tier 2 and 3 suppliers. As such, the production process was dispersed across a number of countries. There are limits to this model, however, as reflected in the fact that this programme was affected by serious coordination problems, along with the similarly organised production of Airbus's A320 airliner, resulting in serious delay (ibid.). By contrast, IT-enabled services are the business sector most amenable to dispersal since their outputs are purely informational rather than material. ICT service companies accordingly engage in elaborate forms of global sourcing, extending to so-called '**homeshoring**' where geographically isolated workers work from home. Whilst this reflects the drive to reduce costs, it has its problems, requiring initiatives to periodically bring workers together to maintain morale and productivity.

Crucially, the relationship between processes of spatial concentration and dispersal should be viewed as dynamic and shifting, with advances in technology often opening additional possibilities for dispersal, although these are often counter-balanced by the human and social benefits of co-location. Interestingly, Moriset and Malecki (2008) suggest that some of the need for proximity at the micro level for individual businesses may be reduced in the future through technological improvement and generational change as 'digital natives' (Prensky 2001) come to dominate the future workforce. At the same time, agglomeration will persist at the macro level due to the enduring benefits of agglomeration economies and the importance of social interaction.

8.3.2 Digital divides

Inequalities in access to ICTs and the internet specifically are a persistent feature of the digital economy, generating a range of policy initiatives to try and address these inequalities. Many studies have shown how access and use are conditioned by income, class, age, gender and race with the poor, the elderly, the undereducated, women and minorities often demonstrating lower levels of access and/or distinctive patterns of use (Warf 2012). Digital inequalities assume different forms, encompassing not only the traditional focus on access to technology, but also differences in the quality and speed of connection and in practices of use and engagement. As internet access has spread (Figure 8.4), differences in the speed and quality of connection and usage are attracting more attention, particularly in developed countries with the highest rates of connection. Accordingly, the rather simplistic idea of a singular 'digital divide' between those who enjoy internet access and those who do not is increasingly giving way to the multifaceted concept of 'digital differentiation', based upon a more nuanced understanding of the differences in patterns and practices of use and engagement between social groups (Longley 2003). These differences can be found at different geographical scales. Here, we focus on the global, intra-national (between cities and regions within countries) and urban (between areas within cities) scales.

Global patterns of internet use show that pronounced divides between developed and developing countries persist (Figure 8.6), despite the rapid growth of the internet in the latter. Just over a third of the population of developing countries use the internet compared to 82 per cent in developed countries (United Nations 2015: 68). In sub-Saharan Africa, less than 21 per cent of the population uses the internet (ibid.). Despite the rapid growth and spread of the internet, particularly in the global South over the past decade, most people in the world remain without access to it (Graham *et al.* 2015: 92). In addition to the uneven provision of internet infrastructure, these global inequalities reflect the relative costs of internet connection (Graham *et al.* 2015). At the same time, however, access to mobile phones has grown dramatically in developing countries, rising from 22.9 per cent of the population in 2005 to 94.1 per cent in 2016, reflecting the greater affordability and flexibility of this technology (ITU 2016).

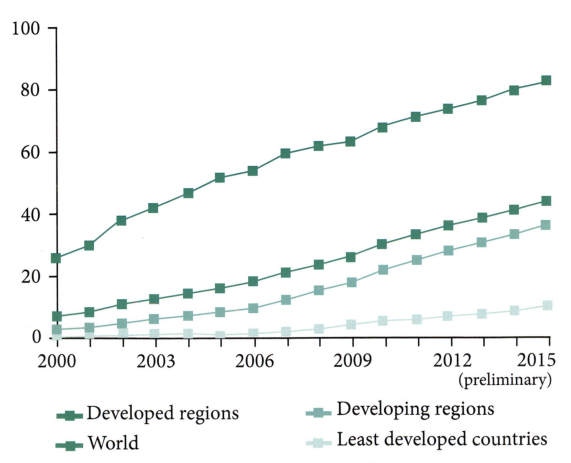

Figure 8.6 Number of internet users per 100 inhabitants 2000–15
Source: United Nations 2015: 68.

Internet access and use is also characterised by significant urban and regional differences with major cities generally enjoying the fastest and most reliable services. By contrast, connections are often slower and less reliable in rural areas, reflecting the negative effects of distance from telephone exchanges and sparsity of population on the quality of transmission and investment in infrastructure respectively (Riddlesden and Singleton 2014). In the US, access to high-speed internet usage is higher in the wealthier, more educated states in the west and north-east and lower in most southern states (File and Ryan 2014). This pattern is also evident at the metropolitan scale. The majority of metropolitan areas with internet access rates of at least 5 per cent above the national average located in the west, Midwest and north-east, while most of the areas in which rates are at least 5 per cent lower are in the South (Figure 8.7). In Europe, rates of usage are higher in northern and western countries than the south and east, reflecting long-standing socio-economic disparities (Warf 2013). The urban-rural divide in internet access is particularly pronounced in the global South, with often stark differences between the major cities

and rural areas in which much of the population live, particularly in sub-Saharan Africa. In general, internet users in the global South tend to be younger, more affluent, educated and male. Internet cafes have grown rapidly in many countries, largely but not exclusively in cities, with their popularity reflecting the high costs of connection relative to average incomes and lower levels of computer ownership.

Whilst highlighting aggregate patterns of internet access and usage, the research outlined above tells us little about how different groups actually engage with internet technologies. Here, qualitative research provides some valuable complementary insights into the relationship between urban inequalities and the use of ICTs. The work of Crang et al. (2006) in Newcastle upon Tyne in northern England, for instance, found that affluent and professional people in middle-class areas used ICTs pervasively and continuously to sustain their relatively privileged lifestyles, although this individualised sense of always being 'online' was associated with increased time pressures and stress. By contrast, working-class residents in lower-income areas experienced a more 'episodic' pattern of connection where

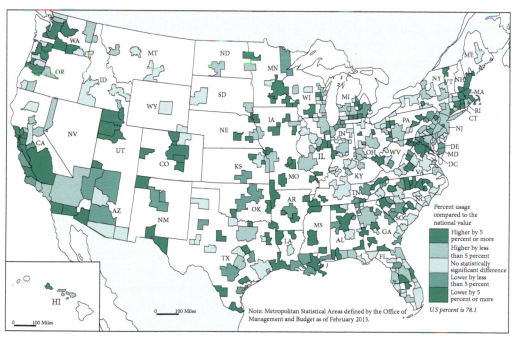

Figure 8.7 High-speed internet use for individuals by Metropolitan Statistical Area 2013
Source: File and Ryan 2014: 14, Figure 7.

ICTs were used for specific, instrumental purposes (for example, booking cheap holidays), often supported by neighbourhood social relations through the sharing of information between family and friends. For low-income residents, this 'episodic' pattern of connection is often punctuated by regular malfunctions and disconnection, reflecting their use of cheaper and less reliable technology and difficulties in paying connection charges and repair bills (Gonzales 2016). This illustrates the crucial interaction between digital inequalities and overlapping, "offline axes of inequality" such as education, employment and income (Robinson *et al.* 2015: 570).

> ### Reflect
>
> In what ways are future technological advances likely to change the relationship between processes of geographical concentration and dispersal?

8.4 The sharing economy

8.4.1 Characterising the sharing economy

There has been an explosion of interest in the concept of the 'sharing economy' in recent years, symbolised by the rapid growth of two Silicon Valley-based platforms: Airbnb and Uber (Box 8.3). Also referred to as the 'digital platform economy' or 'gig economy', the sharing economy refers to forms of economic "exchange that are facilitated through online platforms, encompassing a diversity of for-profit and non-profit activities that all broadly aim to open access to under-utilised resources through what is termed sharing" (Richardson 2015: 121). This definition has three main characteristics. First, the sharing economy is based upon the use of online platforms as digital intermediaries that greatly reduce the costs of connecting large numbers of potential producers and consumers, according to algorithms and reputational rankings controlled by the platform owners (Rabari and Storper 2014). Second, it is peer-to-peer, involving direct transactions between the buyers and sellers of goods and services. Third, it is access-based, enabling participants to buy access to, rather than

ownership of, a resource or service (such as overnight accommodation in the case of Airbnb) for a period of time (Box 8.3). Here, the internet allows for the more efficient use of previously under-utilised resources (for example spare rooms, seats in vehicles, labour skills and time). From a geographical perspective, cities play a key role in fostering the sharing economy, providing the density and critical mass of producers and consumers that underpin the operation of digital platforms (Davidson and Infranca 2016), whilst also increasing economic interactions between cities, permitting residents of one area to access resources and services based in a distant location.

The use of the term 'sharing' is controversial since the common sense definition of sharing excludes forms of exchange in which a monetary benefit accrues to one or more parties (Martin 2016). While the concept of the sharing economy spans both monetary and non-monetary forms of exchange, it often involves companies using the internet to facilitate transactions between buyers and sellers for a fee (O'Connor 2016). The widespread adoption of this label reflects how the wider idea of sharing based on non-monetary forms of collaboration and cooperation has been successfully appropriated by advocates of the 'sharing economy'. While supporters often present the sharing economy as an alternative form of economic organisation that has the potential to empower individuals and disrupt established business models, critics have emphasised its damaging effects such as unfair competition, the casualisation and exploitation of labour, and the promotion of tax avoidance, with Morozov (2013) dubbing it 'neoliberalism on steroids'. Although we continue to use the term 'sharing economy' in this chapter as it remains useful in capturing the breadth and depth of the new forms of economic activity we are concerned with (Richardson 2015: 122), its ambiguous and contested nature should always be borne in mind.

Martin (2016) identifies four main groups of innovations which correspond to key sharing economy sectors: accommodation-sharing platforms; car- and ride-sharing platforms; peer-to-peer employment markets; and, peer-to-peer platforms for sharing and circulating resources (Table 8.2) (Box 8.3). While the first two have captured much attention through the explosive growth of Airbnb and Uber (Box 8.3), online platforms are also being used to organise and allocate work to often freelance workers through platforms such as TaskRabbit, Upwork and Amazon's Mechanical Turk. The inclusion of platforms for sharing and circulating resources highlights the potential for sharing to encompass 'alternative' modes of exchange that promote a more sustainable use of resources.

8.4.2 The growth of the sharing economy

It has proved difficult to estimate the size of the 'sharing economy', which has grown rapidly and is not fully captured by official statistics (Brinkley 2016). Nonetheless, PWC (2015) estimated that it was worth $15 billion in 2015 with the scope to increase to $335 billion by 2025. In a European context, Vaughan and Daverio (2016) estimated that the sharing economy generated revenues of nearly $4 billion and facilitated $28 billion of transactions in 2015. Another survey estimated that it accounted for 1.1 million workers in Great Britain (Balaram *et al.* 2017: 13), while research suggests that much of this 'sharing economy' work is professional and supplementary to full-time work (ibid.; Schor 2017). Focusing on the US, Hathaway and Muro (2016) show that non-employer firms (covering the independent contractors and freelancers who participate in the sharing economy) in ride-sharing and home-sharing grew by 69 per cent and 17 per cent respectively between 2010 and 2014 compared with only 17 per cent and 7 per cent for payroll employment in the same period. This growth occurred mainly in the largest metro areas with 81 per cent of the four-year net growth in non-employer firms in the rides sector taking place in the 25 largest metro areas in the US, reflecting the urban orientation of the sharing economy. As the figures indicate, platform-based freelancing still accounts for only a relatively modest share of overall employment, although continued growth is likely over the next few years (Brinkley 2016).

Table 8.2 Groups of innovations in the sharing economy

Groups of innovation	Broader sector	Examples of sharing economy platforms	Description
Accommodation-sharing platforms	Tourism, ICT	Airbnb	A peer-to-peer marketplace for people to rent out residential accommodation (including their homes) on a short-term basis
		Couchsurfing	An online community of people who offer free short-term accommodation to fellow community members
Car- and ride-sharing platforms	Transport, ICT	Easy car Club and Relayrides	Peer-to-peer car rental platforms
		Lyft and Uber	Peer-to-peer platforms providing taxi and ride-sharing services
		Zipcar	A business-to-consumer vehicle rental platform offering per hour rental of vehicles located within communities
Peer-to-peer employment markets	Employment, ICT	PeoplePerHour and Taskrabbit	Peer-to-peer marketplaces for micro-employment opportunities (i.e. piecemeal contracts or hourly work)
Peer-to-peer platforms for sharing and circulating resources	Waste disposal, production-consumption, ICT	Freecycle	A peer-to-peer platform which enables people to freely and directly give unwanted and underutilised items to others in their local area
		Peerby and Streetbank	Peer-to-peer platforms which enable communities to freely share durable goods, skills and knowledge

Source: Martin 2016: 152.

Box 8.3

The growth of Airbnb

Founded in 2008 in San Francisco, Airbnb is an online platform that links people looking to rent their rooms or homes with people seeking short-term accommodation. It has grown rapidly, providing more than 2 million listings worldwide in over 34,000 cities and 191 countries. Airbnb is now valued at $30 billion (Hook 2016), making it larger than many traditional hotel groups such as Hilton or Hyatt. The platform works by charging a flat commission from hosts (3 per cent) and a transaction fee to guests (between 6 and 12 per cent depending on the booking price). Airbnb's operation as a peer-to-peer digital platform that allows people to purchase access to short-term accommodation exemplifies the three features of the sharing economy emphasised above. Its hosts are generally motivated by generating extra income from their space while guests are often looking to save money and obtain an authentic travel experience (living like a local) (Vaughan and Daverio 2016). Its growth reflects its provision of a generally cheaper alternative to traditional accommodation providers (hotels etc.) with a focus on urban conurbations popular with international tourists, often in non-central districts that are not well served by these traditional providers.

Until recently, Airbnb's growth had been largely untroubled by legal and regulatory restrictions, but it has subsequently faced growing curbs. These reflect the controversy generated by its impact on the rental sector in cities, particularly concerns that holiday lets are reducing the availability of properties for residential letting, and the tax and regulatory advantages it has enjoyed over traditional accommodation providers. In London, for instance, research by the *Financial Times* showed that around one-third of the average £100 saving over the price of an average hotel room is due to tax advantages, reflecting high rates of business property tax and Value Added Tax (VAT) on hotel stays, coupled with generous tax exemptions for owners renting rooms in their home and for small businesses (Houlder 2017) (Figure 8.8). This has generated something of a backlash, prompting regulators to introduce restrictions and fines in several cities, including New York, Berlin, Barcelona and San Francisco. In contrast to the more aggressive stance of Uber, the company has adopted a conciliatory approach, blocking its hosts from exceeding legal restrictions on room renting.

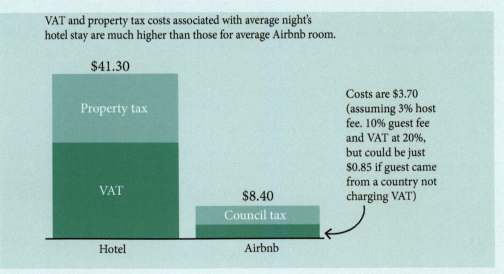

Figure 8.8 Airbnb's tax saving in London
Source: FT research in Houlder 2017.

A number of recent surveys provide some insights into levels of participation in the sharing economy and people's reasons for participating. For instance, one study of 3,000 Americans found that 44 per cent were active on sharing economy platforms, with 42 per cent using services and 22 per cent providing services through at least one platform (De Groen and Maselli 2016: 3). These are fairly evenly spread between different types of services (Figure 8.9). Another survey found that 24 per cent of US adults have earned money from the 'digital platform' economy, with 8 per cent earning this from an online employment platform and 18 per cent from selling something online (Pew Research Centre 2016: 2). A European survey of 'crowd work' found that 8 to 16 per cent of survey respondents across four countries had rented out rooms to paying guests (e.g. Airbnb), while 9 to 19 per cent had undertaken 'crowd work', defined as paid work via an online platform (Huws *et al.* 2016).

Participation tends to be higher among younger age groups, the well educated and relatively high income, with some US studies showing a higher share of racial or ethnic minorities compared to the general population. A study of Uber drivers in the US, for instance, showed that they were younger, more often female and more highly educated than conventional taxi drivers (Hall and Krueger 2015). A study of five European countries, however, indicated that gender differences in participation were relatively limited, with men tending to constitute a small majority of the crowd workforce (Huws *et al.* 2016). This study also showed that crowd workers were also more likely to be from younger age groups, although they could be found in all age groups, confounding the stereotype that such work is confined to the young. In general, crowd work provides a small supplement to total income, contributing only 10 per cent or less of all income for around 45 per cent of survey respondents (ibid.). This is echoed by another US study of three platforms (Airbnb, Relayrides and TaskRabbit) which found that most participants were highly educated people who were using the platforms to increase their earnings, often by doing low-status work (Schor 2017). This suggests that the sharing economy may be increasing income inequality as this crowds out less educated and affluent workers who have traditionally undertaken much of this manual work.

8.4.3 Working in the sharing economy

Digital labour platforms essentially match people who require certain jobs or tasks to be done with people willing to do the job in question. Here, the growth of digital technologies is accelerating a broader shift towards freelance work and independent contracting. These

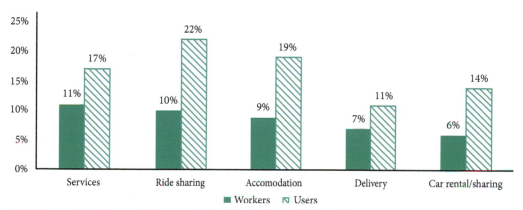

Figure 8.9 Share of US population participating in the collaborative economy
Source: De Groen and Maselli 2016.

labour platforms vary considerably in terms of the skills and employment status of workers, including highly skilled tasks for professionals such as UpWork but also lower-skilled ones like TaskRabbit and ListMinut (a Belgian platform for the provision of local personal services). Their rise has generated considerable controversy about the implications for workers and employment rights. Advocates of the 'sharing economy' emphasise the benefits of flexibility and personal control for workers who can choose the nature and volume of work they do, often fitting this around family and other commitments. Critics, by contrast, highlight low wages and insecurity, alongside a lack of employment rights with workers defined as independent contractors rather than employees, in addition to the potential of the technology to displace labour (Rogers 2015).

On average, Vaughan and Daverio (2016) estimate that over 85 per cent of the revenue generated by sharing economy platforms accrue to the provider rather than the platform, although this is likely to be substantially lower for labour platforms, with ride-sharing services such as Uber, for example, charging a commission of up to 20 per cent. In general, most platforms do not provide sufficient work to generate a comparable income to that earned from full-time jobs, although wages vary considerably between them, with suppliers of physical/local services earning more than suppliers of virtual services and high-skilled performers earning more than low- to medium-skilled ones (De Groen and Maselli 2016). The platforms generally play a key role in setting the conditions of workers, with online ratings and reviews used to allocate tasks and select workers. Some platforms such as Uber require workers to accept a defined proportion of tasks offered and to achieve high average ratings from users. In addition, earnings and commissions are not always made clear to workers in advance and not all work is compensated, particularly not the time spent finding a task on the platform.

The explosive growth of digital platforms in recent years has generally outstripped the capacity of government to regulate their activities, generating a number of controversies over compliance with existing local and national regulations, employment rights and the status of workers, licensing and certification, and taxation (Box 8.3). Uber, for instance, has been criticised for ignoring local taxi regulations and undercutting existing taxi operations, prompting legal challenges to its right to operate and vocal protests from taxi drivers in several cities (Davies 2016). This led Spain and Brazil, for example, to impose national bans on Uber, although the Brazilian ruling was later overturned, while other cities such as London have banned particular Uber services and imposed other legal and licensing restrictions. The growth of Airbnb has also raised the issue of whether room renters should be subject to hotel taxes and regulations with several countries now requiring renters to obtain prior permission from local authorities and to pay tourist taxes (Box 8.3).

On the question of employment status and rights, some workers have begun to mobilise to protest against their self-employed status and demand fuller rights. In a landmark ruling, a UK employment tribunal ruled that Uber drivers were not self-employed and entitled to the national living wage, holiday pay and other standard employment benefits (Osborne 2016). This reflects the fact that drivers are interviewed, recruited and controlled by Uber which sets their routes and fixes their fares. This ruling has huge implications for the 'sharing' or 'gig' economy, which relies on large numbers of independent contractors who often lack basic employment rights as a result of their self-employed status. Given that most online employment platforms are still in their infancy, there are likely to be further disputes between groups of workers and the firms behind the platforms, reflecting an underlying tension between technology and social regulation.

Reflect

Should the 'sharing economy' be subject to the same rules and regulations as other forms of economic activity? Justify your answer.

8.5 Smart cities

8.5.1 New forms of data-driven governance?

According to *The Economist* (2012), the proliferation of ICTs is turning cities into "vast data factories" with the "physical and digital world . . . becoming increasingly intertwined". This reflects the rise of 'big data' in recent years, referring to information that is captured digitally by modem ICTs such as networked sensors, 'smart' objects and devices, the internet and social media (Rabari and Storper 2014: 2). It can take the form of text, web data, tweets, sensor data, audio, video, click streams, log files, social network chatter, mobile phone GPS trails and smart energy meters. This data is often generated by individual users, but generally owned by the private corporations who control and manage the platforms through which it is exchanged. The volume of available data is unprecedented, underpinned by over 10 billion connected consumer devices (ibid.). As a result, private corporations such as IBM, Cisco, Intel, Siemens and Google are becoming key partners in urban governance and management, based upon the idea that the utilisation of 'big data' can promote more effective solutions to a range of long-standing urban problems such as traffic congestion, blight and waste.

The term 'smart cites' emphasises smart cities harnessing this data and the power of ubiquitous computing to address these urban problems. According to the geographer Rob Kitchin (2015), it refers to the use of ICTs to stimulate economic development and data derived from software-embedded technologies to augment urban management. The first element views the urban economy as driven by technological innovation and creativity, based upon the idea that investment in technological infrastructures and programmes, alongside associated human capital, will attract business and jobs, create efficiencies and increase the productivity of government and businesses. As such, the acceptance of economic goals and the notion of urban competitiveness are central to the smart city concept, reflecting its corporate orientation and role in reinforcing and extending the neoliberal direction of urban policy. The second aspect emphasises the use of ICTs to manage and govern cities more effectively, particularly through the provision of more efficient and responsive public services. In particular, the use of technology enables the coordination of previously fragmented urban sub-systems (for example, energy, water, transport, built environment) through processes of digitisation and interconnection, providing a basis for intelligent decision-making. The underlying assumption here is that the application of ICT in cities will benefit everyone with all citizens sharing in the prosperity and wealth generated (Hollands 2015).

The smart city is associated with a new mode of data-driven urban governance, embodied by the emergence of new inter-organisational partnerships and alliances (Shelton *et al.* 2015). In particular, private corporations play a crucial role, reflecting the sense in which they are at the cutting edge of developments in ICTs, possessing valuable technical expertise in areas such as computer programming and data analytics. This means that what Rabari and Storper (2014) call the 'digital skin' of cities is heavily structured by the algorithms, search channels and economic interests of these private actors. The approach requires a transition from the traditional model of silo-based city government with separate functional departments to a more collaborative and integrated model of service delivery (Glasmeier and Christopherson 2015: 4). Yet this emphasis on integration sits very uneasily with the underlying tendency for urban infrastructure and service management to become ever more 'splintered' as a result of the neoliberal 'unbundling', outsourcing and privatisation of urban services (Graham and Marvin 2001). In recent years, private corporations such as IBM and Cisco have begun to use a language of inclusivity and citizen empowerment rather than top-down management to promote their initiatives (Kitchin 2015). This serves to head off potential criticisms of corporate dominance whilst keeping the underlying "mission of capital accumulation and technocratic governance intact" (ibid.: 133).

8.5.2 Smart cities in practice

Smart city projects have been introduced in a number of countries in recent years, although these are often marketing initiatives based on a handful of pilots. These include new cities being built on greenfield sites to 'smart city' specifications. Amongst the best-known examples of these are Masdar City in the United Arab Emirates which is intended to become the world's first sustainable and renewable energy-based clean-tech cluster and Songdo, South Korea's environmentally sustainable, high-tech business city (Box 8.4) as well as Living Plan IT Valley in Portugal. These examples of new cities being built from scratch are the exceptions, however, underlining the need to understand how smart city polices are implemented in a range of established cities (Sheldon *et al.* 2015). In Europe, Barcelona is renowned for its smart city model, hosting successive Smart City Expo World Congresses, while Amsterdam is often held up as an example of how to retrofit a city to improve living and economic conditions and reduce carbon emissions (Hollands 2015: 65). In addition, Helsinki and 'Intelligent' Thessaloniki (Greece) have gained international recognition for efforts in utilising open data and ICT to foster competiveness and sustainability respectively.

Box 8.4

New Songdo City, South Korea

Songdo City is commonly described as the world's first purpose-built smart city (Figure 8.10), based on land reclaimed from the Yellow Sea in north-west Korea, some 64 kilometres (km) west of Seoul and 11 km from Incheon international airport (Kshetri *et al.* 2014). Songdo City is located in the Incheon Free Economic Zone (FEZ) and was designed to attract foreign direct investment, offering a range of benefits including tax breaks, financial support and deregulation, supported by its strategic position on the trade route with China and Japan. Since 2001, the development has been led by a New York-based developer, Gale International, with a 70 per cent stake, and Posco Engineering and Construction, a South Korean *chaebol* or conglomerate which holds the remaining 30 per cent stake. The international corporation Cisco Systems joined the project in 2009 to provide the ICT infrastructure, based upon the provision of ubiquitous computing systems, embodying the underlying principle of the smart city model.

As a high-tech eco-city, the design of Songdo incorporates sensors to

Figure 8.10 New Songdo City, South Korea
Source: www.theguardian.com/cities/2014/dec/22/songdo-south-korea-world-first-smart-city-in-pictures.

Box 8.4 (continued)

monitor temperature, energy use and traffic flow, providing information to both citizens and the local authority. This ICT infrastructure enables the integration of the residential, transport, energy and business information sub-systems in the city. Fibre-optics and high-speed wireless connections will be provided to every home with every resident having a smartcard that will operate as their personal key to a range of services (ibid.). The transport system emphasises safe and carbon-free modes of travel, including a 25 km network of bicycle lanes, supported by the widespread availability of bicycles for rent. The city has been planned around a central park, designed to allow every resident to walk

to work in the central business district (Figure 8.10). The waste disposal system is highly advanced, involving the direct extraction of household waste from kitchens through an extensive underground network of tunnels to waste processing centres. Landmark developments include the Songdo Conversia Convention Centre, completed in 2008 and the North East Asia Trade Tower.

Despite these developments, the realisation of the original masterplan has proved difficult, resulting in several modifications. Completion is behind schedule, having slipped back to 2020 rather than the original date of 2014. The project is about 70 per cent complete with a population of

90,000 in greater Songdo (Shapiro 2015). Fewer than 20 per cent of the commercial offices were occupied in September 2013 and streets, cafes and shopping centres have often been empty (Kshetri et al. 2014). Despite the leading role of Gale International, the project has been unsuccessful in attracting foreign investment, becoming more residential and Korean than originally envisaged, with its high quality of life making Songdo attractive to families from Seoul and the neighbouring region. While the developer plans to use the Songdo model to build other new smart cities in China and India (Hollands 2015), its development has involved considerable improvisation, risk and uncertainty.

Cities' engagement with 'smart' concepts has been promoted by particular corporate and policy initiatives, such as IBM's Smarter Cities Challenges which paired IBM's consultants with city governments to develop technological solutions to urban problems and the UK government's Future Cities Demonstrators Competition which awarded £24 million to the winning city (Glasgow). One of the cities that participated in the IBM programme was Philadelphia, through its Digital On-Ramps project which developed a social media-style workforce education application (app) to prepare low-literacy residents for jobs in the information and knowledge economy (Wiig 2016). This proved unsuccessful, reflecting miscommunication with the target groups, problems with the software and the lack of a clear pathway to defined jobs. This underlines the need for close user interaction in the design phase of projects (Glasmeier and Christopherson 2015). Here, the smart city functioned primarily as a promotional label for the city, highlighting its effort to produce a skilled and competitive workforce, despite the lack of clear results. The failure of this particular project reinforces the wider point about the limitations of narrowly technological solutions to complex and entrenched

urban social problems such as inequality and a lack of employment skills. As the above discussion of 'digital divides' indicates (section 8.3.2), technological applications cannot overcome these problems by themselves. Instead, they remain constrained by issues of uneven access to, and use of, technology that reflect cost and the differential skills and engagement of different social groups. In addition, the algorithms that are used to process and interpret data are controlled by private corporations, remaining largely hidden from residents, generating a new round of information asymmetry and technical complexity for citizens (Rabari and Storper 2014).

The corporate dominance of smart cities is far from complete, however, as digital technologies are employed by a range of citizen movements, NGOs and community groups. These efforts have been central to the growth of the 'open data' movement which has made public datasets available for free use, re-use and redistribution, alongside websites that foster interaction between users and public decision-makers. These groups have successfully adopted smart city technology to engage in public debate and campaign for urban improvements. For the critical urban sociologist Rob Hollands (2015), the most promising initiatives are those that are based on collective ideas and action, using technology to support and strengthen these ideas rather than seeing it as the sole driving force. Examples include the urban crowd-source idea Brickstarter, a Low Impact Living Affordable Community (LILAC) housing project in Leeds and 596 Acres, a project that aims to turn Brooklyn's 596 acres of community-owned land into common use by community groups and individuals. These projects highlight the social, political and cultural dimensions of urban problems such as poverty and inequality, neighbourhood decline, lack of open space and traffic congestion that cannot be solved by simple technological solutions or more sophisticated forms of data-gathering (ibid.). Whilst small-scale, they offer a more inclusive, alternative vision of what smart cities might look like, in contrast to the top-down technocratic models of private companies and city governments.

Reflect

To what extent do you agree with the proposition that the widespread availability of 'big data' provides the basis of more effective solutions to long-standing urban problems?

8.6 Summary

This chapter has focused on increased connectivity between cities through the 'space-shrinking technologies' of transport and ICTs. These technologies underpin processes of globalisation through the faster and cheaper movement of goods, services, information, money and people across distance. They have been widely supported and promoted by policy-makers and business elites for their apparent capacity to promote urban and regional growth and to reduce regional inequalities. Investments in transport infrastructure, for instance, have often been justified not only in terms of fostering greater connectivity, but also in terms of reducing disparities between core and peripheral regions. Even greater emphasis has been placed on the potential of ICTs to overcome the constraints of

geography, fostering claims that they would bring about the 'death of distance' by enabling the instantaneous exchange of goods, services and information from any location through the internet (Cairncross 1997). As a general purpose technology, the internet has permeated virtually all sectors of the economy, underpinning the growth of the digital economy. The recent emergence of the 'sharing economy', based on the proliferation of online platforms, is a deeply urban phenomenon and generates new forms of connection between cities, allowing people to access previously under-utilised resources and services that may be located in other areas. The concept of the 'smart city' refers to the governance dimensions of technology, emphasising the scope for city authorities and their corporate partners to harness the power of big data and ubiquitous computing to develop more integrated and effective solutions to a range of urban problems.

In practice, however, there is little evidence to support inflated claims about the 'death of distance' through the advancement of transport and communications technology. What is apparent, instead, is the persistence of real economic geographies in terms of the continued importance of location, and the perpetuation and, in some cases, the widening of disparities between core and peripheral regions (Rabari and Storper 2014). Rather than widespread dispersal, ICTs have contributed to the renewed concentration of higher-value economic activities in major cities, based upon the multidimensional infrastructure of connectivity they enjoy, alongside the ongoing need for spatial proximity and face-to-face communication in such activities (Castells 2010), though new technological advances and generational change may prompt increased dispersal in the future. Despite the growth and spread of the internet, digital divides based on differential patterns of access to, and use of, the internet remain evident at a range of geographical scales. These differences have shaped participation in the sharing economy and limit the effectiveness of certain 'smart city' initiatives such as Digital On-ramps in Philadelphia. As such, the use of digital technologies continues to be influenced by various "offline axes of inequality" (Robinson et al. 2015: 570) such as income, education, gender, age and race in complex and shifting ways. This reflects their role in amplifying social divisions: not only extending pre-existing differences into the online realm, but also generating distinctive forms of digital disparity (ibid.).

Exercise

Select a key sector of the economy (for example retail, tourism, music, video games) and review how its operation and geography have been transformed by the use of digital technologies.

What are the main effects of digital technologies? To what extent has digitisation fostered the development of new products, outputs or services? How has the organisation of production or the delivery of services changed? Has digitisation altered the relationship between producers and consumers? Have new firms emerged to harness digital technologies? How have incumbent (established) firms sought to respond to technological change and new competitors? In what ways has technological change affected the geography of the sector? What forms of agglomeration and dispersal have emerged? How might this change through further technological advances in the future?

Key reading

Castells, M. (2010) Globalisation, networking, urbanisation: reflections on the spatial dynamics of the information age. *Urban Studies* **47: 2737–45.**
An authoritative and concise summary of the geographical effects of ICTs from one of the world's leading sociologists. Emphasises the renewed importance of clustering in major metropolitan centres, drawing an important distinction between high-level decision-making processes which are based on face-to-face relationships, requiring spatial proximity, and linked processes of decision implementation which are channelled through electronic communication networks.

Graham, M., De Sabbata, S. and Zook, M.A. (2015) Towards a study of information geographies: (im)mutable augmentations and a mapping of the geographies of information. *Transactions of the Institute of British Geographers* **2: 88–105.**
This article explores digital divides at the global scale, framed by a broader discussion of contemporary information

geographies. Presents maps of internet access and participation by country, covering internet users, broadband costs, domain names, coding and Wikipedia editing. Also discusses geographies of key digital representations such as Google searches and OpenStreetMap.

Moriset, B. and Malecki, E. (2008) Organization versus space: the paradoxical geographies of the digital economy. *Geography Compass* **3: 256–74.**
A very accessible review of the changing geographies of the digital economy, emphasising the underlying but dynamic tension between forces favouring agglomeration and those favouring dispersal. The article suggests that while agglomeration will persist, its importance may be reduced by further technological advances and the generational changes in the workforce.

Rabari, C. and Storper, M. (2014) The digital skin of cities: urban theory and research in the age of the sensored and metered city, ubiquitous computing and big data. *Cambridge Journal of Regions, Economy and Society* **8: 27–42.**
A thorough and engaging discussion of the forces behind the rise of smart cities, emphasising the growing importance of 'big data' and ubiquitous computing, supported by a new 'digital skin' of digital objects, networks and communication devices. The article considers the implications of this for both urban research and public policy, with particular reference to the governance of cities.

Richardson, L. (2015) Performing the sharing economy. *Geoforum* **67: 121–9.**
The first sustained analysis of the 'sharing economy' in economic geography. Provides a clear definition of the sharing economy, emphasising how sharing is actually 'performed' through particular forms of community, access and collaboration.

Tomaney, J. and Marques, P. (2013) Evidence, policy, and the politics of regional development: the case of high-speed rail in the United Kingdom. *Environment and Planning C: Government and Policy* **31: 414–27.**
Provides a detailed review of the international evidence on the regional impact of high-speed rail (HSR) in the context of debates on the UK's proposed HS2 line between London and northern England. Emphasises the inclination of governments to favour the construction of HS2 based on the anticipated regional benefits, despite the lack of research evidence to support this position.

Useful websites

www.uic.org/high-speed-database-maps

The high-speed rail section of the website of the International Union of Railways (UIC), the international railway organisation. Provides databases and maps of the growth and extent of HSR globally and by different countries.

http://geography.oii.ox.ac.uk/?page=home

Information Geographies at the Oxford Internet Institute. The website of the Oxford Internet Institute which provides a range of data, visualisations and maps on contemporary information and internet geographies, focusing on three key aspects: access, information production, and information representation.

www.itu.int/en/ITU-D/Statistics/Pages/default.aspx

The statistics section of the website of the International Telecommunications Union (ITU) which collects ICT statistics for 200 economies and over 100 indicators. The site provides free ICT statistics, covering broadband, internet use, mobile-cellular and mobile-broadband.

http://smartcities.ieee.org/

The Institute of Electrical and Electronics Engineers' (IEEE) project website on smart cities. Contains a host of useful articles, resources and links covering key elements of smart cities.

References

Albalate, D. and Bel, G. (2010) High-speed rail: lessons for policy-makers from abroad. WP 2010/3, Research Institute of Applied Economics, University of Barcelona, Barcelona.

Balaram, B., Warden, J. and Wallace-Stephens, F. (2017) *Good Gigs: A Fairer Future for the UK's Gig Economy*. London: Royal Society for the encouragement of Arts, Manufactures and Commerce.

Banister, D. and Berechman, J. (2001) Transport investment and the promotion of economic growth. *Journal of Transport Geography* 9: 209–18.

Brinkley, I. (2016) *In Search of the Gig Economy*. Lancaster: The Work Foundation, Lancaster University.

Cairncross, A. (1997) *The Death of Distance*. Boston: Harvard University Press.

Castells, M. (2010) Globalisation, networking, urbanisation: reflections on the spatial dynamics of the information age. *Urban Studies* 47: 2737–45.

Chen, C.-L. and Hall, P. (2011) The impacts of high-speed trains on British economic geography: a study of the UK's InterCity 125/225 and its effects. *Journal of Transport Geography* 19: 689–704.

Chen, C.-L. and Hall, P. (2012) The wider spatial economic impacts of high-speed trains: a comparative case study of Manchester and Lille sub-regions. *Journal of Transport Geography* 24: 89–110.

Crang, M., Crosbie, T. and Graham, S. (2006) Variable geometries of connection: urban digital divides and uses of information technology. *Urban Studies* 43: 2551–70.

Crescenzi, R. and Rodríguez-Pose, A. (2012) Infrastructure and regional growth in the European Union. *Papers in Regional Science* 91: 487–513.

Darchen, S. and Tremblay, D.G. (2015) Policies for creative clusters: a comparison between the video game industries in Melbourne and Montreal. *European Planning Studies* 23: 311–31.

Davidson, N.M. and Infranca, J.J. (2016) The sharing economy as an urban phenomenon. *Yale Law & Policy Review* 34: 215–79.

Davies, R. (2016) Uber admits defeat in China – but has plenty of fight left for new frontiers. *The Observer*, 7 August, p. 37.

De Groen, W.P. and Maselli, I. (2016) *The Impact of the Collaborative Economy on the Labour Market*. CEPS Special Report No. 138, June. Brussels: Centre for European Policy Studies.

De Prato, G., Feijóo, C., Nepelski, D., Bogdanowicz, M. and Simon, J.P. (2010) *Born Digital/Grown Digital: Assessing the Future Competitiveness of the EU Video Games Software Industry*. Joint Research Centre/Institute for Prospective Technological Studies (JRC-IPTS). Seville: European Commission.

Dicken, P. (2015) *Global Shift: Mapping the Changing Contours of the World Economy*, 7th edition. London: Sage.

Ding, C. (2013) Transport development, regional concentration and economic growth. *Urban Studies* 50: 312–28.

The Economist (2012) The new local, October 27, p. 14.

Eddington, R. (2006) *Transport's Role in Sustaining the UK's Productivity and Competitiveness*. The Eddington Report, Main Report. London: HMSO.

European Union (1996) Council Directive 96/48/EC of 23 July 1996 on the interoperability of the trans-European high-speed rail system. *Official Journal L 235, 17/09/1996 P. 0006–0024*.

File, T. and Ryan, C. (2014) *Computer and Internet Use in the United States: 2013*. American Community Survey Report. Washington, DC: United States Census Bureau.

Fujita, M. (2011) Globalization and spatial economics in the knowledge era. *Research Institute of Economy, Trade and Industry (RIETI) 10th Anniversary Seminar*. Tokyo, Japan: RIETI. At www.rieti.go.jp/en/events/tenth-anniversary-seminar/11011801.html. Last accessed 29 November 2016.

Gibbons, S. (2017) Planes, trains and automobiles: the economic impact of transport infrastructure. *SERC Policy Paper* 12. London: Spatial Economics Research Centre.

Gibbons, S., Lyytikäinen, T., Overman, H. and Sanchis-Guarner, R. (2017) New road infrastructure: the effects on firms. *SERC Discussion Paper* 214. London: Spatial Economics Research Centre.

Gillespie, A., Richardson, R. and Cornford, J. (2001) Regional development and the new economy. *European Investment Bank Papers* 6: 109–31.

Glasmeier, A. and Christopherson, S. (2015) Thinking about smart cities. *Cambridge Journal of Regions, Economy and Society* 8: 3–12.

Gonzales, A. (2016) The contemporary US digital divide: from initial access to technology maintenance. *Information, Communication & Society* 19: 234–48.

Graham, M., De Sabbata, S. and Zook, M.A. (2015) Towards a study of information geographies: (im)mutable augmentations and a mapping of the geographies of information. *Transactions of the Institute of British Geographers* 2: 88–105.

Graham, S. and Marvin, S. (2001) *Splintering Urbanism: Networked Infrastructure, Technological Mobilities and the Urban Condition.* London: Routledge.

Grandadam, D., Cohendent, P. and Simon, L. (2013) Places, spaces and the dynamics of creativity: the video game industry in Montreal. *Regional Studies* 47: 1701–14.

Gwilliam, K. (2003) Urban transport in developing countries. *Transport Reviews* 23: 197–216.

Hall, J.V. and Krueger, A.B. (2015) An analysis of the labor market for Uber's driver-partners in the United States. *Princeton University Industrial Relations Section Working Paper* 587. Princeton, NJ: Princeton University.

Hathaway, I. and Muro, M. (2016) Tracking the gig economy: new numbers. At www.brookings.edu/research/tracking-the-gig-economy-new-numbers/. Last accessed 8 November 2016. Washington, DC: The Brookings Institution.

Hernandez, D.O. and Davila, J.D. (2016) Transport, urban development and the peripheral poor in Colombia – placing splintering urbanism in the context of transport networks. *Journal of Transport Geography* 51: 180–92.

Hollands, R.G. (2015) Critical interventions into the corporate smart city. *Cambridge Journal of Regions, Economy and Society* 8: 61–77.

Hook, L. (2016) Uber and Airbnb business models come under scrutiny. *Financial Times*, 30 December.

Houlder, V. (2017) Airbnb's edge on room prices depends on tax advantages. *Financial Times*, 2 January.

Huws, U., Spencer, N.C. and Joyce, S. (2016) *Crowd Work in Europe: Preliminary Results from a Survey in the UK, Sweden, Germany, Austria and the Netherlands.* University of Hertfordshire: Foundation for European Progressive Studies (FEPS) and UNI Europa.

International Telecommunications Union (ITU) (2016) Key ICT indicators for developed and developing countries and the world (totals and penetration rates). At www.itu.int/en/ITU-D/Statistics/Pages/stat/default.aspx. Last accessed 9 December 2016.

Janelle, D. (1969) Spatial reorganisation: a model and concept. *Annals of the Association of American Geographers* 59: 348–64.

Kim, H. and Sultana, S. (2015) The impacts of high-speed rail extensions on accessibility and spatial equity changes in South Korea from 2004 to 2018. *Journal of Transport Geography* 45: 48–61.

Kitchin, R. (2015) Making sense of smart cities: addressing present shortcomings. *Cambridge Journal of Regions, Economy and Society* 8: 131–6.

Kshetri, N., Alcantara, L.L. and Park, Y. (2014) Development of a smart city and its adoption and acceptance: the case of New Songdo. *Digiworld Economic Journal* 96, 113–28.

Linneker, B. (1997) Transport infrastructure and regional economic development in Europe: a review of theoretical and methodological approaches. *TRP 133.* Sheffield: Dept. of Town and Regional Planning, University of Sheffield.

Longley, P.A. (2003) Towards a better understanding of digital differentiation. *Computers, Environment and Urban Systems* 27: 103–6.

Lucas, K. and Porter, G. (2016) Mobilities and livelihoods in urban development contexts: introduction. *Journal of Transport Geography* 55: 129–31.

Mack, E.A. (2014) Broadband and knowledge intensive firm clusters: essential link or auxiliary connection? *Papers in Regional Science* 93: 1–30.

Martin, C.J. (2016) The sharing economy? A pathway to sustainability or a nightmarish form of neoliberal capitalism? *Ecological Economics* 121: 149–59

Monzón, A., Ortega, E. and López, E. (2013) Efficiency and spatial equity impacts of high-speed rail extensions in urban areas. *Cities* 30: 18–30.

Moriset, B. and Malecki, E. (2008) Organization versus space: the paradoxical geographies of the digital economy. *Geography Compass* 3: 256–74.

Morozov, E. (2013) The 'sharing economy' undermines workers' rights. At http://evgenymorozov.tumblr.com/post/64038831400/the-sharing-economy-undermines-workers-rights. Last accessed 16 January 2017.

O'Brien, R. (1992) *Global Financial Integration: The End of Geography*. London: Pinter.

O'Connor, S. (2016) The gig economy is neither 'sharing' nor 'collaborative'. *Financial Times*, 14 June.

Olvera, L., Plat, D.L. and Pochet, P. (2003) Transportation conditions and access to services in a context of urban sprawl and deregulation: the case of Dar es Salaam. *Transport Policy* 10: 287–98.

Osborne, H. (2016) Uber loses right to classify UK drivers as self-employed. *The Guardian*, 28 October.

Pew Research Centre (2016) Gig work, online selling and home sharing. November 2016. At http://assets.pewresearch.org/wp-content/uploads/sites/14/2016/11/17161707/PI_2016.11.17_Gig-Workers_FINAL.pdf. Last accessed 16 January 2017.

Pratt, A. (2013) Space and place in the digital creative economy. In Handke, C. and Towse, R. (eds) *Handbook of the Digital Creative Economy*. Cheltenham: Edward Elgar, pp. 37–44.

Prensky, M. (2001) Digital natives, digital immigrants. *On the Horizon* 9 (5): 1–6.

Price Waterhouse Coopers (PWC) (2015) *The Sharing Economy*. Consumer Intelligence Series. At pwc.com/CISsharing.

Puga, D. (2008) Agglomeration and crossborder infrastructure. *European Investment Bank Papers* 13: 102–24.

Rabari, C. and Storper, M. (2014) The digital skin of cities: urban theory and research in the age of sensored and metered city, ubiquitous computing and big data. *Cambridge Journal of Regions, Economy and Society* 8: 27–42.

Richardson, L. (2015) Performing the sharing economy. *Geoforum* 67: 121–9.

Riddlesden, D. and Singleton, A. (2014) Broadband speed equity: a new digital divide? *Applied Geography* 52: 25–33.

Robinson, L., Cotton, S.R., Ono, H., Quan-Haase, A., Mesch, G., Chen, W., Schulz, J., Hale, T.M. and Stern, M.J. (2015) Digital inequalities and why they matter. *Information, Communication & Society* 18: 569–82.

Rogers, B. (2015) The social costs of Uber. *The University of Chicago Law Review Dialogue* 82: 82–102.

Schor, J.B. (2017) Does the sharing economy increase inequality within the eighty percent?: findings from a qualitative study of platform providers. *Cambridge Journal of Regions, Economy and Society* 10: 263–79.

Scott, A.J. (2011) A world in emergence: notes toward a resynthesis of urban-economic geography for the 21st century. *Urban Geography* 32: 845–70.

Shapiro, A. (2015) A South Korean city designed for the future takes on a life of its own. At www.npr.org/sections/parallels/2015/10/01/444749534/a-south-korean-city-designed-for-the-future-takes-on-a-life-of-its-own. Last accessed 16 January 2016.

Shelton, T., Zook, M. and Wiig, A. (2015) The actually existing smart city. *Cambridge Journal of Regions, Economy and Society* 8: 13–25.

Sietchiping, R., Permezel, M.J. and Ngomsi, C. (2012) Transport and mobility in sub-Saharan African cities: an overview of practices, lessons and options for improvements. *Cities* 29: 183–9.

Thomas, P. and O'Donoghue, D. (2013) The Channel Tunnel: transport patterns and regional impacts. *Journal of Transport Geography* 31: 104–12.

Tomaney, J. and Marques, P. (2013) Evidence, policy, and the politics of regional development: the case of high-speed rail in the United Kingdom. *Environment and Planning C: Government and Policy* 31: 414–27.

Tranos, E. (2012) The causal effect of the internet infrastructure on the economic development of European city regions. *Spatial Economic Analysis* 7: 319–37.

United Nations (2015) *The Millennium Development Goals Report 2015*. New York and Geneva: United Nations.

Vaughan R. and Daverio, R. (2016) *Assessing the Size and Presence of the Collaborative Economy in Europe*. London: PWC.

Vickerman, R. (1994) The Channel Tunnel, regional competitiveness and regional development. *Applied Geography* 14: 9–25.

Vickerman, R. (2012) High-speed rail – the European experience. In de Urena, J.M. (ed.) *Territorial Implications of High Speed Rail: A Spanish Perspective*. Farnham: Ashgate, pp. 17–32.

Warf, B. (2012) Contemporary digital divides in the United States. *Tijdschrift voor Economische en Sociale Geografie* 104: 1–17.

Warf, B. (2013) *Global Geographies of the Internet*. Dordrecht: Springer.

Wiig, A. (2016) The empty rhetoric of the smart city: from digital inclusion to economic promotion in Philadelphia. *Urban Geography* 37: 535–3.

Chapter 9

Global production networks and regional development

9.1 Introduction

In its *World Investment Report 2013*, the United Nations Conference on Trade and Development (UNCTAD) (2013) estimated that 80 per cent of world trade was conducted through what it termed **global value chains** (**GVCs**) or the production networks of **transnational corporations** (**TNCs**). This involves TNCs investing in cross-border assets and trading inputs and outputs with partners, suppliers and customers on a world-wide basis. It reflects the increased fragmentation of production through the **outsourcing** and interna-tional dispersion of tasks such as manufacturing and assembly, often to lower-cost countries (**offshoring**) (section 3.5). Beyond the traditional trade in goods and services between firms, we are seeing increased 'trade in tasks' or value-added, whereby parts, components and services are traded between firms in **global production networks** (**GPNs**) (Box 9.1). Exports of these parts and components or intermediate goods which are incorporated into finished goods for final consumption now exceed exports of finished goods themselves (UNCTAD 2013: 122). This 'trade in tasks' is facilitated by advances in transportation and information and communications technologies that were examined in Chapter 8. Their role as the organisational structures which coordinate the trade in intermediate goods has led to GVCs or GPNs being described as the "world economy's backbone and central nervous system" (Cattaneo *et al.* 2010: 7).

Box 9.1

The odyssey of the crankshaft

The journey undertaken by the crankshaft used in the BMW Mini – the part of a car that translates the movement of the pistons into the rotational drive to move the vehicle – highlights the significance of the international trade in parts and components as well as finished goods. The crankshaft is made by a supplier based in France before being shipped to BMW's Hams Hall plant in Warwickshire for crafting into shape. The crankshaft is then sent back to Munich where it is inserted in the engine before travelling to the Mini plant in Oxford where the engine is then 'married' with the car (Ruddick and Oltermann 2017) (Figure 9.1). If the car is sold on the continent – as many of the cars produced in the UK are – the crankshaft will cross the English Channel for the fourth time. This reflects the international nature of sourcing in the automotive industry with just 41 per cent of the parts of a car assembled in the UK actually produced there (ibid.). The implications of the UK's 'Brexit' vote on 23 June 2016 threaten to disrupt these supply networks, with tariffs applicable not just to finished goods, but also parts and components. This is a source of great concern for the industry, prompting concerns that car manufacturing could ultimately move away from the UK.

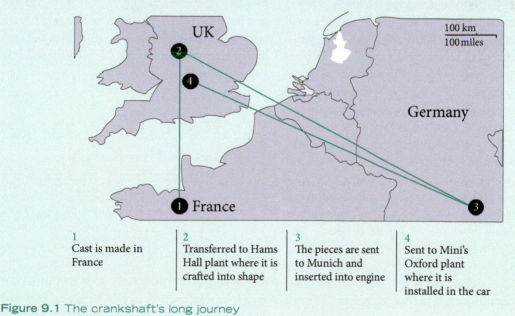

1	2	3	4
Cast is made in France	Transferred to Hams Hall plant where it is crafted into shape	The pieces are sent to Munich and inserted into engine	Sent to Mini's Oxford plant where it is installed in the car

Figure 9.1 The crankshaft's long journey
Source: Ruddick and Oltermann 2017.

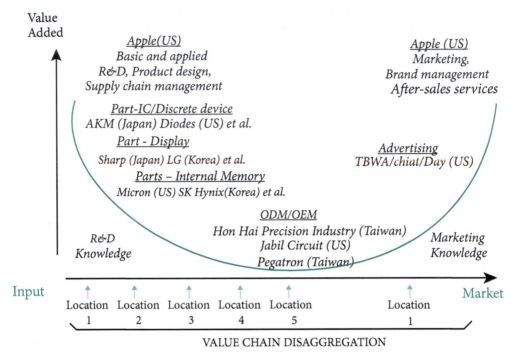

Figure 9.2 Apple's smiling curve and the organisation of global value chains
Source: Grimes and Sun 2016: 99.

The role of GPNs in coordinating this 'trade in tasks' has important implications for urban and regional development, underlining the need for cities and regions (hereafter referred to as regions as a shorthand) to plug themselves into these networks. This reflects the shift from traditional patterns of **regional sectoral specialisation** to increasingly elaborate international divisions of labour (Chapter 3). Rather than producing a particular set of finished goods and all their components internally, before exporting them to international markets, cities and regions now tend to undertake a particular task or stage – for example, research and development (R&D), component manufacture, assembly, distribution – within a broader international division of labour. This involves adding value to the production process. From a regional development perspective, a key issue is the extent to which regions can actually capture this value in terms of retaining the benefits and profits (Coe *et al.* 2004). This will reflect their positon along the so-called 'smiling curve' of value chain organisation with tasks such as R&D, marketing and advertising associated with greater value added and capture than contract manufacturing and assembly (Figure 9.2). This underlines the need to rethink urban and regional development in relational terms as involving cities and regions undertaking particular roles and tasks within GPNs (ibid.) (section 2.5.4).

9.2 The global production networks (GPN) approach

9.2.1 Commodity chains, value chains and global production networks

The development of the GPN approach in economic geography was inspired, in part, by **global commodity chains (GCCs)** research, which originally emerged out of world systems thinking. A commodity chain is defined as "a network of labour and production processes whose end result is a finished commodity" (Hopkins and Wallerstein 1986: 159). Commodity chains involve the transformation of material and non-material inputs into finished goods or services. Four basic stages of inputs, transformation, distribution and consumption have been identified (Figure 9.3), with each stage adding value to the previous stage. The process depends upon a range of technological and service inputs and logistics services and is subject to ongoing forms of regulation, coordination and control from a range of actors, including lead firms, supranational organisations and governments.

Much of the commodity chain literature focused on the issue of **governance**, referring to how chains were organised and coordinated by 'lead' firms. Here, the sociologist Gary Gereffi (1994) drew an important distinction between what he called producer-driven

and buyer-driven commodity chains (Table 9.1). Producer-driven chains are typical of industries which are highly capital and technology intensive such as aircraft, motor vehicles and computers, requiring huge amounts of up-front R&D and investment. The chain is coordinated by large manufacturers who rely on conventional top-down hierarchical links with a range of component suppliers. The level of outsourcing in such chains tended to be moderate, although this has increased in recent years, reflecting wider trends. Buyer-driven chains, by contrast, are coordinated by large, branded retailers who concentrate on design, sales, marketing and finance whilst actual production is outsourced to suppliers in developing countries. This type of arrangement is characteristic of labour-intensive consumer goods industries such as clothing, furniture

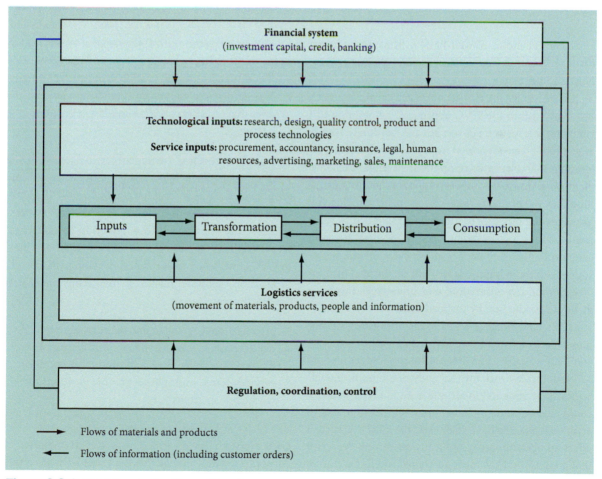

Figure 9.3 A generic production network
Source: Dicken 2007, Figure 1.4c.

Table 9.1 Characteristics of producer-driven and buyer-driven chains

	Form of economic governance	
	Producer-driven	Buyer-driven
Controlling type of capital	Industrial	Commercial
Capital/technology intensity	High	Low
Labour characteristics	Skilled/high wage	Unskilled/low wage
Controlling firm	Manufacturer	Retailer
Production integration	Vertical/bureaucratic	Horizontal/networking
Control	Internalized/hierarchical	Externalized/market
Contracting/outsourcing	Moderate and increasing	High
Suppliers provide	Components	Finished goods
Examples	Automobiles, computers, aircraft, electrical machinery	Clothing, footwear, toys, consumer electronics

Source: Coe *et al.* 2007: 102.

and toys. Examples of well-known firms who operate through buyer-driven chains include Walmart, Ikea, Nike and the Gap.

More recently, the global commodity chain approach has evolved into the GVC approach, maintaining this characteristic focus on governance. Accordingly, Gereffi *et al.* (2005) have identified five types of inter-firm governance: markets, based upon the arm's length exchange of goods and services between suppliers and lead firms; modular value chains whereby suppliers provide full packages and modules to the lead firms; relational value chains requiring frequent communication between lead firms and suppliers; captive value chains where suppliers remain highly dependent on the lead firm for direction and coordination; and hierarchies involving in-house production by the lead firm.

GPNs have become a key focus of research in economic geography and related fields in recent years (Coe and Yeung 2015). The GPN approach provides a broad relational framework for the study of economic globalisation that aims to "incorporate all kinds of network configuration" and to "encompass all relevant sets of actors and relationships" (Coe *et al.* 2008: 272). As such, it offers an open and geographically sensitive perspective that goes beyond the more restricted and linear frameworks offered by the related GCC and GVC perspectives. According to Sturgeon (2001: 10):

A chain maps the vertical sequence of events leading to the delivery, consumption and maintenance of goods and services . . . while a network highlights the nature and extent of the inter-firm relationships that bind sets of firms into larger economic groupings.

The GPN approach is particularly associated with the work of a so-called 'Manchester school' of economic geographers (Bathelt 2006) whereas the GVC framework is associated with sociology and business studies.

The GPN approach directs attention towards networks of firms, spanning a similar range of activities to GVCs, including research and development, design, production, supplier relations, marketing and sales. A GPN is defined as an

organisational arrangement, comprising interconnected economic and noneconomic actors, coordinated by a global firm, and producing goods or services across multiple geographic locations for worldwide markets.

(Yeung and Coe 2015: 32)

GPN research is concerned with the distribution of corporate power in networks, particularly in terms of the relations between lead firms and governments and between such firms and their suppliers (Box 9.2).

A range of firm types can be identified, namely lead firms such as BMW or Apple, strategic partners, specialised suppliers (industry specific or multi-industrial), generic suppliers and customers (Table 9.2). In contrast to the GVC approach, GPN research highlights the role of a broad range of institutions beyond firms – for example, supranational organisations, government agencies, trade unions, employers' organisations, NGOs and consumer groups – in shaping firm activities. While there are important conceptual differences between the GPN and GCC/GVC approaches, with the former having developed out of a critique of the latter (Henderson *et al.* 2002), they share an underlying concern with the changing organisational arrangements through which goods and services are produced, distributed and consumed in the global economy (Figure 9.4).

All three approaches reflect the increased economic importance of TNCs as the 'key movers and shapers' of the global economy (Dicken 2015). TNCs are at the heart of GPNs as the coordinators and controllers of the geographically dispersed networks of partners, contractors and suppliers through which individual goods and services are produced. The geographical mobility of TNCs enables them to shift investment between regions on a global basis, meaning that their decisions to invest, reinvest or disinvest have profound effects on national and regional economies and the livelihoods of individual workers and their families, which tend to be tied to particular places. This has prompted some commentators to suggest they have become more powerful than governments in shaping investment flows and decisions in the global economy (Pike *et al.* 2017: 229).

Production networks are comprised of a number of distinct levels or tiers of suppliers to the global lead firms in Tier 1 (Figure 9.4). In the laptop PC value chain, most assembly operations are subcontracted by the global lead firm (Tier 1) to contract manufacturers and

Table 9.2 Firms as actors in the global production networks

GPN Actors	Role	Value Activity	Examples in Manufacturing	Examples in Service Industries
Lead firms	Coordination and control	Product and market definition	Apple and Samsung (information and communications technology [ICT]); Toyota (automobiles)	HSBC (banking); Singapore Airlines (transport)
Strategic partners	Partial or complete solutions to lead firms	Codesign and development in manufacturing or advanced services	Hon Hai or Flextronics (ICT); ZF (automobiles)	IBM Banking (banking); Boeing or Airbus (transport)
Specialized suppliers (industry-specific)	Dedicated supplies to support lead firms and/or their partners	High value modules, components, or products	Intel (ICT); Delphi and Denso (automobiles)	Microsoft (ICT); Fidelity or Schroders (banking); Amadeus (transport)
Specialized suppliers (multi-industrial)	Critical supplies to lead firms or partners	Cross-industrial intermediate goods or services	DHL (ICT); Panasonic Automotive (automobiles)	DHL (banking); Panasonic Avionics (transport)
Generic suppliers	Arm's-length providers of supplies	Standardized and low value products or services	Plastics in ICT and automobile manufacturing	Cleaning in banking and transport services
Key customers	Transfer of value to lead firms	Intermediate or final consumption	Other lead firms or consumers	Other lead firms or consumers

Source: Yeung and Coe 2015: 45.

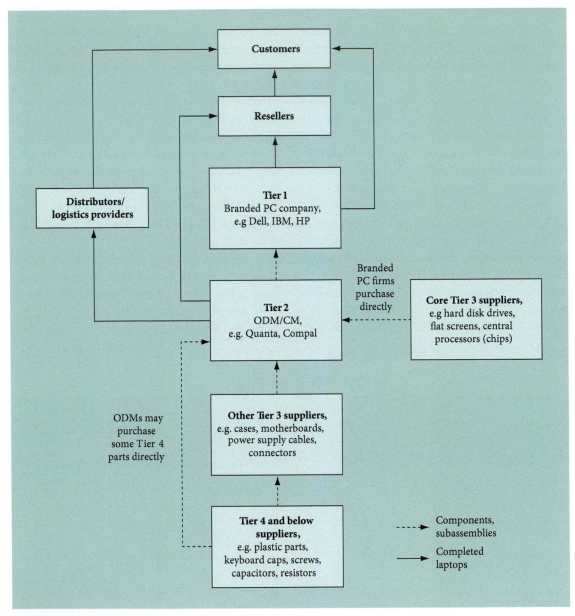

Figure 9.4 The laptop PC production network
Source: Daniels *et al.* 2008: 322.

increasingly original design manufacturers (ODMs) who also contribute to the design process. Tier 2 contractors and ODMs then subcontract the manufacturing of major components and modules to Tier 3 suppliers who subcontract the production of individual components to a fourth tier of suppliers. It should be noted that Figure 9.4 is a highly simplified representation of what is a very complex production process involving hundreds of firms and tens of thousands of workers (Coe 2016: 326).

9.2.2 The GPN approach

The GPN approach is based on three conceptual categories. First, **value**, which attempts to incorporate both Marxist notions of surplus value (see Box 2.4) and

Box 9.2

The global production network of BMW

BMW, the German car manufacturer, provides an interesting example of the evolution of a GPN and its organisation across space (Coe *et al*. 2004). In one sense, the example clearly reflects the characteristics of the automobile sector such as a high level of capital intensity and technology intensity, a globalised form of organisation, a reliance on 'lean production', and the use of just-in-time strategies, whereby manufacturers source supplies as and when they need them, often resulting in the geographical clustering of suppliers around manufacturing and assembly plants. At the same time, BMW remains distinctive as a niche, upmarket, low-volume producer. It retains a strong base in Bavaria with around 46 per cent of shares owned

by the Quandt family. With its headquarters in Munich, BMW has had a major impact on the regional economy of east Bavaria through successive rounds of investment in three plants and supplier parks since the late 1960s. It directly employs some 35,000 people in the region and more in the wider supply chain.

In response to pressures of globalisation and competition within the car industry, BMW has adopted an internationalisation strategy in recent years. It currently manufacturers its products at 30 sites in 14 different countries and on four continents. This is designed to ensure that it can respond rapidly and flexibly to changing customer demands and market requirements, representing a primarily market-seeking form

of internationalisation (Dicken 2015). In addition to facilities in North and South America and South Africa, BMW has recently opened a plant in Araquari, Santa Catarina state, Brazil and is currently constructing a plant in San Luis Potosi in Mexico. In May 2004, BMW opened a plant in Shenyang in Liaoning Province, north-east China, based upon a joint venture with Brilliance Automotive of China (Box 9.3). It also opened a plant in Chennai, India in 2007 to capitalise on the growth of the Indian market. BMW also operates assembly plants in Rayong, Thailand, Kuala Lumpur, Malaysia and Jakarta, Indonesia to serve the South-east Asian market as well as in Kaliningrad, Russia and Egypt.

mainstream economic definitions of economic rent (Henderson *et al*. 2002: 448). It refers to the economic return (profit) or rent generated by the production of commodities for sale, involving the conversion of labour power into actual labour through the labour process. Firms may create value through the control of particular product and process technologies; the development of certain organisational and management capabilities; the harnessing of inter-firm relationships; and the prominence of brand names in key markets. In the face of competition from other firms for market share, there is a need to also enhance value through processes of technological transfer and the upgrading of skills and production capabilities. All this raises the issue of value capture in terms of which actors and locations in the network are able to appropriate and retain value, highlighting questions of ownership and control. Key aspects of this include firm ownership (private-public, domestic-foreign, etc.); government policy in terms of property rights, ownership structures and the repatriation of profits; and systems of corporate

governance in different national contexts (stakeholder-shareholder) (ibid.).

The second conceptual category of interest is **power** (section 1.4.4.), defined primarily as a practice in terms of the capacity to exercise power (Allen 2003). More specifically, Henderson *et al*. (2002) identify three main forms of power. First, corporate power in relation to a lead firm's control of key resources, information, knowledge, skills and brands within a production network. Second, institutional power which is exercised by national and local states (Box 9.3), supranational bodies like the European Union, the 'Bretton Woods institutions' (World Bank, International Monetary Fund and World Trade Organisation), various United Nations agencies (especially the International Labour Organisation) and the major international credit rating agencies. Third, collective power, referring to the actions of various collective actors, including trade unions, employers' organisations and non-government organisations (NGOs).

The third key category is embeddedness (section 2.5.2). Three forms are identified (see Hess 2004)

(Figure 9.5). First, societal embeddedness in a broad sociological sense (Granovetter 1985), invoking Polanyi (Box 1.6) to emphasise how actors are positioned within wider institutional and regulatory frameworks. Second, network embeddedness highlights the social and economic relationships in which a particular actor or firm participates (Henderson *et al.* 2002: 453). Third, **territorial embeddedness** refers to the 'anchoring' of GPNs in different places (Box 9.3). GPNs can become territorially embedded because of lead firms' historic ties to particular locations, often their regions of origin, which may provide particular advantages such as political support through national and local governments, links with key suppliers and access to labour skills. But such embeddedness may be eroded over time as competitive pressures prompt firms to invest in other, often less costly, locations, reflecting the tensions between spatial fixity and mobility that are endemic to capitalism (Harvey 1982).

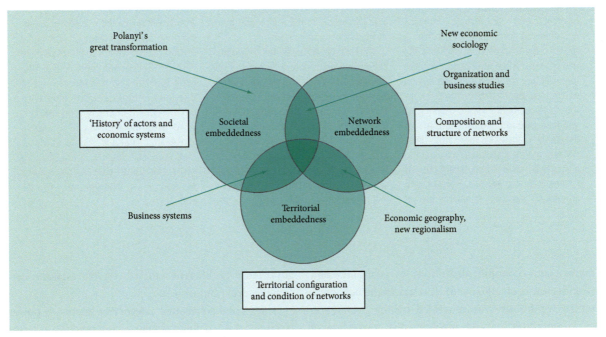

Figure 9.5 Fundamental categories of embeddedness
Source: Hess 2004: 178.

Box 9.3

Obligated embeddedness: TNCs in China

Contrary to the myth of the entirely powerless state in the face of global capital flows, national states remain key actors in the global economy, continuing to control access to their domestic markets. Whereas TNCs seek to maximise their locational flexibility, favouring low levels of political regulation, states aim to capture value within GPNs by trying to embed a TNC's activities as fully as possible in the local or national economy (Liu and Dicken 2006). Here, territorial embeddedness refers to efforts to fix investment in a particular location and the economic benefits and spinoffs (jobs, investment, contracts, etc.) that are derived from this for the host economy in question. Liu and Dicken go on to distinguish between active and obligated forms of embeddedness as two ideal types. The former refers to cases where local assets are widely available and TNCs source and incorporate such assets on an autonomous basis, whilst the latter reflects cases when such assets are not freely available and a particular state can control access to them. In this second set of circumstances,

Box 9.3 (continued)

states will have much greater bargaining power, requiring TNCs who wish to invest in their domestic markets to comply with certain criteria.

China became the fastest-growing automobile market in the world, emerging as the third largest overall after the US and Japan. The industry is concentrated in six main regional clusters within China (Figure 9.6). Its growth followed the selection of automobiles as a strategic growth sector by the Chinese government in the late 1980s. The basic bargaining strategy adopted by the state has been one of trading market access for capital and technology transfer from TNCs. Due to their desire to invest in this huge and rapidly growing market, foreign investors have complied. All the major players in the global automobile industry had invested in China by 2003, and all these investments took the form of joint ventures, as required by the state. This form of obligated embeddedness reflects the unique bargaining power of the Chinese state in controlling access to the world's largest and fastest-growing consumer market, allowing it to reverse the usual scenario of TNCs playing off states against one another into one in which a state could play off different TNCs against one another (ibid.: 1245). Other states are in much weaker positions since they do not control access to such a unique and significant asset as the Chinese market, though larger states or groups of states such as the EU will have more bargaining powers over TNCs than smaller ones.

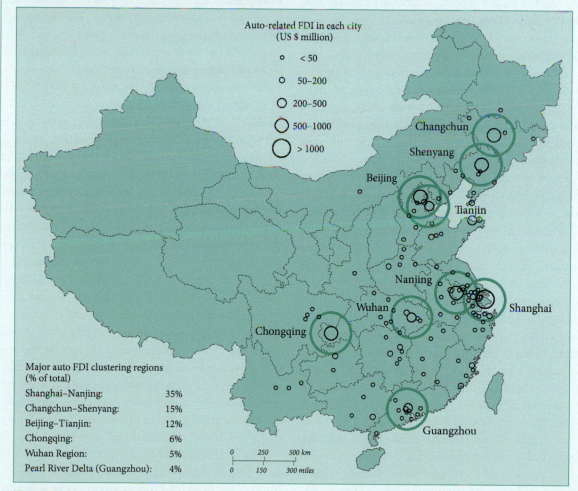

Figure 9.6 China's automobile cluster
Source: Wheelon Co. Ltd 2002, in Liu and Dicken 2006: 1234.

Processes of economic and social **upgrading** have attracted much attention within GVC and GPN research. Economic upgrading can be defined as the improvement of a firm's position in the chain or network through a movement into higher-value activity, based on improved knowledge, technology and skills and leading to increased benefits or profits (Gereffi *et al.* 2005). Four types of upgrading have been identified: process upgrading based on more efficient production; product upgrading (the production of more sophisticated high-value goods or services); functional upgrading which involves moving up the chain by performing higher value-added tasks; and chain or network upgrading whereby a firm moves into a more technologically advanced network, involving a shift into new industries or product markets. By contrast, the parallel process of economic downgrading whereby firms move down the chain to perform lower value-added tasks has received little attention despite its widespread occurrence. According to Blazek (2016), such functional downgrading takes three main forms: passive, based on a decision by a higher-level buyer; adaptive, when firms decide to respond to increased competitive pressures in this way; or strategic, involving a more conscious decision to refocus upon a particular set of core competences or market segments.

Social upgrading is the process of improvement in the rights and status of workers, which enhances the quality of their employment, leading to better working conditions and increased wages, protections and rights (Barrientos *et al.* 2011: 324). It can be divided into two elements: measureable standards such as type of employment, wage levels, social protection and working hours; and enabling rights, such as freedom of association, the right to collective bargaining, non-discrimination, voice and empowerment, which are less easily quantified (ibid.). While it has often been assumed that economic upgrading translates into social upgrading, a number of studies have shown that this is not necessarily the case for all workers, especially women (Barrientos 2014). The example of Hon Hai Precision, a Taiwanese contract manufacturer that has grown from a third-tier supplier of plastic components to become the world's largest provider of electronics manufacturing services through its Foxconn operation in China, illustrates this. Despite the firm's success, a lack of social upgrading through excessive hours, low wages and harsh working conditions culminated in a spate of suicides among the female workforce at Foxconn's factories in China in 2011. More broadly, economic upgrading often involves increased workforce stratification between permanent and relatively well-paid workers, on the one hand, and temporary, poorly paid ones, on the other, in sectors such as clothing (Werner 2016a).

9.2.3 GPN 2.0

In a recent article and book, the leading GPN researchers Neil Coe and Henry Yeung have sought to build on earlier conceptual foundations to develop a more dynamic theory of GPNs (Coe and Yeung 2015; Yeung and Coe 2015). Coe and Yeung term this GPN 2.0, which is designed to move beyond GPN 1.0 concepts of value, power and embeddedness and global value chain typologies of industrial governance. In particular, they have sought to better theorise the relationships between competitive dynamics, firm-specific strategies, value capture trajectories and regional development outcomes.

Yeung and Coe (2015) argue that capitalist dynamics are the *raison d'être* of GPNs, generating actor-specific strategies in different national and regional economies. They identify three main sets of competitive dynamics. First, pressures to optimise cost-capability ratios refer not only to the cost-reduction imperatives associated with the **globalisation** and the outsourcing of tasks in GPNs, but also the need for firms to harness their specific capabilities, referring to the assets, resources and skills they have developed. From this perspective, a firm can be thought of as an organisational and managerial device to optimise the accumulation and deployment of its available resources, defined as its core capability, at the lowest possible cost. Second, sustaining market development emphasises the role of both producers and consumers in creating market structures. This involves firms developing strategies for market research and access, managing the time-to-market for their products and services and seeking to shape customers' behaviour and preferences through a range of advertising and marketing activities. Third, financial disciplines exert another set of pressures through the need for firms to access external finance, often exposing them to the requirements of investors and shareholders. Firms in GPNs are also confronted

Table 9.3 Firm-specific strategies and organisational outcomes in global production networks

| Strategy as Actor Practice | Competitive Dynamics | | | Risks | GPN Structure as Organizational Outcomes |
	Cost-capability Ratio	Market Imperative	Financial Discipline		
Intrafirm coordination (e.g., pharmaceuticals and retail)	Low	High	Low	High	Domestic expansion and/or foreign direct investment and mergers and acquisitions; high level of network integration
Interfirm control (e.g., automobiles and information technology services)	High	Low	High	Medium	Outsourcing but dependent integration of suppliers
Interfirm partnership (e.g., electronics and logistics)	High	High	High	High	Outsourcing, joint development with partners and platform leaders
Extrafirm bargaining (e.g., resources and agrofood)	Medium	High	High	High	Differentiated integration into global production systems

Source: Yeung and Coe 2015: 46.

with the need to manage risks, which may be economic, product-based, regulatory, labour-related or environmental in nature (ibid.).

Coe and Yeung also identify four types of firm strategy in GPNs (Table 9.3). First, intra-firm coordination is based on in-house production within a firm, involving the internalisation and consolidation of value activity within the boundaries of a lead firm, strategic partner or supplier, aiming to reduce costs and inventories, instil greater market responsiveness and develop higher-quality products and services. Second, inter-firm control involves a firm outsourcing a large part of its production activities to independent contractors and suppliers, while exercising tight control over the wider production processes and quality standards. This is often based on a logic of cost reduction by lead firms, seeking to move out of lower value-added activities to concentrate on their higher value-added capabilities. Third, inter-firm partnership involves close cooperation between lead firms and their strategic partners and specialised suppliers, corresponding to relational or modular forms of governance. Fourth, extra-firm bargaining refers to an often contested two-way process of negotiation and accommodation between firms and non-firm actors such as states, supranational organisations, trade unions, consumer and civil society organisations, whose activities can have a significant influence on the evolution of GPNs.

Reflect

Do you agree that the GPN approach offers a conceptual advance over the GCC and GVC approaches from a geographical perspective? Does this have implications for how we might study the organisation of global industries?

9.3 Global production networks and regional development

9.3.1 The process of strategic coupling

The geographical orientation of GPN research underpins its distinctive contribution to the recent rethinking of regional development processes. Compared to the 'new regionalism' of the 1990s, which was preoccupied with social and institutional conditions within regions (see section 2.5.2), the GPN approach signals a new interest in extra-regional relations or the 'outside' of regions whilst seeking to retain certain 'new regionalist' insights, particularly the notion of regional institutions 'holding down the global' (Amin and Thrift 1994). The concern with extra-regional relations is somewhat reminiscent of the Marxist political economy approaches of the 1970s and 1980s (section 2.4), although these relations are conceptualised in a different way. The key point is that regions are not isolated and separate, but closely bound up with the operation of global networks, echoing Massey's **global 'sense of place'** (section 1.2.3) and the concept of the 'relational region' (section 2.5.4). As Dicken *et al.* (2001: 97) argue, this is "a mutually constitutive process; while networks are embedded within territories, territories are, at the same time, embedded into networks".

In particular, the GPN approach aims to 'globalise' regional development through its focus on the link with GPNs, viewing 'the region' as "a porous territorial formation whose notional boundaries are straddled by a broad range of network connections" (Coe *et al.* 2004: 469). Regional development is defined as "a dynamic outcome of the complex interaction between territorialised relational networks and GPNs within the context of changing governance structures" (ibid.). Regional assets in the form of specific kinds of knowledge, skills and expertise provide an important resource for regional development, but must be harnessed by regional institutions to "complement the strategic needs of trans-local actors situated within GPNs" (ibid.: 470). The promotion of local endogenous development based on such assets will not be sufficient to secure increased prosperity. Instead, regional development requires a process of **'strategic coupling'** between regional assets and GPNs.

According to Yeung (2009: 213), ". . . in the context of urban and regional development, strategic coupling refers to the dynamic processes through which actors in cities and/or regions coordinate, mediate, and arbitrage strategic interests between local actors and

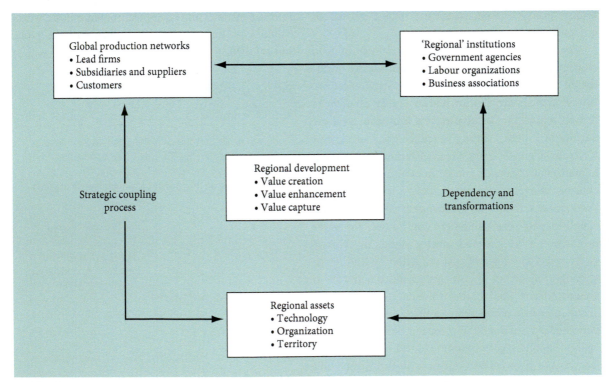

Figure 9.7 The strategic coupling of regions and global production networks
Source: Coe *et al.* 2004: 470.

their counterparts in the global economy". The role of regional institutions is to attract and retain investment by shaping and moulding regional assets to fit the needs of lead firms in GPNs (Figure 9.7). Such coupling processes result in cooperation and the construction of a 'temporary coalition' between groups of actors – the managers of TNC affiliates and regional government officials – who might not otherwise work together in the pursuit of a common objective (ibid.). This conceptualisation can be seen to overlap substantially with an extensive literature on the role of regional institutions in attracting foreign direct investment (FDI) (see MacKinnon and Phelps 2001).

The task for regions is to create, capture and enhance value within GPNs, with regional institutions playing a key role in these processes (Box 9.4) (Coe *et al.* 2004). Value creation involves the creation of supporting conditions for growth by regional institutions through training and education programmes, the promotion of firm start-ups and the provision of venture capital through private-sector investors. The further enhancement of value can occur through knowledge and technological transfer, industrial upgrading, the provision of more advanced infrastructure and the development of specialised skills (Box 9.4). Value capture refers to scenarios in which regions retain the profits generated by economic activities within them, reflecting, in

part, the extent to which key firms are locally owned and embedded in the regional economy. In practice, of course, regions are often involved in a variety of strategic couplings with GPNs in several industries, and each of these will generate different degrees of value creation, enhancement and capture.

In the BMW case (Box 9.2), initial value creation in east Bavaria in the late 1960s and 1970s gave way to value enhancement in the 1980s and 1990s through processes of technological upgrading and the creation of a pool of highly skilled labour. Significant value capture was ensured by the local ownership and control of BMW and by elaborate negotiations between management, the government and labour at the regional and national scales, with, for instance, the regional branch of the national union, IG metal, agreeing to implement a work-shift system that has become a model for the German car industry (Coe *et al.* 2004: 478). In Rayong, Thailand, some value creation occurred through the investment in an assembly plant, supported by the national government, which set up programmes to develop the skills of the workforce and the capabilities of local suppliers (ibid.). There is a lack of real upgrading, however, reflecting limited skills and organisational capacities. At the same time, linkages with the wider regional economy remain very limited and value capture is reduced by continuing dependence on outside investment.

Box 9.4

Value creation, enhancement and capture in 'Silicon Glen', central Scotland

West-central Scotland was one of the key crucibles of industrial capitalism in the mid- to late nineteenth century, with 20 per cent of the world's shipping tonnage built on the River Clyde in 1914 (Devine 1999: 250). In common with many other 'old' industrial regions, west-central Scotland experienced a long decline over the course of the twentieth century, however, culminating in severe deindustrialisation in the 1970s and 1980s. In response, the state sought to attract new industries

to Scotland in order to provide replacement employment, resulting in growing investment in electronics, initially from US and European firms before a new influx from Japan and East Asia in the 1980s and 1990s. This reflected the access that Scotland offered to the European market, a pool of available labour and government incentives.

By the early 1980s, the term 'Silicon Glen' was being used to describe the Scottish electronics industry, invoking the success of 'Silicon Valley'

in California (Box 3.8). The cluster is located in central Scotland, but its geography is relatively dispersed in nature, stretching in a south-westerly line from Dundee to Ayrshire with a particular concentration in West Lothian and Lanarkshire, between the cities of Glasgow and Edinburgh (Figure 9.8). By 2000, electronics had become a very significant sector of the Scottish economy, accounting for over 40,000 direct jobs (and supporting another estimated 29,000

Box 9.4 (continued)

indirectly) and almost half of Scotland's manufactured exports by value
(Scottish Government 2004). This
reflects substantial value creation
through the attraction of investment,

infrastructure provision and workforce
training with the Scottish Development Agency and its successor, Scottish Enterprise, playing a key role in
linking regional assets to the strategic

needs of lead firms within GPNs. Some
value enhancement occurred in the
1990s through additional innovation
and skills development, reflecting
an increasing emphasis on R&D and

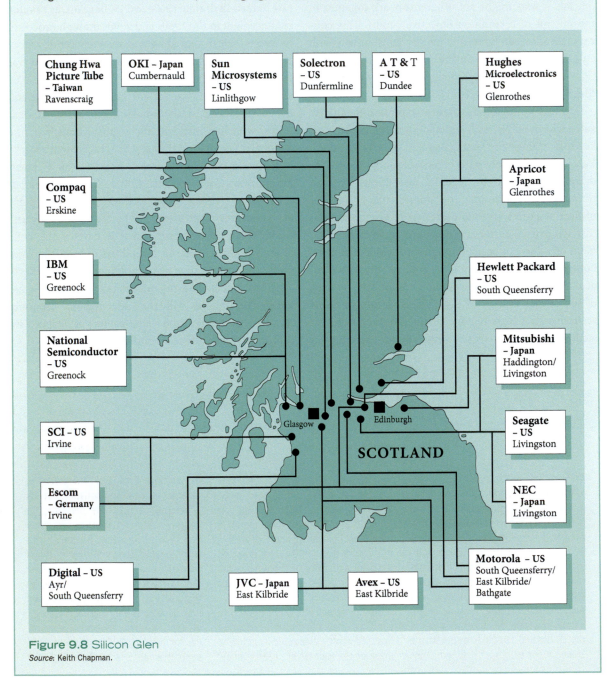

Figure 9.8 Silicon Glen
Source: Keith Chapman.

Box 9.4 (continued)

efforts to move into higher value-added activities.

Such value enhancement has not, however, translated into significant value capture, reflecting a continuing lack of local ownership and control. This was demonstrated by the impact of the sharp downturn in the global electronics sector in the early 2000s, resulting in the Scottish electronics industry contracting by 46 per cent in output terms between 2000 and 2005, compared to 27 per cent for the UK industry, reflecting a continuing specialisation in lower value-added activities (Ashcroft 2006: 6). Closures included Motorola's mobile phone plant in Bathgate in April 2001 with the loss of 3,000 jobs. Competition from Eastern Europe and East Asia where costs are significantly lower has compounded the problems of the Scottish industry with several examples of companies closing down their Scottish operations and moving jobs to Eastern Europe, particularly the Czech Republic and Hungary.

9.3.2 Forms and dynamics of strategic coupling

Based on MacKinnon (2012), the GPN 2.0 approach identifies three forms of strategic coupling between regional assets and GPNs (Yeung 2009):

1. Organic coupling where GPNs develop out of distinctive regional assets over time. This is largely associated with the dynamic growth regions in which lead firms are founded, going on to form GPNs by expanding their operations beyond the region of origin. This form of strategic coupling involves considerable autonomy for TNC operations in the region within the wider GPN. It also generates substantial value capture for the region in question, reflecting the fact that lead firms are owned and based there. Examples of this type of organic coupling include BMW in eastern Bavaria (Box 9.2), Apple or Google in Silicon Valley, California (Box 3.8) and Microsoft and Starbucks in Seattle.

2. Functional coupling is based upon deliberate and mutually beneficial linkages between GPNs and regional actors. These linkages can be derived from domestic firms reaching out to GPNs or from TNCs reaching in to a region from outside. They offer the region in question substantial degrees of both autonomy within wider GPNs and value capture through the activities of TNC affiliates and of suppliers and contractors based in the region. This autonomy and value capture is intermediate between the other two modes of coupling.

Functional coupling is particularly prevalent in Taiwan and Singapore, two of the original East Asian 'tiger' economies that followed Japan. In the former case, value capture is derived from domestic firms serving as strategic partners for global lead firms in the electronics and ICT industries (computer and iPhone production), whereas in the latter it has been based upon FDI by global lead firms in sectors such as electronics, chemicals, finance, transport and logistics.

3. Structural coupling involves lead firms and other actors connecting a region into GPNs. This is characteristic of labour-intensive global industries which are broadly organised according to a **new international division of labour** (NIDL) logic (section 3.5.2). It is characterised by a dependence on TNC investment, a lack of autonomy and limited value capture involving low-skilled work confined to specific, low-value tasks. Here, TNCs are attracted to the regions in question by competitive costs, abundant labour, and financial incentives for TNCs, with the region providing a 'production platform' for the assembly of largely imported components into finished products that are exported to world consumer markets. Examples include investment in the Pearl and Yangtze River Delta regions in China, which has been mediated by strategic partners in Taiwan and Singapore (Box 9.5), Malaysia's Penang region and Thailand's greater Bangkok region as well as the *maquiladora* plants in Mexico and much of 'Silicon Glen' in Scotland (Box 9.4).

Informed by developments in evolutionary economic geography (section 2.5.3), MacKinnon (2012) develops a more dynamic perspective on strategic coupling, emphasising the importance of **strategic decoupling** and **strategic recoupling** processes. Strategic decoupling involves a reduction or rupturing of the relationship between a region and a GPN through disinvestment, the exit of foreign firms and the loss of foreign markets. Strategic recoupling, by contrast, is based upon a renewal of the relationship between regions and GPNs through the recombination of existing regional assets with a new round of investment. It can be continuous or discontinuous in nature. The former refers to repeat investment in existing plants and the latter to investment in new plants and facilities following a period of disengagement or decoupling. The Indian pharmaceutical industry provides an example of the latter (Horner 2014). Here, the Indian Government responded to the limitations of an earlier period of structural coupling between 1947 and 1970, particularly in terms of a dependence upon FDI and a lack of national economic and health benefits, by undertaking a process of strategic decoupling in the 1970s and 1980s, resulting in the growth of a large domestic pharmaceutical industry. This subsequently provided the basis for a period of strategic recoupling with GPNs as part of a broader policy of economic liberalisation from 1991, enabling Indian firms to develop partnerships with foreign TNCs and to form their own GPNs to serve international markets.

9.3.3 Uneven development and regional transformation

Recent research has focused attention on processes of uneven development in GPNs involving periodic disinvestment in certain locations, linked to reinvestment in others (Werner 2016a). Here, there is considerable evidence of heightened inter-regional competition for FDI leading to a partial relocation of production from existing regions to new lower-cost locations. This is apparent in the shift of electronics production from Scotland to Central and Eastern Europe (Box 9.4) and from the Pearl River Delta to other coastal and inland locations in China (Box 9.5). In the garment and textile GPN, changes in trade rules involving the removal of the global quota system, and the erosion of preferential trading arrangements, created new patterns of regional investment and disinvestment in the global South (Werner 2016a). In the case of the Dominican Republic, for instance, this resulted in an exodus of foreign-owned firms from the eastern and capital regions, alongside the consolidation of domestic suppliers in the northern region and growth of subcontracted assembly in Haiti, as well as factory closures (Werner 2016b).

Box 9.5

The shifting geography of Taiwanese PC production in China

Taiwanese firms play a crucial role in the global computer industry, becoming particularly dominant in laptop production, with the top five Taiwanese manufacturers producing 76 per cent of the 46 million laptops sold worldwide in the last quarter of 2014 (Coe 2016: 328). Since the early 1990s, Taiwanese firms have relocated production to mainland China to take advantage of lower labour and land costs with almost 100 per cent of Taiwanese laptops produced in China. Interestingly, the geography of laptop production in China has shifted over time in response to inter-regional competition and changing cost dynamics (Yang 2009).

Earlier desktop production was concentrated in the Pearl River Delta in the south, particularly the city of Dongguan which became the biggest cluster of computer peripherals in the world. This was an 'implicit' or spontaneous process of structural coupling, led by the third and fourth tier of Taiwanese supplier firms, rather than being the result of more explicit regional or national government initiatives (ibid.). Over time, however, Taiwanese manufacturers in the Pearl River Delta began to experience growing shortages and inefficiencies as labour and land costs rose and local officials became less responsive.

From around 2000, Taiwanese investment in laptops became increasingly concentrated in the

Box 9.5 (continued)

Yangtze River Delta around Shanghai (Figure 9.9), which was seen to offer a more attractive environment for expansion. This has involved the clustering of suppliers on special industrial parks. Laptop production has become particularly concentrated in the city of Suzhou which produced over 15 million notebook computers in 2005, around 40 per cent of the global total (ibid.: 390–91). In contrast to Dongguan, this was based upon a more explicit process of strategic coupling in Suzhou driven by regional and local authorities, particularly through the establishment of industrial parks and export processing zones to facilitate the clustering of suppliers alongside the major Tier 2 ODMs (see Figure 9.3). The power relations between Taiwanese investors and local government remain highly asymmetrical, however, with the pressures of inter-regional competition enabling the Taiwanese investors to play regions off against one another through bargaining with local government to meet demands for cost reduction from the lead PC firms who coordinate the wider production network.

The geography of laptop manufacturing investment has shifted again since 2010 through a process of decoupling and relocation from coastal provinces such as the Yangtze River Delta to recoupling with inland provinces. This has been driven by rising costs in the former and China's shift from an export-oriented to a domestic market-driven development model in the post-economic crisis period (Yang 2013). As well as some relocation to the Bohai Rim region around Beijing and Shandong (Figure 9.9), this recoupling process has been focused on the inland Sichuan and Hebei provinces, led by the second-tier contract suppliers. For instance, Foxconn and Hewlett Packard jointly opened two facilities

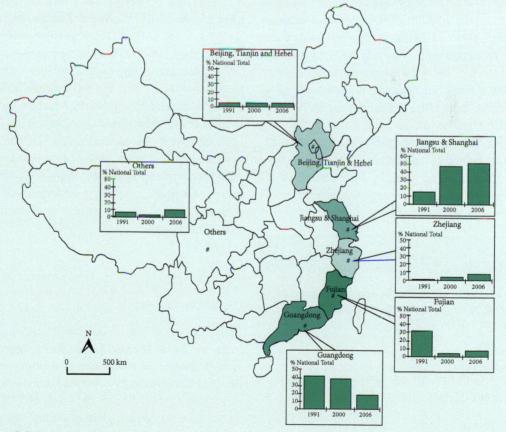

Figure 9.9 Evolving spatial distribution of Taiwanese investment in mainland China, 1991–2006
Source: Yang 2009: 391.

Box 9.5 (continued)

in the city of Chongqing in Sichuan capable of manufacturing 20 million laptops a year (Coe 2016: 330). Such companies were attracted by low labour, land and logistics costs and the city's free trade zone, alongside a discounted corporate tax rate of 15 per cent and the extension of airport facilities to support their logistics operations (ibid.). This reflects how this latest round of inter-regional competition for mobile investment has further increased the bargaining power of Taiwanese investors over local governments.

One limitation of GPN research, however, is that regions are viewed largely as nodes within corporate networks, offering a rather circumscribed view of place which fails to incorporate the broader dynamics of urban and regional development beyond firms and network relations (Kelly 2013). Questions of landscape and environment, households and livelihoods and broader patterns of social differentiation and uneven development tend to be neglected (ibid.). Including these broader dimensions focuses attention on the social distribution of the benefits of incorporation into GPNs, echoing the earlier point about whether economic upgrading is translated into better social outcomes. In his research on the Cavite region of the Philippines, for instance, the economic geographer Philip Kelly (2013) finds that despite dramatic employment growth driven by global manufacturing investment, income and unemployment have actually worsened, reflecting large-scale in-migration from outside the region. Meanwhile, other regions of the Philippines which have not experienced this growth remain outside the scope of GPN research.

back to the 1970s. This reflects the strong emphasis placed on attracting TNCs, particularly in manufacturing, by local and regional development agencies (Pike *et al.* 2017). For many policy-makers and commentators, FDI can help to transform regional economies, bringing new jobs, skills and knowledge as well as helping to connect local economies to wider global networks (Table 9.4). For other researchers, however, the effects are more pernicious, leading to a loss of local control over economic development in return for jobs that are often low skilled and always vulnerable to plant closure if TNCs decide to invest elsewhere (for example, Firn 1975; Massey 1984). In practice, the precise impact of TNC investment on host regions will vary considerably, being contingent upon the relationship between the quality of the investment and the structure and characteristics of the regional economy (Dicken 2015). While more empirically based research on FDI and regional development has often remained divorced from the more conceptually oriented literature on GPNs, the notion of strategic coupling provides a useful lens for refocusing on the recurring question of the impact of FDI on host regions.

Reflect

How useful do you think the concept of 'strategic coupling' is for examining the relationships between regions and GPNs?

9.4 Attracting and embedding FDI as a development strategy

There have been recurring debates in economic geography about the benefits of FDI for host regions going

9.4.1 Branch plant economies

Early studies of FDI in developed economies, often informed by a Marxist perspective, suggested that inward investment could lead to relations of dependency as hitherto locally controlled economies became increasingly subservient to the needs of TNCs. In the 1970s, the term '**branch plant economy**' was coined to reflect how some of the older industrial regions of Western Europe and particularly the UK were being repositioned within the wider space economy of capitalism.

Table 9.4 Advantages and disadvantages of FDI for host regions

Advantages	Disadvantages
Local economy	*Local economy*
Injection of investment and income opportunities Transfer of global knowledge and management techniques – enhancing competitive advantage in context of globalization Improves skills base through training, employment policies Contributes to region's tax base	Development increasingly dependent on external control Greater vulnerability to closure and job loss Deskilling/downgrading of local economy Production enclaves linked to broader global production systems but with few local benefits Marginalization/displacement of other sectors – economy linked to narrow development trajectory
Local firms	*Local firms*
New opportunities to supply inward investors 'Piggy-backing' on MNCs into export markets Learning through imitation of best practice in FDIs	Increased competition in local markets Increased competition for labour, land, capital drives up factor prices Become tied in to dominant client firms as suppliers
Local community	*Local community*
Upgrading of local infrastructure that may not otherwise have occurred (e.g. transport and communication links) FDI 'social investment' (e.g. in local schools, community services)	Disruption/destruction of local culture/society Resource displacement – investment on FDI specific infrastructure rather than community projects/initiatives
Employment	*Employment*
New jobs for local people Wages often in excess of average local wage Training and more progressive approach to human resource management	Introduction of more coercive management strategies Jobs often low skilled, routine Training limited to company-specific skills and not transferable
Political implications	*Political implications*
Influx of powerful non-local actors enhances power of region in context of inter-regional competition for state resources	Regional development agenda becomes hijacked by MNC interests Alternative political discourses marginalized – democratic deficit

Source: Adapted from Pavlinek 2004: 48.

The concept is defined by functional truncation due to a lack of higher-status activities such as strategic planning and R&D, limited linkages between TNC plants and the local economy with many components and materials being imported, and a high degree of external ownership and control (Phelps 1993).

The emergence of branch plant economies in traditional manufacturing regions in the 1970s and 1980s was part of a broader **spatial division of labour** (section 3.5.2). With the decline of traditional industries such as steel production, shipbuilding, textiles and engineering, and the closure of local firms, regional industrial specialisation and local ownership were being replaced by a situation in which regions were being recast and repositioned within corporate production networks. In the UK context, the recession of the early 1980s seemed to bear out many of the critics' concerns as several studies indicated that rates of firm closure and job loss were highest in externally owned plants, with foreign-owned plants experiencing the highest levels of decline (for example, Lloyd and Shutt 1985).

These concerns were reignited by the closure of some East Asian plants since 1998, based upon a new wave of FDI into North America and Western Europe by Japanese and East Asian corporations during the 1980s and early 1990s. In the UK, these closures were linked to the availability of new sources of cheap labour in Central and Eastern Europe (Box 9.4). The collapse of the market for certain electronics components in the late 1990s and early 2000s was felt heavily in the north-east of England, where both the German firm Siemens (1,100 jobs) and the Japanese firm Fujitsu (600 jobs) closed semiconductor plants in 1998 (Dawley 2007). Companies that have relocated from the UK's regions to the Czech Republic include Japanese firm Matsushita (from Wales, costing 1,400 jobs) and US firms Compaq (Scotland, 700 jobs) and Black and Decker (north-east England, 600 jobs) (Pavlinek 2004: 53).

9.4.2 Towards greater regional embeddedness?

The negative branch plant stereotype was questioned by some researchers in the 1980s and 1990s (e.g. Munday *et al.* 1995; Morgan 1997). Compared to the low-skill assembly line activities of previous US investments, the new wave of FDI into the UK and US from East Asian companies was geared towards the creation of locally integrated industrial complexes that were more firmly tied to their host regions. These investments were seen as more strategic and long term in nature. Instead of being purely concerned with the cost advantages of a particular location, they were geared to tapping into local skills and 'know-how', resulting in more locally embedded forms of investment. The flagship Nissan car manufacturing plant set up in the north-east of England in 1986, for example, has subsequently expanded to employ over 6,000 employees directly and support an extensive local supply chain throughout subsequent processes of recoupling involving investment in new models.

This more optimistic scenario hinges around a number of organisational changes taking place in TNCs themselves in response to globalisation. These were characterised as involving a shift away from centralised bureaucratic hierarchies to flatter and more decentralised structures, meaning that decision-making powers and higher-level operations are devolved to local branch plants. Such developments reflect the prerogative of TNCs to be "globally efficient, multinationally flexible, and capable of capturing the benefits of worldwide learning at the same time" (Dicken *et al.* 1994: 30). In these circumstances, it is argued that host regions become

home to key forms of knowledge that are valuable to the firm (Schoenberger 1999). Learning-by-doing and learning-by-using activities are in effect territorialised, through "the everyday experiences of workers, production engineers and sales representatives" (Lundvall 1992: 9), and locationally fixed (at least in the short term) in the sense that they "remain tacit and cannot be removed from [their] human and social context" (Lundvall and Johnson 1994, cited in Morgan.1997: 493).

Such developments provide a basis for functional forms of strategic coupling between regional assets and the global knowledge networks of TNCs, enabling regions to upgrade and improve their competitiveness. For example, Japanese branch plant investments in the Great Lakes region of the US during the 1980s were seen as instrumental in rejuvenating the regional economy, through the transfer of 'best practice' management and production techniques to local firms (Florida 1995). In particular, the high-performance manufacturing model employed by Japanese firms, based upon just-in-time (JIT) production, continuous improvement and team working, was viewed as superior to the assembly line techniques associated with Fordism (see section 3.4.1).

The changing role of urban and regional institutions and policy is also central to these claims of greater 'embeddedness'. Regional institutions have long been involved in seeking to attract FDI to their regions (Table 9.5), relying primarily on the provision of financial incentives in the branch plant era of the 1970s and 1980s. More recently, regional development agencies have emerged as key institutional mechanisms for strategic coupling (Pike *et al.* 2017: 245). They have increasingly adopted a more proactive and sophisticated role as part of the emphasis on territorial embeddedness rather than just attraction. This involves an increasing emphasis on the 'aftercare' of investors through attempts to foster links with local suppliers, universities and research institutes, reflecting the concern with tapping into local skills and knowledge. This is particularly geared to the promotion of forms of recoupling through the attraction of further repeat investment from TNCs, based upon processes of intra-corporate competition (within the same TNC group) between plants in different regions.

Table 9.5 Role and functions of local and regional development agencies in FDI

Policy formation	Liaison and dialogue with parent organisation Guidelines for inward investment policy Assessment of effectiveness of policy Integration with national and regional industrial policies Development of partnership scripts and protocols for joint working
Investment promotion and attraction	Marketing information and intelligence Market planning Marketing operations outside and inside the relevant area Management of overseas agents and offices
Investment approvals	Screening and evaluation of potential projects
Granting of incentives	Consideration of investment offers Incentive application advice and approvals (including direct financial incentives plus training grants, innovation grants, land and buildings, etc.
Providing assistance	Assistance with public utilities (roads, water, electricity, sewerage, telecommunications) Facilities and site Training and recruitment Links with universities and research institutes Supply chain linkage and development
Monitoring and aftercare	Continuation of assistance post-launch Relationship management and liaison (including r- investment projects, upgrading local suppliers etc.).

Source: Pike *et al.* 2017: 246.

9.4.3 Differentiated regional development outcomes

The evidence to support claims of increased embeddedness is mixed with plants remaining vulnerable to closure as corporate strategies and priorities change (Box 9.4). Earlier studies of Japanese branch plants, for example, have found the transfer of high-performance manufacturing to their foreign subsidiaries by Japanese TNCs to be the exception rather than the rule, with activities more often than not characterised by the kind of routine assembly work associated with earlier forms of FDI (Danford 1999). Subsequent research on UK regions found little evidence that firms were setting down longer-term linkages in the regions with a lack of additional investment after the initial start-up, while regional institutions tended to play a relatively limited supporting role (Dawley 2007; Phelps *et al.* 2003). More broadly, Phelps and Waley (2004) suggest that the increased local and regional orientation of TNCs may have been overstated in much of the literature, with the focus on local suppliers and linkages rather at odds with the trend towards centralised and global

sourcing and the consolidation and rationalisation of supplier networks around key international partners and contractors (Werner 2016a). As such, the autonomy of local TNC affiliates and the ability of regional development agencies (RDAs) to influence them often remains constrained by corporate decision-making processes within GPNs. In general, the direct benefits in terms of jobs, investment and income tend to outweigh the broader indirect effects on the local economy through increased links with local suppliers, knowledge spillovers and skills development that have been emphasised by advocates of the increased embeddedness perspective.

In central and Eastern Europe, which experienced an upsurge in FDI from western and predominantly European TNCs in the 1990s and early 2000s, the effects of FDI on host regions has been mixed (Box 9.6). The positive stimulus of jobs, new investment, western management expertise and contributions to the local tax base has been offset by the displacement of local firms, who lose skilled labour to incoming firms who pay higher wage rates. While some local suppliers have managed to upgrade production to meet the quality requirements of incoming TNCs, a lack of domestic capabilities and capital and TNC policies of centralised global sourcing mean that levels of local sourcing have generally remained low (Box 9.6). At the same time, Czech regions have experienced a growth in R&D activities with Volkswagen (VW), for instance, investing in a new R&D centre in 2008 and attracting supplier companies, although these remain foreign-owned rather than domestic (Pavlinek *et al.* 2009). There is also relatively little evidence that incoming firms are becoming firmly embedded within regional economies with many new start-ups serving as final assembly 'turnkey' plants for components shipped in from Western Europe, reducing the scope for local sourcing.

Box 9.6

Local sourcing and linkages in the Czech automotive industry

Automotive production has been partly decentralised from the Western European core to Eastern and Central Europe (ECE) since the 1990s with car assembly almost quadrupling in the latter region between 1990 and 2010 (Pavlinek 2012). This reflects the reintegration of the region into the European and global economies after 1989 and the availability of lower-cost supplies of labour and land. R&D, however, remains concentrated in Western Europe, particularly Germany, despite some limited growth in selected ECE regions. The Czech Republic has been one of the key recipients of FDI in its automotive sector which was comprised of 225 foreign automotive firms employing 135,827 workers in 2008 (Pavlinek and Zizalova 2016: 341). A key event was VW's acquisition of the former state-owned producer, Skoda Auto, which was initiated in the early 1990s and completed in 2000.

VW has subsequently restructured Skoda's supplier base through 'follow sourcing' and the upgrading of domestic suppliers. Follow sourcing involves suppliers following the lead firm to new production sites and setting up operations there. VW encouraged takeovers of Skoda's pre-1991 domestic suppliers to promote technology transfer from foreign suppliers. Economic liberalisation and the influx of FDI exposed domestic firms to international competition, which entailed higher standards for the quality of components and the timing of delivery (ibid.). Domestic suppliers who could not upgrade to these standards were rapidly replaced by foreign firms. As a result, almost two-thirds of pre-1989 Skoda suppliers stopped supplying in the 1990s.

In general, levels of local sourcing in the Czech automotive sector remain low (Table 9.6). Low levels of domestic supply of components to Tier 1 suppliers reflect the lack of firms capable of supplying sophisticated, higher-quality components at the price and quality required,

Table 9.6 The average share of the total volume of automotive suppliers sourced by Czech-based foreign firms (in %)

	Number of firms	Per cent share of total supplies from		
		Domestic firms	Czech-based foreign firms	Abroad
Total	62	12.6	10.9	76.5
Tier 1	14	13.7	10.8	75.5
Tier 2	21	15.4	7.7	76.9
Tier 3	22	7.0	7.9	85.1
Assemblers	5	20.0	32.0	48.0

Source: Pavlinek and Zizalova 2016: 345.

Box 9.6 (continued)

while more standardised low-value Tier 3 components are supplied from low-cost countries, especially China and India. Local sourcing is slightly higher for intermediate Tier 2 components, but suppliers are again subject to cost pressures from lower-cost countries. One key factor is the adoption of centralised sourcing by TNCs. This means that sourcing is controlled by the TNC head office, not the Czech-based affiliate, and that Czech suppliers have to supply the entire TNC group rather than just the local plant, something that many of them lack the capability to do. This underlines the point about the limited autonomy of TNC affiliates within the wider production network. At the same time, however, 40 per cent of foreign firms reported that the proportion of domestic supply had increased since the initial investment, while approximately half indicated that domestic suppliers have improved the quality and sophistication of their operation in response to customer demands (ibid.). This provides some indication of increasing embeddedness over time.

The rise of FDI in services means that assessments of its regional impacts must extend beyond manufacturing. The effects on the leading host regions of FDI in the IT sector such as the Indian city of Bangalore are of particular interest here (section 3.5.3). The Bangalore IT cluster has grown to account for 960,000 jobs and export revenues of US$27.42 billion in 2014–15 (Rao and Balasubrahmanya 2017: 101), while the population of the metropolitan region has expanded to over 10 million. This has been based on the attraction of prominent TNCs such as Microsoft, Hewlett Packard and Cisco and the growth of leading domestic firms, notably Infosys, Wipro and Tata Consultancy Services. These firms began at the bottom of the 'smiling curve' of value creation (Figure 9.2), focusing on 'implementation and testing', i.e. tasks involving coding, data entry, etc. (Lorenzen and Mudambi 2013: 520). The cluster has found it hard to move up the value chain beyond its outsourcing role, with TNCs reluctant to involve Bangalore firms in higher-value tasks due to a lack of domestic capacity and limited intellectual property protections in India (Pike *et al.* 2017: 249). Despite strong knowledge spillovers between TNC affiliates and the main domestic firms, broader local linkages remain limited with the cluster remaining focused on the export of services to TNC clients. In general, the Bangalore IT cluster is characterised by the functional form of coupling, involving international partnerships between local firms and lead TNCs in GPNs, although this operates alongside significant elements of structural coupling.

Reflect

Do you think that the benefits of FDI outweigh the disadvantages for host regions? To what extent does this vary between regions?

9.5 Summary

This chapter has highlighted the complex geographies of GPNs, which now account for the majority of world trade, reflecting widespread processes of outsourcing to suppliers and contractors, leaving lead firms to concentrate on higher value-added tasks such as R&D, design, branding and marketing. In response to the rather restricted and linear approaches of global commodity chains and GVCs in sociology, development studies and management studies, a group of economic geographers have developed a sophisticated and integrated GPN framework which incorporates a range of actors. The original GPN 1.0 formulation focuses attention on processes of value creation, enhancement and capture in GPNs, alongside the power relations between actors and forms of societal, geographical and network embeddedness. The more recent GPN 2.0 approach builds on this earlier framework, developing a more dynamic, evolutionary understanding that stresses the interaction between competitive dynamics, firm-specific strategies, value capture trajectories and regional development outcomes. Processes of economic and social upgrading have been a key focus of interest in GVC and GPN research, with recent research uncovering the complexities of these processes, particularly in terms of the different social outcomes that firm upgrading can have for specific groups of workers.

The GPN approach provides a genuinely multi-scalar framework for research, replacing the local-global binary of the 1990s with a network approach based on following the key actors across various sites and scales. One key scale is that of the region where GPN research has made a major contribution to the rethinking of regional development in relational terms, overcoming some of the limitations of new regionalist research in the 1990s. From this perspective, regional development is a product of the 'strategic coupling' between regional assets and GPNs, emphasising how regions are incorporated into networks and viewing the role of regional institutions as one of matching regional assets to the needs of TNCs. Recent research identifies three forms of strategic coupling between regional assets and GPNs: organic coupling whereby TNCs and their GPNs emerge out of distinctive regional assets; functional coupling based on the development of mutually beneficial relationships between regional actors and GPNs; and structural coupling which is characterised by unequal and dependent relations between TNCs and regional actors.

The attraction of FDI is a long-standing focus of regional development policy, prompting recurring debates over its regional impacts. FDI creates relations of dependency in the eyes of some observers, although offering the opportunities for upgrading according to others. For host regions, vulnerability to competition and TNC relocation seem to be common themes in a range of different circumstances, while recent processes of reorganisation within GPNs present TNCs with greater spatial flexibility. Yet it is important not to view regions as passive victims of corporate restructuring, and regional development outcomes should be viewed as products of the ongoing interaction between regional actors and TNCs within GPNs. FDI retains a very important role as a direct source of employment, investment and income, although there is limited evidence of increased embeddedness through the indirect links to local suppliers, skills and universities.

Exercise

Select a major global industry and examine its geographical organisation and key competitive dynamics. Which major 'lead' or coordinating firms are involved and what are the key strategies they have adopted? What other types of actor are involved in the GPN? How is it governed and coordinated? What is the geography of the network? What forms of strategic coupling can be identified in key regions and how have these evolved over time? What are the main regional development outcomes of such structural coupling?

Key reading

Coe, N. (2016) The geographies of global production networks. In Daniels, P., Bradshaw, M., Shaw, D., Sidaway, J. and Hall, T. (eds) *An Introduction to Human Geography: Issues for the 21st Century*, 5th edition. Harlow: Pearson, pp. 321–42.
A very useful introduction to the GPN approach which defines global production networks and examines their operation across a range of industry contexts, providing some good case studies.

Coe, N. and Yeung, H.W. (2015) *Global Production Networks: Theorising Economic Development in an Interconnected World*. Oxford: Oxford University Press.
An important book by two leading GPN researchers which sets out the GPN 2.0 approach, highlighting the interaction between competitive dynamics, firm-specific strategies, value capture trajectories and regional development outcomes.

Coe, N., Hess, M., Yeung, H.W., Dicken, P. and Henderson, J. (2004) Globalising regional development: a global production networks perspective. *Transactions of the Institute of British Geographers* 29: 464–84.
An important paper which applies the GPN approach to the issue of regional development, using it to recast regions in relational terms and focusing attention on process of strategic coupling between global production networks and regional assets.

Pike, A., Rodríguez-Pose, A. and Tomaney, J. (2017) *Local and Regional Development*, 2nd edition. London: Routledge, chapter 7, pp. 229–53.
An excellent overview of the attraction and embedding of FDI as a regional development strategy. As well as outlining the changing role of TNCs and discussing a GPN and GVC

perspective, the book considers the strengths and weaknesses of the attraction of FDI as an approach to local and regional development. The chapter also considers the changing role of local and regional institutions and policy.

Werner, M. (2016) Global production networks and uneven development: exploring geographies of devaluation, disinvestment and exclusion. *Geography Compass* **10/11: 457–69.**
This paper examines global production networks from the perspective of uneven development, summarising recent research on this issue by GVC and GPN researchers. Expands the scope of research beyond upgrading and downgrading to consider processes of devaluation, regional disinvestment and exclusion from global production networks.

Yeung, H.W. (2015) Regional development in the global economy: a dynamic perspective of strategic coupling in global production networks. *Regional Science Policy & Practice* **7: 1–23.**
An excellent overview of the concept of strategic coupling from a GPN perspective. This is informed by GPN 2.0 thinking, identifying three types of strategic coupling and providing a range of examples as well as discussing decoupling and recoupling processes.

References

Allen, J. (2003) *Lost Geographies of Power*. Oxford: Blackwell.

Amin, A. and Thrift, N. (1994) Living in the global. In Amin, A. and Thrift, N. (eds) *Globalisation, Institutions and Regional Development in Europe*. Oxford: Oxford University Press, pp. 1–22.

Ashcroft, B. (2006) Outlook and appraisal. *Quarterly Economic Commentary* 30 (4): 1–9.

Barrientos, S. (2014) Gendered global production networks: analysis of cocoa–chocolate sourcing. *Regional Studies* 48: 791–803.

Barrientos, S., Gereffi, G. and Rossi, A. (2011) Economic and social upgrading in global production networks: a new paradigm for a changing world. *International Labour Review* 150: 319–40.

Bathelt, H. (2006) Geographies of production: growth regimes in spatial perspective 3 – towards a relational view of economic action and policy. *Progress in Human Geography* 30: 223–36.

Blazek, J. (2016) Towards a typology of repositioning strategies of GVC/GPN suppliers: the case of functional upgrading and downgrading. *Journal of Economic Geography* 16: 849–69.

Cattaneo, O., Gereffi, G. and Staritz, C. (2010) Global value chains in a post-crisis world: resilience, consolidation and shifting end markets. In Cattaneo, O., Gereffi, G. and Staritz, C. (eds) *Global Value Chains in a Post-Crisis World: A Development Perspective*. Washington, DC: World Bank, pp. 3–20.

Coe, N. (2016) The geographies of global production networks. In Daniels, P., Bradshaw, M., Shaw, D., Sidaway, J. and Hall, T. (eds) *An Introduction to Human Geography: Issues for the 21st Century*, 5th edition. Harlow: Pearson, pp. 321–42.

Coe, N. and Yeung, H.W. (2015) *Global Production Networks: Theorising Economic Development in an Interconnected World*. Oxford: Oxford University Press.

Coe, N., Dicken, P. and Hess, M. (2008) Global production networks: realising the potential. *Journal of Economic Geography* 8: 271–95.

Coe, N., Hess, M., Yeung, H.W., Dicken, P. and Henderson, J. (2004) Globalising regional development: a global production networks perspective. *Transactions of the Institute of British Geographers* 29: 464–84.

Coe, N.M., Kelly, P.F. and Yeung, H.W. (2007) *Economic Geography: A Contemporary Introduction*. Oxford: Wiley Blackwell.

Danford, A. (1999) *Japanese Management Techniques and British Workers*. London: Mansell.

Daniels, P., Bradshaw, M., Shaw, D. and Sidaway, J. (eds) (2008) *An Introduction to Human Geography*, 3rd edition. Pearson Education.

Dawley, S. (2007) Fluctuating rounds of inward investment in peripheral regions: semiconductors in the North East of England. *Economic Geography* 83: 51–73.

Devine, T.M. (1999) *The Scottish Nation*. London: Penguin.

Dicken, P. (2007) *Global Shift: Reshaping the Global Economic Map in the 21st Century*, 5th edition. London: Sage.

Dicken, P. (2015) *Global Shift: Mapping the Changing Contours of the World Economy*, 7th edition. London: Sage.

Dicken, P., Forsgren, M. and Malmberg, A. (1994) The local embeddedness of transnational corporations. In Amin, A. and Thrift, N. (eds) *Globalisation, Institutions and Regional Development in Europe*. Oxford: Oxford University Press, pp. 23–45.

Dicken, P., Kelly, P.F., Olds, K. and Yeung, H.W. (2001) Chains and networks, territories and scales: towards a relational framework for analysing the global economy. *Global Networks* 1: 89–112.

Firn, J. (1975) External control and regional development: the case of Scotland. *Environment & Planning* A 7: 393–414.

Florida, R. (1995) Toward the learning region. *Futures* 27: 527–36.

Gereffi, G. (1994) The organisation of buyer-driven commodity chains: how US retailers shape overseas production networks. In Gereffi, G. and Korzeniewicz, M. (eds) *Commodity Chains and Global Capitalism*. Westport, CT: Greenwood Press, pp. 95–122.

Gereffi, G., Humphrey, J. and Sturgeon, T. (2005) The governance of global value chains. *Review of International Political Economy* 12: 78–104.

Granovetter, M. (1985) Economic action and social structure: the problem of embeddedness. *American Journal of Sociology* 91: 481–510.

Grimes, S. and Sun, Y. (2016) China's evolving role in Apple's global value chain. *Area Development & Policy* 1 (1): 94–112.

Harvey, D. (1982) *The Limits to Capital*. Oxford: Blackwell.

Henderson, J., Dicken, P., Hess, M., Coe, N. and Yeung, H.W. (2002) Global production networks and economic development. *Review of International Political Economy* 9: 436–64.

Hess, M. (2004) 'Spatial' relationships? Towards a reconceptualisation of embeddedness. *Progress in Human Geography* 28: 165–86.

Hopkins, T. and Wallerstein, I. (1986) Commodity chains in the world economy prior to 1800. *Review* 10: 157–70.

Horner, R. (2014) Strategic decoupling, recoupling and global production networks: India's pharmaceutical industry. *Journal of Economic Geography* 14: 1117–40

Kelly, P. (2013) Production networks, place and development: thinking through global production networks in Cavita, Philippines. *Geoforum* 44: 82–92.

Liu, W. and Dicken, P. (2006) Transnational corporations and 'obligated embeddedness': foreign direct investment in China's automobile industry. *Environment and Planning A* 38: 1229–47.

Lloyd, P. and Shutt, J. (1985) Recession and restructuring in the North West region 1974–82: the implications of recent events. In Massey, D. and Meegan, R. (eds) *Politics and Method*. London: Methuen, pp. 16–60.

Lorenzen, M. and Mudambi, R. (2013) Clusters, connectivity and catch-up: Bollywood and Bangalore in the world economy. *Journal of Economic Geography* 13: 501–34.

Lundvall, B.-A. (1992) *National Systems of Innovation*. London: Pinter.

Lundvall, B.-A. and Johnson, B. (1994) The learning economy. *Journal of Industry Studies* 1: 23–43.

MacKinnon, D. (2012) Beyond strategic coupling: reassessing the firm-region nexus in global production networks. *Journal of Economic Geography* 12: 227–45.

MacKinnon, D. and Phelps, N.A. (2001) Devolution and the territorial politics of foreign direct investment. *Political Geography* 20: 353–79.

Massey, D. (1984) *Spatial Divisions of Labour: Social Structures and the Geography of Production*. London: Macmillan.

Morgan, K. (1997) The learning region: institutions, innovation and regional renewal. *Regional Studies* 31: 491–504.

Munday, M., Morris, J. and Wilkins, B. (1995) Factories or warehouses? A Welsh perspective on Japanese transplant manufacturing. *Regional Studies* 29: 1–17.

Pavlinek, P. (2004) Regional development implications of foreign direct investment in Central Europe. *European Urban and Regional Studies* 11: 47–70.

Pavlinek, P. (2012) The internationalization of corporate R&D and the automotive industry R&D of East-Central Europe. *Economic Geography* 88: 279–310.

Pavlinek, P. and Zizalova, P. (2016) Linkages and spillovers in global production networks. *Journal of Economic Geography* 16: 331–63.

Pavlinek, P., Domanski, B. and Guzik, R. (2009) Industrial upgrading through foreign direct investment in Central European automotive manufacturing. *European Urban and Regional Studies* 16: 43–63.

Phelps, N.A. (1993) Branch plants and the evolving spatial division of labour: a study of material linkage change in the North East of England. *Regional Studies* 27: 87–101.

Phelps, N.A. and Waley, P. (2004) Capital versus the districts: the story of one multinational company's attempts to disembed itself. *Economic Geography* 80: 191–215.

Phelps, N.A., MacKinnon, D., Stone, I. and Braidford, P. (2003) Embedding the multinationals? Institutions and the development of overseas manufacturing affiliates in Wales and North East England. *Regional Studies* 37: 27–40.

Pike, A., Rodríguez-Pose, A. and Tomaney, J. (2017) *Local and Regional Development*, 2nd edition. London: Routledge.

Rao, P.M. and Balasubrahmanya, M.H. (2017) The rise of IT services clusters in India: a case of growth by replication. *Telecommunications Policy* 41: 90–105.

Ruddick, G. and Oltermann, P. (2017) The crankshaft's odyssey and how Brexit could undo an industry. *The Guardian*, 4 March, p. 29. At www.theguardian.com/business/2017/mar/03/brexit-uk-car-industry-mini-britain-eu.

Schoenberger, E. (1999) The firm in the region and the region in the firm. In Barnes, T.J. and Gertler, M. (eds) *The New*

Industrial Geography: Regions, Regulation and Institutions. London: Routledge, pp. 205–24.

Scottish Government (2004) *Scottish Economic Report, March 2004.* Edinburgh: The Scottish Government.

Sturgeon, T. (2001) How do we define value chains and production networks? *IDS Bulletin* 32: 9–18.

United Nations Conference on Trade & Development (UNCTAD) (2013) *World Investment Report 2103: Global Value Chains, Investment and Trade for Development.* New York: UNCTAD.

Werner, M. (2016a) Global production networks and uneven development: exploring geographies of devaluation, disinvestment and exclusion. *Geography Compass* 10/11: 457–69.

Werner, M. (2016b) *Global Displacements: The Making of Uneven Development in the Caribbean.* Oxford: Wiley Blackwell.

Yang, C. (2009) Strategic coupling of regional development in global production networks: redistribution of Taiwanese personal computer investment from the Pearl River Delta to the Yangtze River Delta, China. *Regional Studies* 43: 385–408.

Yang, C. (2013) From strategic coupling to recoupling and decoupling: restructuring global production networks and regional evolution in China. *European Planning Studies* 21: 1046–63.

Yeung, H.W. (2009) Transnational corporations, global production networks and urban and regional development: a geographer's perspective on multinational enterprises and the global economy. *Growth and Change* 40: 197–226.

Yeung, H.W. and Coe, N.M. (2015) Towards a dynamic theory of global production networks. *Economic Geography* 91: 29–58.

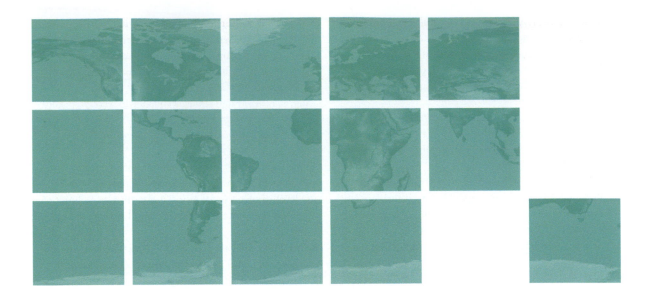

Chapter 10
Urban agglomeration, innovation and creativity

Topics covered in this chapter

➤ Influential recent arguments about the increasing agglomeration of economic activity in cities and the benefits of such agglomeration.

➤ The widespread emphasis on knowledge and innovation as key drivers of economic development.

➤ Processes of innovation and clustering in cities.

➤ The concept of the creative classes and its adoption by urban and regional policy-makers.

➤ The significance of wider global knowledge linkages for the development of cities and regions.

10.1 Introduction

A resurgence of interest in cities and regions as economic units has been apparent from both academics and policy-makers over the past couple of decades. At first sight, this renewed focus on cities and regions seems paradoxical, given the prevailing emphasis on globalisation as perhaps *the* political and economic force of the last 15 or so years. As we argue in this book, however, globalisation is an uneven process, leading to the concentration of economic activity in particular places and creating increasingly close linkages between the urban/regional and global scales of activity. National economic coherence has been undermined since the 1970s as states have lost control over increasingly globalised flows of investment. The abandonment of Keynesian policies of demand management and full employment has exposed cities and regions to the effects of international competition. This has focused attention on the need for interventions at the sub-national scale if cities and regions are to be able to shape their own development prospects in a climate of rapid technological change and increased capital mobility (Amin and Thrift 1994).

Particularly strong claims about the increased urban **agglomeration** of economic activity and prosperity have been made in recent years (see Glaeser 2011), making cities the main focus of this chapter. Spatial agglomeration or concentration, referring to the tendency for industries to cluster in particular places, can be contrasted with the opposite process of **spatial dispersal** where firms move out of existing centres into other, often less developed regions. As we saw in Chapter 3, the balance between agglomeration and spatial dispersal changes over time, reflecting prevailing forms of economic development and organisation. Thus, for example, the decline of Fordist manufacturing industries was associated with dispersal in the late 1970s, while the rise of new growth sectors like ICT, life sciences and financial services fostered renewed agglomeration from the late 1980s. As such, agglomeration, which generates regional divergence, is associated with the growth of new industries. By contrast, dispersal, which promotes regional convergence, is characteristic of mature and declining economic sectors (Storper *et al.* 2016).

The renewed emphasis on urban agglomeration reflects the increased importance of knowledge and skills in economic development (Lundvall 1994). Recent theories of agglomeration emphasise the role of knowledge 'spillovers' between firms, involving firms working together to solve problems and to develop new products and services, as well as the concentration of human capital in the form of educated and skilled workers. Skilled people are viewed as increasingly mobile between places, and some theorists have argued that cities should focus on developing amenities such as parks, museums, art galleries and landmark buildings in order to attract skilled people rather than firms *per se* (Florida 2002). The capacity of cities and regions to generate and harness knowledge and attract skilled labour is central to their capacity to adapt to economic change by moving into new and related economic activities in response to wider processes of economic restructuring. Whilst much of the literature on agglomeration has centred on large metropolitan cities, such renewal and reinvention has proved a particular challenge for post-industrial cities in Europe and North America that lost much of their traditional economic base through deindustrialisation in the 1970s and 1980s (section 3.4.3).

10.2 Cities, agglomeration and knowledge

10.2.1 Agglomeration and urban economies

According to many influential commentators, the contemporary period is one in which population, productive activity and wealth are increasingly concentrated in cities (Storper and Scott 2016). This reflects the fact that over half the world's population now lives in cities, coupled with the proliferation of a number of mega-cities of over 10 million people (Pike *et al.* 2017: 5). The NEG (section 2.3) and its conceptual cousin, **urban economics**, emphasise the economic benefits generated by the scale and density of economic activity and wealth within cities – agglomeration. For the leading urban economist Edward Glaeser, cities are engines of **innovation**, based upon the increasing returns to knowledge that are generated by people in close proximity to each other (Box 10.1) This argument has become increasingly influential with policy-makers, including the World Bank and other international organisations as well as many national governments, fostering an approach that seeks to work with the grain of the market forces that generate **agglomeration economies** (Cheshire *et al.* 2014).

Box 10.1

Triumph of the City

Triumph of the City is the title of a best-selling 2011 book by Edward Glaeser, a leading Harvard economist who is a regular contributor to public debate through newspaper and policy articles. Glaeser is a product of the renowned Chicago School of economics, home of the neoclassical counter-revolution against Keynesianism in the 1970s, and his work in urban economics has sought to apply microeconomic techniques to the analysis of the competitive city from a free market Chicago-style approach. As Peck (2016: 2) notes, the book represents the "economically rationalist strain" of what Brendan Gleeson (2012) calls the 'new urbanology', emanating from North America, which celebrates the economic and social vitality and potential of cities, with a particular focus on urban entrepreneurialism.

For Glaeser (2011: 6) "cities are the absence of physical space between people and companies. They are proximity, density, closeness." The central paradox of the modern city is that proximity has actually become more important and economically valuable as the costs of connecting and transporting goods over long distances have fallen. This, of course, reflects the impact of globalisation, falling transport and communication costs and the growth of a **knowledge-based economy**. In particular, the notion of cities as engines of innovation reflects how "technological change has increased the returns to the knowledge that is best produced by people in close proximity to other people" (ibid.). Cities thrive when they have many small firms and skilled citizens (ibid.: 8), reflecting Glaeser's underlying attachment to market competition. Based upon the importance of human capital to urban growth, Glaeser contends that cites should focus on attracting skilled people through low taxes, limited regulation and high-quality education facilities and cultural amenities (see Peck 2016).

While this celebratory account of cities as "centres of ideas creation and transmission" (Glaeser 2000: 83) acknowledges the reality of urban poverty, Glaeser's view is described as strongly sanguine by Gleeson (2012: 931). Peck (2016: 2) summarises Glaeser's politics as "pro-market, pro-development and anti-regulation". He has become closely associated with the staunchly conservative Manhattan Institute for Policy Research, where he is a senior fellow. From this perspective, Glaeser (2011: 11) observes that "there's a lot to like about urban poverty", based on the assertion that cities attract poor people rather than making people poor, meaning that they move to cities in search of advancement and prosperity. For Glaeser and other new urban economists and new economic geographers, policies should aim to help poor people not poor places, particularly through education and the development of skills. This distinction between people and places is, of course, something of an anathema to geographers who emphasise the multiple connections between people and places. At the same time, Glaeser does support increased state funding for places with higher levels of poverty (ibid.: 259).

Scott and Storper (2015: 6) describe agglomeration as "the basic glue" that holds cities together as complex ensembles of human activities. It is driven by an underlying division of labour in economic life whereby the production of goods and services and related activities is organised through networks of specialised but complementary units of activity (firms and other economic actors) (Storper and Scott 2016: 1116). The need for frequent interaction and communication between these units explains their concentration in cities, overcoming the 'friction of distance'. This concentration of economic activities in cities is not at odds with wider economic relationships and flows. In particular, "trade enables cities to specialise and sell their outputs in exchange for the specialised outputs of other places. The economic viability of cities and the growth of long-distance trade are therefore complementary and mutually-reinforcing phenomena" (Scott and Storper 2015: 6). In the contemporary period, the increased agglomeration of economic activities is closely associated with the process of globalisation. **Clusters** of firms in cities often serve global markets as the increased volume of external connections reinforces the need for geographical proximity and localised interaction. Of course, as Walker (2016) argues, agglomeration is not the sole function of cities which are also places in which the economic surplus from trade and production is concentrated and invested in the built environment. This concentration of wealth is associated with social inequality and exploitation. As such, the city also functions as an arena for the display of wealth and power, symbolising the prevailing social order (ibid.).

10.2.2 Knowledge and agglomeration

Renewed interest in agglomeration reflects the broader shift towards an increasingly knowledge-based economy since the early 1990s. According to the Scandinavian economist Lundvall, capitalism has entered a new stage in which "knowledge is the most important resource and learning the most important process" (Lundvall 1994). Knowledge should be seen as distinct from information, referring to the broader frameworks of meaning through which information or data about real-world events and trends is processed and understood. As Nonaka *et al.* (2001: 15) observe, "information becomes knowledge when it is interpreted by individuals and given a context and anchored in the beliefs and commitments of individuals".

A distinction is often drawn between codified and tacit forms of knowledge (ibid.). **Codified** or explicit **knowledge** refers to formal, systematic knowledge that can be conveyed in written form through, for example, programmes or operating manuals. **Tacit knowledge**, on the other hand, refers to direct experience and expertise, which is not communicable through written documents. It is a form of practical 'know-how' embodied in the skills and work practices of individuals or organisations. Traditionally, in industries such as construction, practical skills were acquired through apprenticeships where new entrants learnt on the job by shadowing and assisting established tradesmen. Contemporary understandings of agglomeration are based on the assumption that codified knowledge has become increasingly global in organisation and reach, whilst tacit knowledge remains local, relying on geographical proximity to foster communication and interaction between firms (Maskell and Malmberg 1999).

Firm learning and innovation play a crucial role in the agglomeration process. A key factor in determining a firm's success in knowledge generation and innovation is its '**absorptive capacity**', referring to the ability to recognise, assimilate and exploit knowledge, derived from either internal or external sources (Cooke and Morgan 1998: 16). The absorption of knowledge, in turn, depends upon the existence of a common corporate culture and language, meaning that everybody shares the same broad outlook and sense of the company's overall purpose and objectives. In the absence of a common language and outlook, it becomes far more difficult to exploit tacit knowledge without it leaking out to competitors (Box 10.2).

Box 10.2

'Fumbling the future': Xerox, Apple and the personal computer

A group of scientists working at Xerox's Palo Alto Research Centre in Silicon Valley discovered key elements of the personal computer (PC) in the 1970s. This included not only the processing equipment that sits beneath the desk, but also the screen desktop of icons, folders and menus that make up systems such as Windows, Macintosh and the World Wide Web.

Yet the development of the PC did not benefit the firm which initially created the technology. This reflected divisions within Xerox between scientists based in Palo Alto, development engineers in Dallas, Texas and the company management in Stanford, California. Not only were the different groups separated geographically, there was also an absence of the common language

and outlook required to exploit emerging technologies commercially. The engineers found the scientists naïve and unrealistic whilst the scientists viewed employees in other divisions as 'toner heads' who were interested only in photocopiers (Brown and Duguid 2000: 151).

Such internal divisions meant that the knowledge embodied in the emerging computer technology

10.2.3 Agglomeration, skills and productivity

Urban economics understandings of spatial agglomeration identify three key mechanisms or microfoundations: sharing, matching and learning (Duranton and Puga 2004). While these echo Marshall's much earlier formulation in many respects (section 3.3.1), they are more theoretical in nature, based on the identification of specific mechanisms rather than the looser distinction between the different types of resources and materials (infrastructure, suppliers and labour). Sharing refers to the sharing of certain indivisible facilities such as infrastructure; the gains from a variety of input providers; the benefits of specialisation; and risk.

Matching involves the pairing of people and jobs, with agglomeration improving either the quality or quantity of the match, based upon the existence of large pools of firms and workers. Learning refers to the dense flows of knowledge and information between proximate firms and workers, fostering various forms of innovation. Here, contemporary theories of agglomeration go beyond the traditional Marshallian approach by stressing not only cost reduction, but also the dynamic benefits of clustering in facilitating innovation and learning processes (Malmberg and Maskell 2002).

Urban economics research emphasises the role of human capital in creating and sustaining agglomerations. Here, the positive relationship between the initial level of human capital in a city and its later growth is

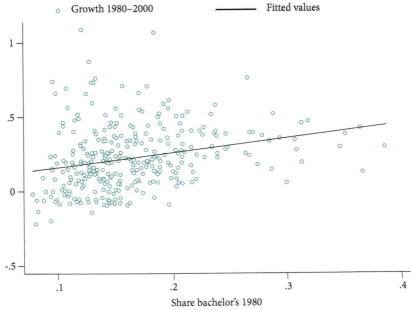

Figure 10.1 Metropolitan area growth, United States (1980–2000) and human capital (share with bachelor's degree 1980)

Source: Glaeser and Saiz 2003, Figure 1.

a central finding (Figure 10.1), with the evidence indicating that skills induce growth rather than vice versa (Glaeser and Saiz 2003). Relatedly, some studies have found there to be a relationship between productivity, usually measured in terms of output per worker, and city size in population terms (Glaeser and Resseger 2010). This follows from urban economics theory as agglomeration effects and increasing returns should be most pronounced in the largest cities (Martin *et al.* 2016). Furthermore, output is itself "subject to agglomeration economies, whereby people are more productive when they work in densely populated areas", whilst 'human capital spillovers' mean that concentrations of educated people raise both the level and growth rate of productivity (Glaeser and Gottlieb 2008: 155).

These arguments based on US data have not gone undisputed, however, particularly in terms of the relationship between city size and productivity. Other studies in Europe and the US, for instance, have indicated that doubling local employment or population density only increases productivity by a small amount, 2–6 per cent (Ciccone 2012). In the UK context, the relationship between city size and productivity is weak beyond London with many of the larger cities having productivity levels below many smaller-sized cities (Martin *et al.* 2016). More broadly, there is no simple relationship between city size and growth among European cities with the performance of both larger and small- to medium-sized cities proving mixed in practice (Dijkstra 2013). At the same time, larger cities tend to have higher levels of income inequality and poverty (Lee *et al.* 2014; Royuela *et al.* 2014).

From a developmental perspective, the capacity of cities and regions to generate and utilise knowledge is central to their economic prospects. Large metropolitan regions are seen to be particularly advantaged in this respect, based upon the concentration of large numbers of firms in close proximity, fostering knowledge exchange and innovation between them (Doloreux and Shearmur 2012). But smaller cities and more peripheral regions also need to try and compete within an economy driven by knowledge, focusing attention on initiatives to promote the agglomeration of firms in similar and related sectors. These attempts to foster local 'buzz' through increased interaction between firms in smaller cities have not proved particularly successful, however,

suggesting that efforts to encourage such firms to build knowledge linkages or 'pipelines' with partners beyond the city or region in question may have a greater chance of success (Rodríguez-Pose and Fitjar 2013).

10.3 Cities, innovation and clusters

10.3.1 Models of innovation

Innovation is a key theme of knowledge-based economic development. It can be defined as the creation of new products and services or the modification of existing ones to gain a competitive advantage in the market. The commercial exploitation of ideas is crucial, distinguishing innovation from invention. Traditionally, a distinction was drawn between product and process innovations, with the former referring to new outputs (mobile phones, for instance) and the latter to new methods of making or doing things (the moving assembly line, for example). This was derived from studies of manufacturing industries, with innovation in service industries often involving a particular service being delivered or packaged in new ways. Another distinction is often made between radical and incremental forms of innovation, referring to major innovations such as the development of the internet and smaller improvements in the design and operation of particular products and services respectively (Freeman 1994).

Understandings of innovation as a process have changed over time from a linear to an interactive approach. The traditional **linear model of innovation** was focused on large corporations, breaking it down into a series of well-defined stages running from the research laboratory to the production line, marketing department and retail outlet (Figure 10.2). Sometimes known as the science, technology and innovation model, it views formal research and development, involving advanced scientists and engineers, operating separately from other divisions of the company (Rodríguez-Pose and Fitjar 2013). In recent years, this has been replaced by an **interactive approach**, viewing innovation as a circular process based on cooperation and collaboration between manufacturers or services providers, users (customers), suppliers, research institutes, development

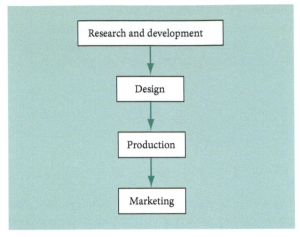

Figure 10.2 The linear model of innovation
Source: D. MacKinnon.

agencies, etc. (Figure 10.3) (Box 10.3). The metaphor of the firm as a laboratory draws attention to the experimental nature of innovation, which is often based on trial and error, involving the adoption of existing practices and the trying out of new combinations (Cooke and Morgan 1998: 47–53). This second approach has recently been termed the doing, using and interacting model (Rodríguez-Pose and Fitjar 2013).

An extensive literature on the geography of innovation has emerged since the 1990s, focusing particularly on how to conceptualise the innovation process from an urban and regional perspective and the implications for economic development policy. Much of this work is underpinned by the interactive model of innovation, based on the idea that interaction between firms and

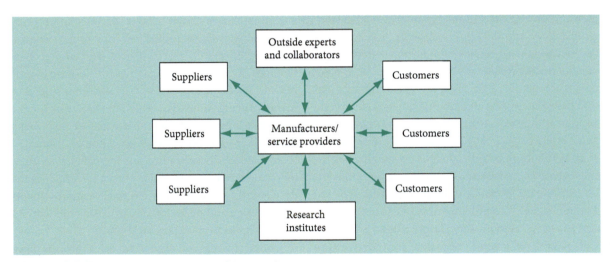

Figure 10.3 The interactive model of innovation
Source: D. MacKinnon.

The development of Apple's iPod

Apple's iPod rapidly became the most popular and fashionable digital music player in the marketplace, following its launch in October 2001 (Figure 10.4). It is based on the recombination of several existing components, involving collaboration between a number of different companies. Rather than stemming from a revolutionary new invention, the iPod began with a sense in early 2001 that Apple could develop a better product than any of its rivals within the emerging market for MP3 players (Hardagon 2005). The original idea came from an independent contractor, Tony Fadell, who was hired by Apple to develop the

Box 10.3 (continued)

product. The platform design came from Portal Player and the operating system from Pixo (both Silicon Valley start-ups) whilst the hard disk was developed in collaboration with Toshiba and the lithium battery was obtained from Sony (Nambisan 2005). By integrating these diverse components during an intensive eight-month design period, Apple was able to produce the most portable, user-friendly and fashionable digital music player which combines small size and ease of use with a large storage capacity (holding over 1,000 songs) thanks to the hard drive developed by Toshiba. Sales grew rapidly, equalling Apple's computer sales in two years and reaching 14 million in the three months up to 31 December 2005, as the iPod became a 'must-have' Christmas gift for many people (*The Economist* 2006).

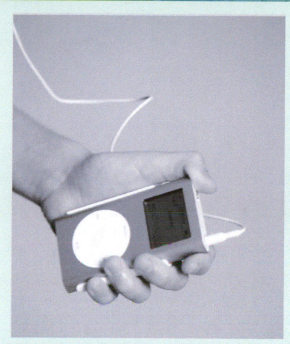

Figure 10.4 Apple's iPod
Source: © Dana Hoff/Beatworks/Corbis.

associated organisations is facilitated by spatial proximity. This supports the transmission of tacit knowledge through face-to-face contact and allows firms to learn directly from one another (Lee and Rodríguez-Pose 2013). As such, innovation is fostered by the agglomeration and clustering of firms in cities and regions. Since the 1990s, a range of urban and regional development policies have sought to support innovation and learning processes, often focusing on the softer intangible factors of interaction, networking and collaboration (Morgan 1997).

10.3.2 Territorial innovation models and clusters

A number of related theories of how agglomeration and clustering support innovation and learning in cities and regions have been advanced. These **territorial**

innovation models have many elements in common (Table 10.1) (Moulaert and Sekia 2003), based upon theories of agglomeration between firms in similar and related industries. In addition to the conception of innovation as an interactive process and the economics of agglomeration, territorial innovation models focus attention on the role of institutions, the process of regional development, the importance of culture, particularly with reference to the generation of trust and reciprocity between economic actors, the types of relations between agents, and the types of relations with the environment (Table 10.1). A number of models are identified, namely innovative milieu, industrial districts (section 1.4.2), regional innovation systems, new industrial spaces (section 3.4.3), clusters and learning regions.

One of the most influential models of territorial innovation is the Harvard business economist Michael Porter's theory of clusters. According to Moulaert and

Table 10.1 Territorial innovation models

Model

Features of innovation	Milieu innovateur (innovative milieu) (MI)	Industrial district (ID)	Regional innovation system (RIS)	New industrial spaces (NSI)	Clusters	Learning regions
Core of innovation dynamics	Capacity of firms to innovate through the relationships with agents of the same milieu	Capacity of actors to implement innovation in systems of common values	Innovation as an interactive, cumulative and specific process of research and development (path dependence)	A result of R&D and its implementation; application of new production methods (JIT etc.)	Based in interaction between demand conditions, supported and related industries, factors conditions and firm strategy and rivalry	As for RIS but stressing co-evolution of technology and institutions
Role of institutions	Very important role of institutions in the research process (universities, firms, public agencies, etc.)	Institutions are 'agents'; and enabling social regulation, fostering innovation and development	Definitions vary according to authors, but they all agree that the institutions lead to a regulation of behaviour both inside and outside organisations	Social regulation of the coordination of inter-firm transactions and the dynamics of entrepreneurial activity	Same as for ID, but with focus on role of governance	As in RIS, but with a stronger focus on the role of institutions
Regional development	Territorial view, based on MI and on agent's capacity of innovating in a cooperative atmosphere	Territorial view based on spatial solidarity and flexibility of districts; this flexibility is an element of this innovation	View of the region as system of 'learning by interacting' and by 'steering regulation'	Interaction between social regulation and agglomerated production systems	Effects of geographical concentration in enhancing innovation and productivity	Double dynamics: technological and techno-organisational dynamics; socio-economic and institutional dynamics
Culture	Culture of trust and reciprocity links	Sharing values among IS agents; trust and reciprocity	The source of learning and interacting	Culture of networking and social interaction	Role of local social-culture context in development	Strong focus on interaction between economic and social-cultural life
Types of relations between agents	The role of the support space: strategic relations between the firm, its partners, suppliers and clients	The network is a social regulation mode and a source of discipline. It enables a coexistence of both cooperation and competition	The networks is an organisation mode of 'interactive learning'	Inter-firm transactions	Inter-firm and inter-institution networks	Networks of agents (embeddedness)
Types of relations with the environment	Capacity of agents in modifying their behaviour according to the changes in their environment. Very 'rich' relations: third dimension of support space	The relationships with the environment impose some constraints and new ideas; must be able to react to changes in the environment; 'rich' relations; limited spatial view of environment	Balance between inside-specific relations and environment constraints; 'rich' relations	The dynamics of community formation and social reproduction	Close to MI	As in RIS

Source: Adapted from Moulaert and Sekia 2003: 294.

Sekia (2003), this is both the most practice-oriented and the most market-based of the territorial innovation models, reflecting its influence on policy and emphasis on competition and productivity respectively (Table 10.1). For Porter, clusters are defined as

> geographical concentrations of interconnected companies, specialised suppliers, service providers, firms in related industries, and associated institutions (for example universities, standards agencies and trade associations) that compete but also co-operate.
>
> (Porter 1998: 197)

There are two key elements in this definition (Martin and Sunley 2003: 10). First, the firms in the cluster must be linked in some way, for example by the supply of specialised inputs or services. Second, a cluster is defined by geographical concentration with proximity creating a commonality of interests between firms and encouraging frequent interaction.

Porter's **diamond model** is based on the contention that clusters enhance competitiveness and productivity by fostering the interaction between four sets of factors (Figure 10.5). Demand conditions refer to the tendency for successful clusters to serve global markets with 'leading edge' local customers – firms who sell to global markets – playing a key role in encouraging innovation among suppliers. Second, a concentration of supporting and related industries creates the critical mass to support advanced skills, training and infrastructure. The third dimension of the model relates to factor conditions, referring to the main factors of production. Here, the availability of capital, the presence of a skilled workforce and close links between universities/research institutes and leading firms play a crucial role. The final element is firm strategy, structure and rivalry. Successful clusters require a high rate of new firm formation through mechanisms such as corporate spin-off. At the same time, rivalry between firms is based upon spatial proximity which enables them to monitor the activities of their neighbours, encouraging innovation in order to keep up and remain competitive (Box 10.3).

A number of examples of dynamic, innovative clusters have been emphasised in the literature. Particular attention has been directed towards successful, high-tech cities and regions: for example, Cambridge, England, Silicon Valley, California and Baden-Wurttemberg, Germany. In these iconic cases, spatial

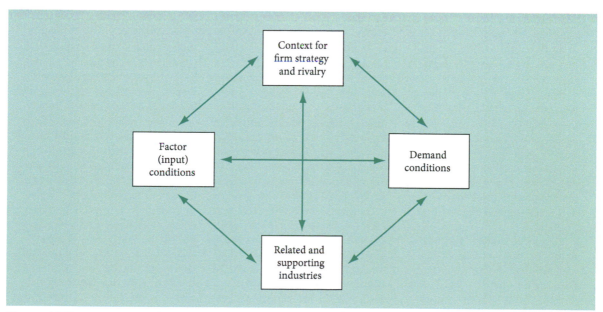

Figure 10.5 Porter's diamond model
Source: Porter 2000.

proximity supports interactive innovation and knowledge exchange through processes of inter-firm networking, labour mobility and firm spin-off (Gertler 2003). This propensity towards innovation and learning provides not only the underpinnings of growth in existing industries, but also a basis for economic diversification over time as firms branch into related and emerging industries and technologies (Box 3.8). In Cambridge, for example, early investment in IT, especially electronic equipment and instruments, paved the way for subsequent rounds of growth in biotechnology, IT applications and software. As such, these cities and regions have been able to reinvent and renew their economies in response to competition and technological

change. This renewal is based upon cultures of innovation and entrepreneurship, supported by the venture capital derived from earlier episodes of investment and growth (Kenney and von Burg 2001).

In general, this process of continuous innovation and renewal has proved easier for so-called 'sunbelt' cities distant from the older centres of heavy industry than for the old industrial cities scarred by deindustrialisation and economic decline since the 1970s (Simmie *et al*. 2008). Detroit remains the poster child for prolonged and catastrophic urban decline. There are cases, however, of the successful renewal of post-industrial cities such as Boston which exhibited some of the classic symptoms of 'rustbelt' decline in the 1980s, but was

Box 10.4

Economic renewal and innovation in Boston

Boston, Massachusetts was a declining city in 1980, with its population having fallen from a peak of just over 800,000 in 1950 to 563,000 (Pike *et al*. 2017: 268). It was characterised by many of the problems that affected other 'rustbelt' cities in the context of economic restructuring and deindustrialisation, including industrial decline, racial tension, white flight and fiscal crisis (ibid.). In short, there was little reason to suspect that it would be any more successful than other declining cities such as Buffalo, New York or Cleveland, Ohio over the next few decades. Yet by the early 2000s, Boston had become the eighth-richest metropolitan area by income and the fourth most expensive metropolitan area outside the New York and San Francisco regions (Glaeser 2005a).

Boston's recovery and growth was driven by skills and innovation in high-tech industry and the knowledge economy from the early 1980s. The city had a high-skills base relative to other rustbelt cities in 1980,

reflecting its wealth of prestigious educational institutions. In addition, the technological shift towards microcomputing and defence-based contracting fuelled the growth of a high-tech computer industry along Route 128 in the Boston suburbs (Saxenian 1994), contributing to the 'Massachusetts miracle' of the 1980s. At the same time, the city experienced rapid growth in financial and business services. The city's growth was far from smooth, however, with rapid recovery giving way to a severe recession in the early 1990s as the defence industry contracted following the end of the Cold War; the region lost its lead in the computing industry to Silicon Valley; and real estate values collapsed (Pike *et al*. 2017: 271). This was followed by renewed growth from the mid-1990s, generated by knowledge-intensive technology and financial services, though punctuated by the 'dot.com' crash of 2001. Whilst adversely affected by the financial crisis of 2008 and subsequent Great Recession, the Boston economy has

demonstrated considerable resilience in recovering from it.

"Boston's transformation from a dying factory town to a thriving information city" (Glaeser 2005a: 120) reflects the abundance of a number of social and economic assets relative to other 'rustbelt cities'. First, as emphasised above, the city's recovery reflects its high levels of human capital, based upon a concentration of world-leading universities and research organisations which supplied graduates and fostered technological innovation. Second, unlike other industrial cities, Boston retained a level of industrial diversity with substantial employment in financial and business services prior to the 1980s. Third, the city was able to attract rather than lose residents as a consumer city with high-quality urban amenities in which people want to live (ibid.). Finally, the state also played a key role, as a result of investments in defence and infrastructure and the leadership role played by the mayor and state government (Pike *et al*. 2017: 273).

subsequently able to turn this around (Box 10.4). More generally, many post-industrial cities in the US and Western Europe have achieved partial renewal through economic diversification and downtown regeneration strategies, but have often found it more difficult to secure the kind of broader economic renewal and reinvention achieved by Boston. The city of Pittsburgh, for instance, has achieved a degree of international renown as an example of a 'turnaround' city through a rebirth strategy oriented toward innovation in high-tech industries linked to the city's universities and hospitals ('eds and meds'), although it continues to be characterised by entrenched inequality, racial disparities and poor job quality (Rhodes-Conway *et al.* 2016).

10.3.3 Differentiated geographies of innovation

While much of the literature has assumed that cities have an innovation advantage over rural areas on the basis of density and agglomeration (Crescenzi *et al.* 2007), some recent research has sought to develop a more nuanced view of this relationship, distinguishing between different kinds of innovation. As expected, in a UK study, Lee and Rodríguez-Pose (2013) found that both product and process forms of innovation were higher for urban than rural firms, but also that urban firms were disproportionately likely to introduce process innovations that are only new to the firm, rather than wholly original. This suggests that proximity allows urban firms to monitor and mimic the processes of neighbouring firms which are easier to introduce and

less subject to protection through patents and intellectual property rights. There was less evidence that urban firms undertake more original innovations, resonating with other research that indicates that such innovations can take place in more remote locations (Fitjar and Rodríguez-Pose 2011). Another study of knowledge-intensive business services (KIBS) in Quebec found that while product innovation (introducing a new service) was higher in metropolitan areas, reflecting its interactive nature, other forms of process and managerial innovation were found to be higher in rural areas (Doloreux and Shearmur 2012). This was explained by the requirement for firms in rural areas to continuously adapt to the needs of a small number of clients, compared to the scope for KIBS firms to develop specialised niches in urban areas, which can be exploited through the same set of practices.

These studies are part of a renewed concern for the wider geographical organisation of innovation beyond urban environments and local linkages (Gertler 2003), reflecting, in part, the influence of relational thinking in economic geography (section 2.5.4). One important contribution to this research was Boschma's (2005) identification of a number of other forms of proximity. These are social, organisational, cognitive and institutional in addition to geographical proximity (Table 10.2). These other forms of proximity can support and underpin processes of innovation that occur through extra-local linkages, often alongside more localised relationships (Rodríguez-Pose and Fitjar 2013).

This multifaceted understanding of proximity feeds into the concept of '**global pipelines**', developed by a group of Scandinavian researchers (Bathelt *et al.* 2004).

Table 10.2 Different forms of proximity	
Geographical proximity	The lack of spatial or physical distance between economic actors
Social proximity	Based upon micro-level ties of trust and friendship
Organisational proximity	The existence and operation of organisational arrangements, either within or between organisations
Cognitive proximity	The existence of a shared vocabulary and conceptual framework to communicate, understand, absorb and process new information effectively
Institutional proximity	Shared rules, regulations, habits and values

Source: Adapted from Boschma 2005.

The key claim is that, in addition to engaging in processes of localised learning within a cluster, firms seek to build channels of communication or pipelines with selected partners outside the cluster (Figure 10.6). Such strategic partnerships offer access to knowledge and assets not available locally, although their number and scope is limited by the cost and time involved in building them. Successful establishment of global pipelines requires firms to develop a shared organisational context which enables them to learn and solve problems together. In a survey of firms in Norwegian cities, Fitjar and Rodríguez-Pose (2011) found that firms which had managed to build pipelines outside Norway were more innovative than those that were solely reliant on local linkages. This indicates that such pipelines, underpinned by social, organisational and relational forms of proximity are a critical source of innovation for firms in remoter cities and regions.

At the same time, such relationships tend to complement and enhance local linkages, rather than being a substitute for them. Whilst a firm's **territorial embeddedness** in a cluster provides automatic and routine access to a range of information and knowledge, pipelines provide access to more specialised

forms of knowledge that are not locally available. This specialised knowledge may relate to the development of new technologies or new market opportunities. Bathelt *et al.* (2004) suggest that wider links are particularly important during the early stages of cluster formation, providing access to markets and knowledge before critical mass is achieved locally. Maintaining such links as clusters mature is also seen as important to avoid introversion if local linkages become too close and rigid, leading to 'lock-in' as firms fail to respond to change. As such, pipelines can contribute to the adaptation and renewal of clusters, although tensions may arise between a growing reliance on extra-local networks and the need to maintain and replenish local knowledge as a distinctive source of competitive advantage.

Reflect

What measures do you think cities should adopt to stimulate innovation and learning, particularly in terms of encouraging collaboration between firms?

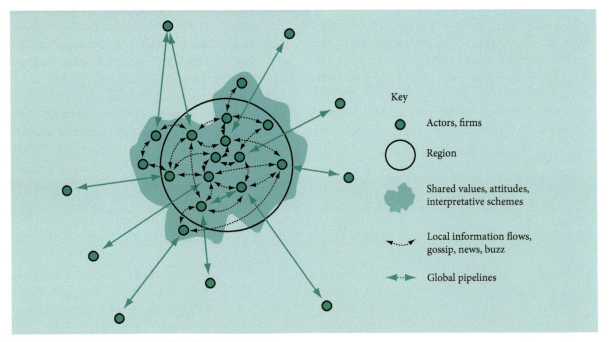

Figure 10.6 Local buzz and global pipelines
Source: Bathelt *et al.* 2004: 46.

10.4 Creative cities

The attraction of skilled people has become a central element of recent urban and regional development policy, reflecting the close connection between skills or human capital and urban growth (Figure 10.1). As such, urban agglomeration is based upon the concentration of skilled people (workers, entrepreneurs, consultants, researchers, etc.) as well as firms. As emphasised above, this has become increasingly important in the contemporary knowledge-based economy. As Glaeser (2005b: 594) puts it, ". . . the productive advantage that one area has over another is driven mostly by the people. Urban success comes from being an attractive 'consumer city' for high skill people". Here, the consumer city refers to the increased importance of cities as centres of consumption, focusing around retail, entertainment and culture, as their role as centres of industrial production has declined.

10.4.1 The creative cities approach

This emphasis on skills as a key source of urban innovation and prosperity underpins the creative city approach that became highly influential from the mid-2000s (Figure 10.7). This is closely associated with the work of Richard Florida, an economic geographer based at the University of Toronto. His 2002 book, the *Rise of the Creative Classes*, is based on the idea that urban prosperity has become increasingly dependent on the attraction and retention of the so-called **creative classes**, highly skilled and educated workers who have distinct lifestyle preferences. The basis of Florida's creative class is economic, consisting of people who add economic value through their creativity, engaging in work "whose function is to create meaningful new forms" (2002: 68). As such,

> the super-creative core of this new class includes scientists and engineers, university professors, poets and novelists, artists, entertainers, actors, designers, and architects, as well as the thought leadership of modern society: nonfiction writers, editors, cultural figures, think-tank researchers, analysts and other opinion-makers.
>
> (Ibid.: 34)

This super-creative core is estimated to account for 15 million workers in the US, or 12 per cent of the workforce. In addition, the concept of the creative classes consists of creative professionals in high-tech sectors, business and finance, law, health care and management who engage in complex problem solving that involves independent judgement, requiring a high level of education or human capital. As a whole, Florida argues, the creative class has grown markedly to incorporate 30 per cent of the US workforce, increasing tenfold since 1900. This is a very broad definition, however, including scientists, engineers, researchers and analysts alongside the more traditional creative professions of the arts, entertainment, architects, etc. In practice, there are substantial differences between these occupations in terms of their working practices, locational preferences and political outlooks (Box 10.5)

The creative class thesis has clear implications for economic development, involving cities and regions competing to attract mobile 'creatives'. According to Florida, the new economic geography of creativity is based on what he calls the 3 Ts: technology, talent and tolerance. Technology is defined in terms of the presence of high-tech industry, measured according to its size and importance to the city's or region's economy. Talent refers to high levels of human capital, and is measured largely in terms of educational attainment, particularly the percentage of the population with a bachelor's (undergraduate) degree. Tolerance is the most novel of Florida's categories, referring to the openness and diversity of a particular place, measured though various indexes such as the gay index (the proportion of gay people in the population) and the 'bohemian index' (the percentage of writers, designers, musicians, actors and directors, painters and sculptors, photographers and dancers).

The principal method of the so-called urban creativity industry associated with Florida (2005) is the ranking of cities according to these various indexes and correlation analyses of the relationships between the different measures (Table 10.4) (Figures 10.9, 10.10). Cities such as San Francisco, Austin and San Diego, already celebrated for their high-tech industries and cultural attractions, emerge as the big-city winners of the urban creativity race (Table 10.4), followed by the

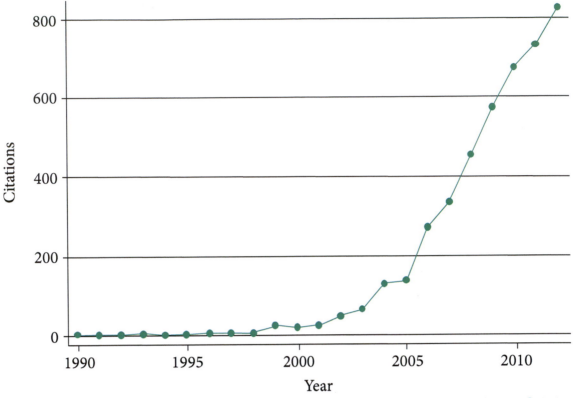

Figure 10.7 Annual number of citations of the term 'creative cities' as recorded by Google Scholar, 1990–2012
Source: Scott 2014: 568.

Box 10.5

Unpacking the creative class: artists in Minneapolis–St Paul

One study which does shed considerable light on the workings of creativity in the economy is Ann Markusen's study of artists in the twin cities of Minneapolis and St Paul in the US Midwest, reflecting the need to break down the very broad and aggregative concept of the creative class into specific creative occupations (Markusen 2006). Markusen's definition of artists encompasses four sub-groups: writers, musicians, visual artists and creative artists, accounting for 1.4 million jobs in the US. Artists are more likely to be self-employed than the labour force as a whole, seeming to fit the characterisation of the creative class as 'footloose' and being prone to choose locations on the basis of place rather than employment. In the 1990s, artists became more concentrated in the three 'superarts' metropolitan areas of Los Angeles, New York and San Francisco which increased their leads over second-tier arts-specialised metros of Washington, DC, Seattle, Boston,

Box 10.5 (continued)

Minneapolis–St Paul and San Diego (Table 10.3). Such urban economies both attract and 'homegrow' artists, with educational institutions and cultural organisations playing a key role. The forces that attract artists to particular cities are complex, but include agglomerations of prospective employers in media, advertising and the arts, in addition to lower costs of living, recreation and environmental amenities and supportive cultural conventions and activities.

Markusen's study found that artists do make discretionary 'creative class' location choices which are semi-independent of employers. At the same time, the relationship between artists and high-tech industry is far from clear with little evidence of them clustering together in the same cities. Within cities, they tend to gravitate towards dense, inner city neighbourhoods, often relatively seedy, transitional ones which provide access to arts schools, performance and exhibition spaces, affordable live/work and studio space, training agencies, artists' centres and amenities like nightlife and recreational opportunities (Figure 10.8). Artists tend to be politically progressive, voting and campaigning for left and Democratic candidates in elections. In general, they also support more decentralised, neighbourhood-based theatres, galleries and other artist-centred spaces, which they believe to be under-supported, often involving work with coloured and minority groups. Whilst often regarded as agents of gentrification in taking over abandoned buildings in run-down neighbourhoods, they are only one player in this broader process which is fundamentally structured by developers and wider zoning and land use practices. In respect of both their progressive politics and the factors shaping their location within cities, artists have very little in common with other groups of the 'creative class', reinforcing the need to disaggregate this overarching category.

Table 10.3 Artistic specialisations, selected metros

	Location quotient		
	1980	1990	2000
Los Angeles	2.39	2.31	2.99
New York	2.60	2.42	2.52
San Francisco	1.79	1.60	1.82
Washington, DC	1.76	1.63	1.36
Seattle	1.59	1.40	1.33
Boston	1.51	1.49	1.27
Orange County, CA	1.15	1.26	1.18
Minneapolis–St Paul	1.20	1.27	1.16
San Diego	1.24	1.15	1.15
Portland	1.18	1.24	1.09
Atlanta	1.31	1.08	1.08
Chicago	1.03	1.09	1.04
Cleveland	0.82	0.83	0.79

Source: Markusen 2006: 1929.

Box 10.5 (continued)

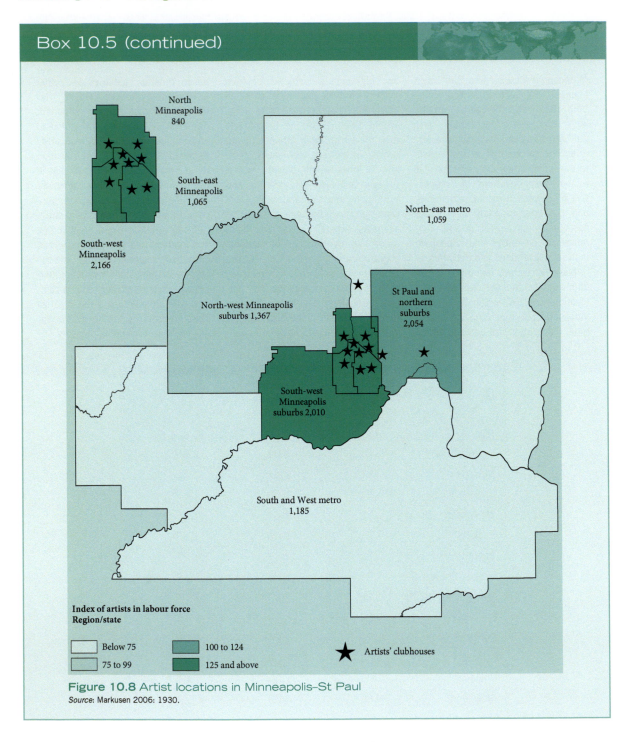

Figure 10.8 Artist locations in Minneapolis–St Paul
Source: Markusen 2006: 1930.

likes of Albuquerque, New Mexico, Albany, New York and Tucson, Arizona for the mid-sized cities group and Madison, Wisconsin, Des Moines, Iowa, and Santa Barbara, California for smaller cities (Peck 2005: 747).

According to the European Commission (2017: 23), the top five 'creative economy' cities in Europe are Paris, Stuttgart, Munich, Copenhagen and Eindhoven, measured by jobs and innovation. When this is narrowed to

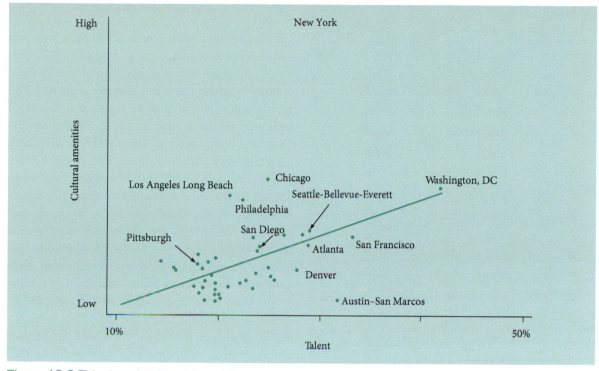

Figure 10.9 Talent and cultural amenities
Source: Florida 2005: 98.

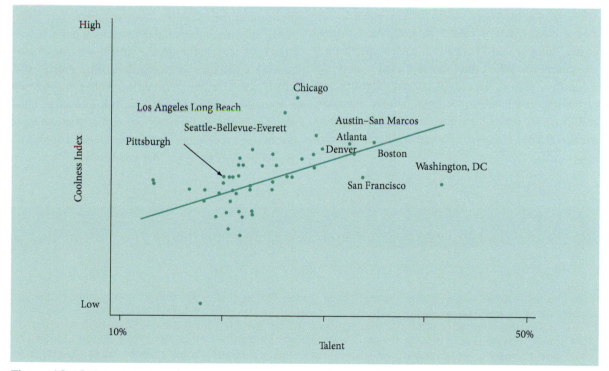

Figure 10.10 Talent and coolness
Source: Florida 2005: 98.

Table 10.4 The creativity index

Rank	Region	Score
1	San Francisco	1057
2	Austin	1028
3	Boston	1015
4	San Diego	1015
5	Seattle	1008
6	Raleigh-Durham	996
7	Houston	980
8	Washington, DC	964
9	New York	962
10	Dallas–Fort Worth	960
11	Minneapolis–St Paul	960
12	Los Angeles	942
13	Atlanta	940
14	Denver	940
15	Chicago	935

Source: Florida 2005: 157.

10.4.2 Creative cities policy

The creative cities policy agenda is that cities must promote tolerance, viewed as the 'crucial magnet' in Florida's formulation, encouraging openness and diversity, in order to attract talent and generate prosperity. This agenda was widely adopted by cities across the world with one prominent advocate observing that creative cities have been spreading "like a rash" (Landry 2005, quoted in Peck 2012: 482). Here, Florida can be seen as a leading guru of the 'new urbanology' (Box 10.1), whose role is to "translate academic ideas into digestible policy formulations" for city governments and managers (Peck 2012: 480). Famously, the mayor of Denver ordered multiple copies of *The Rise of the Creative Class* as bedtime reading for his senior staff, while developing a strategy to rebrand the city as a creative centre. Many other cities bought into Florida's ideas on creativity, reinforcing their rapid spread and influence, including cities as diverse as Abu Dhabi, Amsterdam, Berlin, Beijing, Melbourne and Mumbai (ibid.: 482). The concept has proved particularly popular in Asia, especially China where cities such as Beijing, Shanghai, Guangzhou, Chongqing and Wuhan have been active in asserting their creative economy credentials (Scott 2014). There are substantial differences, though, in how cities have adopted the creative cities thesis, particularly between cities such as Amsterdam, on the one hand, which were already renowned as creative and cultural centres (Box 10.6) and declining post-industrial cities such as Detroit, on the other, where it arrives as a more distinct new economy narrative that is seized upon with greater urgency by urban mangers (Peck 2012).

jobs in culture, arts and entertainment, Paris, Copenhagen, Bologna and Perugia lead their respective city size groupings. When the focus is shifted to new jobs in creative sectors, including creative start-ups and knowledge-based industries, Paris, Vilnius, Bratislava and Umea are top of their size categories, highlighting the significance of Eastern European cities in this area (ibid.: 68–9).

Box 10.6

The adoption of creative city policies in Amsterdam

According to the economic geographer Jamie Peck (2012: 464), the city of Amsterdam has acted as one of many 'translation spaces' between the general model of urban creativity and specific regimes of urban governance in Europe. The city has adapted the model to its own circumstances and has also been itself held up as something of an exemplar or model for other cities to emulate and follow. While the Dutch debate on creative cities was initiated by the publication of the book *Creative Stieden* (Creative Cities) by the Ministry of Housing and Spatial Planning in early 2002, it was the visit of Richard Florida in 2003 (for a fee of $50,000 for

Box 10.6 (continued)

one day) that was the "game chang-ing moment" (ibid.: 466). Amster-dam was of course already a cultural and creative centre of international renown, with associated traditions of openness and tolerance. Although the urban economy experienced some-thing of a crisis in the 1980s, evi-dent in terms of high unemployment, large numbers of welfare claimants, increased crime and fiscal austerity, it experienced a widespread recovery in the 1990s through the growth of the knowledge and service economy, paralleling the trajectory of other European and North American cities (Box 10.4). Nonetheless, income and value-added were low in the cre-ative economy, compared to other 'uncreative' sectors, with 'hard core'

creatives accounting for as little as 7 per cent of the workforce as groups such as artists struggled to make a living (ibid.: 470) (see Box 10.5).

This history and recent economic trajectory ensured that Amsterdam had "long possessed the kind of funky, mix and don't match urban cul-ture that is celebrated in the generic creative-cities script" (ibid.: 465). The close fit between Florida's message and the city meant that creativity was a label "that just works" for Amsterdam (ibid.: 467). Its effect has largely been to rebadge and reframe existing policies, providing a kind of overarching meta-narrative or story of urban policy that has allowed the municipal administration to bracket several disparate economic, cultural

and housing initiatives together. Partly in response to early accusa-tions of elitism and controversies over gentrification, the city has adopted a 'relaxed' and laid-back approach (ibid.: 470) that views creativity as a symbolic overarching brand rather than as a series of substantive policy interventions to reshape the social and economic fabric of the city. This grants Amsterdam's adoption of the creative script a welcome degree of flexibility, allowing it to mean different things to different groups and allowing space for criticism, even from urban managers themselves, and the accommodation of edgier elements of the inner city which provide a degree of authenticity.

Creating an attractive and 'funky' environment for the creative classes typically involves the redevelopment of neighbourhoods to create a sense of "hipster chic" (ibid.: 462), based upon authentic historical buildings, converted lofts, walkable streets, coffee shops, arts and live music spaces. As such, it has tended to reinforce and accelerate existing processes of gentrification through which former industrial and working-class inner city areas have been redeveloped and colonised by middle-class professionals and hipsters. This can lead to the dis-placement of established lower-income residents who can no longer afford the increased rents resulting from gentrification. As a result the creative cities approach has been widely criticised for its elitism and contribution to the gentrification process (Grodach 2017).

The widespread adoption of the creative cities poli-cies reflects its close fit with established modes of **urban entrepreneurialism** and competition:

the contemporary cult of urban creativity has a clear genealogical history, stretching back at least as far as the entrepreneurial efforts of reindustri-alised cities. The script of urban creativity reworks

and augments the old methods and arguments of urban entrepreneurialism in politically seductive ways. The emphasis on the mobilisation of new regimes of local governance around the aggressive pursuit of growth-focused development agendas is a compelling recurring theme. The tonic of urban creativity is a remixed version of this cocktail: just pop the same basic ingredients into your new-urbanist blender, add a slug of Schumpeter-lite for some new-economy fizz, and finish it off with a pink twist.

(Peck 2005: 766)

In this context, urban entrepreneurialism, which has been prevalent in North America and Western Europe since the early 1980s, is based on competition between cities for investment and visitors, requiring the estab-lishment of public-private partnerships between local government and business, place promotion and the redevelopment of neighbourhoods and creation of new cultural and leisure facilities. The underlying discourse of **competitiveness**, based on the idea that cities and regions compete with one another for market share in the global economy, has shaped the thinking of local politicians and business leaders over the last couple of decades, helping to explain their receptiveness to Florida's ideas. In this respect, the creative class approach can be seen as adding a new dimension or layer to the idea of urban competition, particularly through the innumerable rankings of cities which present policy-makers with a very direct and tangible indication of their city's performance relative to its competitors.

10.4.3 Criticisms of the creative class thesis

The creative class thesis has proved highly controversial, with its eager adoption by many policy-makers and consultants countered by a number of academic criticisms and political objections. Interestingly, Florida has been castigated by both the political right and left for advocating big government-style spending programmes and progressive liberal policies over low taxes and 'family values', on the one hand, and for favouring elitist policies that increase social polarisation by privileging the needs of educated, high-income groups over working-class residents, on the other hand (Malanga 2004; Peck 2005).

In academic terms, a number of problems have been identified. First, the creative class is a rather fuzzy concept, presenting a number of generalisations that are 'proven' by correlations between particular measures of creativity. These are based on rather thin US data, raising questions as to whether the findings would hold in other parts of the world (Musterd and Gritasi 2012). From an urban economics standpoint, Glaeser (2005b) argues that creativity reflects nothing more than the effects of human capital (skills), with the Bohemian effect actually driven by two cities, Las Vegas and Sarasota, Florida, while the gay index actually has a negative impact on growth in Glaeser's regressions.

Second, several analysts have taken issue with the idea that creatives are attracted to cities through the provision of cultural and other amenities rather than employment, with jobs following rather than driving migration. While Florida argues that tolerance attracts talent and technology, it is at least equally plausible to suggest that the direction of causation runs the other

way, whereby centres of high-tech industry attract talent and generate diversity as a by-product of growth, as suggested by the leading position of cities such as San Francisco, Austin, San Diego and Boston (Markusen 2006). In a European study, Musterd and Gritsai (2012) find that the most important groups of factors in shaping the decisions of creative knowledge workers to settle in a particular city were: first, personal networks, including place of birth, family connections, having studied there and proximity to friends; second, 'hard' conditions linked to employment and wages; and third, 'softer' conditions such as amenities, diversity, openness and tolerance. More generally, as Storper and Scott (2009: 156) argue, "it strains credulity to suppose that members of the creative class move about the economic landscape as though they were principally in search of amenity based gratification".

Third, while the thesis that creativity is an increasingly important driver of economic development is widely accepted, the liberal policy agenda of attracting bohemian types to cool and 'funky' downtown areas is far more contentious. By contrast, Glaeser (2005b) argues that creative people prefer the traditional American staples of big suburban houses and car-based commuting, alongside safe streets, good schools and low taxes. This contention is equally over-generalised and contentious, however, with the research of Markusen (2006) and others suggesting that different creative groups and occupations have different locational and lifestyle preferences. Nonetheless, there is little doubt that the specific policy prescription of attracting bohemian, hipster types by building trendy cultural quarters is under-determined by the underlying thesis of the increased economic importance of creativity, which supports only the general policy of attracting creative people.

Finally, Florida presents a celebratory account of economic change and competition that not only neglects problems of urban inequality and poverty, but is also likely to reinforce them through the policies it advocates. For instance, as indicated above, the redevelopment and gentrification of selected urban neighbourhoods in order to attract talent is likely to result in the further marginalisation and potential displacement of 'non-creative' working-class residents. Here, critics on the left have argued that the most creative cities and regions, as measured by Florida's rankings, also tend to be the most socially polarised and unequal, reflecting how creative strategies have become part of the broader neoliberal policy agenda of urban competition and entrepreneurialism (Peck 2005). In his recent work, Florida (2017) has himself acknowledged a 'new urban crisis' of gentrification, rising inequality, and increasingly unaffordable housing, requiring investment in affordable housing, upgrading low-wage service jobs, providing infrastructure to connect people and places to employment centres, and the creation of a stronger social safety net.

Reflect

Explain the popularity of the creative cities model among urban and regional policy-makers.

10.5 Summary

As this chapter has demonstrated, knowledge and skills have been widely viewed by academic researchers and policy-makers as the key factors shaping urban and regional development in recent years. The focus on knowledge and innovation has revived interest in the topic of spatial agglomeration as knowledge-based processes have been stressed over the traditional emphasis on cost minimisation. From an urban economics perspective, agglomeration is rooted in the three processes of sharing, matching and learning (Duranton and Puga 2004). Two specific academic concepts have attracted much attention from policy-makers. First, as the most practice-oriented and market-based of a family of territorial innovation models, Porter's clusters model emphasises the main advantages of geographical concentration in terms of the competitive diamond of demand conditions, related and supporting industries, factor conditions and firm strategy and rivalry. Second, Florida's theory of the creative classes argues that urban and regional prosperity depends upon the attraction and retention of highly skilled and educated workers, who have distinct lifestyle preferences, favouring diverse and tolerant places. While Porter highlights the actions of firms as the basis of urban and regional

prosperity and growth, in common with earlier theories of agglomeration, Florida's approach focuses attention on the role of skilled individuals (talent).

While the clusters approach in particular retains a bounded notion of the region, and the creative cities model views cities and regions as entities competing with one another to attract 'talent', contemporary theories of regional development have become increasingly concerned with the wider linkages and flows that connect regions to the global economy, underpinning the notion of the 'relational region' (section 2.5.4). As such, it is important to appreciate that wider global linkages – beyond those connecting firms to product markets – play an important role in the development of cities and regions. As the emphasis on local buzz and global pipelines suggests, localisation and globalisation can be seen as complementary rather than contradictory forces, reflecting the central paradox of the modern city that spatial proximity has become more economically valuable as the costs of transporting goods have fallen. Agglomeration is a dynamic process, however, and changes in the opportunities for profits offered by different locations mean that the balance between agglomeration and **spatial dispersal** will change over time. The mobility of capital as it 'see-saws' between places means that some established clusters experience decline (automobiles in Detroit or shipbuilding in north-east England), whilst others continue to attract investment (world cities like London and New York) and new ones emerge (Silicon Valley, Cambridge).

Exercise

Select a long-established cluster of economic activity and review its development over time. Appropriate resources to draw on here are academic articles, reports by regional development agencies or local authorities, economic statistics, materials or websites produced by business associations, newspaper articles and media reports.

What were the origins of the cluster? To what extent was its development based on the three processes of sharing, matching and learning? What markets did it serve? What were its sources of competitive advantage? Was it based on a few large firms or a network of smaller firms? What were the key supporting institutions? What was the role of local linkages between firms and institutions? What were the main external linkages (global pipelines)?

Did the cluster remain competitive or experience a process of decline as it matured? How would you explain this outcome? How important were internal conditions (capital, skills, knowledge, infrastructure, attitudes/practices) and external factors (competition, technology, markets)? If the cluster remained competitive, what processes of adaptation were involved? If it experienced decline, why did this occur?

Key reading

Bathelt, H., Malmberg, A. and Maskell, P. (2004) Clusters and knowledge: local buzz, global pipelines and the process of knowledge formation. *Progress in Human Geography* **28: 31–57.**

An important paper which adopts a broader relational perspective on agglomeration and clusters, viewing 'local buzz' and 'global pipelines' as complementary rather than conflicting or incompatible processes.

Florida, R. (2005) *Cities and the Creative Class.* **London: Routledge.**

One of the key texts in Florida's creative classes series in which he consolidates and extends the underlying concepts and applies them to cities in North America. *Cities and the Creative Class* is particularly instructive in illustrating the methods of the creative class approach through a profusion of tables and charts which rank cities according to various creative criteria.

Markusen, A. (2006) Urban development and the politics of a creative class – evidence from a study of artists. *Environment and Planning A* **38: 1921–40.**

A very interesting case study of the locational preferences and cultural politics of artists in the twin cities of Minneapolis and St Paul in Minnesota. Markusen marshals her research very effectively to question key aspects of the creative class thesis, stressing the need to break it down through the study of specific occupations.

Peck, J. (2005) Struggling with the creative classes. *International Journal of Urban and Regional Research* **29: 740–70.**

An excellent critical assessment of the creative class approach which explains its popularity among policy-makers in terms of its fit with existing discourses of urban entrepreneurialism and competitiveness.

Porter, M.E. (2000) Locations, clusters and company strategy. In Clark, G.L., Feldman, M. and Gertler, M. (eds) *The Oxford Handbook of Economic Geography.* **Oxford: Oxford University Press, 253–74.**

An accessible account of the clusters concept from its founder. Porter defines the concept and explains the key mechanisms of growth within clusters, drawing on a range of examples. The discussion is set within a framework of firm strategy, focusing on the notion of competitiveness which he has done so much to popularise.

Scott, A.J. and Storper, M. (2015) The nature of cities: the scope and limits of urban theory. *International Journal of Urban and Regional Research* **39: 1–15.**

An important paper which argues that agglomeration processes represent the underlying basis of cities, giving rise to an associated nexus of locations, land uses and human interaction. The paper reflects the renewed emphasis on agglomeration in recent urban economic and economic geography research.

Useful websites

www.creativeclass.com/

The website of the Creative Class Group, featuring one Richard Florida as 'author and thought leader'. The site contains background information on the underlying approach, an introduction to services offered by the Creative Class Group, biographical information on members of the group, summaries of creative class community initiatives and a forum for exchange.

www.isc.hbs.edu/

The website of the Institute for Strategy and Competitiveness at Harvard Business School, led by Michael Porter. Contains a wealth of information on clusters, competitiveness and business strategy.

References

Amin, A. and Thrift, N. (1994) Living in the global. In Amin, A. and Thrift, N. (eds) *Globalisation, Institutions and Regional Development in Europe.* Oxford: Oxford University Press, pp. 1–22.

Bathelt, H., Malmberg, A. and Maskell, P. (2004) Clusters and knowledge: local buzz, global pipelines and the process of knowledge formation. *Progress in Human Geography* 28: 31–57.

Boschma, R. (2005) Proximity and innovation: a critical assessment. *Regional Studies* 39: 61–74.

Brown, J.S. and Duguid, P. (2000) *The Social Life of Information*. Cambridge, MA: Harvard University Press.

Cheshire, P., Nathan, M. and Overman, H.G. (2014) *Urban Economics and Urban Policy*. Cheltenham: Edward Elgar.

Ciccone, A. (2012) Agglomeration effects in Europe. *European Economic Review* 46: 213–27.

Cooke, P. and Morgan, K. (1998) *The Associational Economy: Firms, Regions, and Innovation*. Oxford: Oxford University Press.

Crescenzi, R., Rodríguez-Pose, A. and Storper, M. (2007) On the geographical determinants of innovation in Europe and the United States. *Journal of Economic Geography* 7: 673–709.

Dijkstra, L. (2013) Why investing more in the capital can lead to less growth. *Cambridge Journal of Regions, Economy and Society* 6: 251–68.

Doloreux, D. and Shearmur, R. (2012) Collaboration, innovation and the geography of innovation in knowledge intensive business services. *Journal of Economic Geography* 12: 79–105.

Duranton, G. and Puga, D. (2004) Micro-foundations of urban agglomeration economies. In Henderson, V. and Thisse, J.-F. (eds) *Handbook of Urban and Regional Economics*, vol 4. Amsterdam: North-Holland, pp. 2063–177.

The Economist (2006) Podtastic, Apple. 14 January, p. 68.

European Commission (2017) *The Cultural and Creative Cities Monitor, 2017 Edition*. Brussels: European Commission.

Fitjar, R. and Rodríguez-Pose, A. (2011) When local interaction does not suffice: sources of firm innovation in urban Norway. *Environment and Planning A* 43: 1248–67.

Florida, R. (2002) *The Rise of the Creative Classes*. New York: Basic Books.

Florida, R. (2005) *Cities and the Creative Class*. London: Routledge.

Florida, R. (2017) *The New Urban Crisis: How Our Cities Are Increasing Inequality, Deepening Segregation, and Failing the Middle Class-and What We Can Do About It*. New York: Basic Books.

Freeman, C. (1994) The economics of technical change. *Cambridge Journal of Economics* 18: 463–514.

Gertler, M. (2003) A cultural economic geography of production. In Anderson, K., Domosh, M., Pile, S. and Thrift, N.J. (eds) *The Handbook of Cultural Geography*. London: Sage, pp. 131–46.

Glaeser, E.L. (2000) The new economics of urban and regional growth. In Clark, G., Feldmann, M. and Gertler,

M. (eds) *The Oxford Handbook of Economic Geography*. Oxford: Oxford University Press, pp. 83–98.

Glaeser, E.L. (2005a) Reinventing Boston: 1630–2003. *Journal of Economic Geography* 5: 119–53.

Glaeser, E.L. (2005b) Review of Richard Florida's *The Rise of the Creative Class*. *Regional Science & Urban Economics* 35: 593–6.

Glaeser, E.L. (2011) *Triumph of the City*. London: Macmillan.

Glaeser, E.L. and Gottlieb, J.D. (2008) The economics of place-making policies. *Brookings Papers on Economic Activity* 1: 155–253.

Glaeser, E.L. and Resseger, M.G. (2010) The complementarity between cities and skills. *Journal of Regional Science* 50: 221–44.

Glaeser, E.L. and Saiz, A. (2003) The rise of the skilled city. *NBER Working Papers*, Working Paper 10191. Cambridge, MA: National Bureau of Economic Research. At www. nber.org/papers/w10191.

Gleeson, B. (2012) The Urban Age: paradox and prospect. *Urban Studies* 49: 931–42.

Grodach, C. (2017) Urban cultural policy and creative city making. *Cities* 18: 82–91.

Hardagon, A. (2005) Technology brokering and innovation: linking strategy, practice and people. *Strategy and Leadership* 33: 32–6.

Kenney, M. and von Burg, U. (2001) Paths and regions: the creation and growth of Silicon Valley. In Garud, R. and Karnøe, P. (eds) *Path Dependence and Creation*. London: Lawrence Erlbaum, pp. 127–48.

Lee, N. and Rodríguez-Pose, A. (2013) Original innovation, learnt innovation and cities: evidence from UK SMEs. *Urban Studies* 50: 1742–59.

Lee, N., Sissons, P., Hughes, C., Green, A., Atfield, G., Adam, D. and Rodríguez-Pose, A. (2014) *Cities, Growth and Poverty: A Review of the Evidence*. York: Joseph Rowntree Foundation.

Lundvall, B.A. (1994) The learning economy: challenges to economic theory and policy. Paper presented at the *European Association for Evolutionary Political Economy Conference*, October, Copenhagen.

Malanga, S. (2004) The curse of the creative class. *City Journal*, Winter issue: 1–9.

Malmberg, A. and Maskell, P. (2002) The elusive concept of localisation economies: towards a knowledge-based theory of spatial clustering. *Environment and Planning A* 34: 429–49.

Markusen, A. (2006) Urban development and the politics of a creative class – evidence from a study of artists. *Environment and Planning A* 38: 1921–40.

Martin, R. and Sunley, P. (2003) Deconstructing clusters: chaotic concept or policy panacea? *Journal of Economic Geography* 3: 5–35.

Martin, R., Sunley, P., Tyler, P. and Gardiner, B. (2016) Divergent cities in post-industrial Britain. *Cambridge Journal of Regions, Economy and Society* 9: 269–99.

Maskell, P. and Malmberg, A. (1999) The competitiveness of firms and regions: 'ubiquitification' and the importance of localised learning. *European Urban and Regional Studies* 6: 9–25.

Morgan, K. (1997) The learning region: institutions, innovation and regional renewal. *Regional Studies* 31: 491–504.

Moulaert, F. and Sekia, F. (2003) Territorial innovation models: a critical survey. *Regional Studies* 37: 289–302.

Musterd, S. and Gritasi, O. (2012) The creative knowledge city in Europe: structural conditions and urban policy strategies for competitive cities. *European Urban and Regional Studies* 20: 343–59.

Nambisan, S. (2005) How to prepare tomorrow's technologists for global networks of innovation. *Communications of the Association for Computing Machinery* (ACM) 48 (5): 29–31.

Nonaka, I., Toyama, R. and Konno, N. (2001) SECI, *ba* and leadership: a unified model of dynamic knowledge creation. In Nonaka, I. and Teece, D. (eds) *Managing Industrial Knowledge: Creation, Transfer and Utilisation*. London: Sage, pp. 13–43.

Peck, J. (2005) Struggling with the creative classes. *International Journal of Urban and Regional Research* 29: 740–70.

Peck, J. (2012) Recreative city: Amsterdam, vehicular idea and the adaptive spaces of creativity policy. *International Journal of Urban and Regional Research* 36: 462–85.

Peck, J. (2016) Economic rationality meets celebrity urbanology: exploring Edward Glaeser's city. *International Journal of Urban and Regional Research* 40: 1–30.

Pike, A., Rodríguez-Pose, A. and Tomaney, J. (2017) *Local and Regional Development*, 2nd edition. London: Routledge.

Porter, M.E. (1998) Clusters and the new economics of competition. *Harvard Business Review*, December: 77–90.

Porter, M.E. (2000) Locations, clusters and company strategy. In Clark, G.L., Feldman, M. and Gertler, M. (eds) *The Oxford Handbook of Economic Geography*. Oxford: Oxford University Press, pp. 253–74.

Rhodes-Conway, S., Dresser, L., Meder, M. and Ebeling, M. (2016) *A Pittsburgh That Works for Working People*. Madison, WI: COWS, University of Wisconsin-Madison.

Rodríguez-Pose, A. and Fitjar, R. (2013) Buzz, archipelago economies and the future of intermediate and peripheral areas in a spiky world. *European Planning Studies* 21: 355–72.

Royuela, V., Venen, P. and Ramos, P. (2014) Income inequality, urban size and economic growth in OECD regions. *OECD Regional Development Working Papers, 2014/10*. Paris: OECD.

Saxenian, A.L. (1994) *Regional Advantage: Culture and Competition in Silicon Valley and Route 128*. Cambridge, MA: Harvard University Press.

Scott, A.J. (2014) Beyond the creative city: cognitive–cultural capitalism and the new urbanism. *Regional Studies* 48: 565–78.

Scott, A.J. and Storper, M. (2015) The nature of cities: the scope and limits of urban theory. *International Journal of Urban and Regional Research* 39: 1–15.

Simmie, J., Martin, R., Carpenter, J. and Chadwick, A. (2008) *History Matters: Path Dependence and Innovation in British City Regions*. London: National Endowment for Science, Technology and the Arts.

Storper, M. and Scott, A.J. (2009) Rethinking human capital, creativity and urban growth. *Journal of Economic Geography* 9: 147–67.

Storper, M. and Scott, A.J. (2016) Current debates in urban theory: a critical assessment. *Urban Studies* 53: 1114–36.

Storper, M., Kemeney, T., Makarem, N.P. and Osman, T. (2016) On specialisation, divergence and evolution: a brief response to Ron Martin's review. *Regional Studies* 50: 1628–30.

Walker, R.A. (2016) Why cities? A response. *International Journal of Urban and Regional Research* 40: 164–80.

PART 4

Reordering economic life

Chapter 11
Consumption and retail

11.1 Introduction

Understanding **consumption** as a distinct, yet integral, part of the workings of global capitalism has become an increasingly important aspect of economic geography. Ultimately, the success of all economic activity depends on the ability of firms to sell products and services to customers. Consumption and the effort expounded around selling, bound up in retail activities, but also a range of other activities such as marketing and advertising, is therefore fundamental and inextricably linked to the sphere of production, rather than a separate economic realm. Consumption and retail geographies have therefore become a flourishing sub-discipline of economic geography over the past two decades (e.g. Wrigley and Lowe 2002; Mansvelt 2013; Pike 2015; Crewe 2017).

As an object of critical analysis, the sphere of consumption extends beyond narrow economics to a more complex set of social and cultural practices and identities. Although mainstream economists still tend to see consumption rather simplistically as about atomised individuals exercising preferences to maximise their utility in competitive markets, a bit of serious thought reveals a more complicated reality. How and what we consume, and how this positions us in relation to other groups, and even classes of people, is fundamental to our sense of being and self, as well as our economic health and well-being. A critical geographical contribution to the literature has been to emphasise the growing volume and intensity of global networks and flows that connect places which still retain their distinctiveness in a highly uneven and variegated world of consumption.

As outlined in Chapter 9, the production of commodities involves a complex chain of actors (Figure 9.3) spread across different countries and regions. While consumption has conventionally been thought of by geographers as a discrete stage in the production of commodities, its role is far more important than just "the expression of production"; representing, instead, "a critical part of how it [production] proceeds, influencing the constitution of space, material and social life and ways places themselves are imagined and consumed" (Mansvelt 2013: 381).

11.2 Understanding consumption, consumers and commodities

11.2.1 Unpacking consumption as a set of economic and cultural practices

Consumption has become a major focus of interest for geographers and other social scientists over the last couple of decades. Much of this is a response to the previous neglect of consumption, seen as very much secondary to, and derivative of, the more fundamental process of production. Three broad premises underpin this recent concern with consumption (Slater 2003). First, it is seen as central to the reproduction of social and cultural life, referring to people's everyday actions in supporting themselves and their families, involving feeding, clothing, sheltering, socialising, etc. Second, modern market societies are said to be characterised by a 'consumer culture' organised around the logic of individual choice in the marketplace. Third, studying consumption enables us to better comprehend the importance of culture in shaping economic processes and institutions, representing "the site on which culture and economy most dramatically converge" (Slater 2003: 149).

Consumption is ultimately about the way that people procure goods and services to meet their economic, social, cultural needs and desires, and, as such, is critical to social reproduction. It can relate to basic needs required to survive: the use and consumption of food, water, energy, clothing, housing, etc.; but it can also relate to what mainstream economics calls 'luxury goods': non-essential items consumed for pleasure or leisure purposes. While consumption can be reduced to simple economic relations, its operations reveals far more about the social, cultural and even political relations that underpin it. For most people, it is a central, taken-for-granted part of life in contemporary society. Watching television, eating a burger, going to the cinema, shopping and clubbing are all acts of consumption. Whilst such everyday activities may seem mundane and trivial, they are of considerable economic and cultural significance (Cook and Crang 2016).

All societies have to contend with how consumption is organised and resources distributed but under capitalism the imperatives of profit and competitive markets lead to the exhortation for consumption to grow well beyond meeting basic needs. The sale of commodities becomes fundamental in enabling firms to generate revenue and profits, fuelling the process of economic growth. The importance of consumption as a driver of economic growth is emphasised through regular media references to patterns of consumer spending and levels of consumer confidence. In developed economies such as the UK and US, consumption typically accounts for 70 per cent of GDP, while in rapidly developing economies, such as China, commentators urge policymakers to increase domestic consumption to reduce export dependence and create more balanced growth models (Wolf 2016). However, there are growing concerns around whether the present growth-driven economic model is sustainable, both socially and environmentally (section 12.2). With regard to the former, a growing proportion of consumption is linked to spiralling household debt (section 4.6.3), given the decline in real wages and the growing dependence of consumers on credit to fuel their spending patterns. In terms of the latter, for instance, global consumption of meat has quadrupled in half a century, partly fuelled by the geographical spread of westernised fast food habits, which puts enormous strains on the environment, in the amount of land used for cattle rearing. This can in turn lead to desertification and depleted water supplies through the increase in crops required for animal feed as well as massively increasing gases such as methane and CO_2 (Worldwatch Institute 2015).

11.2.2 Consumption, consumers and identity

Two main perspectives on consumption can be identified. First, a number of influential social critics, including Karl Marx and Herbert Marcuse, have viewed it as signalling the triumph of market exchange and industrial society over deeper human qualities and meanings. As such, consumption marks the process through which culture is colonised by economic forces (Slater 2003: 150). In capitalist societies, needs and wants are artificially created and manufactured, inducing people to consume far more than they actually need. A number of studies have focused on the role of advertising in stimulating demand for products. Whilst there is certainly a sense in which market demand is induced through processes like advertising, this approach invariably tends to cast the consumer in a passive role as a 'cultural dupe' or 'dope' manipulated and controlled by corporations and the media. Some versions of postmodernism (Box 2.7) have reproduced this theme of the passive consumer, emphasising his/her powerlessness in the face of an infinite universe of abstract signs and meanings.

A second view emphasises the active role of consumers in utilising things for their own ends. Earlier work in this vein viewed consumption and leisure practices as a source of social status and distinction, as captured in the institutional economist Thorstein Veblen's notion of 'conspicuous consumption' and the sociologist Pierre Bourdieu's more recent work on cultural capital and taste. Contemporary studies have moved away from this concern with social status and distinction to stress how consumers creatively rework the products they buy, generating new meanings in the process (Cook and Crang 2016). Rather than reading off the process of consumption from production and corporate strategies, as critical approaches have tended to do, one has to understand the social and cultural relations in which it is entangled. From this perspective, consumption is seen as a relatively fertile arena for the expression of individuality and creativity, compared to the world of work, which involves considerable drudgery and monotony for many people.

An example of the wider social and cultural relations in which active consumption is rooted are those of family and friendship. According to the anthropologist Danny Miller, instead of being driven by individual greed and hedonism, consumption is based on acts of love and devotion involving the purchase of commodities for partners, children and friends (Miller 1998). Of particular significance here is gift-buying, focused around rituals such as Christmas and birthdays (Box 11.1). Much of this work on active and creative consumption has been ethnographic in nature, attempting to understand the meaning and significance that people attach to consumption through detailed fieldwork and observation.

Whilst much work on consumption has been cultural in orientation, and sometimes rather celebratory in tone, some political economists have sought to overcome **commodity fetishism** by reasserting the relations of production. This involves uncovering the working conditions and regimes through which particular commodities are produced and distributed before being sold to consumers (Harvey 1989). The downside of this is the tendency to reduce consumption to production. One framework which helps to overcome the opposite extremes of commodity fetishism and productionism is the **commodity chain** approach (Box 1.4). This traces how commodities link together diverse actors performing various roles such as farmers, labourers, haulage companies, shipping companies, retailers and consumers. The increasing globalisation of the economy means that such actors are often located in separate countries and continents, emphasising how commodity flows create a range of linkages between people, things and places (Mansvelt 2005).

Box 11.1

Linking the economy and culture: the example of Christmas

The annual festival of Christmas offers an instructive example of close links between the economy and culture. For most people in developed countries particularly, Christmas is a time of leisure and consumption, associated with a holiday from work, usually spent with friends and family (Figure 11.1). As a cultural festival, Christmas draws together traditions from different countries: the Christmas tree from Germany, the practice of filling stockings from the Netherlands, the idea of Santa Claus or Father Christmas from the US and the Christmas card from Britain (Miller 1993, cited in Thrift and Olds 1996). As part of the global spread of a western consumer culture, these practices are spreading to emerging economies such as China.

The social customs associated with Christmas support the practice of gift-buying. This provides the economy with a crucial stimulus in the final quarter of the retail year, with the Christmas shopping season (December) typically accounting for 12 per cent of total sales (British Retail Consortium 2016). In 2016 the British population spent a total of £77.5 billion on Christmas with the US comparable figure being £492 billion (Centre for Retail Research: www. Retailresearch.org/shoppingforchristmas.php). Such expenditure benefits manufacturers as well as retailers, with production often located in areas distant from the main centres of global consumption. Hong Kong has been an important centre for toy production since the 1970s, but has been increasingly eclipsed by China which is estimated to produce around 80 per cent of the world's toys, particularly in the southern province of Guangdong, centred in cities such as Shenzhen, Dongguan, Guangzhou (HKTDC Research 2017).

Figure 11.1 Christmas consumption
Source: Franco Zecchin, Getty Images.

11.2.3 Consumption, commodities and retail

'The consumer' or 'the market' is often invoked as the reason for producing goods and organising services in particular ways. For example, bananas must be of a particular size and quality to satisfy consumer expectations whilst services such as banking should be provided through the internet rather than face-to-face because this is what 'the consumer demands' (Crang 2005: 126). The implications of this can be serious, leading to the impoverishment of Caribbean banana growers who cannot meet these standards (Box 1.5) or the closure of bank branches. In this sense, 'the consumer' has become a kind of 'global dictator' (Miller 1995), with the demands of affluent northern consumers in particular determining how goods are produced and services delivered throughout the world economy.

Emphasising these connections between consumption and production, it is useful to locate consumption within broader global commodity chains (Box 1.4) as a means of uncovering the complex spatial relations of power that underpin it, linking up people and places in diverse but often unequal ways. This draws attention also to the leading role played by larger retailers and branded manufacturers who, hitherto, have been successful in dominating and orchestrating supply chains. In this respect, the sociologist Gary Gereffi distinguishes between producer-driven and buyer-driven commodity chains (see Table 9.1). A growing proportion of global economic activity accounted for the latter, whereby the chain is coordinated by large, branded retailers who concentrate on design, sales, marketing and finance whilst actual production is outsourced to suppliers in developed countries. Examples of well-known firms who operate through buyer-driven chains include Walmart, Ikea, Nike, the Gap and Adidas.

The production and distribution of commodities reproduces patterns of uneven development between places with low-value activities often confined to poorer regions in the global South whilst higher-value ones are typically located in wealthier places in the global North, reflecting, to a considerable extent, the legacy of colonialism. The process of commodity fetishism means that consumers in the global North are typically concerned with the price and appearance of the goods that they buy, obscuring the relations of production and distribution associated with these goods. This commodity fetishism is often reinforced by advertising.

By contrast, research by economic geographers, sociologists and anthropologists has been concerned to overcome commodity fetishism by uncovering the complex geographies of commodity production and distribution, revealing webs of interdependencies that connect different places within the global economy. As the following quotation from David Harvey illustrates, even routine, everyday activities like having breakfast rely on complex sets of spatial relations:

Consider, for example, where my breakfast comes from. The coffee was from Costa Rica, the flour that made up the bread probably from Canada, the oranges in the marmalade came from Spain, those in the Orange juice came from Morocco and the sugar came from Barbados. Then I think of all the things that went into making the production of these things possible – the machinery that came from West Germany, the fertiliser from the United States, the oil from Saudi Arabia . . . it takes very little investigation for the map of where my breakfast came from to become incredibly complicated. I also find that literally millions of people all over the world in all kinds of different places were involved just in the production of my breakfast. The odd thing is that I don't have

to know that in order to eat my breakfast. Nor do I have to know it when I go shopping in the supermarket. I just lay down the money and take whatever it will buy.

(Harvey 1989: 93)

It is these kinds of complex connections that the geographer Michael Watts (2005: 530) is referring to when he describes the commodity as a bundle of social relations. By uncovering these relations as suggested by the Harvey quote above, it is possible to trace the 'life' or 'biography' of a commodity.

The growing power of retailers has caused one group of commentators to suggest that firms such as Walmart, Carrefour and Tesco are becoming the "key organisers of the global economy" (Hamilton *et al.* 2011: 3) through their ability to exercise great power over manufactures and suppliers in their supply chains as efforts to reduce consumer prices are translating into lower prices for suppliers, enabling retailers to capture a greater share of the value added in a commodity chain (see Box 1.5).

> ## Reflect
>
> How far would you agree that retailers are now the dominant actors in the global economy? What role do individual consumers play: dupes or dictators?

11.3 Changing patterns of consumption

Modern consumer culture took shape in the second half of the nineteenth century, symbolised by the rise of the department store, particularly in the major cities of North America and Western Europe such as New York, London and Paris. Described as "the quintessential consumption site of the late nineteenth and early twentieth century" (Lowe and Wrigley 1996: 18), department stores presented "the most visible, urban manifestation of consumer culture and the economics of mass production and selling" (Domosh 1996: 257)

(Figure 11.2). A vast array of goods was placed on public display in the store and prices fixed for standardised goods. Whilst we take this for granted today, prior to the development of modern stores goods were not displayed, with customers having to ask to view them, and prices were negotiated between the customer and shop owner or clerk (Domosh 1996: 264). In the new department stores, shoppers were directly able to compare the prices and qualities of different goods as shopping became a knowledgeable and skilled activity, encouraged by the retailers and advertisers.

Shopping also became highly gendered as the department stores targeted middle-class women in particular (Box 11.2). In this way, a crucial set of links were forged between gender, class and culture in the late nineteenth century which continue to shape retail and consumption today. Shopping became a major part of women's work, with the purchase of goods replacing the domestic production of food and clothing, creating a market for manufacturers and retailers. The realm of consumption was defined as feminine, associated with leisure and self-indulgence, in contrast to the masculine domain of production, governed by the work ethic and associated notions of self-denial and self-discipline

(ibid.: 262). Department stores provided spaces in which women could be taught to shop through the display of goods as spectacle, the use of advertising and demonstration and the assistance of specialist staff (Hudson 2005: 147). The introduction and manipulation of fashion was a key mechanism for increasing demand, meaning that frequent changes in style were required to keep up. Shopping became an important female duty with store owners cultivating associations between women, fashion and religion, referring to their stores as 'cathedrals' and goods as 'objects of devotion' (Domosh: 1996: 266).

While **mass consumption** became established in the nineteenth century, it was consolidated and reinforced during the period of **Fordism** from the 1940s to the 1970s. As we have emphasised (section 3.4), Fordism was a system of industrial organisation based on a balance between mass production and mass consumption. The key link here was higher wages for workers, received in exchange for increased productivity. Fordism involved the mass production of consumer durables such as automobiles, fridges and washing machines, produced in standard forms. Rising wages meant that more and more workers were able to afford such goods,

Figure 11.2 Macy's: a famous New York department store with nineteenth-century origins
Source: D. MacKinnon.

Box 11.2

Gender and consumption in nineteenth-century New York City

Whilst the rapid and massive expansion of New York's retail district in the nineteenth century reflected the spectacular growth of the city's economy, its form was shaped by the shopping habits of the middles classes, particularly women. Shopping in the city's burgeoning department stores was an important activity for middle-class women in New York, becoming almost a daily ritual for some (Domosh 1996). The very wealthy defined the styles and

fashions which middle-class women sought to emulate, allowing the latter to define their own tastes and, by extension, their social status (ibid.: 261). Rather than being entirely confined to the domestic sphere, women made frequent trips from their suburban homes to downtown Manhattan, meaning that they played an important role in shaping this and other American downtowns. Whilst shopping was a frequent activity, they also met for

church prayer groups, lectures and concerts, and to pay bills or make social calls (ibid.: 258).

A distinct retailing area developed in late nineteenth-century Manhattan, focused on Fifth Avenue between Union and Madison Squares, extending west to Sixth Avenue and east to Broadway (Figure 11.3). This was an "urban landscape designed specifically for consumption", made up of "ornamental architecture and grand

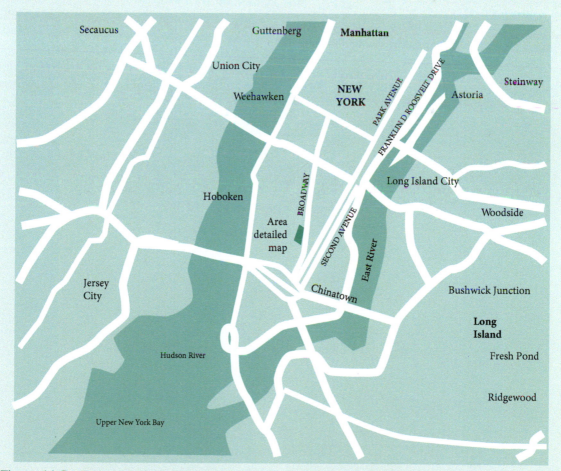

Figure 11.3a The original retail district in nineteenth-century New York
Source: Constructed from area map from Multimap, at www.multimap.com/.

Box 11.2 (continued)

boulevards, of restaurants and bars, and of small boutiques and large department stores" (ibid.: 263–4). Improvements in urban transportation through the development of a rapid transit network allowed middle-class women to travel downtown to shop in the morning, returning to their uptown residences for lunch before proceeding back downtown in the afternoon. Wide, paved streets, lit by gas and electricity, made the retailing district feel safe and congenial for women.

The city's first department store, Stewart's, opened on Broadway in 1846. Its four storeys, devoted entirely to retailing, and its white marble façade were unprecedented in the city (ibid.: 264). The interior of the store was arranged in order to maximise the display of goods and to create an appropriate atmosphere for women. It included large mirrors, chandeliers, a gallery and parlour. A later store was opened by Stewart's further uptown on Broadway in 1862,

becoming an important tourist attraction. Its six storeys contained a main floor and five encircling balconies, lit by a skylight which provided natural light. Reflecting the analogies drawn between consumption and religion, a commentator compared Stewart's with nearby Grace Church. By the end of the nineteenth century, the domestic sphere had become more fully incorporated into stores in the form of tea rooms, restaurants, art galleries and grand architectural displays.

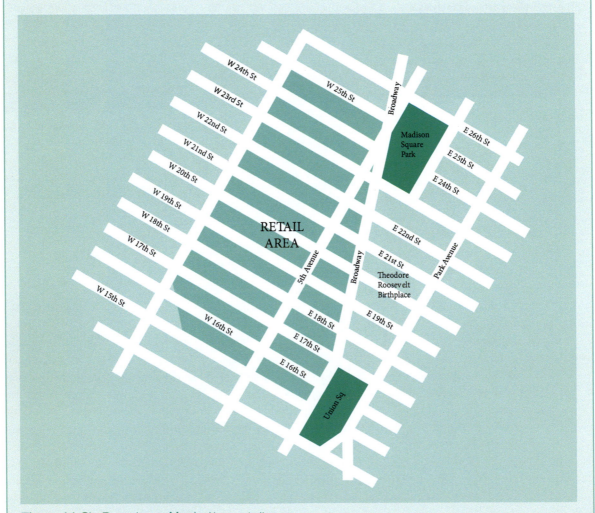

Figure 11.3b Downtown Manhattan retail area

ensuring a larger market for manufacturers and retailers. A key trend in the geographical organisation of society was the growth of suburbs, particularly in North America, facilitated by state investment in infrastructure such as roads and electricity. Suburban lifestyles became closely associated with mass consumption patterns, with every household requiring its car, washing machine and lawnmower (Goss 2005).

Mass markets for standardised goods became increasingly saturated from the late 1960s, however, as economic growth slowed and the Fordist system experienced growing problems. As a result, mass consumption has been eclipsed by the rise of **post-Fordist** patterns of consumption since the 1970s. These are defined by flexibility as markets have become fragmented into distinct segments and niches. Accordingly, patterns of consumption are defined individually rather than collectively. Consumer choice and identity has become increasingly important with individual consumers regarding the purchase and consumption of commodities as expressions of their lifestyles and aspirations (Mansvelt 2013). Individualised patterns of consumption oriented towards identity and lifestyles are a key component of postmodern culture, characterised by an emphasis on flexibility, difference and diversity (Box 2.7). The rapid circulation of ideas, images and signs, fuelled by the advertising industry and the media, is another central aspect of postmodern consumer culture.

Producers and retailers have become increasingly consumer oriented, striving to tailor goods and services to the demands of individuals and specific groups of consumers. The growth of ICTs has allowed retailers to store, process and convey data about changing patterns of consumer demand. The introduction of point-of-sale terminals in the 1980s allowed retailers to rapidly transmit information about consumption trends from shops to company headquarters from where it was passed on to designers and suppliers. More recently, the advent of online shopping, emergence of big data and analytics and the increasing sophistication with which companies can use electronic information about consumers' past spending habits to selectively market and advertise products to different market segments promises to further transform consumption relations.

> **Reflect**
>
> ➤ To what extent do you view consumption as an important means of expressing your individuality and identity?

11.4 Spaces and places of consumption

Consumption is implicated in the production and reproduction of geographic space at a range of scales, from the local to the global (Cook and Crang 2016). It shapes the local spaces in which everyday life takes place, with contemporary urban landscapes, for instance, organised to facilitate consumption through the construction of shopping centres, retail parks and the like, often in out-of-town locations accessible by car. Consumption has been viewed as increasingly important to the fashioning of global space too through the creation of a **global consumer culture** centred upon **brands** like McDonald's, Coca-Cola and Nike. For many commentators, this is erasing the distinctiveness of local places and cultures, heralding the 'end of geography' (Ritzer 2004).

11.4.1 Global cultures and local variations

The notion of a homogenous global consumer culture has become popularised through the media. The central image here is of the erasure and dissolution of distinctive local cultures in the face of global corporations and brands (Figure 11.4) (Slater 2003: 157). International tourism is seen as a key agent of this kind of cultural imperialism, subordinating local cultures to the dominance of western consumer norms. As an increasing number of studies have shown, however, this cultural homogenisation argument is highly simplistic, resting on a number of problematic assumptions (Crang 2005). Not least amongst these is that it reintroduces the discredited notion of the passive consumer, with non-western populations powerless to resist western

Figure 11.4 McDonald's in Beijing
Source: D. MacKinnon.

consumer norms propagated by powerful corporate interests (Slater 2003). At the same time, the authenticity of non-western cultures is seen as ultimately dependent on an underlying purity and absence of contamination by external forces and influences. In reality, however, cultures are a product of the relationships and connections between places, blending elements from different sources (Massey 1994). Think, for example, of the importance of drinking tea or eating curry to contemporary British culture.

A number of recent studies have shown that global consumer cultures assume locally specific forms, blending with pre-existing local cultures in particular ways. Research on how McDonald's is consumed in East Asian countries, for example, found that the chain has been localised and incorporated into local practices, with restaurants functioning also as "leisure centres, where people can retreat from the stresses of urban life" (Crang 2005: 368). The consumption of such non-local products can actually be seen as part of the production of identifiable local cultures, with people appropriating such goods for their own ends, as shown by Miller's research on the consumption of Coca-Cola in Trinidad (Box 11.3).

Box 11.3

'Coca-Cola: a black sweet drink from Trinidad'

Coca-Cola is usually regarded as one of the pre-eminent global brands, central to the creation of a global consumer culture which is actively marginalising and subordinating more authentic local cultures. As several commentators have observed, however, this view is highly simplistic (Crang 2005). One study which demonstrates this is the economic anthropologist Danny Miller's work on the consumption of Coca-Cola in the Caribbean island of Trinidad. Rather than representing the dominance of western consumer culture, Miller shows that Coca-Cola is consumed in locally specific ways in Trinidad, having become absorbed into local

Box 11.3 (continued)

cultures and traditions. This leads him to term it 'a black sweet drink from Trinidad'.

Instead of being viewed as an imported western drink, Coke is seen as authentically Trinidadian. It is regarded as a basic necessity and the common person's drink (Miller 1998: 177–8). For local consumers, the categories which frame and inform their choice of product are not those used by the producer and advertisers, but the distinctly local concepts of the 'black' sweet drink and the 'red' sweet drink. The latter is a traditional category particularly associated with the Indian population. The former is summarised

in the centrality of a 'rum and coke' as the most popular alcoholic drink on the island, although the consumption of 'black' sweet drinks without alcohol is equally common. Coke is probably the most popular of these 'black drinks', becoming associated with the black African population. These links are historical associations rather than describing actual patterns of consumption with possibly a higher proportion of Indians than blacks drinking Coke, identifying with its modern image, while many Africans consume red drinks, associating it with an image of Indianness that is an essential part of their Trinidadian identity.

These local cultural specificities and complexities impose real limits on the marketing strategies of the producers, indicating that consumption is shaped by a wide range of locally specific factors beyond their control. More broadly, the case study shows how the consumption of a prominent global brand is dependent on locally specific cultural practices and traditions. In this way, **mass consumption** practices effectively transform global products into locally specific forms, suggesting that capitalism should be viewed as a diverse collection of practices rather than as a set of overarching economic imperatives.

There is a clear sense in which modern consumer culture actively embraces cultural and geographical difference with the globalisation of food, for instance, presenting consumers in western countries like the UK and US with the 'world on a plate' through a choice of ethnic cuisine from different cultural regions (Cook and Crang 1996). In any large British city, for example, consumers are presented with a wide choice of ethnic restaurants, stretching beyond the popular Indian, Chinese, Mexican and Italian to include specialties such as Lebanese, Thai and Vietnamese, and an array of goods of diverse geographical origins in supermarkets. Rather than eradicating geographical difference, consumption produces new geographies, presenting us with particular representations of the global and the local, the foreign and the domestic, etc. (Crang 2005).

11.4.2 The geographies of brands

The emergence of brands and the branding of products has become a critical aspect of the shift towards a more consumption-based form of capitalism. The ability of the sellers of a product to differentiate it

from a mass of market competitors has become a critical source of capturing value. Brands are recognised as having considerable value in themselves, beyond the intrinsic value of the product itself. Top brands are often estimated to account for a sizeable proportion of the originating company's market value and in some cases can exceed a corporation's sales turnover (see Table 11.1)

Around the rise of the brand, a whole new set of creative and cultural economic activities such as advertising, marketing, consultancy and design have emerged with significant employment- and revenue-generating effects. One study in the UK, for example, estimated that employment in advertising and marketing had grown by 6 per cent annually between 1981 and 2006 to reach around 200,000, although growth has plateaued over the past decade, partly through the impact of the financial crisis and subsequent recession (see: www.thecreativeindustries.co.uk/uk-creative-overview/facts-and-figures/employment-figures). As with consumption more generally, brands have a much broader cultural and social significance. Kornberger has described the emergence of a "brand society" where brands can provide "ready-made identities" that become "so mashed up with our social world that

Table 11.1 Brand values for selected global brands

Ranking	Company	Brand value ($m)	Brand value as % of market capitalization	Brand value as % of total sales	Country of ownership
1	Coca-Cola	67 525	64	290	United States
2	Microsoft	59 941	22	138	United States
3	IBM	53 376	44	54	United States
4	GE	46 996	12	28	United States
5	Intel	35 588	21	93	United States
6	Nokia	26 452	34	68	Finland
7	Disney	26 441	46	82	United States
8	McDonald's	26 014	71	128	United States
9	Toyota	24 837	19	14	Japan
10	Marlboro	21 189	15	22	United States
11	Mercedes	20 006	49	12	Germany
12	Citi	19 967	8	22	United States
13	Hewlett-Packard	18 866	29	22	United States
14	American Express	18 559	27	57	United States
15	Gillette	17 534	33	157	United States
16	BMW	17 126	61	31	Germany
17	Cisco	16 592	13	67	United States
18	Louis Vuitton	16 077	44	102	France
19	Honda	15 788	33	19	Japan
20	Samsung	14 956	19	26	South Korea

Source: Pike 2015: 31, Table 2.3.

they have become a life-shaping force" (cited in Pike 2015: 9). This is evident with the role that global icons such as Apple, Coca-Cola, Levi's, etc. have in constructing people's sense of self. Because of this power, brands are seen as a negative force by some writers – notably Naomi Klein in her celebrated book *No Logo* (Klein 2010) – in seducing and controlling people in an increasingly information-driven knowledge economy.

While talk of global brands gives a sense of place-lessness, there are important spatial dimensions underpinning brands (Pike 2015). Global brands themselves are heavily dominated by the US in their place of origin, reflecting the broader strength of American capitalism and its corporations. Beyond this superficial level,

Pike suggests three critical aspects to the geography of brands: first, they have particular geographical "connections and connotations" (ibid.: 32) in their source of origin and their success in marketing; second, they take specific spatial forms and flows between places in how their economic value is constructed and maintained; and third, they are unevenly distributed in the ways they capture value and benefit different economic and social actors and places. There is also a ceaseless spatial dynamic to brands and their evolution, seeking new markets and consumers in ways that serve to deepen the commodification of economic and social life. As Pike puts it: "Brand actors invest much time, effort and resources grappling with social and geographical differences and specifically how they can be used and perpetuated to create and realise meaning and value in spatial circuits" (ibid.: 36).

Pike (2015) uses the concept of 'origination' to identify the way actors construct and use geographical meanings and associations to enhance brands. This happens in a range of different ways that defies easy conceptualisation: for example, Nike's success in branding its sportswear in association with Brazil and its footballing success; Guinness's association with a romantic sense of Irishness; or the way numerous brands successfully identify themselves with New York as an iconic global city. Other strategies involve certification and legally enforceable regulation of brands associated with particular territories for products such as Scotch whisky, French champagne, Parma ham or Roquefort cheese. However, some geographical branding also displays "spatial discontinuity" where there is a deliberate obscuring of "geographical associations" (ibid.: 41). This happens, for example, in the way that many clothing and fashion firms deploy the label 'Made in Italy', signifying a high-quality Italianate craft threshold, whilst outsourcing production to low-cost foreign locations.

11.4.3 Places of consumption

Another major focus of attention has been particular **places of consumption**. Key sites include the department store, the mall, the street, the market and the home as well as a host of more inconspicuous sites of consumption (for example, charity shops and car boot sales) (Mansvelt 2005). Tourist regions and heritage parks are also sites of consumption, albeit of landscapes and experiences rather than material goods and services. In recent decades, department stores have been reinvented in the form of the 'flagship store', which incorporates its own labels, building on concepts introduced by chains like Habitat. Harrods and Harvey Nichols are good examples of such chains, which claim to be selling 'lifestyles' rather than simply goods, combining "designer interiors, rituals of display and leisure, sexuality and food" (Lowe and Wrigley 1996: 25).

The mall or shopping centre has attracted a lot of interest from geographers and other consumption researchers, representing perhaps the most visible and spectacular kind of retail environment. It consists of a range of shops and entertainment facilities within an enclosed space that is usually privately owned and managed (Mansvelt 2005: 61). Malls are widely viewed as the iconic space of contemporary retailing, representing the 'urban cathedrals' of contemporary capitalism (Goss 1993). The world's first fully enclosed mall was opened in Southdale, Minneapolis in 1956, becoming the prototype for thousands of others over the succeeding decades.

Shopping centres are designed in order to maximise the exposure of consumers to goods, with those of the 1960s and 1970s designed as 'machines for shopping'. More recently, planners and developers have sought to provide spectacular places that people want to spend time in, thus maximising spend, incorporating entertainment facilities, food courts and visual features as well as shops (Crang 2005: 373). A number of regional malls were opened in Britain in the 1980s such as the Metro Centre in Gateshead or Meadowhall in Sheffield. Further developments occurred through the development of mega malls in North America in the 1980s and 1990s such as the West Edmonton Mall and the Mall of America (see Box 11.4). Again, people's use of such spaces is not wholly determined by the intentions of developers and chains. They provide certain groups with a place to socialise or 'hang out' instead of shop, with teenagers, for instance, often coming into conflict with centre management (Crang 2005).

More recently, the focus of attention has moved away from malls as representing the grand and spectacular towards more mundane and everyday sites of consumption such as the street, the home and the likes of car boot sales. Streets provide a great variety of environments for consumption, with particular types of shop often tending to congregate in particular districts. Inconspicuous consumption spaces such as car boot sales, charity shops and retro-vintage clothes shops involve the valuing and purchase of second-hand commodities (Crewe and Gregson 1998). Domestic space has also been re-examined as a site of consumption with research examining how a range of consumer goods are utilised within the home. The role of home-based shopping has also been examined in terms of catalogues, classified adverts and Tupperware, for example. Food and cooking represents another strand of research, with

Box 11.4

The Mall of America

The Mall of America (MoA) was opened in 1992 in Bloomington, Minnesota. It is probably the most spectacular example of the modern mall in the world today, representing "the largest fully enclosed retail and family entertainment complex in America" (MoA, 1997, quoted in Goss 1999 45). It receives between 35 and 40 million visitors a year, more than the Grand Canyon, Disneyland and Graceland combined. The Mall contains over 520 stores, over 50 restaurants, 14 movie screens, the largest indoor family theme park in the USA, a 1.2-million-gallon aquarium and a range of other attractions.

A key narrative (story) is that of authenticity, stressing the Mall's rootedness in the local environment and culture. An important source of this is the site itself, previously the Metropolitan Stadium: home of local sports teams the Minnesota Twins (baseball) and the Minnesota Vikings (American football). The connections with nearby Southdale are also stressed. As such, the site is a strong symbol of local identity, representing a natural place of congregation, and conveying a local 'sense of place' as rooted and authentic. Notions of travel and tourism are incorporated into the fabric of the mall itself, divided into districts based on imagined tourist-retail destinations (West Market, North Garden, etc.).

A close inspection of certain displays of goods provided some glaring examples of the process of commodity fetishism in terms of how the descriptions of goods served to obscure their actual origins. Most notably perhaps, a stuffed bear in the shop 'Love From Minnesota' had a tag in its right ear saying "'Minnesotans who live deep in the northwoods among the loons, wolves and scented pine trees, listen to the gentle lapping of the waves of the shoreline while they handicraft unique memories of our homeland, like this one, to share with you'; a tag on the other ear said, 'Bear made by the Mary Meyer Corp., Townsend, Vermont . . . Made in Indonesia'" (Goss 1999: 54–5).

In terms of time, the developers of the Mall sought to mobilise meaning and a sense of magic through the four key themes of nature, primitivism, childhood and heritage. In this way, a strong sense of nostalgia is evoked, building a collective dream of authenticity which becomes attached to the products sold within the Mall. According to Goss (1999: 72), ". . . it [the Mall] must promote the spontaneity of crowds in order to evoke [the] natural commerce of the marketplace". The obvious point underpinning all of this is that elaborate narratives of authenticity are constructed to sell products, creating an aura of mystery in order to overcome the perceived meaninglessness and superficiality of modern life.

studies focusing on this as an expression of social relations, particularly those of gender, within the household. The body can also be seen as an important site of consumption, being central to the creation of identity through appearance and image, underpinning the consumption of items such as clothes and cosmetics.

11.4.4 The reconfiguration of retail spaces

There is a sense in the first decades of the twenty-first century that we are seeing a reconfiguration of the relations between space and consumption with the emergence of an increasing amount of online shopping. In what one commentator has described as the "great retail apocalypse" (Rusche 2017), retail stores are coming under increasing pressure from the trend towards online shopping, which promises once again to remake relations between consumption, space and identity. Traditional department stores and retail outlets have seen massive declines in their share of overall sales since 2000 while online sales have grown dramatically (see Figure 11.5). One 2017 study suggested that there had been no major shopping malls built in the US in the previous three years, and around half of the 1,200 existing malls would close within five years (ibid.). The spectacular rise of Amazon and other digital platforms, and the growth of home shopping, seems to be threatening existing consumer spaces quite dramatically and potentially transforming the whole sociality of consumption, potentially leading to more atomised and individualist sets of relations.

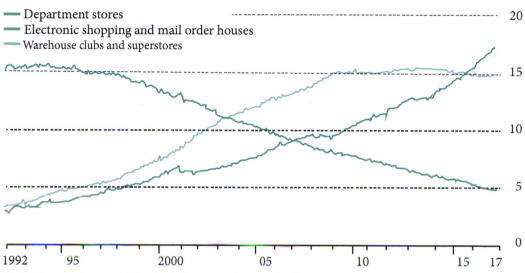

American online retailing is growing rapidly

% of core retail sales

— Department stores
— Electronic shopping and mail order houses
— Warehouse clubs and superstores

Figure 11.5 The decline of US retail stores and growth of online shopping
Source: Financial Times, 16 July 2007, at www.ft.com/content/d34ad3a6-5fd3-11e7-91a7-502f7ee26895.

The pernicious effects of the decline in retail shopping have been felt heavily in some areas where there has been a growing discourse and concern about the 'death of the High Street' (Hubbard 2017). In the wake of the financial crisis, added to the longer-term drop in retail sales, and the growth of out-of-town retail parks, the UK's high streets have suffered from a process of ongoing decline, with a growing number of empty shops and vacant premises (see Figure 11.6). In 2014 it was estimated that 14 per cent of retail premises in the country's high streets were vacant (ibid.: 16). What were once thriving centres of sociability and conviviality are increasingly characterised as places in crisis with considerable public anxiety and concern about the negative antisocial and potentially criminal effects of these developments (ibid.). As Hubbard adroitly puts it:

The fear is that these on-going processes of decline, left alone, will precipitate a mass abandonment that will leave many local shopping streets threadbare and irrelevant in the context of increasingly privatized and segmented cities where much economic and social life is online, not grounded

in the physical spaces in which urban life has long been conducted.

(Hubbard 2017: 17)

In many towns, the closure of well-established retail chain stores, alongside the disappearance of local butchers, bakers and other small independents has been accompanied by the growth of retail outlets that cater for poorer and more marginalised groups, which include discount 'pound' stores, charity shops and pawnbrokers. Similar trends are at work in the US where chains like Dollar General are increasing sales rapidly while the mainstream retail chains experience decline (Authers and Leatherby 2017).

Reflect

➤ What are the key trends transforming retailing in the contemporary era and what are their social and spatial consequences?

Figure 11.6 The decline of the UK high street
Source: Alamy.

Box 11.5

The death of the British high street?

The demise of many British high street shopping areas through the closure of retail outlets and the growing number of vacant properties has been the cause of much public concern and media commentary. Hubbard contrasts the decline of traditional urban centres – such as Margate on the south-east coast of England, which earned the unwanted moniker of Britain's most empty high street in 2010 – with the growth of new retail spaces such as Westfield Stratford City in east London. The latter, a new shopping centre of 2 million square feet constructed in a regenerated area near to the site of the city's 2012 Olympics stadium, housing both up-market brands as well as the more mainstream mass market stores and fast food outlets, is viewed as a new kind of public space where the diverse social groups of a world city can mingle. According to Hubbard,

> Margate and Westfield of course represent extremes: one, a contrived and carefully planned retail spectacle benefitting from its proximity to London's Olympic Park, the other, a 'gap toothed' High Street in a seaside town largely forgotten and unloved by a British public that has moved on from its love affair with the 'kiss me quick' bucket-and-spade holiday.
>
> (Hubbard 2017: 17–18)

Clearly we are witnessing a dramatic change in shopping habits but also in the ways that people interact with each other in public spaces. For Hubbard, the decline of high streets is a symptom of fracturing cities, a growing divide between the affluent and the more impoverished and marginalised elements of society. He criticises British media discourse and official government policy, which he says portrays a nostalgia for an older and often white, racialised vision of high streets from the 1940s and 1950s that does not match the multiracial, multicultural societies of the twenty-first century. Middle-class consumer culture and tastes, in his view, are also coming to dominate retail visions over the needs of poorer groups. High-fashion brands, retailers and more pristine and manicured spaces of collective consumption are driving out more organic, diverse and chaotic spaces of multi-cultural and cross-class coexistence.

11.5 The restructuring and internationalisation of retail

Alongside the globalisation of consumer culture, the retail sector itself has undergone considerable **internationalisation** in recent times. Although retailing has traditionally been seen as a very nationally specific economic activity, in common with many other services it has experienced a dramatic internationalisation in recent times. The period since the late 1990s has witnessed what Coe and Wrigley (2007: 342) term a 'deluge of retail FDI', focused largely on emerging markets in East Asia, Central and Eastern Europe and Latin America. The most recent data suggests that between 2005 and 2015, international retail sales (as a proportion of total retail sales) increased from 14.4 per cent to 22.8 per cent (Coe *et al.* 2017) while the average number of countries in which multinational retailers operate increased from 5.9 to 10.1.

11.5.1 Trends in retail internationalisation

An exploration of the world's leading retailers (Table 11.2) reveals a mixed picture in terms of internationalisation with some of the larger US firms displaying relatively low levels of foreign activity. For those firms that have internationalised, there is a tendency to adopt a regional rather than fully global strategy, focusing on developing and consolidating their activities in a limited number of markets in selected regions such as East Asia rather than seeking to 'collect countries' on a global basis (Dawson 2007).

Retail internationalisation is clearly driven by market access with these emerging markets offering the prospect of sustained future growth, in contrast to saturated and highly competitive domestic markets. It was facilitated by access to low-cost capital in the growth years of the late 1990s and early 2000s through debt and equity financing and the substantial liberalisation of the retail sector in many emerging markets (Coe and Wrigley 2007). Retail TNCs' expectations of growth in emerging economies were derived not only from the prospect of rapid economic development, but also from the underdevelopment of the modern retail sector and the preponderance of traditional outlets in these economies. Three types of strategies are evident: more global strategies pursued by transnational food retailers such as Carrefour and Walmart; highly localised 'franchise' approaches of retailers such as the Netherlands-based corporation Ahold, who operate through networks of fairly autonomous and self-contained

Table 11.2 World's top ten retailers by revenue 2015					
Rank	Company	Country of origin	Retail Revenue	Countries of operation	% retail revenue from foreign operations
1	Walmart	US	482,130	30	25.8
2	Costco	US	116,119	10	27.4
3	The Kroger Co	US	109,830	1	0.0
4	Schwarz Unter. KG	Germany	84,448	26	61.3
5	Walgreen Boots Alliance	US	89,631	10	9.7
6	The Home Depot	US	88,519	4	9.0
7	Carrefour	France	84,856	35	52.9
8	Aldi Einkauf GmbH	Germany	82,164	17	66.2
9	Tesco	UK	81,019	10	19.1
10	Amazon	US	79,268	14	38.0

Source: Derived from Deloitte 2017: 14.

Box 11.6

The complex relational geographies of Tesco's internationalisation process

Expansion into overseas markets has been a major component of Tesco's growth strategy since the mid-1990s, helping it to become the third-largest food retail TNC in the world by the mid-2000s. The company moved into 13 overseas markets, comprised of five countries in Eastern Europe and the Republic of Ireland, six in Asia and the US. As such, its strategy has been highly regionalised, rather than global, focused on building up market share in a relatively small number of strategic markets.

The company's 'strategic localisation' strategy (Coe and Lee 2006) appeared to be giving it an advantage over some of its rivals in the 2000s, particularly in the competitive Asian markets where it pursued partnership relations with local firms or in the case of India a franchise agreement with a local corporation. Its most successful international venture was viewed as its South Korean operation launched in 1999 through a merger with the Samsung Corporation's distribution unit, creating Samsung-Tesco. The stores were branded as Homeplus, further emphasising the local. This was designed to respond to a consumer culture that is noted for its suspicion of foreign brands and a preference

for established national products and brands (Coe and Lee 2006). The overwhelming majority of staff were local, including the Chief Executive who sought to create a hybrid organisational culture, melding Korean notions of staff loyalty and identification with the more 'rational' business practices of Tesco as the largest UK food retailer (ibid.). Local sourcing was also emphasised, involving the establishment of direct procurement channels with local producers and manufacturers.

The South Korean venture was viewed as the 'jewel in the crown' in Tesco's international operations, expanding from 2 stores employing 500 workers in 1999 to 139 stores employing 25,972 by 2014 (Coe et al. 2017). It was something of a surprise therefore when Tesco announced that it was selling Homeplus to an Asian private equity concern in 2015 for £4.2 billion (ibid.). To explain the decision, it is necessary to understand the complex relational geographies of Tesco, bound up in the linkages between its home market in the UK and its overseas operations. Although its international operations had been expanding, its domestic market remained dominant (see Table 11.3). Faced with the

aftermath of the financial crisis, challenging domestic market conditions and growing competition (including from new low-cost foreign entrants such as Aldi and Lidl), Tesco recorded the largest ever loss by a UK retailer in 2015. It was also hit by a scandal following the over-recording of profits. Losing the confidence of its investors, and with problems in some of its other international operations – notably a loss-making operation in China and the withdrawal from a failed attempt to enter the highly competitive US market – Tesco needed to shore up its domestic core business.

Tesco was also facing problems in South Korea where after early growth, its fortunes were faltering. Falling consumer satisfaction and a growing number of campaign groups protesting at alleged malpractices by Homeplus combined with growing resistance by local retailers at the entry of foreign retailers into their traditional markets. The company was also facing growing collective action by workers demanding living wages and more secure conditions of employment (ibid.). Faced with such pressures at home and abroad and with the need to satisfy its institutional investors, the sale of Homeplus becomes more understandable.

Table 11.3 The changing orientation of Tesco's international operations 2000–17 (% share of total turnover)

	2000	2005	2010	2017
UK	90	80	68	76
Rest of Europe	7	11	16	15
Asia	3	9	22	9
US			1	0

Source: Derived from Coe and Lee 2006: 72; Tesco 2017: 167.

subsidiaries; and a 'third way', pursued by firms such as UK retailer Tesco, of 'strategic localisation' (Coe and Lee 2006) (Box 11.6)

11.5.2 Complex spatial mosaics of retail TNCs

In general, the retail sector is subject to high levels of **territorial embeddedness**, reflecting several underlying characteristics of retail as an industry (Coe and Lee 2006: 68). First, firms are closely connected to the property markets and planning systems of host countries since they require an extensive network of stores. Second, retailers need to be highly attentive to local patterns of consumption, which reflect distinct cultural preferences, tastes and attitudes, raising a host of socio-cultural questions beyond the purely economic. Third, retail firms typically source a wider range of products from local suppliers, alongside the growing emphasis on global and regional sourcing. Such embeddedness can be contrasted with the situation in high-volume manufacturing sectors in which TNC plants often have few links with the host economy, relying on a host of imported materials which are assembled for export.

The process of retail internationalisation has not been a story of inexorable global expansion and integration, but what has been described as a more "complex mosaic of success and failure" (Coe *et al.* 2017: 2743). On the one hand, retail firms continue to be heavily dependent on their host country environment, both for market share and, increasingly importantly since the financial crisis, for investor capital to deliver their strategies (ibid.). Additionally, the continuing territorial embeddedness of the retail sector, and the difficulties of transcending highly localised consumer markets and cultures in penetrating host retail markets mean that there have been many examples of unsuccessful attempts to internationalise over the past decade. In one well-known example, Walmart struggled to transplant its low-cost deregulated retail model to the more regulated German market, eventually withdrawing in 2006 (Christopherson 2007).

Growing host country opposition to international retailers has also been evident where "domains of resistance" (Coe *et al.* 2017: 2756) are emerging on a range of fronts. Facing pressures from smaller domestic retailers threatened by the low-cost models of incoming foreign corporations, many governments have begun to introduce new regulations and even taxes on the large retail sector. The Polish government recently tried to introduce a new sales tax on large and predominantly foreign retailers, including Carrefour and Tesco, while exempting smaller local firms. The European Commission, however, ordered the Polish government to withdraw the tax on the grounds that it would infringe EU state aid rules, giving smaller countries an unfair competitive advantage (Shotter 2017). Additionally, growing labour resistance, union organisation and demands for higher wages often drive foreign retailers out of host markets. In practice, therefore, international retail strategies tend to be marked by processes of capital switching and ongoing spatial adjustment from different markets and countries as economic conditions change (Coe *et al.* 2017).

As we have already noted, the dominance of conventional retailers is also being challenged by new online and digital consumption patterns and platforms and the emergence of major new corporate players such as Facebook, Netflix and Amazon. A recent analysis (Authers and Leatherby 2017) suggests that Amazon may be poised to overtake Walmart in terms of total sales in its US markets. While some retailers are able to transition to an online presence, including so far Walmart, it is likely that the growth of online platforms will lead to considerable disruption of established retail operations and business models over the next few years.

What this will mean for the spaces and places of consumption is an open question but it is likely that, rather than a displacement of real material spaces for more virtual ones, we will see an increasingly interwoven geography of consumption and retail where as one observer shrewdly puts it: "the precise impact of the Internet [on consumption's] geographies and practices is complex and still evolving, and will continue to be in part shaped by the spatial configuration of existing value chains, power relations, and network structures by virtue of the remediation effects at work" (Crewe 2013: 776). This may well mean that it helps to consolidate existing patterns of uneven development while at the same time rupturing others and creating new forms of relations between places.

11.6 Summary

This chapter has highlighted the importance of consumption to the contemporary global economy. The increasing focus around the mass consumer, commodities and brands represents an underlying shift away from the productivist and industrialised capitalism of the early twentieth century towards a modern consumer-oriented capitalism in the second half, with a concomitant shift in many advanced economies towards services. However, rather than viewing consumption as displacing production, an emphasis upon commodity chains reveals how the two are enmeshed and mutually constituted within the global economy, bound up in uneven and differentiated geographical flows, circuits and connections between places.

Consumption is also bound up in cultural practices and social identities, which in turn mediate economic processes. While the emergence of globalised consumption practices is seen by many as creating more homogenised cultures, the chapter has emphasised continuing spatial divergence and the geographical uniqueness of place. Changing modes of consumption help to remake the economic landscape, not least of major cities, in common ways, through the emergence of increasingly ubiquitous urban phenomena such as department stores and out-of-town malls. But local cultures and practices also co-produce the global consumer economy in spatially distinctive ways, as evident from our examples of Coca-Cola, brand origination and the strategic localisation approaches of international retailers. The contemporary ongoing shift towards online consumption, digital technologies and the internet promises to further transform the global economy, consumer identities and practices. However, rather than eradicating space – as predicted by some arch-globalising discourses – it is likely to produce new forms of geographical variety, spatial connections and flows.

Exercise

Select a major conventional global retail chain and assess how it organises both (i) its own retail organisation across space and (ii) its commodity chains across a few of its key product markets. Critically evaluate the robustness of its existing spatial strategies to the threats posed by growing online consumption and the emergence of competitor digital platforms. How is the retailer responding to these threats? Evaluate its strengths and weaknesses for the new digital age.

Key reading

Crewe, L. (2017) *The Geographies of Fashion: Consumption, Space and Value,* London: Bloomsbury.
An analysis of fashion and the clothing industry from a critical geographical perspective. In particular, it uncovers the dense and complex relations around the production and consumption of high-end clothing, exploring both the economic and culture values bound up in the fashion industry and their spatial connections.

Gereffi, G. (1994) **The organisation of buyer-driven commodity chains: how US retailers shape overseas production networks.** In Gereffi, G. and Korzeniewicz, M. (eds) *Commodity Chains and Global Capitalism.* Westport, CT: Greenwood Press, pp. 95–122.
A key founding statement in the commodity chains literature. It usefully conceptualises the growing role of retailer and brand manufacturer power in driving global supply chains, with particular reference to clothing.

Hubbard, P. (2017) *The Battle for the High Street: Retail Gentrification, Class and Disgust.* Basingstoke: Palgrave Macmillan.
A highly provocative and readable account of the issues facing conventional high streets and retailers in austerity-driven Britain. It usefully critiques different narratives and policy discourses around retailing and the 'death of the high street' from spatial, social and class perspectives.

Mansvelt, J. (2013) Consumption-reproduction. In Cloke, P. *et al. Introducing Human Geographies*, **3rd edition.** London: Taylor & Francis, pp. 378–90.

A useful introduction to debates around the economic and cultural significance of consumption. In particular, it emphasises multi-dimensional aspects of consumption, in terms of its central role within global economic relations but also its cultural aspects and the way it shapes identities.

Pike, A. (2015) *Origination: The Geographies of Brands and Branding.* Oxford: Wiley-Blackwell.

An important and original contribution to thinking about the significance of brands and their geographies. It develops the concept of origination to think about the way that brands are produced and reproduced through place-based and spatialised relations.

Useful websites

https://followtheblog.org/author/iancooketal/
Important and innovative blog on consumption matters by leading cultural geographer Ian Cook.

References

Authers, J. and Leatherby, L. (2017) In charts: how US retailers fared as Amazon powered ahead. *Financial Times,* 22 November.

British Retail Consortium (2016) *Festive FAQs: 2016.* London: British Retail Consortium. Available at: https://brc.org.uk/media/105790/brc-festive-faqs-2016.pdf.

Christopherson, S. (2007) Barriers to 'US style' lean retailing: the case of Wal-Mart's failure in Germany. *Journal of Economic Geography* 7(4): 451–69.

Coe, N. and Lee, Y.-S. (2006) The strategic localization of transnational retailers: the case of Samsung-Tesco in South Korea. *Journal of Economic Geography* 82 (1): 61–88.

Coe, N. and Wrigley, N. (2007) Host economy impacts of transnational retail: the research agenda. *Journal of Economic Geography* 76: 341–71.

Coe, N., Lee, Y.-S. and Wood, S. (2017) Conceptualising contemporary retail divestment: Tesco's departure from South Korea. *Environment and Planning A* 49 (12): 2739–61.

Cook, I. and Crang, P. (1996) The world on a plate: culinary culture, displacement and geographical knowledges. *Journal of Material Culture* 1: 131–53.

Cook, I. and Crang, P. (2016) Consumption and its geographies. In Daniels, P., Bradshaw, M., Shaw, D., Sidaway, J. and Hall, T. (eds) *Introduction to Human Geography*, 5th edition. London: Pearson Education, pp. 379–96.

Crang, P. (2005) Consumption and its geographies. In Daniels, P., Bradshaw, M., Shaw, D. and Sidaway, J. (eds) *Human Geography: Issues for the Twenty First Century*, 2nd edition, London: Pearson, pp. 359–79.

Crewe, L., (2013) When virtual and material worlds collide: democratic fashion in the digital age. *Environment and Planning A* 45 (4): 760–80

Crewe, L. (2017) *The Geographies of Fashion: Consumption, Space and Value.* London: Bloomsbury.

Crewe, L. and Gregson, N. (1998) Tales of the unexpected: exploring car boot sales as marginal spaces of consumption. *Transactions, Institute of British Geographers* NS 23: 39–53.

Dawson, J.A. (2007) Scoping and conceptualising retailer internationalisation. *Journal of Economic Geography* 7: 373–97.

Deloitte (2017) *Global Powers of Retailing*, Annual Report.

Domosh, M. (1996) The feminised retail landscape: gender, ideology and consumer culture in nineteenth-century New York City. In Wrigley, N. and Lowe, M. (eds) *Retailing, Consumption and Capital: Towards the New Retail Geography.* Harlow: Longman, pp. 257–70.

Goss, J. (1993) The 'magic of the mall': an analysis of form, function and meaning in the contemporary retail built environment. *Annals of the Association of American Geographers* 83 (1): 18–47.

Goss, J. (1999) Once upon a time in the commodity world: an unofficial guide to the mall of America. *Annals of the Association of American Geographers* 89: 45–75.

Goss, J. (2005) Consumption geographies. In Cloke, P., Crang, P. and Goodwin, M. (eds) *Introducing Human Geographies*, 2nd edition. London: Arnold, pp. 253–66.

Hamilton, G.G., Senauer, B. and Petrovic, M. (2011) *The Market Makers: How Retailers are Reshaping the Global Economy.* Oxford: Oxford University Press.

Harvey, D. (1989) Editorial: a breakfast vision. *Geography Review* 3: 1.

HKTDC Research (2017) *China's Toy Market*, 11 September. Hong Kong: Hong Kong Trade Development Council.

Hubbard, P. (2017) *The Battle for the High Street: Retail Gentrification, Class and Disgust.* Basingstoke: Palgrave Macmillan.

Hudson, R. (2005) *Economic Geographies: Circuits, Flows and Spaces.* London: Sage.

Klein, N. (2010) *No Logo*, 4th edition. New York: Fourth Estate.

Lowe, M. and Wrigley, N. (1996) Towards the new retail geography. In Wrigley, N. and Lowe, M. (eds) *Retailing, Consumption and Capital: Towards the New Retail Geography*. Harlow: Longman, pp. 3–30.

Mansvelt, J. (2005) *Geographies of Consumption*. London, Sage.

Mansvelt, J. (2013) Consumption-reproduction. In Cloke, P., Crang, P. and Goodwin, M. (eds) *Introducing Human Geographies*, 3rd edition. London: Taylor & Francis, pp. 378–90.

Massey, D. (1994) A global sense of place. In Massey, D. (ed.) *Place, Space and Gender*. Cambridge: Polity, pp. 146–56.

Miller, D. (1995) Consumption as the vanguard of history: a polemic by way of introduction. In Miller, D. (ed.) *Acknowledging Consumption: A Review of New Studies*. London: Routledge, pp. 1–57.

Miller, D. (1998) Coca-Cola: A black sweet drink from Trinidad. In Miller, D. (ed.) *Material Culture: Why Some Things Matter*. Chicago: University of Chicago Press, pp. 169–88.

Pike, A. (2015) *Origination: The Geographies of Brands and Branding*. Oxford: Wiley-Blackwell.

Ritzer, G. (2004) *The Globalisation of Nothing*. Thousand Oaks, CA, and London: Pine Forge Press.

Rusche, D. (2017) Big, bold . . . and broke: is the US shopping mall in fatal decline? *The Guardian*, 23 July.

Shotter, J. (2017) European Commission rules Poland retail tax unfair. *Financial Times*, 30 June.

Slater, D. (2003) Cultures of consumption. In Anderson, K., Domosh, M., Pile, S. and Thrift, N. (eds) *Handbook of Cultural Geography*. London: Sage, pp. 146–63.

Tesco (2017) *Annual Report and Financial Statement*. Tesco plc, Cheshunt, Hertfordshire.

Thrift, N. and Olds, K. (1996) Refiguring the economic in economic geography. *Progress in Human Geography* 20: 29–42.

Watts, M. (2005) Commodities. In Cloke, P., Crang, P. and Goodwin, P. (eds) *Introducing Human Geographies*, 2nd edition. London: Arnold, pp. 527–46.

Wolf, M. (2016) China's great economic shift needs to begin. *Financial Times*, 19 January.

Worldwatch Institute (2015) *Vital Signs*, vol. 22. Washington, DC: Worldwatch Institute.

Wrigley, N. and Lowe, M. (2002) *Reading Retail: A Geographical Perspective*. London: Arnold.

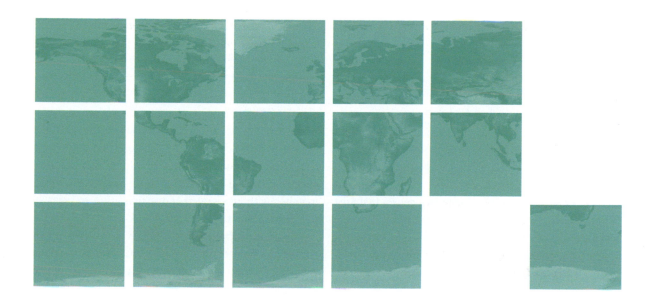

Chapter 12
Economic geography and the environment

Key topics covered in this chapter

➤ Nature, resources and economic development.

➤ The political economy of climate change.

➤ Different economic geography perspectives on economy–environment relations.

➤ Sustainability transitions: the multi-level perspective and beyond.

➤ The emerging economic geography of energy transition.

12.1 Introduction

Economic geographers have long had an interest in the role of natural environmental factors in shaping the location of economic activity, going back to the origins of German location theory in the 1920s and even earlier (section 2.2). But until relatively recently, the environment was always regarded in a more passive sense as a source of raw materials feeding into location models as another factor of production alongside labour, capital, market proximity and transport costs. Analysis of the environment per se as a subject on its own terms largely happened beyond the boundaries of the sub-discipline. While other areas of human geography, such as development studies, cultural, rural and urban geography have begun to pay attention to the environment and nature, not just as critical spheres for thinking spatially, but also as more active dimensions in shaping human society, **economic geography** has tended to lag behind.

This has changed with the increasing sense of urgency around tackling **climate change** and its economic, social and spatial implications. There are two critical elements of this that are the focus of this chapter. First, conceptually, how does an environmental lens reshape our understanding of the spatial economy, complicating our existing concepts and theories? Second, what are the implications of taking the environment more seriously, and in particular, the growing policy agenda to combat climate change, for existing practices and processes of local and regional economic development and the economic relations between places?

As is fairly self-evident, geography, the environment and the looming threat of climate change is a vast area of study, and there is not space here to do justice to the diverse traditions and perspectives that have developed. Instead, we focus on the renewed interest in environmental and especially energy issues within economic geography in recent years, particularly those associated with climate change.

12.2 Nature, climate change and economic development

12.2.1 Resources, nature and climate change

As we have already noted, resources are fundamental to any attempt to grapple with the workings of the economy. For example, Penrose's theory of the firm (section 3.2.1) is underpinned by the basic idea that a firm's competitive advantage ultimately stems from the way it develops effective competences to use the resources available to it. The resources of nature – land, water, soil, minerals, etc, – and their accessibility have long been key ingredients in economic geography's explanations. While it might be overstating the point to describe them as largely passive elements in explanation of patterns of uneven development, it is fair to say that there had been little attempt, until recently, to think more critically about resources or nature more broadly within the sub-discipline of economic geography.

The increased threat of climate change for people and the planet, and the role of economic development linked to fossil fuel extraction in accelerating **global warming**, can, however, no longer be sidestepped. The term '**Anthropocene**' is increasingly used to identify

what the United Nations' Intergovernmental Panel on Climate Change (IPCC) defines as the 'industrial era' denoting the effect of the burning of fossil fuels for rising temperatures, increased CO_2 emissions and other greenhouse gases responsible for global warming (see Figure 12.1). Not only is there now a 95 per cent certainty that industrialisation has caused global warming, but the evidence (see Box 12.1) suggests that the process has actually speeded up in recent years as the long-term effects of industrialisation in the global North come together with accelerated industrialisation in parts of the global South since the late 1950s.

Although certain vested interests – particularly among the powerful global oil **TNCs** and automobile producers – continue to dispute the evidence for global warming, the vast majority of climate scientists are now convinced both of its existence and its human and industrial causes. There is also near unanimity among scientific opinion that drastic policy action is required now to arrest the most harmful effects of global warming. A recent IPCC report could not have been clearer in its assessment:

> Anthropogenic greenhouse gas emissions have increased since the pre industrial era, driven largely by economic and population growth, and are now higher than ever. This has led to atmospheric concentrations of carbon dioxide, methane and nitrous oxide that are unprecedented in at least the last 800,000 years. Their effects, together with those of other anthropogenic drivers, have been detected throughout the climate system and are extremely likely to have been the dominant cause of the observed warming since the mid-20th century.
>
> (IPCC 2014: 4)

12.2.2 Climate change as another dimension to global uneven development

Geographically, most of the blame for global warming sits with the developed economies of the global North, with Europe and North America accounting for 70 per cent of total carbon emissions since 1850 (ibid.). This raises a critical dilemma for global economic development. Less developed economies from the global

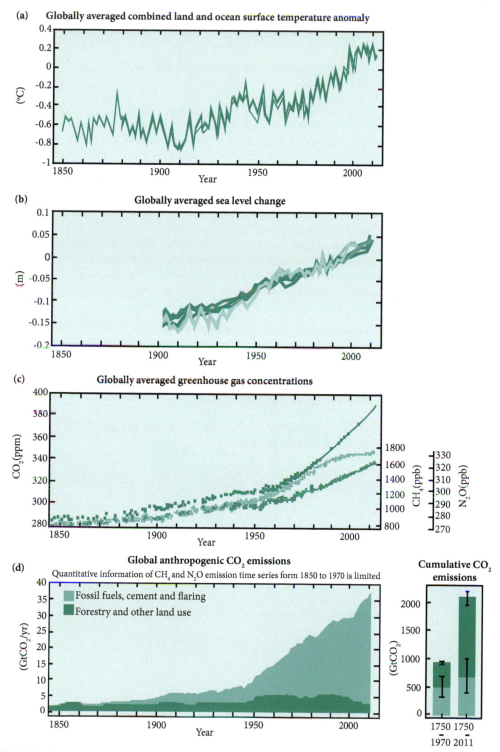

Figure 12.1 The link between industrialisation, fossil fuels and global warming

Source: IPCC (2015) Climate Change 2014: Synthesis Report, Intergovernmental Panel on Climate Change, Geneva, p. 3.

Box 12.1

The *Stern Review*, climate change, economics and beyond 'business as usual' (BAU)

One of the most thorough analyses of climate change from an economic perspective is the *Stern Review* (Stern 2006), commissioned by the UK Government and named after its chair, Lord Stern. It has estimated that 24 per cent of CO_2 emissions come from power generation – i.e. the burning of fossil fuels like coal, oil and gas – with transport, industry and agriculture being the other main culprits (Figure 12.2). The implications are clear, as the *Stern Review* puts it: "The scientific evidence points to increasing risks of serious, irreversible impacts from climate change associated with business-as-usual (BAU) paths for emissions" (Stern 2006: iii). In other words, there needs to be a complete rethink in the way the economy operates and a dramatic transition away from fossil fuels towards a low- or, better still, a no-carbon future.

As the report goes on to say, a BAU scenario will have serious detrimental effects for human (as well as animal and plant) life on the planet. The trebling of greenhouse gases, which will occur without a change of economic trajectory, is likely to raise average temperatures by 5 degrees Celsius that would make much of the planet uninhabitable. "Such changes would transform the physical geography of the world. A radical change in the physical geography of the world must have powerful implications for the human geography – where people live, and how they live their lives" (ibid.: iv).

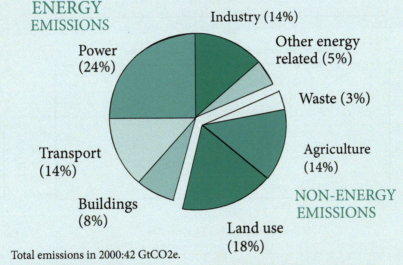

ENERGY EMISSIONS

Industry (14%)

Power (24%)

Other energy related (5%)

Waste (3%)

Transport (14%)

Agriculture (14%)

NON-ENERGY EMISSIONS

Buildings (8%)

Land use (18%)

Total emissions in 2000:42 GtCO2e.

Energy emissions are mostly CO_2 (some non-CO_2 in industry and other energy related). Non-energy emissions are CO_2 (land use) and non-CO_2 (agriculture and waste).

Figure 12.2 Producers of greenhouse gas emissions by sector
Source: Stern 2006: iv.

South, seeking to reach the same standards of wealth and prosperity enjoyed in the global North, cannot replicate similar trajectories of industrial development and resource use. Not surprisingly, this is causing some consternation amongst rapidly developing economies, with a sense of injustice arising in global climate change discussions. Why should development in the global South be constrained by problems emanating in the North? Despite such tensions, countries from the South have signed up to international climate change mitigation, including the recent Paris Accord to take action to limit global warming. China, in particular, has been promoting renewable energies strongly in recent years.

The worst effects of climate change will be felt in the global South. While rising sea levels will result in increased flood risks for coastal areas across the world, including major world cities such as New York and London, some of the most devastating consequences will be in some of the world's poorest countries. In one of the most severe cases, it is estimated that one-fifth of Bangladesh's land area could be under water by 2100. Climate change is also causing an increase in extreme weather events, such as hurricanes and floods; for example, the flooding caused by the early monsoon in Nepal, Bangladesh and India in September 2017 where over 1,000 people died and 41 million people were affected by loss of home and livelihood according to Oxfam (see Figure 12.3).

More broadly, climate change is felt through the effects of the melting of glaciers in the Himalayan region and Andes for flood risk and reduced water supplies on the Indian subcontinent, China and South America; declining crop yields and the increased threat of hunger and famine for millions in Africa; and the increased threat of water-borne diseases such as malaria in tropical and sub-tropical regions. Because so many developing world countries remain reliant on agriculture, the threat to livelihoods produced by greater climatic instability could be particularly ruinous for the livelihoods of many rural communities.

In summary, there is an ecological and climate debt owed by the countries of the global North to those of the global South, which forms an important part of the demands of environmental and social movements for what is termed 'climate justice' (Chatterton *et al.* 2013). There is what the late Doreen Massey has termed a 'geography of responsibility' around climate change (Massey 2004), whereby one set of places is the source for broader spatial processes that have devastating effects elsewhere.

12.2.3 The political economy of climate change: going beyond 'business as usual' (BAU)?

While the broad consensus among academics and informed expert opinion is that some sort of transition

Figure 12.3 The devastating effects of the early monsoon in Bangladesh, September 2017
Source: Getty Images.

in economic practices is required to counteract climate change, there is considerable debate over whether this means reforms within capitalism or a transition to a very different economic system (see Chapter 13). The *Stern Review*, for example, suggests that, "despite the historical pattern and the BAU projections, the world does not need to choose between averting climate change and promoting growth and development" (Stern 2006: xi).

Others are far more critical of this position, which seems to advocate a moderated BAU agenda around markets and competition with some better regulation and the better 'pricing' of environmental costs arising from existing forms of economic development. They point to the way capitalism as a system inevitably leads to environmental destruction. The journalist and activist Naomi Klein insists that it requires a change in the economic and social system away from a rapacious capitalism, driven by growth, resource extraction and destructive competition when she talks of climate change as "a battle between capitalism and the planet" which at present "capitalism is winning hands down" (Klein 2014: 22). For some, the very term, Anthropocene, abrogates the responsibility of capitalism as an environmentally destructive social system, with the term 'Capitalocene' used by one geographer (Moore 2017) to put the blame for climate change and broader environmental crises at the door of capitalism's inexorable search for profit and accumulation of resources.

There are strong arguments that a return to growth and BAU is incompatible with tackling climate change and the ecological sustainability of a finite planet (Jackson 2009). Radical ecological economists have called for a new set of economic relations around 'de-growth' (Latouche 2003). While there is an appreciation that de-growth does not mean negative growth, but rather a fundamental transformation in the "array of analysis, propositions and principles guiding the economy" (Martinez-Alier *et al.* 2010: 1742), there is as yet little consensus around what kinds of practices and

mechanisms will be required to establish such an economy, or how we get there.

12.3 A critical economic geography of the environment

In grappling with these issues, critical geographical approaches to economy–environment interactions have broadly tended to be of three kinds (see Bakker and Bridge 2006). The first is an applied tradition of environmental management from the 1970s onwards with a remit to achieve better and more sustainable forms of management of the earth and its resources (Table 12.1). This tradition tends to regard nature passively as something to be managed and controlled and reflects a mainstream concern with how existing market mechanisms and thinking can better utilise natural resources for economic development. Although there is increased attention to sustainability issues, these tend to be framed around an environmental economics tradition of costing the 'externalities' produced by climate change. Here, we focus largely on the second and third perspectives, linking to key theoretical perspectives in economic geography (Chapter 2).

12.3.1 Thinking politically about the economy–environment dialectic

A second set of approaches comes variously under the umbrella of **political economy** and **political ecology** perspectives (Smith 1990; Robbins 2012; Castree 2015; Perreault *et al.* 2015) (section 2.4). These have a more 'political' sense of nature-economy-society relations, recognising that "the world around us is always already political: no ecologies may be understood outside of politics" (Sundberg and Dempsey 2014: 175). Two key themes unite these approaches. First, a concern with how processes of capitalist accumulation produce uneven spatial effects that are detrimental for both society and the environment. Second, in response to this, the articulation of an alternative politics to tackle the inequalities produced by capitalism. This brings an

---

Reflect

To what extent does tackling climate change effectively require a different kind of global economic system?

Table 12.1 Main contemporary approaches in geography to economy-nature-society relations

	Environmental management	Political economy/ecology	Socio-environment hybridities
Perspective on environment	Nature and resources passive entities to be managed and produced	Largely passive but capitalist dynamics can create catastrophic natural responses (e.g. global warming)	Environment, resources, materials actively shape interaction with economy and society
Objects of study	Better management of resources Protection of environment Exploitation of resources	Over-exploitation of nature for capital accumulation Creation of spatial-temporal regimes for capital accumulation Effects of industry and modernisation on traditional cultures, knowledge and environment Advocating an alternative politics of the environment and economy	Identifying hybrid eco-social formations Attention to emergent and imminent phenomena Focus on material and discursive aspects Emphasis on non-linearity, feedback loops and post-structural understandings
Spatial concerns	Mapping environmental destruction and resource depletion Identifying location of resources for economic development	Consequences of uneven development for people and environment Spatial production of nature through capital accumulation processes Effects of mass urbanisation Disruption of traditional cultures and traditions	Identifying new spatial assemblages and formations forged through social-environment interactions Emphasis on temporary and fluid spatial constellations Focus on deterritorialisation and reterritorialisation

additional concern with ecological, or, more recently climate, justice to augment the long-standing emphasis on social justice in Marxist approaches (Chatterton *et al.* 2013). Political ecology research has also tended to focus upon social movements and struggles against capitalist industrialisation and state modernisation processes, particularly in the global South. A growing body of research is concerned with critical resource geographies, with a particular focus on energy, water and food (e.g. Routledge *et al.* 2018).

A particularly influential political economy-inspired account is the late Neil Smith's production of nature thesis (Smith 1990). Smith introduced to geography the idea that nature itself, rather than being something external to capitalism, is itself produced, exploited and often despoiled by market forces and the pursuit of profit (see also Harvey 1996). As Smith points out, heavily implicit in Marx's work is the idea that as society evolves over time, from relatively primitive forms to industrial capitalism, nature (like labour) is increasingly enrolled into capitalist processes of value creation. Indeed, it is work and human labour itself that transforms nature as society develops more advanced and complex forms. "So completely do human societies now produce nature, that a cessation of productive labour would render enormous changes in nature, including the extinction of human nature" (Smith 1990, p. 36).

There are two important aspects of this from an economic geography perspective. The first, more obvious point, is that natural resources – such as water, forests, land, energy resources – are incorporated into production processes, not in a neutral fashion but often in a way that damages and ultimately can destroy nature, as the exchange values of capitalism are often in tension with the ecological values needed to sustain particular ecosystems. This positions climate change, and the threatened extinction of human life on the planet through global warming, within the broader process of the human transformation of nature, involving the depletion of non-sustainable resources by capitalism's rapacious thirst for growth.

Second, nature itself becomes socialised with particular landscapes becoming increasingly constructed through capitalist processes. Even the most advanced urban landscape is an example of the production of nature, through the way that land, materials and resources are absorbed in the production of the built environment in the pursuit of exchange value. This results in new forms of uneven development from which social exclusions can arise, even in cases that seem to create new pleasant green spaces for general public consumption in the city (see Box 12.2).

Box 12.2

Central Park, the production of nature and uneven development in the city

Generations of tourists have enjoyed the tranquil setting of Central Park in the middle of Manhattan, as an oasis of green space, trees and water providing respite from the hustle and bustle of the New York metropolis (Figure 12.4). Yet, some might have a few qualms if they knew the divisive history around its construction. Like many

Figure 12.4 Central Park amidst skyscrapers
Source: Getty Images.

Box 12.2 (continued)

urban green spaces created in the nineteenth century, amidst a broader landscape of rapid urbanisation and industrialisation, the aim of the park's architect, Frederick Law Olmstead, was to preserve a 'natural' landscape of rolling meadows with rock outcrops for Manhattan's citizens.

Central Park was intended to be a public space where all groups and classes could mingle in an open and democratic way. However, like many such well-intentioned attempts to blend the urban with the rural, the social with the natural, in urban planning, its production required the displacement of poorer and more marginal social groups within the city who did not fit with Olmstead's view

of civilising the industrial city. In his own words:

When purchased by the city, the southern portion of the site was already a part of its straggling suburbs, and a suburb more filthy, squalid and disgusting can hardly be imagined. A considerable number of its inhabitants were engaged in occupations which are nuisances in the eye of the law, and forbidden to be carried on so near the city. . . . During the autumn of 1857, three hundred dwellings were removed or demolished by the Commissioners of the Central Park, together with several factories, and numerous

'swill-milk' and hog-feeding establishments.

(Quoted in Taylor 1999: 439)

It is an early example of gentrification, where low-value neighbourhoods at the fringes of the city are improved and regenerated for urban 'improvement'. Although the motives of Olmstead and others were to enhance the overall quality of public space, green spaces like Central Park often have the effect of driving up land and property rents around them because of the increased desirability of such neighbourhoods, pushing out poorer groups in the process. Manhattan now has some of the highest property values on the planet.

12.3.2 Socio-economic hybridities

A third set of approaches are what we label here 'socio-environmental hybridities' (e.g. Whatmore 2002). These give equal weighting to the agency of nature with humans in the co-construction of the economic landscape. An important influence is the work of Bruno Latour (1993), which itself comes out of the sociology of science tradition (e.g. Bijker *et al.* 2011) and is interested in how power is exercised in concrete settings from a broadly **post-structuralist** perspective rather than the political economy concern with uncovering deeper underlying power relations and structures (Allen 2003). Research in energy geography that draws on these insights has emphasised the importance of the multiple interactions between society and the environment using key concepts such as materiality (Bakker and Bridge 2006) and assemblage (Haarstad and Wanvik 2016).

A key focus is in how material things – such as infrastructure, commodities like oil or natural resources like water, or even phenomena such as global warming,

nuclear disasters or earthquakes – actively co-produce landscapes. A recent pertinent example would be the 2011 disaster at the Fukushima Daiichi nuclear power plant in Japan (Figure 12.5). Here, naturally occurring phenomena – an earthquake and associated tsunami – interacted with human agency in the form of nuclear technology and management failure. The disaster was in effect co-produced by the earthquake, the failure of the company concerned (Tokyo Electric Power Company) to meet the required safety and assessment standards, which in turn led to overheating of the plant's cooling system (i.e. second-order non-human agency), and the release of dangerous radioactive material. Although there were no immediate deaths, it has been estimated that over 300 people may die from related cancers over the next decade (Ten Hoeve and Jacobson 2012). This event has had significant broader political effects, notably in Japan, but also in Germany, where governments in both countries have initiated a policy shift away from nuclear power to renewables.

Human geographers working within this broad field have a concern with how combinations of the human and non-human come together to create temporary

Figure 12.5 Aerial view of the Fukushima Daiichi Nuclear Power Plant
Source: Getty Images.

spatial fixes. In the words of a leading geographical theorist, this is "a way of thinking the social, political, economic or cultural as a relational processuality of composition and as a methodology attuned to practice, materiality and emergence" (McFarlane 2011: 652). This means that 'natural' or environmental entities are viewed as sets of ongoing material processes subject to transformation and change, eschewing any attempt to create large-scale and more stable concepts of human–environment interaction over time. Such insights remind us of the importance of taking the influence of material objects seriously but also of avoiding overly structural interpretations of more fluid and heterogeneous realities. In one account, drawing insights from the oil industry, emphasis is placed upon how sudden dramatic and unanticipated changes in the oil price can have massive disruptive and transformative effects on the global economy itself – particularly the period of oil shocks in the 1970s (Haarstad and Wanvik 2016). The same is true of the renewable energy sector (see section 12.5 below).

12.3.3 An open political economy perspective on nature and the environment

While we are sympathetic to approaches that challenge overly deterministic accounts of human–nature interactions, being particularly critical of tendencies to imply 'lock-in' to particular technical and developmental trajectories without a sense of the potential for rupture and upheaval in environmental systems (ibid.), we are at the same time wary of neglecting the extent to which relatively stable political and economic governance regimes of 'structured coherence' (Jessop 2006) emerge and endure over time. Whether we are at a critical (and even terminal) branching point for capitalism, because of its growing social and ecological contradictions, is another matter.

But too much focus on immanence and rupture and a failure to distinguish between underlying structural processes and different forms of agency seems to us a weakness of much of this type of theorising. For that reason, we are drawn to accounts that attempt to integrate transition dynamics in economic geography within our broader open political economy perspective. This recognises the importance of agency, uncertainty and contingency but is also attentive to the interaction between agency and inherited structures and processes (see for example Box 12.3 below).

Box 12.3

Carbon democracy and the spatial politics of energy transition

The US-based sociologist Tim Mitchell (2009; 2011) has developed an influential account of energy transition that situates an agency-sensitive account of the political and technical aspects of transition within a long-term political economy of energy. His account stresses how society and economy co-evolve with technology and key materials and resources (such as oil in this case). 'Following the carbon' is important in connecting up how resource extraction is organised, what social relations, materials, networks and connections are brought together, and what the outcomes are in terms of the organisation and distribution of social and economic benefits.

Mitchell recognises the central importance of the spatialities of energy to understanding its political economy. He emphasises the flows, concentrations and networks through which energy is produced and what these reveal about how particular groups of people and places become connected. His account emphasises both the agency of non-human materials and also the changing political and economic relations associated with how different energy regimes evolve over time with different social, political and spatial configurations (Table 12.2).

Until 1800 almost all energy to sustain human society came from renewable sources: sun for crops and grain; grassland and woodlands for firewood; conversion of wind and water for transport and machinery.

Human society was spatially quite dispersed with settlement located along waterways and close to the best agricultural land with access to adequate woodland being the main energy imperative. Much of the world beyond Europe and North America remained like this until the middle of the twentieth century.

Mass industrialisation from 1850 onwards, linked to the transition from wood to coal and latterly oil (hence the carbon era), fundamentally changed the spatial and political relations of economy and society creating what he terms the era of '**carbon democracy**'. The development of coal as the main source of power, plus the establishment of infrastructure required to extract and distribute it to the main sources of demand in growing metropolises and industrial regions across Western Europe and North America, was important politically as well as economically.

In particular, it empowered certain working-class groups who had the ability to disrupt capitalism through strikes and other forms of actions that could halt the flow of coal to factories and homes. This is what is sometimes referred to as the 'structural power' of workers (Wright 2001) and refers essentially to political capacities given to certain workers through their location at critical junctures in the geography of production and distribution networks. Reflecting this, most of the strikes that occurred in the late nineteenth and early twentieth centuries were in the mines and

transport sectors rather than manufacturing (Silver 2003). A critical aspect of Mitchell's argument here is that workers' power comes as much from the actual spatiality and materiality of coal as their own organisation. As he puts it:

Great quantities of energy now flowed along very narrow channels. Large numbers of workers had to be concentrated at the main junctions of these channels. Their position and concentration gave them, at certain moments, a new kind of political power . . . from the extraordinary concentrations of carbon energy whose flow they could now slow, disrupt or cut off.

(Mitchell 2009: 403)

Carbon democracy therefore emerges in the form of the mass democracies of Europe and North America in the twentieth century, as workers are able to challenge established elites, demanding a redistribution of wealth and resources. Notably, this was a restricted form of democracy in the global North running alongside imperialism and colonisation of much of the global South.

By the 1920s a new regime based on oil (which is gradually replacing coal as the primary fuel in the global economy) is emerging which disrupts and transforms these power relations. The emergence of major oil multinationals, the development of the Middle Eastern oil fields, the

Box 12.3 (continued)

Table 12.2 Carbon democracy and energy transition

Energy regime	Intrinsic properties of energy	Social + political configurations	Geographical features
Pre-carbon	Widely available solar, water and woodlands but low energy capacity	Low levels of industrialisation, agriculture dominant and traditional feudal undemocratic power relations	Economic activity widely dispersed but concentration close to resources such as water, fertile agricultural land and woodlands
Carbon democracy 1 (coal)	Highly concentrated underground stores of energy. Requires considerable effort and labour to extract and distribute. Empowers labour	Growth of coal and transport-based trade unions 'Triple Alliance' Popular demands produce social reforms and welfare states	Heavily concentrated in particular regions Extraction and distribution require new infrastructures and geographies linking coalfields and growing industrial + metropolitan regions though still predominantly national production and distribution systems
Carbon democracy 2 (oil)	Underground concentrations but fluid and more portable, requiring less labour to extract than coal, more mechanised + technology driven	Easier for managerial supervision and control. Less easy (though still possible) for labour and social disruptions at key hubs (terminals, refineries). New geopolitical tensions emerge	New world regions becoming prominent (e.g. Middle East, South America, West Africa, North Sea). Globalised distribution networks of pipelines and shipping
Post-carbon	More diverse materials, less concentrated and more networked forms, requiring new decentred infrastructures	Threat to carbon interests (oil companies, trade unions and oil producing countries). Shift from geopolitical tensions to localised conflicts over renewable resources (e.g. water, wind, marine)?	Networked decentralised and dispersed geographies have the potential to re-emerge around more localised forms

Source: Derived from Mitchell 2009.

geopolitical extension of US and allied interests into these areas after the Second World War all lead to a different form of carbon democracy. The power of the coal and transport workers is much reduced – indeed US support for oil in Western Europe is aimed at neutering radical coal-based trade unions – although the growth of unions in the automotive industry in major cities creates new forms of worker power.

New political tensions and crises developed reflecting the particular materialities of oil's global supply networks and new sources of vulnerability to political action at key junctions and hubs. The 1956 Suez Crisis, following the nationalisation by Egypt of the Suez Canal, and the 1970s oil crisis, arising in part from conflict between Israel and Arab states in the Middle East, are emblematic of these new geopolitical realities. Mitchell's

Reflect

How does a serious consideration of the agency of nature complicate our understandings of climate change and its social and economic effects?

12.4 Sustainability transitions: the multi-level perspective and beyond

12.4.1 The multi-level perspective (MLP)

Enacting a sustainability transition is becoming an important area of concern for economic geographers, who recognise the possibility that a low-carbon economy is likely to lead to new spatial configurations of economic activity and a recasting of relations between cities and regions (e.g. Truffer and Coenen 2012; Essletzbichler 2012; Gibbs and O'Neill 2017). One very influential framework is the **multi-level perspective** (**MLP**) (Geels 2002), which focuses upon how technological change comes about and what mechanisms and processes shape the transition between technological regimes (e.g. from sail boats to steam ships or typewriters to computers). As one of its leading proponents says: "technology, of itself, has no power, does nothing. Only in association with human agency, social structures and organisations does technology fulfill functions" (Geels 2002: 1257).

Interpreting transition is not just about the replacement of one technology by a more efficient and productive one, but understanding how technological

processes interact with social ones, which leads to changing social customs, practices and behaviour and their regulation. In this respect, three key levels of analysis are identified (Figure 12.6): niches (micro), defined as protected spaces which act as test beds for innovative ideas and technologies; **technological regimes** (meso), comprised of established institutions, technologies, rules and practices; and wider landscapes (macro) structured by social values, cultural norms, natural systems and macroeconomic frameworks (ibid.).

New technological pathways can only be realised by actors capable of overcoming numerous barriers and blockages, which represent path-dependent existing ways of being and operating. These barriers can be both material – in the sense of existing technologies and their materialities and infrastructures and the ways these prevent transition – and also socio-political, in the sense of inherited social practices, routines, customs and existing modes of practice that can be resistant to change and innovation.

Resistance may also reflect vested interests that benefit from existing socio-technical regimes evident in Mitchell's analysis, above. While human-induced climate change and its devastating consequences for the future of the planet are widely acknowledged, and combating these effects is an accepted policy priority at global, national and local scales, there are still important social and political interests seeking to block transition. Powerful vested interests associated with established technological regimes – most notably the multinational oil industry, firms and workers operating in coal mining and related sectors – exist that are capable of influencing public opinion and government policy to shape transition pathways and even block transition in some cases.

In the US, for example, the Koch brothers, whose family business made its fortune from fossil fuels, have

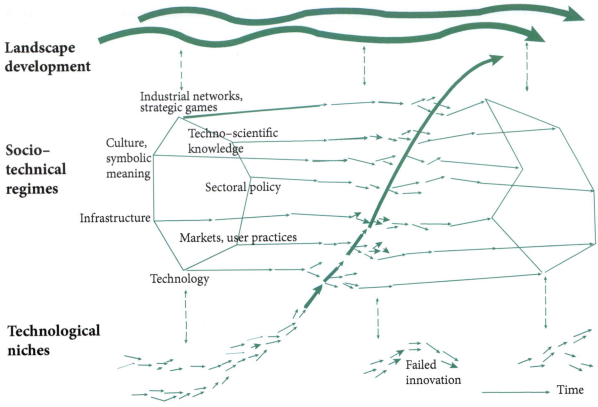

Landscape
development

Industrial networks,
strategic games

Techno–scientific
knowledge

Socio-
technical
regimes

Culture,
symbolic
meaning

Sectoral policy

Infrastructure

Markets, user practices

Technology

Technological
niches

Failed
innovation

Time

Figure 12.6 The multi-level transitions perspective
Source: Geels 2002: Figure 5.

spent over $100 million dollars funding climate change denial groups and foundations since the late 1990s to influence media and popular opinion (see Greenpeace: www.greenpeace.org/usa/global-warming/climate-deniers/koch-industries/). Such activities have helped to fuel a growing scepticism in the US despite the overwhelming scientific evidence, with the proportion of Republicans believing that global warming is an increasing threat actually falling from just half in the early 2000s to less than 30 per cent by 2010 (McCright and Dunlap 2011).

Such developments helped to provide the deeper structural context for the arrival of President Trump in the White House, and his refusal to commit the US to the Paris Agreement. Even in countries such as Germany, where national policy has shifted decisively towards a post-carbon transition, there are still setbacks occasionally as entrenched interests attempt to block or subvert green initiatives. The most recent example is the

success of the coal lobby to block a federal government proposal to impose a higher carbon levy on older coal-fired power stations because of objections by the main coal producing regions of North Rhine-Westphalia, Brandenburg and Saxony (Cumbers 2016).

Effecting a sustainability transition needs more than just confronting powerful established interests; it also requires developing new institutional, regulatory and infrastructural mechanisms that shift both individual behaviours and established societal norms and practices away from a dependence on fossil fuels. Most notably, this means moving from economies centred upon individualised consumption habits, symbolised by the gas-guzzling automobile and suburban westernised lifestyles, towards very different modes of living with dramatic spatial consequences across a range of different areas from energy consumption, housing, transportation and even the way we produce and consume food (see Box 13.5).

12.4.3 Facilitating sustainability transitions

However, incumbent interests under one socio-technical regime can successfully reposition and realign themselves with new insurgent technologies and practices; indeed this may be a crucial element in developing momentum towards a successful transition. An interesting example here would be the growing number of automobile companies announcing their intention to shift completely away from petrol and diesel engine vehicles to electrical ones – most recently the Swedish car maker Volvo which announced that all its cars would be electric powered from 2019 onwards.

The MLP approach is suggestive of the tensions that exist between creating the conditions necessary to sustain economic development in a stable manner, and the necessity for innovation, adaptation and diversification if economy, society and indeed the natural environment are to successfully evolve as conditions change (Hodgson 1999). On the one hand, for technologies to become broadly established, there is a need for commonly accepted norms, rules and patterns of behaviour to exist, while successful evolution requires space and protection for niches of experimentation. In a healthy and diverse economy, 'hopeful monstrosities' (Geels 2002) or 'mindful deviation' (Garud and Karnoe 2003) should not just be permitted but positively encouraged.

Shifting trajectories also requires the kinds of niche experimentation with alternative technologies and practices that are outside the mainstream, and state support for these against powerful and establishment incumbent technologies and actors. An interesting success story in this respect is the Danish wind turbine sector, which emerged to be a world leader in the 1980s and 1990s (Box 12.4), capturing 50 per cent of the world market for wind turbine manufacture (Cumbers 2012). More broadly, the country has been a global leader in the transition away from carbon towards renewable energy. A combination of state support, grassroots political mobilisation against nuclear power, and a cooperative approach to economic development meant that the country has led the way in a transition from a dependence on oil and gas – imported foreign oil met 90 per cent of its energy needs in the 1970s (DEA 2010) – to a situation in which renewable energy now accounts (as of 2014) for 28.5 per cent of total energy consumption (State Of Green, undated). The country has ambitious plans to be completely independent of fossil fuels by 2050.

Box 12.4

How did Denmark's wind turbine sector become a world leader? A triumph of 'bricolage' over 'breakthrough' in innovation processes

As a small northern European country of 5.7 million people, Denmark is an unlikely global leader in the green energy sector. Yet, it has experienced dramatic success in outperforming much larger and better-resourced countries, notably in the growth of its wind turbine sector. How in particular did it establish a significant technological lead over the much larger and financially superior US industry during the 1980s and 1990s to capture a 50 per cent share of world markets and 20,000 jobs (DEA 2010)?

Different phases and actors can be identified. Following the MLP approach, early wind turbine experiments were undertaken by a few enthusiasts in rural areas of western Denmark from the 1950s through to the 1970s (Andersen and Drejer 2008). This attracted the attention of agricultural machinery manufacturers like Vestas who identified potential market diversification opportunities, particularly as the broader global energy situation began to change.

In the wake of the 1970s crisis, concerns about the country's foreign oil dependence and grassroots political mobilisation against nuclear power, Danish government policy shifted the socio-technical regime dramatically in favour of encouraging and subsidising renewable energy (Cumbers 2012: chapter 9). Strong government support through the 1980s and 1990s – including 30 per cent funding for investment in wind turbines and a feed-in tariff (FIT) forcing electricity distribution

Box 12.4 (continued)

companies to purchase renewable energy – provided an important and stable operating environment for the fledgling industry.

While these were all important factors, the key trigger for success according to one influential account was a system of localised collective learning and innovation that underpinned the development of wind turbine technology and manufacture (Garud and Karnoe 2003). In contrast to the US which pursued a high-tech approach to wind turbine development, driven by engineering science

and aerospace concepts, the Danish approach was remarkably low-tech, and driven by more incremental innovation, as designers adapted equipment from agricultural practices. A crucial difference was the way Danish innovation was embedded in strong collaborative relationships between wind turbine users, designers and producers with strong ongoing feedback processes helping to amend and adapt.

This trajectory has been referred to as a "bricolage" process by Garud and Karnoe (2003: 284), which they

contrast with the US "breakthrough" strategy of focusing policy support on high-end technologically advanced designs without similar collaborative processes between users and producers that would incorporate the important process of market testing and product revision, or the kinds of 'infant industry' state support apparent in Denmark. US firms did not respond well to setbacks and failures, lacking the collective learning traditions of the Danish wind power sector, while not benefiting from the Danish state's **infant industry** strategy (section 5.3.3).

12.4.4 Putting space into sustainability transitions

One of the key criticisms of the MLP is that it lacks a critical spatial focus, with an implicit emphasis upon the national level (Gibbs and O'Neill 2017). This has partly been addressed by recognising the importance of the local scale for niche experimentation that can then be scaled up to national and even international scales (e.g. Truffer and Coenen 2012). There is considerable interest, in particular, in the role that cities might play in **sustainability transitions** (e.g. Bulkeley and Betsill 2005; Castan Broto *et al.* 2010; Rutherford and Coutard 2014). Given that the local is the scale of everyday life, urban political actors can play an important role in tackling climate change through their ability to develop an integrated approach to key infrastructure such as housing, transport, energy and water.

In some contexts, where national-level governance frameworks are lagging in tackling climate change, or frustrated by vested interests, city actors often develop their own autonomous strategies. For example, in the United States a new coalition of cities such as Los Angeles, Pittsburgh, Atlanta and others and states such as California, New York and Washington committed to developing their own targets on greenhouse gas emissions as part of the UN process and the Paris Climate

Accord (Tabuchi and Fountain 2017). But city and local initiatives remain embedded within and partially constrained by broader multi-scalar governance frameworks, which can be critical in facilitating or frustrating sustainability strategies if national state actors fail to effectively provide support (see section 12.5).

A more fundamental criticism of the MLP is that the approach emphasises the temporal nature of transition while not appreciating the extent to which transition is a fundamentally geographical process (Bridge *et al.* 2013). Most evidently, shifting beyond a carbon-based economy towards a renewable one is likely to need dramatic changes in the spatial organisation of energy, requiring more decentralised and dispersed networks and infrastructures than the more centralised forms of carbon energy (see Box 12.3). It might also lead to broader spatial reconfigurations of the economy and society to adjust and adapt to the requirements of transition across a diverse range of sectors and competences (ibid.). As with all technological shifts, there will be massive disruption associated with the destruction of carbon-based industrial landscapes, and the movement towards new green economy industries, infrastructure and materials. This will involve both scalar recalibration of the relations between the local, national and global, as well as new forms of uneven development between places, themes we explore below in section 12.5.

The transition literature is also silent on the different forms that a sustainable green economy should take (Gibbs and O'Neill 2017). As noted earlier, there are significant differences in perspective between business-as-usual arguments which emphasise how the green economy generates new sources of local and national competitive advantage, and de-growth arguments based upon an entirely different set of economic values (see Table 12.3).

Reflect

How does the MLP perspective advance our understanding of sustainability transitions? What additional value would a geographical sensibility bring to the MLP?

12.5 The emerging economic geography of energy transition

The transition to a low-carbon economy is likely to have significant geographical repercussions. Despite the recent decision by US President Donald Trump to pull out of UN agreements to reduce carbon emissions, almost every other major country, including importantly the fastest-growing developing economies of the global South, China and India, is committed to the shift to a low-carbon economy. This means that there will be significant government policy pressure and incentives to shift the economy and business actors towards renewable energy, increased energy efficiency and other forms of innovation and new technology development to facilitate the transition. The threats from global warming, outlined earlier, and related problems in major urban and industrial areas, such as severe pollution arising from carbon emissions, as well as growing health concerns around emissions of other gases from diesel- and petrol-fuelled automobiles, are facilitating a major policy shift, despite the incumbent vested interests in the existing system.

12.5.1 The emerging renewable energy landscape

To illustrate, we focus here on the energy sector and the shift towards renewable energy. In this respect, it is possible to detect some emerging global trends and new industrial spaces associated with the green economy. The first key point is that renewable energy is still in its infancy and continues to be dwarfed by carbon-based and nuclear forms of energy (Figure 12.7), both of which are still increasing in some parts of the world.

Table 12.3 Discourses of the green economy

Frequently articulated in policy ↔ Rarely articulated in policy
Incremental change ↔ Transformative change
Fit and conform ↔ Stretch and transform

Conventional pro-growth/almost business as usual	Selective growth/greening the economy	Limits to growth/socioeconomic transformation
Greening as an investment opportunity	Resource efficiency	Steady-state economy
Restarting market economies	Low-carbon growth	Prosperity without growth
Green Keynesianism	Decoupling	Degrowth
Green job creation	Clean-technologies	Social well-being
Green New Deal policies	Ecological modernization	Alternative food networks
	Cleantech clusters	Eco-housing developments
	Makerspaces	

Source: Gibbs and O'Neill 2017: 164, Table 1.

This suggests that the geographical restructuring of the energy landscape is just beginning.

Wind and solar energy still only account for less than 4 per cent of global electricity production combined. However, there is likely to be significant growth in the years ahead, given the UN-agreed targets to reduce carbon-based energy and convert to renewables. Solar alone, which currently accounts for less than 1 per cent of global electricity supply, is expected to rise to 16 per cent by 2050 (Ball *et al.* 2017). Even if the progress is slower than what is required to meet the UN's goals, the push towards renewables will lead to significant market expansion and therefore potential opportunities for those places that are successful in securing competitive advantage.

There has been significant growth in investment in renewables over the past decade and a half (Figure 12.8). So far, Europe has dominated as a market for green energy, partly as a result of political commitments to tackle climate change, notably in Germany as the largest economy, but across the European Union more generally where high targets for carbon reduction, and subsequently considerable subsidy and government support for renewable energy, have been established.

More recently, other parts of the world are also seeing growth in renewable energy investment, particularly China where the national government has become increasingly committed to expanding solar and wind power in its energy mix. China has overtaken Germany, as both the leading market and the dominant global manufacturer, in the production of solar photovoltaic panels. One set of observers described its market dominance (71 per cent share of solar module production in 2016), as a mixture of "aggressive government support and rough-and-tumble entrepreneurialism" (Ball *et al.* 2017: 29).

Government policy and political shifts can have massive effects in both encouraging and deterring investment, reminding us of the ability of powerful state actors to effectively make markets and play leading roles in the process of economic development more generally (see section 5.3.3). It is particularly evident in the renewable energy sector where only effective state policy – particularly its ability as a strategic economic actor to encourage nascent technologies and invest for the longer term (section 5.3.3) – can overcome the cost advantages of existing technologies such as oil, coal and nuclear.

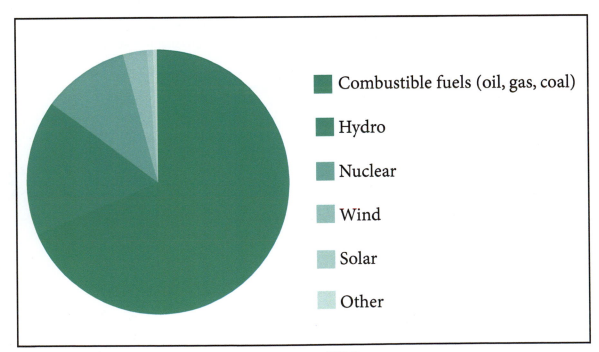

Figure 12.7 Global electricity production by fuel type 2014
Source: UN Energy Statistics.

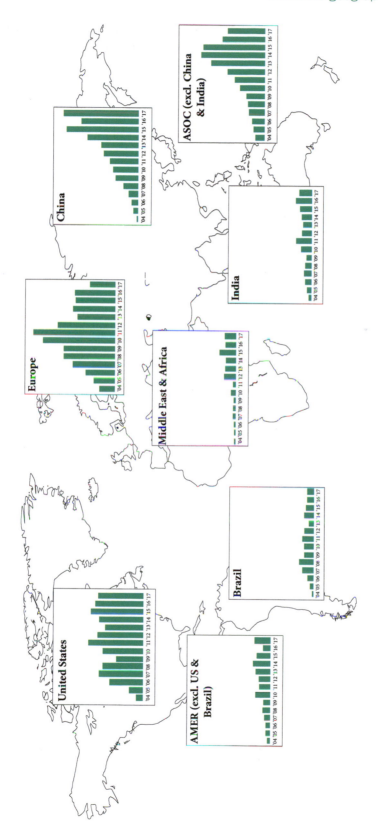

New investment volume adjusts for re-invested equity. Total values include estimates for undisclosed deals

Figure 12.8 Global new investment in renewable energy by region, 2004–16 ($Bn)

Source: Frankfurt School 2017: 22.

12.5.2 Creating local and regional competitive advantage from the post-carbon transition

While cities and regions have been identified as key actors in encouraging sustainability transitions, some will be better placed than others to benefit from a shift towards a green economy. Not all places will have the resources, skills, capacity or political agency to successfully adapt their local economies to broader transition dynamics. A key issue becomes the way in which local and regional development actors are not only successful in responding to broader external changes (Truffer and Coenen 2012), but also have the capacity to actually shape those changes themselves in creating new development directions (Dawley *et al.* 2015).

A particularly crucial nexus regards the relations between local and national governance. This is most evident in China, which by any standards has been remarkably successful in stimulating renewable industries and fostering local development. Having prioritised solar energy as a key growth sector from 2005 onwards as part of its five-year plan, central government set policy goals and encouraged competition between local governments for resources to develop industrial clusters (Chen 2016). The strategy was effective initially in allowing Chinese companies to enter and dominate global markets in the 2000s. Financial investment from both government and private foreign capital allowed the development of large Chinese solar panel manufacturers, competing on the basis of massive economies of scale compared to foreign competitors, and taking advantage of policy-led market growth in both Europe and North America.

However, the limits of this growth-led strategy became evident following the financial crisis and the collapse of European markets. China was faced with a massive oversupply, leading to the bankruptcy of two of its largest firms, Suntech and LDK in 2013. This policy approach encouraged a type of local **developmental state**, where regional governments encouraged their own industry champions and vertically integrated complexes, independent of a broader national industrial strategy (Chen 2016). This maximised job

generation and the growth of a local tax base, replicating broader Chinese success in penetrating export markets by setting up basic processing and assembly operations at low cost. Concerns are now being raised about the flexibility and adaptability of this approach as market conditions change. The Chinese government has ambitious plans to stimulate more research and development and high-tech activities in the sector, at the same time as trying to expand domestic markets. So far, there has been little coherence between industrial strategy and energy policy, with many regional governments continuing to expand nuclear and even coal-fired power stations.

As alluded to earlier, Denmark represents an interesting contrast with China (see Table 12.4). The country has a thriving green economy, with 59,000 employees, or an estimated 2.8 per cent of total employment (one of the highest in the world), accounting for 7 per cent of total Danish exports, of which renewable energy in the form of wind turbines represents 86 per cent (Danmarks Statistik 2014). Driven by both government policy and an openness to market competition, the economic geography of Denmark's green economy has a pronounced **spatial division of labour** in which the capital Copenhagen tends to dominate the service-based activities while the manufacturing sector (especially wind turbine production) is concentrated on the mainland on the Jutland peninsula with a particularly pronounced clustering in the central region.

Within this, there is a further spatial division of labour with the headquarters and research and development functions of major players such as Vestas and Siemens located in the second-largest city, Aarhus, on the east coast, and some of the basic production facilities in remoter rural areas. A downturn in the sector in the wake of the financial crisis produced some restructuring for the latter areas with the closure of four manufacturing plants and the loss of 3,000 jobs (Cornett and Sorensen 2011). While Vestas continues to prosper, and has developed its own global production network, maintaining its lead in world markets despite growing competition from other countries (Figure 12.9), there are fears for the sustainability of the more peripheral regions in Denmark and routine

Table 12.4 The emergent geographical political economy of sustainability transition: a comparison of Danish and Chinese success

	Denmark	China
Sectoral success	Dominance of wind turbine manufacture	Dominance of market for pv solar panels
Key actors	Central state, specialist technical/ scientific communities, locally embedded entrepreneurs, collaborative business institutions	Central state, local state + established private manufacturing capital
Multi-scalar dynamics	Localised and grassroots entrepreneurialism; effective political mobilisation and coalition building around green technologies, aided by state legislative and institutional market-making support infrastructure	Top-down developmentalist state allied to considerable local governance autonomy to pick winners; lack of overall systemic governance coherence
Competitive advantage	Strong, dispersed but collective learning tendencies; tolerance of divergent and disruptive innovation	Low cost + efficient economies of scale; availability of finance for large-scale investment, patient capital
Key to initial success	Effective and stable state support + tradition of open knowledge sharing in local entrepreneurial ecosystem	Strong state capacities to set targets + existing export-driven manufacturing base
Longer-term development issues – emergent tensions	Retaining core position within innovation and technology networks without financial power + capacity of larger states; reliance on larger states' regulatory and market access regimes; ability to create new sources of employment through retaining innovation lead	Over-dependence on export markets + other countries' regulatory and open trade policies; lack of effective 'exit' strategies to produce smart productivities; over-reliance on cost-based competition v ability to move up supply chain to knowledge-based competitive advantage; ability to successfully generate domestic demand for its manufacturing sectors
Relational positioning vis-à-vis global economy	Weak global market power, dependent on continuing EU market and regulatory support for renewables, importance of securing good continuing market access to US, China and emerging markets in Africa, Asia + Latin America	Strong global market power but needs to successfully manage trade tensions with other powerful blocs; needs to create successful and stable relations with emergent markets

manufacturing jobs may move to lower-cost locations in key markets such as southern Europe and China in the years ahead.

It remains an open question whether Denmark can maintain its position of global leadership in the face of strong competition from much larger and more powerful states such as Germany and the US, and in the global South from China and India, which are both intent on maximising the economic development potential of their own renewable sectors. Of the other

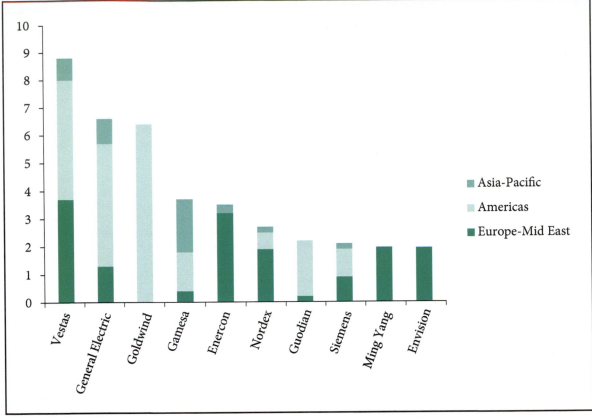

Figure 12.9 Top ten onshore wind manufacturers with region of manufacture
Source: Bloomberg New Energy: https://about.bnef.com/blog/vestas-reclaims-top-spot-annual-ranking-wind-turbine-makers/.

companies in the top ten, one is American (General Electric); three are German (Enercon, Nordex and Siemens); one Spanish (Gamesa); and the remaining four Chinese, including the third largest, Goldwind. The dominance of German and Chinese firms in particular may signal the shape of things to come in terms of the geography of global renewables manufacturing.

In their different ways, the success of China, Denmark and even Germany in establishing a significant presence in terms of local clusters, employment and leading firms has been predicated on the state at the national level playing a critical developmental role. A common feature in all three is the way national governments provide long-term and sustained policy support and subsidies for fledgling 'infant industries' such as solar and wind power – FIT schemes that guarantee

a price and market share are effectively creating market spaces. The situation can be contrasted with the UK, where successive governments' commitment to the rapid development of renewable energy markets has not yet been matched by the growth of indigenous industrial capacity in sub-sectors such as offshore wind (see Box 12.5).

Reflect

What strategies are most likely to be successful in creating sustainable local renewables clusters? What development tensions are evident in multi-scalar governance dynamics?

Box 12.5

Market-led path creation: the UK's offshore wind experience

As an island off the north-west coast of Europe with a changeable and wet climate, the UK has immense natural advantages in renewable energy from its marine, tidal and wind resources. Committed to achieving ambitious decarbonisation targets and increasing renewable energy to 30 per cent of electricity demand by 2030, there has been a strong policy push towards this goal. The UK's offshore wind resources are particularly vast, with the country generating more offshore wind power than anywhere else in the world and accounting for 46 per cent of Europe's total installed capacity in 2015 (MacKinnon *et al.* forthcoming: 29).

The sector has experienced dramatic growth and investment: from virtually nothing in 2005 to 5.3 per cent of total electricity consumption by 2017 with a figure of over 10 per cent expected by 2020 (Department for Business, Energy and Industrial Strategy 2017: 57). This has, to date, involved a massive £9.5 billion investment programme with an additional £18 billion expected by 2021 (Labour Energy Forum 2017: 6).

Based on the country's rich natural resources, the UK government has stressed the economic opportunity represented by onshore wind for the generation of electricity. The strategic vision of government in the 2008–11 period was one of large-scale growth, supported by the Renewables Obligation (RO) as the principal support mechanism. This placed a mandatory requirement on suppliers to source an increasing proportion of electricity from renewable sources, rising to 14.4 per cent by 2015–16. Following the introduction of 'banding' in 2009, the doubling of the support level for offshore wind in response to its higher

costs of development compared to onshore wind stimulated the rapid expansion of the sector. After 2010, however, concerned about the high costs of the RO in a climate of austerity, the Conservative–Liberal Democratic Coalition Government embarked upon a process of Electricity Market Reform (EMR) which saw the RO replaced by a competitive auction system, representing a liberalisation of the support regime (MacKinnon *et al.* forthcoming). Whilst the uncertainty associated with this process slowed the growth of the sector, it has been highly successful in reducing costs, with the awarded price for offshore wind projects falling by 50 per cent in the 2017 round compared to the 2015 round. This is raising the prospect of subsidy-free development in future years, reflecting a broader fall in the costs of renewable energies compared to established fossil fuel sources as technology develops and knowledge advances.

In contrast to the rapid development of its offshore wind market, the UK has lagged behind in industrial development, becoming dependent on attracting inward investment across the whole manufacturing and deployment chain. This can be contrasted with Germany and Denmark which have developed much greater industrial capacity in offshore wind, reflecting the different national varieties of capitalism (e.g. the UK's post-industrial service economy versus Germany's manufacturing-based export economy). With over 80 per cent of the value of some existing UK installations having been sourced from outside the UK, mostly by Siemens and Vestas in Germany and Denmark, the publication of an Offshore Wind Industrial Strategy in 2013 by

the UK government was an attempt to improve the levels of domestic content and better "complete the economic circle of benefits to the UK" (OWIC 2014: 3). This included the strategic aspiration of securing 50 per cent domestic content from the offshore market for future projects.

There is little sign as yet that this is bearing fruit, with a recent report suggesting that UK firms were accounting for only 18 per cent of local content and only 5 per cent of the rest of the EU offshore wind market, and with only 6,800 jobs created nationally (Labour Energy Forum 2017: 14). In manufacturing, this has fostered a dependence on the attraction of inward investment. Almost the entire capacity of the UK market has been installed using turbines produced and assembled in Germany and Denmark with no domestic turbine manufacturer. Particularly in the expansion period of 2009–12, a number of UK regions were competing to attract the 'holy grail' of a turbine manufacturer (Dawley *et al.* 2015). Here, the Humber region on England's east coast has been successful in attracting a Siemens blade plant to Hull, which opened in late 2016 providing around 1,000 jobs. This is likely to be the only turbine manufacturing plant in the UK given Siemens's dominance of the offshore market and the more limited size of the UK market relative to what was envisaged in 2009–12. UK firms have a limited presence in the production of towers and foundation structures, but have had more success in diversifying into substations and cables. With the exception of the Siemens plant, most employment takes place in deployment activities, including construction and installation, and operation and maintenance.

12.6 Summary

The environment is becoming an increasingly critical issue for economic geographers to study. While economic geography has from its early beginnings (see Chapter 2) distinguished itself from economics in its concern for the relationships between firms, people and the external environment within which the economy is located, it is only recently that economic geographers have developed more critical theoretical insights into these relations. A key element of this is to recognise the mutually constituted relationship between nature and the economy, and the social and ecological relations that this entails. Marxist accounts, notably that of the late Neil Smith, with his emphasis upon the way processes of uneven development fundamentally reshape natural environments as well as social reality, were an important step forward in the 1980s. Growing concern with the environmentally destructive nature of capitalism as a mode of production has been an important element of spatial understandings of nature–society relations. Other research in geography places increased emphasis upon the agency and materiality of nature and the environment itself, helping to enhance a relational approach (section 2.5.4).

Economic geography, in comparison with other areas of human geography, has been slow off the mark in integrating environmental dimensions into its critical analysis of the changing economic landscape. However, recent work, informed by the concern with the threat to human survival on the planet represented by climate change, has begun to pay closer attention to the spatial, social and ecological implications of capitalist forms of economic development. As we have detailed in this chapter, this is leading to a growing focus on issues surrounding sustainability transitions, a questioning of mainstream forms of economic development centred on growth, an interest in alternative forms of economy (see Chapter 13), and concern with the social and spatial consequences of a transition to a low-carbon economy. From a local and regional development perspective, areas of future study will need to address the way that sustainability transitions will recast global economic geography. This has implications for broader processes of uneven development with the likely emergence of new green industrial spaces and new strategies by economic actors to capture a share of emergent low-carbon economy sectors. As the chapter has demonstrated through the cases of China and Denmark, new economic geographies are emerging in the post-carbon economy that reflect differences in the way that places are able to embed themselves in broader global economic networks. Whether the post-carbon transition will lead to a more fundamental reshaping of political and economic relations in the global economy, and new spatial divisions of labour, with new winners and losers, remains to be seen.

Exercise

With reference to an emerging green economy sector (e.g. solar power, offshore wind, biomass, electric cars, advanced batteries), attempt to map its current economic geography and spatial division of labour. How is this economic geography organised in terms of the relations between manufacturing and services, research and development, high-tech activities and more routine manufacturing and service operations? Who are the lead actors in emergent global production networks and what are the geographical connections of power and decision making? Identify the ways in which different countries and regions within countries are becoming positioned within this emergent division of labour.

Key reading

Garud, R. and Karnoe, P. (2003) Bricolage versus break-through: distributed and embedded agency in technology entrepreneurship. *Research Policy* **32: 277–300.**
Insightful and illuminating perspective on the competitive success of the Danish wind turbine sector, which blends a business-school perspective with a socio-technical explanation.

Klein, N. (2014) *This Changes Everything: Capitalism versus the Planet.* **New York: Simon and Schuster.**
Activist and journalist Naomi Klein's polemic against the rapacious and environmentally destructive tendencies of capitalism and the need for an alternative political economy.

Martínez-Alier, J., Pascual, U., Vivien, F.-D. and Zaccai, E. (2010) Sustainable de-growth: mapping the context,

criticisms and future prospects of an emergent paradigm. *Ecological Economics* 69: 1741–7.

A foundational text for the de-growth perspective, which stresses the elements for a more socially and environmentally sensitive mode of economic development.

Mitchell, T. (2009) Carbon democracy. *Economy and Society* **38: 399–432.**

Highly innovative sociological account of energy transition which combines a political economy analysis with a more socio-technical framing to emphasise the way energy and capitalist landscapes are co-produced by human agency, political and social as well as natural and material effects.

Useful websites

http://gwec.net/
Global wind energy association which has useful updates and statistics.

www.iea.org/
Official energy arm of the OECD which produces authoritative reports on world energy trends.

www.foei.org/
www.sierraclub.org/
www.greenpeace.org.uk/
Leading environmental NGOs committed to fighting climate change.

References

Allen, J. (2003) *Lost Geographies of Power*. Oxford: Blackwell.

Andersen, P.H. and Drejer, I. (2008) Systemic innovation in a distributed network: the case of Danish wind turbines. *Strategic Organization* 6: 13–46.

Bakker, K. and Bridge, G. (2006) Material worlds? Resource geographies and the matter of nature. *Progress in Human Geography* 30 (1): 5–27.

Ball, J., Reicher, D., Sun, X. and Pollock, C. (2017) *The New Solar System: China's Evolving Solar Industry and Its Implications for Competitive Solar Power in The United States and the World*. Stanford, CA: Stanford Law and Business Schools.

Bijker, W.E., Hughes, T.P., Pinch, T. and Douglas, D.G. (2011) *The Social Construction of Technological Systems: New Directions in the Sociology and History of Technology*. Cambridge, MA: MIT Press.

Bridge, G., Bouzarovski, S., Bradshaw, M. and Eyre, N. (2013) Geographies of energy transition: space, place and the low-carbon economy. *Energy Policy* 53: 331–40.

Bulkeley, H. and Betsill, M. (2005) *Cities and Climate Change: Urban Sustainability and Global Environmental Governance*. London: Routledge.

Castan Broto, V., Hodson, M., Marvin, S. and Bulkeley, H. (2010) *Cities and Low Carbon Transitions*. London: Routledge.

Castree, N. (2015) Capitalism and the Marxist critique of political ecology. In Perreault, T., Bridge, G. and McCarthy, J. (eds) *The Routledge Handbook of Political Ecology*. Abingdon, UK: Routledge, pp. 279–92.

Chatterton, P., Featherstone, D. and Routledge, P. (2013) Articulating climate justice in Copenhagen: antagonism, the commons, and solidarity. *Antipode* 45 (3): 602–20.

Chen, T.-J. (2016) The development of China's solar photovoltaic industry: why industrial policy failed. *Cambridge Journal of Economics* 40: 755–74.

Cornett, A. and Sorensen, N.K. (2011) Regional economic aspects of the Danish Windmill Cluster: the case of the emerging off shore wind energy cluster on the west coast of Jutland. Paper delivered at *Cluster Development and Regional Transformation in an Economic Perspective*, 14th Uddevalla Symposium June 16–18, Bergamo, Italy.

Cumbers, A. (2012) *Reclaiming Public Ownership: Making Space for Economic Democracy*. London: Zed.

Cumbers, A. (2016) Remunicipalization, the low-carbon transition, and energy democracy. In Worldwatch Institute, *State of the World Report 2016*. Washington, DC: Worldwatch.

Danmarks Statistiks (2014) *59.000 Grønne Arbejdspladser*. Bulletin available at: https://dst.dk/da/Statistik/nyt/NytHtml?cid=22249.

Dawley, S., MacKinnon, D., Cumbers, A. and Pike, A. (2015) Policy activism and regional path creation: the promotion of offshore wind in North East England and Scotland. *Cambridge Journal of Regions, Economy and Society* 8 (2): 257–72.

DEA (2010) *Danish Energy Policy 1970–2010*. Copenhagen: Danish Energy Agency.

Department for Business, Energy and Industrial Strategy (2017) *Energy Trends December 2017*. London: Department for Business, Energy and Industrial Strategy.

Essletzbichler, J. (2012) Renewable energy technology and path creation: a multi-scalar approach to energy transition in the UK. *European Planning Studies* 20: 791–816.

Frankfurt School (2017) *Global Trends in Renewable Energy Investment 2017*. Frankfurt am Main: FS-UNEP Collaborating Centre. Available at: http://fs-unep-centre.org/sites/default/files/publications/globaltrendsinrenewableenergyinvestment2017.pdf.

Garud, R. and Karnoe, P. (2003) Bricolage versus breakthrough: distributed and embedded agency in technology entrepreneurship. *Research Policy* 32: 277–300.

Geels, F.W. (2002) Technological transitions as evolutionary reconfiguration processes: a multi-level perspective and a case-study. *Research Policy* 31: 1257–74.

Gibbs, D. and O'Neill, K. (2017) Future green economies and regional development: a research agenda. *Regional Studies* 51 (1): 161–73.

Haarstad, H. and Wanvik, T.I. (2016) Carbonscapes and beyond: conceptualizing the instability of oil landscapes. *Progress in Human Geography*. Online first, DOI: 10.1177/0309132516648007.

Harvey, D. (1996) *Justice, Nature and the Geography of Difference*. Oxford: Blackwell.

Hodgson, G.M. (1999) *Economics and Utopia: Why the Learning Economy is Not the End of History*. London: Routledge.

IPCC (2014) *Climate Change 2014: Synthesis Report*. Contribution of Working Groups I, II and III to the Fifth Assessment Report of the Intergovernmental Panel on Climate Change [Core Writing Team: Pachauri, R.K. and Meyer, L.A. (eds)]. Geneva: IPCC.

Jackson, T. (2009) *Prosperity without Growth? The Transition to a Sustainable Economy*. London: Earthscan.

Jessop, B. (2006) Spatial fixes, temporal fixes, and spatio-temporal fixes. In Castree, N. and Gregory, D. (eds) *David Harvey: A Critical Reader*. Oxford: Blackwell, pp. 142–66.

Klein, N. (2014) *This Changes Everything: Capitalism versus the Planet*. New York: Simon and Schuster.

Labour Energy Forum (2017) Who Owns the Wind Owns the Future. Labour Energy Forum. Available at: https://labourenergy.org/policy-briefings-reports/.

Latouche, S. (2003) Pour une société de décroissance. *Le monde diplomatique*, pp. 18–19. At www.monde-diplomatique.fr/2003/11/LATOUCHE/10651.

Latour, B. (1993) *We Have Never Been Modern*. Cambridge, MA: Harvard University Press.

MacKinnon, D., Dawley, S., Steen, M., Menzel, M.P., Karlsen, A., Sommer, P., Hansen, G.H. and Normann, H.E.

(forthcoming) Path creation, global production networks and regional development: a comparative analysis of the offshore wind sector. *Progress in Planning*.

Martínez-Alier, J., Pascual, U., Vivien, F.-D. and Zaccai, E. (2010) Sustainable de-growth: mapping the context, criticisms and future prospects of an emergent paradigm. *Ecological Economics* 69: 1741–7.

Massey, D. (2004) Geographies of responsibility. *Geografisker Annaler B* 86: 5–18.

McCright, A.M. and Dunlap, R. (2011) The politicization of climate change and polarization in the American public's views of global warming, 2001–2010. *The Sociological Quarterly* 52: 155–94.

McFarlane, C. (2011) The city as assemblage. *Environment and Planning D: Society and Space* 29: 649–71.

Mitchell, T. (2009) Carbon democracy. *Economy and Society* 38: 399–432.

Mitchell, T. (2011) *Carbon Democracy: Political Power in the Age of Oil*. New York: Verso.

Moore, J. (2017) The Capitalocene, Part I: on the nature and origins of our ecological crisis. *The Journal of Peasant Studies* 44 (3): 594–630.

OWIC (2014) *The UK Offshore Wind Supply Chain: A Review of Opportunities and Barriers*. London: Offshore Wind Industry Council (OWIC).

Perreault, T., Bridge, G. and McCarthy, J. (eds) (2015) *The Routledge Handbook of Political Ecology*. London: Routledge.

Robbins, P. (2012) *Political Ecology: A Critical Introduction*. Oxford: Wiley Blackwell.

Routledge, P., Cumbers, A. and Derickson, K. (2018) States of just transition: realizing climate justice through and against the state. *Geoforum* 88: 78–86.

Rutherford, J. and Coutard, O. (2014) Urban energy transitions: places, processes and politics of socio-technical change. *Urban Studies* 51 (7): 1353–77.

Silver, B. (2003) *Forces of Labor: Workers' Movements and Globalisation since 1870*. Cambridge: Cambridge University Press.

Smith, N. (1990) *Uneven Development*. Oxford: Blackwell.

State of Green (undated) *Basic Facts about Denmark*. At https://stateofgreen.com/en/pages/facts-about-denmark. Last accessed October 2017.

Stern, N. (2006) *Stern Review: The Economics of Climate Change*. London: HM Treasury.

Sundberg, J. and Dempsey, J. (2014) Political ecology. In Cloke, P., Crang, P. and Goodwin, M. (eds) *Introducing Human Geographies*, 3rd edition. London: Routledge.

Tabuchi, H. and Fountain, H. (2017) Bucking Trump, these cities, states and companies commit to Paris Accord. *New York Times*, 1 June. At www.nytimes.com/2017/06/01/climate/american-cities-climate-standards.html. Last accessed 16 August 2018.

Taylor, D.E. (1999) Central Park as a model for social control: urban parks, social class and leisure behavior in nineteenth-century America. *Journal of Leisure Research* 31: 420–77.

Ten Hoeve, J.E. and Jacobson, M.Z. (2012) Worldwide health effects of the Fukushima Daiichi nuclear accident. *Energy and Environmental Science* 5: 8743–57.

Truffer, B. and Coenen, L. (2012) Environmental innovation and sustainability transitions in regional studies. *Regional Studies* 46: 1–21.

Whatmore, S. (2002) *Hybrid Geographies: Natures Cultures Spaces*. London: Sage.

Wright, E.O. (2001) Working-class power, capitalist-class interests, and class compromise. *American Journal of Sociology* 105: 957–1002.

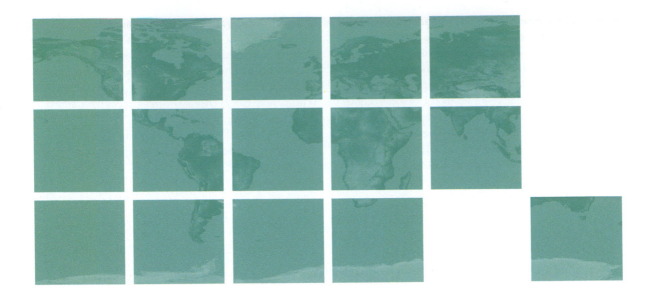

Chapter 13
Alternative economic geographies

13.1 Introduction

Given the nature of much media debate and public discourse, it would be easy to assume that **capitalism** has colonised our 'lifeworld' (Habermas 1981). At one level this is evident in the emergence of an increasingly integrated global economy dominated by capitalist social relations, commodity exchange, intensified competition and the pursuit of profit.

While some proponents of **neoliberalism** would like to extend market logics and competition into all walks of life (Mirowski 2013), human and economic relations are characterised by much more diverse values and practices. As we have stressed throughout this book, capitalism itself is constituted by a complex set of economic and geographical relations. One set of complications reflects the continued diverse nature of capitalist economies, as reflected in the **varieties of capitalism** literature (Hall and Soskice 2001). Another arises because capitalism coexists with a plethora of other forms of social relationships in shaping the economy. Whilst we think it is important to recognise the emergence of an increasingly interconnected global capitalist economy, not all the relations within the economy are capitalist, and capitalism itself takes different forms in different places.

Furthermore, the evidence that inequality, poverty and alienation are features of global capitalism has resulted in a growing interest in alternative forms of economic development that might privilege non-capitalist social relations, replacing a competitive individualism with more cooperative and collaborative forms of economic organisation. The search for alternative and more equitable forms of economic development is not new, but has been a feature of capitalism from its very early phases, spawning a wide range of movements and philosophies, from anarchism to communism to 'deep green' environmentalism. However, there has been a resurgence of interest in alternatives with the growth of an alternative 'bottom-up' **counter-globalisation movement** (section 1.2.1).

13.2 Capitalism and its alternatives

The history of industrial capitalism (since its inception in the late 1700s) is replete with resistance and oppositional movements from those dispossessed by its accumulative thrust (Harvey 2003). What equally characterises the history of capitalism, right up to the present day, is the proposal of alternatives, both within capitalism (e.g. social democracy) and by those who completely reject it and attempt to create new modes of production outside capitalist social relations (e.g. anarchist, communist, socialist, etc.).

Attempts by wealthy elites to enclose common land in England in the sixteenth and seventeenth centuries for the purpose of making profits spawned resistance movements such as the Levellers and the Diggers who were committed to more democratic and egalitarian forms of government and economy. In the nineteenth century, the development of an industrialised capitalism and factory system, and the growth of massive new urban centres to house an emergent industrial working class – or proletariat in the words of Engels and Marx – produced new forms of exploitation and alienation. Out of these conditions grew new social movements, most notably trade unions, but also other individuals or groups committed to radical social reform, concerned both with improving the lives of workers and also with constructing new forms of society, based upon non-capitalist principles of equality and solidarity.

13.2.1 The cooperative movement and alternative values to capitalism

One of the earliest reformers to argue for an alternative economics to capitalism was Robert Owen, a Welsh factory owner, who envisaged a society where workers owned and ran their own companies, rather than being exploited by a class of 'bosses'. Owen developed a series of utopian schemes from 1800 onwards seeking to put his principles into practice (Box 13.1).

Box 13.1

Robert Owen and the Cooperative Movement

The idea of employees owning and controlling their work as an alternative to the exploitation and alienation apparent under capitalist employment relations is almost as old as industrial capitalism itself. Awareness of the harsh economic and social consequences of early industrial capitalism for the workers led to experiments by social reformers such as Robert Owen in forms of employee ownership and mutualism. As an early socialist, Owen

believed firmly in the principles of social cooperation as a replacement for the socially destructive competition unleashed by capitalism. Owen had witnessed the harsh and oppressive realities of the capitalist workplace as a manager of a cotton spinning mill in Manchester. In 1800, he began to put his ideals into practice in the purchase of a cotton mill at New Lanark in Scotland. New Lanark combined more humane working conditions

with new efficient methods of production, although it was ultimately a moral order imposed from above, rather than an example of a grassroots initiative. Owen subsequently attempted to establish a number of cooperative communities in Scotland and the United States although they ultimately foundered due to a lack of funds and community support.

Nevertheless, Owen's ideals spawned a new movement, known as Owenism,

Box 13.1 (continued)

which believed in the principles of a society premised on "equal exchange" (Pollard 1971: 106) rather than exploitation. These ideas were to influence the cooperative movement, but can also be found in the more recent emergence of a Fair Trade movement. Although cooperative ideas have proved very popular and enduring, the more radical intentions of Owen and others that a cooperative society should replace a competitive one have not materialised. In practice, capitalism exists alongside cooperative movements, often absorbing them: where the latter are in competition with capitalist-run organisations, low profits have often meant the collapse or absorption of cooperative movements into the economic mainstream.

Figure 13.1 New Lanark in 1799
Source: RCAHMS Enterprises: © Royal Burgh of Lanark Museum Trust. Licensor www.scran.ac.uk.

Over the past 200 years, Owen's cooperative principles have developed into a global movement, both in the global North and South. A recent report estimated that cooperative employment now accounts for around 250 million people and 12 per cent of the employed population in the G20 countries (Roelants *et al.* 2014: 9). **Cooperatives** are a worldwide phenomenon and include some very well-established companies (see Table 13.1), and are active in virtually every major sector of the economy, from agriculture where farmers' cooperatives play a major role in many countries in the global North and South, to retailing, to financial services and even in house-building and ownership.

One of the most successful and frequently cited examples of cooperative operation is the Basque corporation Mondragon, which operates now as a transnational. Mondragon has 30,000 worker-owners, over 100 separate cooperatives in industrial, service and retail sectors, plus its own bank which is tasked with financing more cooperatives (Gibson-Graham 2006). Founded in 1941 by a catholic priest, Father Arizmendiarrieta, who was influenced by Robert Owen and

early Spanish anarchist cooperatives, its mission was to develop "democratic economic and social arrangements that might benefit all in the community and give a strong footing for postwar society" (Gibson-Graham 2006: 223).

Cooperatives are formed for different reasons, and not all sign up to the ideals espoused by Owen and Father Arizmendiarrieta. Many do not operate that differently to privately owned or corporate sector firms. Whilst most have the aim of providing their members with security from the vicissitudes of free markets (whether this is in regulating prices for farmers and consumers, or providing decent wages and employment safeguards for workers), the criticism that is often levelled at them is that they still operate within the wider capitalist economy and are therefore not insulated from the same competitive pressures facing other firms. While they may be able to cushion the effects of capitalism, they cannot work against it.

This is to take a rather deterministic view of capitalism and markets. From a more agency-centred perspective, one might argue that cooperatives – such as

Table 13.1 Top ten cooperatives by income 2014

Organisation	Country	Sector	Income ($ billion)
Groupe Crédit Agricole	France	Financial services	90.21
BVR	Germany	Financial services	70.05
Groupe BPC	France	Financial services	68.96
ENH Nonghyup	Republic of Korea	Agriculture/food	63.76
State FarmKaiser	United States	Insurance	63.73
Permanente	United States	Insurance	62.66
ACDLEC – E.Leclerc	France	Wholesale/retail	58.40
Groupe Crédit Mutuel	France	Financial services	56.54
ReWe Group	Germany	Wholesale/retail	56.42
Zenkyoren	Japan	Insurance	54.71

Source: Roelants *et al*, 2014.

the example of Mondragon – evoke different values and alternative perspectives of the economy which provide a challenge to the more rapacious forms of capitalism, associated, for example, with contemporary forms of neoliberalism. They can, in turn, shape the economy towards a different set of ethics and values. To take one example, banking cooperatives were less adversely affected by the financial crisis than the sector as a whole (Birchall and Ketilson 2009; Birchall 2013). Because cooperative banks tended not to engage in the more risky and speculative activity of other banks in real estate and securitisation (see section 4.4) prior to the financial crisis, but remained focused upon retail banking, their revenues remained relatively stable through the financial crisis and its aftermath (Groeneveld 2017).

13.2.2 Developments in anti-capitalist thought and practice

From a very early stage, there was a basic division between those seeking to reform capitalism in favour of a more socialised model (a social democratic tendency) and those seeking to overthrow capitalism through revolution. This schism emerged first in what was known, from the 1890s onwards, as the 'Second International' (of the international working class movement, the First International having been founded by Marx and others in the 1860s).

These opposing tendencies were most evident in the largest Marxist-inspired movement of the time, the German Social Democratic Party (SPD), which was split between revolutionary theorists such as Rosa Luxembourg and Karl Liebnecht, on the one hand, who argued for violent class struggle to overthrow capitalism, and more reformist positions taken by intellectuals such as Karl Kautsky and Eduard Bernstein, on the other, who argued for reform from within.

After the First World War, these tensions split the SPD and the wider international labour movement, shaping the history of class struggle and economic alternatives during the twentieth century. In the Soviet Union (after the Bolshevik Revolution in 1917), Eastern Europe (after 1945), China (after 1949) and Cuba (post-1959) successful revolutions tried to instil alternative communist-inspired models to capitalism, whilst in other countries, especially in northern Europe and Scandinavia, labour unions and social democratic parties were successful in developing more socially oriented and egalitarian forms of capitalism (at least between 1945 and the mid-1970s). By the late 1980s, the collapse of the Soviet bloc and the opening up of China to global capitalism effectively ended these alternative experiments, although most observers would accept that the more egalitarian and democratic impulses of these movements were thwarted much earlier.

With the collapse of the Soviet Union, the period after 1989 led to a phase of theorising around the concept of a single global economy, where market freedoms and individual liberty were triumphing over other older sets of social relations (Fukuyama 1992). Attempts to reform capitalism from within were also dealt a setback in the 1980s as **globalisation** and the growing dominance of neoliberal ideas in economic policymaking meant that most states retreated from social democracy in the face of increasing competition from the global South and especially China. Business elites were successful in projecting their own views onto the public policy agenda with the result that alternative and more egalitarian views of society were in retreat and competitiveness agendas in the ascendant. Indeed, as the US writer Robert McChesney (1999) puts it: "Neoliberalism's loudest message is that there is no alternative to the status quo, and that humanity has reached its highest level". Originally associated with Margaret Thatcher, British Prime Minister during the 1980s, the phrase 'There is no alternative' or the politics of TINA became a handy slogan to those advocating neoliberal free market principles in the context of globalisation.

Since the late 1990s, however, growing concerns about increased social inequality at the global and local levels and the environmental destruction of capitalism have given rise to a new set of counter-globalisation movements (section 1.2.1, Box 1.3). Through the growth of the World Social Forum (WSF) and various regional and local forums, this movement has developed an important momentum and role in promoting alternative ideas with its slogan 'Another World is Possible'. At the Mali WSF in Bamako in 2006, the movement promoted a manifesto for an alternative globalisation (see Table 13.2).

As we noted in Chapter 1, there are considerable divisions within this counter-globalisation movement, once again exhibiting the tensions between those who wish to reform the current system of economic globalisation and those 'anti-capitalists' who aim to overthrow the current order and develop a new society. Even within the latter, there are divisions between those who want to rebuild from below, through a diversity of local forms, and those that hold to a singular model of communism or socialism (Routledge and Cumbers 2009,

---

Table 13.2 The Bamako Accord: a manifesto for an alternative globalisation

1. The cancellation of the debt of countries in the global South.

2. Implementation of the Tobin Tax on financial speculation.

3. The dismantling of tax havens.

4. The implementation of basic rights to employment, welfare and a decent pension and equality in this regard for men and women.

5. Rejection of free trade and implementation of fair trade and environmentally sound trade principles.

6. Guarantees of national sovereignty over agricultural production, rural development and food policy.

7. Outlawing of knowledge patenting on living organisms and privatisation of common goods, especially water.

8. Fight by means of public policies against all kinds of discrimination, sexism, xenophobia, anti-semitism and racism.

9. Urgent action to address climate change, including the development of an alternative model for energy efficiency and democratic control of natural resources.

10. Dismantling of foreign military bases except those under UN supervision.

11. Freedom of information for individuals, the creation of a more democratic media and controls on the operation of major conglomerates.

12. Reform and democratisation of global institutions incorporating institutions such as the World Bank and IMF under the control of the UN.

Source: Derived from Routledge and Cumbers 2009: 195.

chapter 7). The Bamako Manifesto itself was the subject of considerable controversy between its supporters who wanted to promote a more global vision and those who maintained the primacy of local autonomy over alternatives.

13.2.3 Variety, diversity and uneven development

Although many writers correctly talk of the predatory and competitive forces that drive capitalism in the form of 'primitive accumulation' or 'accumulation by dispossession' (Harvey 2003), as it seeks to impose itself as a system over other forms of social relations, this tendency is at the same time a rather schizophrenic one. One of the hallmarks of capitalism is its ability to coexist with, and even adapt, other forms of social relations into its operation, resulting in the development of diverse forms of economy emerging in practice. In this regard, we would agree with the institutional economist Geoff Hodgson in recognising the principle of variety that is at work in shaping economic life (Hodgson 1999).

The literature on varieties or national systems of capitalism (section 5.4) provides a conceptual handle on this variation, demonstrating how sets of national institutions structure economic performance and development (for example Amable 2003; Hall and Soskice 2001), though economic geographers have been slow to engage with this work in comparative political economy and economic sociology. Such theories emphasise the difference between coordinated-market economies (for example, Germany and Japan) with their 'long-term, structural relationships', and the decentralised and short-termist liberal-market economies such as the US and UK (Peck and Theodore 2007: 736). The later varieties of capitalism approach favoured by Hall and Soskice (2001), amongst others, positions economies on a continuum between coordinated-market economies and liberal-market economies in which institutions adapt to complement one another. At the same time, the 'varieties' approach is characterised by some serious limitations in terms of how it conceptualises institutional variation within capitalism. These include its preoccupation with the national scale and neglect of

regional and local institutions and its tendency to conceive of change as a process of convergence to a pre-defined (neo) liberal market model (Peck and Theodore 2007).

As we have emphasised throughout this book, another key feature of capitalism is **uneven development** (section 1.2.2). According to the Marxist geographer Neil Smith (1984), there are two opposing tendencies of equalisation and differentiation at work here. The former involves the attempt to incorporate people, resources and spaces into the orbit of capitalism in the search for new spatial fixes. The latter, by contrast, refers to the rapid development of certain favoured locations and the abandonment of others, involving the devaluation or writing off of unwanted and unprofitable assets (in the form of unemployment, redundant factories and devastated old industrial landscapes) from previous rounds of accumulation (Harvey 1982). Both spaces that have yet to be fully incorporated and devalued ones can be seen as outside of the central engine of capitalist accumulation, often creating a reliance on the development of alternative economic practices.

> ### Reflect
>
> With reference to both the de-growth arguments in Chapter 12 and the anti-capitalist sentiments reflected in this chapter, assess the extent to which realistic alternatives to capitalism at the systemic level might arise. What would be the characteristics of an alternative global economy?

13.3 Alternative economic spaces

The recognition of diversity and variety, alongside the continued operation of non-capitalist spaces, within the economy has become the focus of a growing body of work in economic geography. Two of the leading thinkers in the emergent '**alternative economic spaces**' tradition are the feminist geographers Kathy Gibson and the late Julie Graham, who write together under the moniker, J.K. Gibson-Graham. Their approach is both radical and controversial in regarding the economy not as a single system dominated by capitalist imperatives, but rather as "a zone of cohabitation and contestation among multiple economic forms" (Gibson-Graham 2006: xxi).

13.3.1 Gibson-Graham and diverse economies

In a series of writings over two decades, notably their two monographs *The End of Capitalism (As We Knew It): A Feminist Critique of Political Economy* (1995) and *Post-Capitalist Politics* (2006), Gibson-Graham challenge what they view as the 'capitalocentric' views of the economy held both by mainstream economists and Marxists. Drawing upon feminist, post-colonialist and post-structuralist philosophies, their work seeks to de-centre capitalism as the predominant set of processes to highlight the varied, overlapping and competing forms and practices that constitute the economy. Though still framed by a political economy understanding of the world, class is just one of the axes through which economic power relations are played out; others being gender, race, age, caste and even more primitive social forms such as slavery and feudalism.

A key concept is **diverse economies** (section 1.3.1) whereby capitalist economic relations represent "just one particular set of economic relations situated in a vast sea of economic activity" (2006: 70). Most vividly demonstrated in their 'iceberg model' (Figure 13.2), this draws attention to the other forms of activity that exist in the economy outside of wage labour in capitalist enterprises (see Box 13.2). Indeed, it has been estimated by feminist economists that 30–50 per cent of economic activity globally is non-capitalist in the form of unpaid work in the home or non-market transactions (Gibson-Graham 2008; Ironmonger 1996), a good example being the diverse economy of childcare (Box 13.2).

Gibson-Graham are motivated by a progressive left politics which views a fixation with 'capitalocentric' (ibid.: 72) discourses as a barrier to developing more egalitarian and democratic ways of organising the economy. Their aim is to produce "a discourse of economic difference as a contribution to a politics of economic innovation" (Gibson-Graham 2008: 615). For

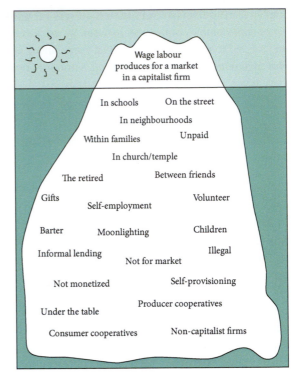

Figure 13.2 The iceberg model
Source: Drawn by Ken Byrne (Gibson-Graham 2006: 70).

them, subjugation under capitalism is as much about language, discourse and identity as it is about the hard material realities of exchange and surplus value. Recognition of the diversity that already exists in the economy opens us to anti-capitalocentric readings (ibid.: 72) where we can choose between alternative forms of social relations (collective over individualistic, cooperative over competitive) that can produce fairer economic outcomes. Their concept of the **community economy**, where markets are regulated by ethical concerns around **fair trade** and exchange, and private ownership is replaced by forms of common ownership, represents an alternative to capitalist market economics.

There are important geographical implications in their approach which emphasises openness and difference in economic processes and outcomes, rather than an underlying over-determined capitalist logic informing the relations between places. For them, the economic landscape is a mosaic of "specific geographies, histories and ethical practices" (ibid.: 71). Rather than capitalism spreading and colonising the globe in an

Box 13.2

The diverse economies and geographies of childcare

A good example of the diverse forms of social relations which exist is childcare (Table 13.3). While often undervalued (in terms of economic rewards), childcare is fundamental to the operation of the economy, when thought of as reproducing the future labour force. Yet, most of the ways that the work of childcare is organised have little to do with market relations or mechanisms (Gibson-Graham 2006: 73), although, as an earlier generation of feminists argued, this does not mean that exploitative relations are absent. The intersection of patriarchal and capitalist forms of power meant that historically, women's work in the sphere of social reproduction (such as childcare and housework) has been excluded from paid employment.

In the contemporary economy, childcare exists in both capitalist and non-capitalist forms. There are paid childcare workers who work in the home, or in private childcare centres (e.g. after-school clubs, work-based nurseries and kindergartens), but the majority of childcare takes places through other forms of exchange. The most usual is 'nonmarket' forms, through parents and grandparents sharing childcare responsibilities, friends and family offering babysitting and babysitting clubs or cooperatives set up by groups of parents. In many non-western societies, there are also traditions of children being raised by other family members than their parents. A range of alternative non-capitalistic market mechanisms such as childcare trading schemes (as part of wider local non-monetary trading networks) and informal economy payments to babysitters can also be identified. State-run and community schemes would also feature in this group where the price for childcare is not driven by the profit motive.

The diverse economy of childcare is, in turn, producing an increasingly diverse geography of places and flows. Traditional forms of childcare centred upon the home and family have given way to a diversity of spaces, including the home, the specialist childcare centre, workplace facilities and state-run nurseries. Focusing upon the space of flows through which work and social interaction is organised also draws attention to diverse geographies. Paid childcare in the home often involves the use of a transnational migrant workforce (typically on low pay and often in exploitative conditions) (Pratt 1999). Trans-local and transnational flows of labour might also feature in other forms of childcare; for example, in the use of transnational relations of care within the extended family networks of migrant communities. Other spaces – such as childcare cooperatives and parent support networks – tend to be more locally bounded. State nurseries often have strict territorial boundaries, which in common with school catchment areas can have important knock-on effects in driving up house prices in local economies that are perceived to have better-performing childcare services.

Table 13.3 Diverse identities of childcare and their geographies

	Market	Alternative market	Nonmarket
Forms of relation	- Domestic service (e.g. nanny, au-pair, housekeeper) - care worker in childcare centre - workplace	- childcare trading schemes - informal economy (e.g. babysitters)	- parents sharing - grandparents - friends, neighbours - extended family - volunteer, cooperative forms
Spatial dimensions	- Home - Workplace - Transnational migrant flows	- domestic home - community centres - state nursery - local flows of reciprocal exchange - bonded labour	- domestic home - family/friends' home - cooperative space - local and non-local family/communal networks

Source: Derived from Gibson-Graham 2006: 73.

unproblematic fashion, their approach reinforces the importance of place in shaping economic relations.

13.3.2 Alternative geographies of the corporation

Contrary to the economic geography of the 1970s and 1980s which tended to regard the geographical effects of corporate restructuring as driven by underlying economic and political imperatives, an alternative approach views spatial decision making as a more open and contested process which is shaped by competing discourses (O'Neill and Gibson-Graham 1999). Even within privately owned or joint-stock capitalist firms, Gibson-Graham argue that we cannot assume that the pursuit of surplus value dominates to the exclusion of all other motivations. Instead, there is considerable diversity in practice. Transnational corporations, for example, do not have singular dominant strategies, but are home to competing and sometimes contradictory discourses between different groups (O'Neill and Gibson-Graham 1999). In essence, their approach insists that any economic relationship has multiple sets of identities and meanings rather than being reducible to a narrow profit maximisation agenda.

From this perspective, corporate strategies can be challenged and even transformed for more ethical and progressive ends. Campaigns for corporate social responsibility and ethical trade seek to intervene in broader corporate discourses to challenge and shape executive decision making for more progressive ends. Although these campaigns have grown in prominence since the late 1990s, their effectiveness in promoting a more progressive globalisation is open to debate, with most of the evidence suggesting business as usual in terms of how TNCs function worldwide. In addition, the increasingly complex networks and webs through which corporations operate, using many different suppliers and subcontractors, allows them to render many of their environmental and labour abuses less visible. At the same time, however, the availability of the internet, which enhances activists' and campaigners' ability to highlight corporate malpractice, means that TNCs are now being subjected to a greater level of scrutiny than ever before.

Within corporations, alternative discourses are being developed by shareholder groups who are effectively developing networks of common concern between the ordinary members of pension schemes – who have dual identities as workers and shareholders – and workers and communities affected by TNC activities. Pension activism operates through intervening in the complex and contradictory spaces of corporations (O'Neill and Gibson-Graham 1999), challenging existing narratives in pursuit of more social and environmentally conscious forms of development. Whether these kinds of initiatives have much effect on the broader economic landscape is a matter of much debate. As with much of the broader issue surrounding corporate social responsibility, there are limitations to what can be done to challenge the actions of firms that operate across borders. Nevertheless, it is a reminder of different and alternative value discourses that can be generated even within the most hardened profit-seeking companies.

It is not only private pension funds that have been subject to campaigns for more ethical strategies. The Norwegian Government's Global Pension Fund, which invests the state's profits from its oil activities and is the largest pension fund in Europe – recently estimated at over £700 billion or 1 per cent of total world equity stocks (www.nbim.no/en/the-fund/, last accessed March 2017) – set up an advisory Council on Ethics which has resulted in the withdrawal of funding from a range of companies, especially those in the arms and tobacco industries. It has five criteria upon which it bases its decisions. These are:

➤ Serious or systematic human rights violations including forms of torture, murder and child exploitation;

➤ Violations of individual rights relating to wars and conflict;

➤ Environmental abuses;

➤ Larger-scale evidence of corruption;

➤ Other violation of ethical norms.

The Council has recently (as of March 2017) taken the decision to disinvest from companies that use fossil fuels as part of power generation with a new exclusion of companies that derive more than 30 per cent of their business from coal (Norges Bank 2017) (see Table 13.4). The approach is not without its contradictions,

Table 13.4 Coal-based exclusions from Norwegian Pension Fund, 2017

Company	Country
CEZ AS	Czech Republic
Eneva SA	Brazil
Great River Energy	United States
HK Electric Investments & HK Electric Investments	Hong Kong
Huadian Energy Company	China
Korea Electric Power Corp	South Korea
Malakoff Corp Bhd	Malaysia
Otter Tail Corp	United States
PGE Polska Grupa Energetyczna SA	Poland
SDIC Power Holding Companies Ltd	China

Source: Norges Bank 2017.

Box 13.3

US students divestment campaign against fossil fuels

Student and activist groups in the US have been waging an impressive campaign across university campuses against corporations in, or associated with, the production of fossil fuels, notably oil and coal. The first fossil fuel divestment (FFD) campaign, aimed at encouraging universities and higher education institutions to divest their considerable investments away from fossil fuel sectors took place in Swarthmore College, Pennsylvania with the support of local groups in the Appalachian mountains protesting against the environmental and health effects of open-cast mining. Subsequently, there have been over 400 campaigns in the US and over 500 worldwide with the movement spreading to Europe (Grady-Benson and Sarathy 2016). Success has been mixed with 25 universities divesting as a result of effective student campaigns with around the same number refusing to divest (ibid.).

Opponents of these campaigns usually claim that divestment could cause serious financial risk and that there are hypocrisies involved, given the extent to which universities still use fossil fuels for their own operations. There is typically an unwillingness by many university authorities to become involved in campaigns that they regard as overtly 'political'. However, even many of the most ardent opponents of divestment often subsequently develop their own sustainability initiatives, notably Harvard University, which, having set its face against divestment, became the first US university to sign the UN's Principles for Responsible Investment (ibid.).

What is critical about the campaign is the broader agenda of attempting to create an alternative set of values around environment, society and the economy. The student campaign is as much about challenging individualised conceptions of environmental problems that emphasise changing consumer behaviour as it is about tackling broader societal structures and values (ibid.). Animated by writers such as Naomi Klein and Bill McKibben and the environmental action group 350.Org, the emphasis is upon collective and socially just transitions away from a carbon-based economy. On its website, the Divest Student Network evokes a set of principles that not only challenge the status quo but argue for an alternative and more ethical set of economic relations, recognising that "The path to ecological sustainability requires a moral and material transformation in our relationships to land, labor and one another. Transitioning away from fossil fuels means transitioning towards justice." (www.studentsdivest.org/about, last accessed May 2017). One of the important aspects of the student campaigns against universities is to hold the latter accountable to their own non-economic values of promoting ethical and sustainable societies and creating engaged citizens.

however, for a country where oil and gas still accounts for around 55 per cent of total exports (see World Bank data at http://wits.worldbank.org/CountrySnapshot/en/NOR/textview, last accessed May 2017).

More broadly, it has been argued that divestment campaigns may not always deliver the outcomes they intend (e.g. MacAskill 2015). One objection is that divestment from concerned shareholders, or as a result of pressure from citizens' initiatives, might not necessarily lead to damage to the corporation's profitability, particularly if there are other potential investors available. For example, campaigns against fossil fuel corporations, such as that by the US student movement (see Box 13.3), have had mixed effects. Coal sectors can be badly hit because of the absence of investors for a sector that is becoming uncompetitive against other energy alternatives, whereas oil and gas companies have often found willing investors from the Middle East and China (MacAskill 2015). At the same time, targets can be highly selective and not necessarily the most appropriate companies. Because most campaigns tend to emanate from richer, western countries, typically animated by predominantly white and middle-class activist groups, they can represent a rather partial 'geography of responsibility' (Massey 2004), which at its worst represents another form of colonialism, albeit a well-intentioned one. Notwithstanding such criticisms, the advocates of these campaigns suggest that, even where they do not have a direct effect on corporate behaviour, they do raise the profile of a more responsible and ethical capitalism in important ways and can lead to longer-term changes in corporate norms and values (Grady-Benson and Sarathy 2016).

13.3.3 Alternative local economic spaces

Following Gibson-Graham's lead, a body of work by economic geographers has developed since the mid-1990s interested in alternative spaces of local economic practice (e.g. Gibson-Graham *et al.* 2013; Gomez and Helmsing 2008; Leyshon *et al.* 2003; North 2005, 2014, 2015; Seyfang 2006). The focus of these studies is on how the exclusion of some places from the capitalist economy results in the creation of alternative

local economic spaces of exchange and circulation, where relations of economic reciprocity – rather than commodification – are created to provide essential goods and services outside the free market. The philosophy in such local economic spaces runs directly counter to the integrating dynamics of globalisation; instead, the emphasis is upon creating endogenous power and capacities that allow individuals to carry out sustainable economic activities at the local level, detached from broader circuits of capital (Gomez and Helmsing 2008). The idea of decoupling local and regional economies from the broader global economy is not new (e.g. Stohr and Todling 1978), but interest has grown in response to the pressures and threats of globalisation and more recently the financial crisis, austerity policies and their punitive effects on the more vulnerable citizens (see section 5.5).

Local currencies are viewed as among the more radical initiatives in alternative local economics; they are aimed at 'short-circuiting' the global economy, creating spaces of closure that prevent leakages out of the local economy (Gomez and Helmsing 2008; North 2010). Seen as a direct response to increasing globalisation and the loss of power of local communities over their economies, local or community currencies in particular are viewed by many radical and green economists as a way of challenging the disciplinary power of national and international monetary regimes. An additional argument that has become more prominent since the financial crisis and the problems associated with the Euro (see section 4.5.2) is the idea that large single-currency regimes that range over large national or continental spaces can all too easily lead to money flowing from "poorer to richer places, condemning less favoured regions to structural poverty, or to crises like those in contemporary Ireland, Spain, Portugal and Greece" (North 2014: 249). There has therefore been something of a resurgence of the idea of more localised and contained currencies in recent years. It has been estimated that there are as many as 2,700 local currencies active in 56 countries.

One of the most celebrated examples of local currencies, known as the Red de Trueque (RT), occurred in Argentina in the wake of the 2001–02 financial crisis. Although there was a number of very small-scale local currency schemes operating prior to the crisis, it was this event that marked a critical acceleration in their usage. As the economy plummeted as a result of the broader financial crisis in Asia and Latin America, with GDP falling by 25 per cent between 1999 and 2002, growing numbers of people were made unemployed or destitute (among traditionally more prosperous middle-class groups as well as the poor). One of the government's desperate policies was to freeze bank accounts in a bid to prevent capital leaving the country. In these circumstances, people turned in desperation to alternative forms of economic exchange. Resources released from the capitalist economy, land, labour and technology were taken up in what one set of observers terms an "emergency circuit of low productivity and small-scale production" (Gomez and Helmsing 2008: 2495). Out of this situation, alternative currencies and barter networks developed across the country (North 2005). One such was the RT which at its height in 2002 had over 3 million participants (Figure 13.3).

The desperate conditions of the mainstream economy meant that many millions of people participated in this alternative economy. A system of credits (*creditos*) developed to allow people to barter goods and services at informal street markets that sprung up across the country. These markets operated through a system of 'nodes' which met at organised times in a range of places, including church halls, disused factories and car parks. North (2005) suggests that the largest of these nodes – in the western city of Mendoza – had over 36,000 members at its height. A wide variety of food and clothing, as well as services such as haircuts, were on offer in these nodes. Nodes were not territorially confined but were overlapping with people able to travel across cities and regions to exchange goods and services. Without a central coordinating body, the system somehow allowed people to exchange credits from one part of the country to another as long as the person they traded with considered their credits to be valid (North 2005). Although attempts were made to develop alternative currencies at a national scale, such as the Arbolito, many activists remained avowedly localist, seeing any attempt to 'scale up' monetary regulation as an erosion of the local participatory democracy through which local currencies could be monitored and controlled, and as prone to inflationary pressures. Over time, interest in the alternative currency declined (Figure 13.3), due to not only a number of corruption scandals involving those administering the alterative currencies, but also the recovery of the national economy (Gomez and Helmsing 2008; North 2005). There are also suggestions that business owners and the Peronist Government sought to undermine the currencies, fearful of the long-term effects on the economic mainstream (ibid.).

Credit unions are another form of alternative economic institution. They effectively operate as financial cooperatives that are owned and run by their members. Credit unions are often linked to particular community

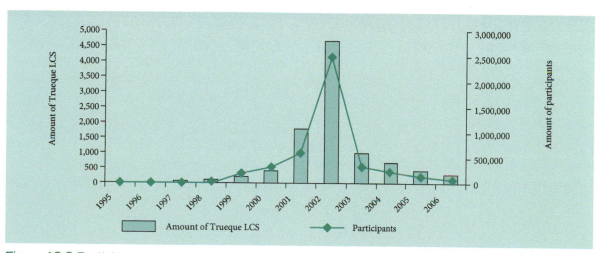

Figure 13.3 Participants and amount of Trueque local currency scheme
Source: Gomez and Helmsing 2008: 2496, Graph 1.

groups or organisations, sometimes based along ethnic or religious lines. To access credit, individuals usually must have built up a minimum of their own savings; often a ratio of three to one, in terms of loans to savings, is the maximum.

Credit unions offer the opportunity for hard-pressed individuals to save or borrow money on better terms than those of mainstream banks or 'loan sharks' of the underground economy, which are frequently the only option for people in poorer communities to access credit. They usually operate at the local or 'community' level, although they can operate at broader economic scales and often represent a response to the withdrawal of 'mainstream' economic actors such as banks and building societies from disadvantaged areas, resulting in **financial exclusion**. Economic geographers have suggested that credit unions can strengthen communities and create autonomous economic spaces for local people (Lee 1999).

The credit union movement originated in Germany in the 1850s, but by the year 2015 had reached over 60,000 worldwide in 109 countries with over 220 million members (Table 13.5). Its geography varies dramatically by world region from North America, where its penetration rate is over 40 per cent (number of credit union members divided by the economically active population), to continental Western European countries where membership is negligible, although this is probably because many people are already members of

long-established local and regional banking cooperatives, such as the Sparkasse network in Germany, which fulfils a similar role. Ireland is a notable exception with a 70 per cent penetration rate for credit unions.

13.3.4 The 'social economy' and the limits to alternative local economic practice

Although many schemes set out with the more radical intention of creating alternative models to capitalism, in practice activists find it difficult to maintain their autonomy from the mainstream economy and broader processes of uneven development. Many credit unions, for example, were incorporated into **social economy** policy agendas pursued by many centre-left governments in the 1990s and 2000s as a way of dealing with socially excluded regions and communities in North America and Western Europe. The social economy – comprised of organisations such as cooperatives, NGOs and charities that are not part of the private or public sectors of the mainstream economy – was perceived as being a way of delivering local economic and social regeneration in communities left behind by globalisation, by governments otherwise committed to neoliberal economic policies. As Amin *et al.* (2003: 48–9) noted, policies were only adopted in those "localities, communities and neighbourhoods where conventional

Table 13.5 Key facts and figures on credit unions worldwide, 2015

Number of countries involved	109
Number of credit unions	60,657
Number of members	223 million
Financial statistics:	
- savings	$1.5 billion
- loans	$1.2 billion
- assets	$1.823 billion
Penetration of credit unions by world region:	
Africa	6.8%
Asia	2.9%
Caribbean	19.9%
Europe	3.4%
Latin America	8.1%
North America	48.3%
Oceania	18.9%

Source: Derived from World Credit Union website, available at: www.woccu.org/documents/2015_Statistical_Report_WOCCU, last accessed March 2017.

economic practice is deemed to be no longer possible, practical or desirable", often serving to reinforce inequalities between places rather than address them.

Even for those schemes determined to maintain their autonomy from wider government programmes and initiatives, activists are inevitably drawn into wider economic structures and processes beyond their control. For example, Fuller and Jonas (2003: 70) found in their study of credit unions in the UK that these institutions became involved in filling the gaps left by the withdrawal of mainstream financial institutions. This has been a major concern in many western economies following the financial crisis as the onset of austerity policies (section 4.5.3, section 5.5) has led to the effective withdrawal of the state from supporting many disadvantaged communities and places. In this context, social economy activities often seem to provide a poor substitute for the state provision of essential needs and services.

One such debate has developed in recent years over the proliferation of urban agriculture and community gardening initiatives in deprived urban areas. A recent study in the US found that about one-third of all urban households are engaging in food production (McClintock 2014). For half of these, producing their own food to save money was an important motivation. More broadly, a 2012 survey in the US recorded more than 9,000 community gardens with almost 40 per cent set up in the previous five years (Lawson 2012). Some see these as having the potential to create alternative sets of more ethical, socially progressive relations around food that might also contribute to tackling climate change by creating shorter food supply chains while others see the same kinds of 'lifeboat' (i.e. a small-scale response to increasing poverty) and 'incorporation' (into the mainstream capitalist economy) dangers as in other areas of the social economy (see Box 13.4).

Box 13.4

Community gardens: cultivating new spaces of mutual self-help or doing the dirty work of neoliberalism?

The resurgence of urban food growing has been a particular phenomenon of deprived urban areas, from Detroit to Glasgow to Berlin, suggesting that it represents an important grassroots response to processes of deindustrialisation and urban decay (Crossan et al. 2016). Many old industrial cities in the US in particular are the site of innovative policies around food and community building. Places such as Cleveland, Detroit and Milwaukee have established land trusts, supported training in food growing and distribution and encouraged new community consumer outlets with the aim of creating more sustainable and healthier local food networks (Tornaghi 2014).

In a broader political climate of austerity and state retrenchment, community gardens are often presented as a progressive step in regenerating communities and places that have been left behind by both the state and capital. Finding new ways of using community resources and physical assets such as land that is derelict or vacant has been seen as important to the revival of many declining cities (Hospers 2014). Proponents even suggest that they could be part of a more fundamental shift against the social and environmental destructions wrought by capitalism by developing more progressive collective and sustainable values around food (Crossan et al. 2016; Cumbers et al. 2018). One leading commentator on the phenomenon, Kevin Morgan, has even gone so far as to identify a new urban food paradigm (Morgan 2015).

Some researchers are more sceptical about the claims being made on behalf of community gardens. Rosol, from her research in Berlin, describes them rather scathingly as "private activity in the public realm" (Rosol 2011: 249), with the implication that community gardeners are merely helping to turn unused urban space back into land that can subsequently be appropriated by capital for profit-seeking activities, paving the way for gentrification and privatisation of urban neighbourhoods. This chimes with David Harvey's deeper structural argument made earlier in the chapter about the way that land and communities are thrown out of capital accumulation processes when they become devoid of opportunities for value creation, only to be reincorporated into circuits of capital following regeneration (Harvey 2003).

Box 13.4 (continued)

Figure 13.4 (a) and (b) Alternative modes of urban living in Glasgow's community gardens
Source: Blair Cunningham.

Box 13.4 (continued)

Another criticism is that much community gardening work, while positive in helping to regenerate some more impoverished neighbourhoods, only serves to reinforce dominant neoliberal practices because such work effectively lets the local state 'off the hook' in managing and maintaining certain city spaces for collective use. Ghose and Pettigrove's study of Milwaukee's Harembee community gardens highlights the exploitative nature of community gardens which rely on volunteer labour, rather than properly paid jobs, from citizens who are already among the most impoverished in the city (Ghose and Pettigrove 2014).

Other writers take a more nuanced view, recognising that, depending on the local context, community gardens have both progressive and regressive dimensions in dealing with urban problems. While they might have unanticipated consequences in some places such as gentrification, elsewhere they can help with community generation, not least by giving back a sense of political agency and a collective ethos of self-help and mutual trust and responsibility to places that have been forgotten by the mainstream economy (McClintock 2014).

Moreover, small-scale alternatives, which are typically dependent upon a handful of activists and enthusiasts for their functioning and organisation, often encounter problems of sustainability and regeneration over time. For example, the Manchester Local Exchange Trading Scheme (LETS) studied by North (2005) dwindled from 485 members at its height in 1995 to 125 by 2001 with most of the original participants having left as a result of changing personal circumstances – such as having children, becoming unemployed and being unable to afford the participation fees or having less time to commit because of working in regular schemes. As North (2005: 227) puts it, "the resistances of everyday life meant that [people] did not maintain membership, so that, rather than being the precursor of a new localised experiment, LETS became episodic".

As the example of the Argentinian barter economy demonstrates, even where alternative practices get scaled up to the national level through local initiatives coalescing into broader solidarity networks, they are at present more useful as escape valves for people and communities when the mainstream economy enters periodic crises. In this sense, North (2005: 222) describes alternative economic spaces as "lifeboats against globalization developed by the marginal in spaces suffering from uneven capitalist development". The decline of alternative currencies in the wake of the recovery of the Argentinian national economy would appear to bear this out, although it should be noted that alternative forms of economic organising continue there to a greater degree than in other countries.

Reflect

Discuss the extent to which local alternatives can formulate different values to the capitalist mainstream. How might such alternatives be scaled up to create broader social reform?

13.4 Alternative global networks of trade and development

13.4.1 Fair trade initiatives

A different – but often complementary – approach to the more localist alternatives identified above is provided by movements that seek to intervene in global economic networks to promote fairer and more ethical forms of social relations. Since the 1970s, concerned consumers in the more developed economies of Western Europe and North America have campaigned to improve the working and living conditions of farmers and workers who are involved in global **commodity chains**. This has resulted in the growth of movements such as the Fair Trade network, the Clean Clothes Campaign – a Netherlands-based alliance of trade unions and NGOs, which exists to improve the rights of workers in the clothing and textile sectors – and the Ethical Trade Initiative, a similar UK initiative that also involves some of the major transnational retail chains.

Because many developing countries remain dependent upon one or two commodities for export, producers can be dramatically affected by sudden price fluctuations in global markets over which they have little control. Many farmers in the global South, for example, are in highly dependent power relations with transnational food or retail chains over which they have virtually no control. In contrast, ethical or fair trade networks seek to develop different and less exploitative sets of relations between producers and consumers in global trade networks. The political aspects of fair trade are evident in the following quote: "When you buy Fair Trade you are supporting our democracy" (Guillermo Vargas, Costa Rican farmer, talk to UK House of Commons 2002, cited in Morgan et al. 2006: 1).

The Fair Trade Initiative describes itself as follows:

Our mission is to connect disadvantaged producers and consumers, promote fairer trading conditions and empower producers to combat poverty, strengthen their position and take more control over their lives.

Our vision and mission will be reflected in the values by which we work as an organisation so that we ourselves set an example for the changes we seek in others. Therefore we will work collaboratively and seek to empower those who wish to be partners in our mission. Trust is a crucial factor in our work and we will be mindful of our responsibilities to those who place their trust in us. Embracing transparency and stakeholder participation is an important way that we will be accountable for our work.

(Website of the Fairtrade Labelling Organizations International, www.fairtrade.net/our_vision.html, last accessed 8 July 2010).

Thus, at one level it can be seen as quite revolutionary, going beyond Marx's famous critique of the mainstream economy's **commodity fetishism** to reveal the true nature of relations between people and places as a way of developing more equitable connections between consumers and producers.

Making the geographies of production, consumption and circulation more visible is a key theme of much recent writing in economic geography (e.g. Morgan et al. 2006; Cook et al. 2010; Crang and Cook 2012; Mansvelt 2013), and there have been detailed criticisms of the limitations of fair trade initiatives (see the excellent discussion of these issues in Cook et al. 2010). One criticism is that one set of dependency relations – between transnationals and producers – is replaced by another set, this time between producers and middle-class consumers in the West, who are able to afford to pay higher prices for fair trade products. Other research suggests that there is considerable variation in the practice of fair and ethical trade, often highly dependent upon relations in developed countries between consumers and retailers involved in ethical initiatives (Hughes et al. 2008).

There is no doubt that fair trade has expanded dramatically in recent years with sales increasing, despite the recent global crisis and recession, from $3.4 billion in 2009 to around $7.88 billion in 2016 (Fair Trade International 2017). Similarly to the critique of the alternative local economic initiatives outlined earlier, doubts are sometimes raised about the real economic significance of these schemes. For example, the above figures can be put into perspective somewhat by comparing them to the global sales of the British retail TNC, Tesco, which amounted to £50 billion in 2017 (Tesco 2017: 12).

13.4.2 The moral economy and geographies of responsibility

The sociologist Andrew Sayer (2000) uses the term **'moral economy'** in advocating an approach which replaces capitalist economic values and the exploitation of one group by another, with the recognition that everyone is entitled to certain basic economic rights (such as decent housing, food and clothing, standard of living, education, etc.), alongside a recognition of economic responsibility to others. Of course, such principles are nothing new; they have been the concern of theorists as diverse as Smith, Marx, Keynes and Polanyi. A global set of economic rights has also existed in principle since the end of the Second World War and the United Nations' Declaration of Human Rights:

> Everyone has the right to a standard of living adequate for the health and well-being of himself and of his family, including food, clothing, housing and medical care and necessary social services, and the right to security in the event of unemployment, sickness, disability, widowhood, old age or other lack of livelihood in circumstances beyond his control.
>
> (UN Declaration of Human Rights 1948, clause 25)

However, the practice of contemporary globalisation – as shaped by powerful actors such as transnational corporations, leading states and global governance institutions – has departed markedly from these ideals. The values espoused by fair trade groups are clearly motivated by alternative ideals that connect with Sayer's moral economy. With regard to the consequences of global economic processes and connections, Doreen Massey (2004: 7) similarly argues for a **geography of responsibility** in which we recognise how the "lived reality of our daily lives, invoked so often to buttress the meaningfulness of place, is in fact pretty much dispersed in its sources and its repercussions".

Drawing upon her earlier insights with regard to a 'global sense of place', Massey's point is that a more ethical and responsible globalisation is one in which we recognise how our own economic actions impact upon distant others through increasingly dense global connections. This also requires contesting certain forms of economic development in particular places if they have harmful effects on others. So, for example, developing an alternative and more egalitarian financial globalisation would mean challenging the role of London as a corporate and financial centre that is home to a global decision-making elite, and arguing for an alternative politics of the local as a means of addressing broader processes of uneven development.

13.4.3 Reconnecting places through responsible geographies

In their work on the global food sector, Morgan *et al.* (2006) distinguish between two types of social relations and geographies that are practiced within global food commodity chains: a regime of 'hard power' related to increasingly 'placeless' global production networks dominated by larger transnational retailers and agri-food companies such as Walmart, Tesco, Aldi, Cargill and Kraft Foods; and an emergent and more ethical regime, characterised by fair trade networks, a growing concern with food quality and attempts to recover 'place' by reconnecting ethical consumers with the producers of their food by making visible the geographies of food production and distribution (see Box 13.5).

Box 13.5

Making responsible connections through Fair Trade Towns

The Fair Trade concept was extended through the establishment of Fair Trade Towns, an initiative first started by an Oxfam member, Bruce Crowther. Crowther spent eight years attempting to sell the fair trade concept to cafes and restaurants in his home town of Garstang, a market town in rural Lancashire with a population of 5,000 people. A key turning point for Crowther was watching a demonstration by local farmers protesting at falling prices for their own dairy products. As he noted: "When I saw dairy farmers marching down Garstang High Street with banners saying 'we want a fair price for our bottle' I realised that we couldn't just campaign for developing countries."

Crowther recognised that local farmers in a developed country such as the UK were connected to the same relations of exploitation as developing country peasants and farmers in global food commodity chains. Subsequent support from local farmers' groups was crucial in making Garstang the first Fair Trade Town in April 2000. The movement has subsequently spread to include 632 towns in the UK and 2,016 worldwide in 30 countries.

Box 13.5 (continued)

Fair Trade Towns sign up to 5 key principles:

➤ the local council passes a resolution supporting fair trade, and agrees to serve fair trade products (for example, in meetings, offices and canteens);
➤ a range of fair trade products are available locally (targets vary from country to country);
➤ schools, workplaces, places of worship and community organisations support fair trade and use fair trade products whenever possible;

➤ media coverage and events raise awareness and understanding of fair trade across the community;
➤ a fair trade steering group representing different sectors is formed to coordinate action around the goals and develop them over the years.

There is now a European network of Fair Trade Towns, with support from the European Union, which host conferences that bring together members to communicate activities and share best practice. The ethos shared by Fair Trade Towns of encouraging their residents to use their everyday local consumption choices to benefit and bring about positive changes in the livelihoods of food producers globally connects strongly with Massey's concept of a geography of responsibility.

Source: Fair Trade Towns website: www.fairtradetowns.org/about/what-is-a-fairtrade-town/, last accessed December 2017.

Table 13.6 Mainstream and alternative economic geographies of globalisation

	Mainstream global economy	Moral/alternative global economy
Dominating economic values	Competitive advantage, free trade	Solidarity relations, fair trade, consumer quality
Underlying rationalities	Profit maximisation	Ethics of care and responsibility
Relations of power	Hard power, hierarchical and coercive	Soft power, collaborative and discursive
Organisational forms	Private ownership (though usually TNC dominated), subcontractors, home workers, child workers	Family firms, SMEs, cooperatives
Geographies	Placeless, encouraging mobility and low-cost competition between places	Re-envisaging place through politics of responsibility and reconnection
Economic outcomes	Highly unequal distribution of income with value captured by larger actors (e.g. TNCs, global retailers)	Income more evenly allocated through commodity chains with producers in global south provided with living incomes

Source: Derived from Massey 2004; Morgan *et al.* 2006; Sayer 2000.

We think that these principles can be applied more generally to theorising an alternative geography of the global economy where the profit maximisation rationalities of mainstream or neoliberal capitalism can be juxtaposed against Sayer's moral economy and Massey's geographies of responsibility (see Table 13.6). An alternative economic geography of counter-globalisation would emphasise solidarity between places within commodity chains, and fair trade, contrasting with the mainstream espousal of free trade and competitive advantage as providing the optimum economic outcome. It would also favour organisational forms that are likely to evoke alternative ethical and moral values (e.g. family firms, cooperatives) to private firms and transnational corporations driven by narrow monetary values.

The hierarchical and uneven power relations that characterise global commodity chains in the mainstream economy would be replaced by 'softer' and more decentred forms of power that would involve more collaboration and dialogue between actors to organise commodity chains and production networks. The 'placeless' landscape of many contemporary global commodity chains, in which local actors are played off against each other and are easily substitutable in a competitive race to the bottom, is replaced by an approach that re-emphasises place. Economic actors in commodity chains develop awareness of the varied and entangled geographies of their commodity chain and forge closer and more intimate connections by recognising how their own actions (as consumers, producers, traders) have impacts on distant others.

showing how the global economy is shaped by non-capitalist relations and practices, both within firms and local communities. Important work on alternative economic spaces – such as community gardens (McClintock 2014; Crossan et al. 2016) – has highlighted not only the diversity and variety which exist in the economy, but also the political limits of such alternatives. Critiques of the limits to more localist alternatives emphasise the dangers of incorporation into the mainstream capitalist economy, while others such as Sayer suggest capitalism itself can be reformed along more moral lines. In this respect, fair trade initiatives attempt to evoke more ethical relations and a more responsible set of geographies (Massey 2004) that question uneven development and the spatial imbalances that exist between places in the global economy.

Reflect

How might the concept of a geography of responsibility be applied to the relationship between places that are economically connected?

Exercise

Using a case study of a particular **fair or ethical trade**, assess how far its practice evokes an alternative economics to mainstream commodity chains. Pay attention to the connections and relations within all aspects of the commodity chain and theorise how far a geography of responsibility is or can be established in the face of broader processes of uneven development. How does the selected initiative seek to monitor and regulate processes of trade and exchange? And, how far is it successful in replacing dependency relations within mainstream networks with new and more egalitarian sets of relations? How are the relations between places reconfigured as a result?

13.5 Summary

Capitalist economic practices have never become completely dominant over other forms of social relations. The evolution of capitalism as a global economic system has always been accompanied by resistance and the articulation of alternatives. During the 1980s and 1990s, however, globalisation, the emergence of neoliberal economic policies, the collapse of concrete alternative systems in the Soviet Union and Eastern Europe and the entry of China into the global capitalist economy seemed to herald the end of history for many commentators (Fukuyama 1992). The onset of new capitalist crises in the 1990s and 2000s has exposed the limits to such thinking, whilst the emergence of alternative and more localist forms of experimentation shows the increased potential for reframing the economy away from capitalist values.

Geographers such as Gibson-Graham have also been at the forefront of critiquing capitalocentric thinking,

Key reading

Cook I. et al. (2010) Geographies of food: 'Afters'. Progress in Human Geography 35 (1): 104–20.
A really useful discussion based on an online blog discussion between Cook and others that debates the ethics and politics of alternative food production systems. A very accessible and informative introduction to the issues in a novel format.

Gibson-Graham, J.-K. (2006) Post-Capitalist Politics. Minneapolis: University of Minnesota Press.
A highly influential book advocating ways of rethinking the economy that go beyond mainstream capitalist values

to consider the more social and community-based forms of economy that exist and might be extended.

Massey, D. (2004) Geographies of responsibility. *Geografisker Annaler B*: **86 (1): 5–18.**
Massey's 2004 article is a useful summary of her arguments for a geography of responsibility between places and is a useful way of thinking about the relationality of places under globalisation and the different power geometries at work in the global economy.

McClintock, N. (2014) Radical, reformist, and garden-variety neoliberal: coming to terms with urban agriculture's contradictions. *Local Environment* **19 (2): 147–71.**
An interesting and nuanced analysis of the growing phenomenon of community gardening and its potential to offer alternative spaces for community development in the city. The article considers both the arguments that community gardens become incorporated into broader neoliberal-driven gentrification of the city as well as more positive views that see their emancipatory potential.

Useful websites

www.communityeconomies.org.
The website of the Community Economies Collective: a group of activists and scholars inspired in large part by the work of J.-K. Gibson-Graham but incorporating researchers and community groups in Australia, North America, Europe and South-east Asia.

https://thenextsystem.org/
A website and research project based in the US which is articulating a range of visions for a more solidaristic and democratic economy beyond capitalism and heavily influenced by the work of the writer and activist Gar Alperovitz.

www.fairtrade.net/
The main website for the global fair trade network.

www.woccu.org
The main website for the global credit union federation.

References

Amable, B. (2003) *The Diversity of Modern Capitalism*. Oxford: Oxford University Press.

Amin, A., Cameron, A. and Hudson, R. (2003) The alterity of the social economy. In Leyshon, A., Lee, R. and Williams, C.C. (eds) *Alternative Economic Space*. London: Sage, pp. 27–54.

Birchall, J. (2013) The potential of cooperatives during the current recession: theorizing comparative advantage. *Journal of Entrepreneurial and Organizational Diversity* 2 (1): 1–22.

Birchall, J. and Ketilson, L.H. (2009) *Resilience of the Cooperative Business Model in Times of Crisis*. Geneva: ILO.

Cook, I. *et al.* (2010) Geographies of food: 'Afters'. *Progress in Human Geography* 35 (1): 104–20.

Crang, P. and Cook, I. (2012) Consumption and its geographies. In Daniels, P., Bradshaw, M., Shaw, D. and Sidaway, J. (eds) *Introduction to Human Geography*, 4th edition. London: Pearson Education.

Crossan, J., Cumbers, A., McMaster, R. and Shaw, D. (2016) Contesting neoliberal urbanism in Glasgow's community gardens: the practice of DIY citizenship. *Antipode* 48: 937–55.

Cumbers, A., Shaw, D., Crossan, J. and McMaster, R. (2018) The work of community gardens: reclaiming place for community in the city. *Work Employment and Society* 32 (1): 133–49.

Fair Trade International (2017) *Creating Innovations, Scaling Up Impact, Annual Report 2016–2017*. Fair Trade International. Available at: https://annualreport16-17.fairtrade.net/en/.

Fukuyama, F. (1992) *The End of History and the Last Man*. London: Penguin.

Fuller, D. and Jonas, A. (2003) Alternative financial spaces. In Leyshon, A., Lee, R. and Williams, C.C. (eds) *Alternative Economic Space*. London: Sage, pp. 55–73.

Ghose, R. and Pettigrove, M. (2014) Urban community gardens as spaces of citizenship. *Antipode* 46 (4): 1092–112.

Gibson-Graham, J.-K. (2006) *Post-Capitalist Politics*. Minneapolis: University of Minnesota Press.

Gibson-Graham, J.-K. (2008) Diverse economies: performative practices for other worlds. *Progress in Human Geography* 32: 613–32.

Gibson-Graham, J.-K., Cameron, J. and Healy, S. (2013) *Take Back the Economy, Any Time, Any Place*. Minneapolis: University of Minnesota Press.

Gomez, G. and Helmsing, A.H.J. (2008) Selective spatial closure and local economic development: what do we learn from the Argentine local currency systems? *World Development* 36: 2489–511.

Grady-Benson, J. and Sarathy, B. (2016) Fossil fuel divestment in US higher education: student-led organising for climate justice. *Local Environment* 21: 661–81.

Groeneveld, H. (2017) *Snapshot of European Cooperative Banking 2017*. Tilburg: Tilburg University, TIAS School for Business and Society.

Habermas, J. (1981) *The Theory of Communicative Action. Volume 1*. Boston: Beacon Press.

Hall, D. and Soskice, D. (eds) (2001) *Varieties of Capitalism: The Institutional Foundations of Comparative Advantage*. Oxford: Oxford University Press.

Harvey, D. (1982) *The Limits to Capital*. Oxford: Blackwell.

Harvey, D. (2003) *The New Imperialism*. Oxford: Oxford University Press.

Hodgson, G. (1999) *Economics and Utopia: Why the Learning Economy is not the End of History*. London, Routledge.

Hospers, G.J. (2014) Policy responses to urban shrinkage: from growth thinking to civic engagement. *European Planning Studies* 22: 1507–23.

Hughes, A., Wrigley, N. and Buttle, M. (2008) Global production networks, ethical campaigning and the embeddedness of responsible governance. *Journal of Economic Geography* 8: 345–67.

Ironmonger, D. (1996) Counting outputs, capital inputs and caring labour: estimating gross household product. *Feminist Economics* 2: 37–64.

Lawson, L.J. (2012) *Community Garden Research Survey Preliminary Report*. Presented at the Greater and Greener: Re-imagining Parks for 21st Century Cities. New York: City Parks Alliance.

Lee, R. (1999) Local money: geographies of autonomy and resistance. In Martin, R. (ed.) *Money and the Space Economy*. Chichester: Wiley, pp. 207–24.

Leyshon, A., Lee, R. and Williams, C.C. (2003) *Alternative Economic Spaces*. London: Sage.

MacAskill, W. (2015) Does divestment work? *The New Yorker*, October.

Mansvelt, J. (2013) Consumption-reproduction. In Cloke, P. et al. (eds) *Human Geography*, 3rd edition. London: Taylor & Francis, pp. 378–90.

Massey, D. (2004) Geographies of responsibility. *Geografisker Annaler B* 86 (1): 5–18.

McChesney, R.W. (1999) Noam Chomsky and the struggle against neoliberalism. *Monthly Review* 50: 40–47.

McClintock, N. (2014) Radical, reformist, and garden-variety neoliberal: coming to terms with urban agriculture's contradictions. *Local Environment* 19 (2): 147–71.

Mirowski, P. (2013) *Never Let a Serious Crisis Go to Waste: How Neoliberalism Survived the Financial Meltdown*. London: Verso.

Morgan, K. (2015) Nourishing the city: the rise of the urban food question in the global north. *Urban Studies* 52 (8): 1379–94.

Morgan, K., Marsden, T. and Murdoch, J. (2006) *Worlds of Food: Place Power and Provenance in the Food Chain*. Oxford: Oxford University Press.

Norges Bank (2017) *Grounds for Decision – Product-based Coal Exclusions*. At www.nbim.no/.

North, P. (2005) Scaling alternative economic practices? Some lessons from alternative currencies. *Transactions of the Institute of British Geographers* 30: 233–5.

North, P. (2010) *Local Money: How to Make it Happen in Your Community*. Dartington: Green Books.

North, P. (2014) Ten square miles surrounded by reality? Materialising alternative economies using local currencies. *Antipode* 46 (1): 246–65.

North, P. (2015) The business of the Anthropocene? Substantivist and Diverse Economies perspectives on SME engagement in local low carbon transitions. *Progress in Human Geography* 4 (2): 437–54.

O'Neill, P. and Gibson-Graham, J.-K. (1999) Enterprise discourse and executive talk: stories that destabilize the company. *Transactions of the Institute of British Geographers* 24: 11–22.

Peck, J. and Theodore, N. (2007) Variegated capitalism. *Progress in Human Geography* 31: 731–72.

Pollard, S. (1971) Introduction. In Pollard, S. and Salt, J. (eds) *Robert Owen, Profit of the Poor*. London: Redfern Percy.

Pratt, G. (1999) From registered nurse to registered nanny: discursive geographies of Filipina domestic workers in Vancouver BC. *Economic Geography* 75: 215–36.

Roelants, B., Hyungsik, E. and Terrasi, E. (2014) *Cooperatives and Employment: A Global Report*. Brussels: International Organisation of Industrial and Service Cooperatives.

Rosol, M. (2011) Community volunteering as neoliberal strategy? Green space production in Berlin. *Antipode* 44 (1): 239–57.

Routledge, P. and Cumbers, A. (2009) *Global Justice Networks*. Manchester: Manchester University Press.

Sayer, A. (2000) Moral economy and political economy. *Studies in Political Economy* 61 (1): 79–103.

Seyfang, G. (2006) Sustainable consumption, the new economics and community currencies: developing new institutions for environmental governance. *Regional Studies* 40: 781–91.

Smith, N. (1984) *Uneven Development.* Oxford: Blackwell.

Stohr, W. and Todling, F. (1978) Spatial equity – some antitheses to current regional development doctrine. *Papers of the Regional Science Association* 38: 33–54.

Tesco (2017) *Annual Report.* Tesco Plc. Available at: www.tescoplc.com/media/392373/68336_tesco_ar_digital_interactive_250417.pdf.

Tornaghi, C. (2014) Critical geography of urban agriculture. *Progress in Human Geography* 38 (4): 551–67.

PART 5

Prospects

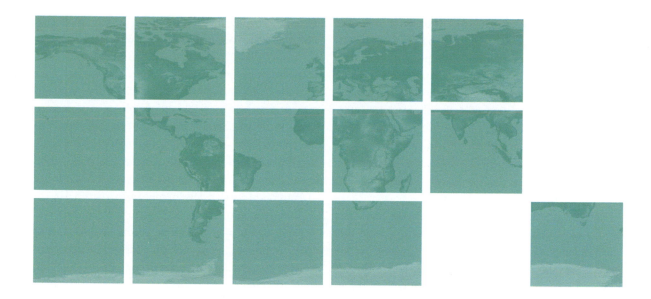

Chapter 14
Conclusions

In this short concluding chapter, we summarise the key themes of the book, and offer some brief reflections on post-financial crisis concerns around slow economic growth, **climate change**, rapid technological change, social and spatial inequality and the rise of economic populism.

14.1 Summary of key themes

The three interrelated themes running through the book are **globalisation**, **uneven development** and **place**. Globalisation should be viewed as an ongoing and unfolding process, not a final outcome or 'end state', which is shaped by the actions of a wide range of individuals, organisations, firms and governments (Dicken 2015). States should be viewed as active facilitators of globalisation, rather than its passive 'victims', through the adoption of **neoliberal** policies including the abolition of exchange and capital controls, the reduction of trade barriers and the implementation of **privatisation** programmes. Instead of economic integration creating a 'flatter' or more equal world, through the reduction of international inequalities (Friedmann 2005), globalisation is an uneven process, creating prosperity in some places whilst others become increasingly marginalised. On a global scale, this is symbolised by the dramatic growth of East Asia, especially China, and continued poverty across much of sub-Saharan Africa. At the same time as between-country inequality has fallen, levels of inequality within countries have generally risen since the late 1980s, not only in rapidly growing economies such as China, India and Russia, but also in most developed countries (Horner and Hulme 2017). These inequalities have generated a strong populist backlash against globalisation in recent years, symbolised by the election of President Trump in the US and the UK's Brexit vote. The rise of populism and economic nationalism raises the prospect of de-globalisation or a reversal of key aspects of globalisation through growing opposition to multilateral trade agreements, the dismantling of trade blocs like the EU and a general reassertion of national economic interests. If it were to happen, a process of de-globalisation would be likely to have profound and uneven effects on the geography of the global economy, reflecting how different countries and regions are positioned in relation to global trade and investment flows.

Uneven development represents a basic characteristic of **capitalism** as a mode of production, reflecting the tendency for capital and labour to move to the areas where they can secure the highest returns (profits and wages respectively). In this way, growth becomes concentrated in core regions though a process of 'cumulative causation' (Myrdal 1957) (Box 3.4). Surrounding regions often get left behind, becoming subordinate peripheries supplying resources and labour to the core. Patterns of uneven development are not static, however, with capital and labour moving between locations in search of higher profits and wages. Over time, the process of economic growth in a particular region tends to undermine its own foundations, generating dis-economies of **agglomeration** such as rising land costs and congestion. At the same time, lower costs in underdeveloped regions may attract capital and labour. This 'see-sawing' of capital between locations creates shifting patterns of uneven development at different geographical scales (Smith 1984). This process of uneven regional growth and decline is apparent from the decline of so-called 'rustbelt' regions (northern England, the American Midwest) and the rise of 'sunbelt' ones (the south and west of the US, southern Germany).

Patterns of uneven development are underpinned by **spatial divisions of labour**, where different parts of the production process become located in different types of region. On a global scale, **TNCs** moved routine manufacturing and assembly operations in industries such as textiles, footwear and electronics to certain developing countries in the 1960s and 1970s. This relocation process was driven by the availability of large surpluses of low-cost labour in developing countries, and facilitated by the increasing division of labour in manufacturing and new transport and communication technologies. It created a **NIDL** where routine assembly and manufacturing was increasingly carried out in low-wage countries whilst higher-level functions like strategic management and research and development remained based in developed countries. More recently, the offshoring of IT-enabled services to developing countries like India has prompted talk of a 'second global shift', creating a new NIDL in services (Bryson 2016).

Patterns of uneven development are structured by the opposing forces of spatial agglomeration and dispersal. The question of which of these predominates depends upon a range of factors, including the type of economic activity in question, the level of available technology, the development of the (technical) division of labour and the size and location of the market being served. ICTs have been a key force reshaping the economic landscape, facilitating the relocation of both manufacturing and IT-enabled services on a global scale. This does not herald the 'end of geography' (O'Brien 1992), however, as shown by the increased agglomeration of corporate headquarters and advanced financial and business services in world cities. This brings us to the crucial observation that high-value activities show a marked tendency towards agglomeration or concentration in large urban areas whilst lower-value activities are more susceptible to **spatial dispersal** to lower-wage locations. While there are always exceptions, we regard this as a key underlying 'rule' or 'stylised fact' in economic geography.

The role of place represents the third key theme of the book, partly reflecting the effects of wider processes of economic development in creating distinctive forms of production in particular places. These general processes interact with pre-existing local conditions and practices (for example, resource bases, employment patterns, skills, income levels, cultural values, institutional arrangements, political orientations) to create new geographies of production and consumption. Rather than simply reflecting geographical isolation, the distinctiveness or uniqueness of places is actually reproduced through this process of interaction (Johnston 1984). From this perspective, place can be regarded as a key node constructed out of wider social relations and connections (Massey 1994). Such relations and connections span the spheres of production, consumption and circulation as transport and communication networks and financial flows bind places together. The volume and intensity of these spatial flows and connections have increased markedly under globalisation, inspiring Massey's effort to develop a **global sense of place**.

The assumption that globalisation is erasing the distinctiveness of place, making distant localities appear increasingly similar, is only true at a very superficial level. In terms of consumption, for instance, the spread of global brands like McDonald's and Coca-Cola is often cited, together with the profusion of large retail malls and centres. Yet the associated idea of a single **global consumer culture** is crude and simplistic, masking a more complex process whereby different local cultures have become increasingly mixed and entangled (Cook and Crang 2016). Local sites of consumption remain highly significant, not least in terms of the growth of tourism and heritage in which place becomes a central object of consumption.

In the sphere of production too, place remains important in the global economy, although many of the specialised industrial regions of the past have disappeared in the face of wider processes of economic restructuring. Nonetheless, particular places continue to be associated with distinctive forms of economic activity. Numerous examples have been cited and discussed through the text, from the densely clustered financial districts of Wall Street and the City of London, the high-tech centre of Silicon Valley to the post-industrial regions of Wales and north-east England and the mining districts of the Andes. While some places have prospered under globalisation, others have become marginalised and impoverished, reflecting the process of uneven development within capitalism. The impact of processes of uneven economic development upon particular places cannot be easily predicted or modelled in advance, since this will depend upon a range of contingent factors and relationships. This means that the shaping of the economic landscape is characterised by a basic openness and unpredictability, providing much of what makes economic geography such an interesting subject to study.

14.2 Contested futures: globalisation, inequality and populism

The geographical **political economy** approach that we have adopted in this book emphasises the dynamic and unstable nature of capitalism as an economic system. Over the long term, it has been characterised by periods of growth and stagnation, punctuated by moments of crisis. It is this dynamism that underpinned Schumpeter's famous notion of **creative destruction**, which emphasises how the periodic emergence of new technologies undermines and destroys established industries and skills. This dynamism and instability has been reinforced by the widespread adoption of **neoliberalism** as the dominant economic policy framework since the late 1970s. Policies of **deregulation** and **liberalisation** have unleashed renewed market forces, removed social protections and eroded the ability of states to control global flows of capital.

The instability of capitalism has been underlined by the experience of the financial crisis of 2008–09, and the prolonged period of economic stagnation and instability that has followed it, reflecting the particularly slow and weak nature of the recovery from the Great Recession. Whilst triggered by failures in the 'sub-prime' mortgage market in the US, the increasing global integration of the financial system ensured that the crisis soon enveloped the world economy, since banks and financial institutions from a range of countries had become heavily involved in the repackaging and trading of US mortgage liabilities (section 4.5). From 2013, the crisis spread to developing countries and the **BRICS** economies, driven by falling commodity prices. Renewed signs of recovery in the global economy became apparent from early 2017, however, based on strong growth in the US, economic revival in Europe and only a limited moderation of growth in China (Economist Intelligence Unit 2018). It remains to be seen, however, whether this apparent recovery will prove economically sustainable given the fragility of previous post-crisis upturns and the inability to address many underlying structural problems (Chapter 12).

The other source of crisis to affect the global economy in recent years is, of course, environmental. Concerns about climate change have grown markedly, supporting an increased commitment to reducing emissions and fostering a low-carbon economy, although this continues to generate opposition from vested interests, symbolised by President Trump's recent decision to withdraw the US from UN emission agreements (section 12.3). The issue of climate change is shot through with questions of uneven development and justice as the impact will be felt most severely in developing countries due to their greater dependence on natural resources and ecosystems to meet their everyday material needs. Even in the developed world, however, the presumption that economic development can proceed without any adverse effect on the environment is no longer valid with resources likely to come under increasing pressure. More broadly, the era of cheap energy and low transport costs that underpinned processes of economic globalisation may be coming to

an end, raising questions about the scope for the re-localisation of supply chains (Cumbers 2016).

Questions of social and spatial inequality have attracted increased attention in recent years, reflecting a prolonged squeeze of wages and living standards in the aftermath of the 2008–09 crisis and Great Recession. This has fuelled the populist backlash against globalisation. The principal beneficiaries of globalisation have been the very richest people in the world, together with the emerging global middle class in developing countries, whilst those who have lost out have been the very poorest and the middle class in the global North (between 75th and 90th percentiles of the global income distribution), as represented by the 'elephant graph' below (Figure 14.1) (Milanovic 2016). The percentiles of the global income distribution who did best from 1988 to 2008 were 90 per cent comprised of Asians, while 86 per cent of the least successful were from the developed countries of the global North (Horner and Hulme 2017: 13–14). This reflects the eastwards shift

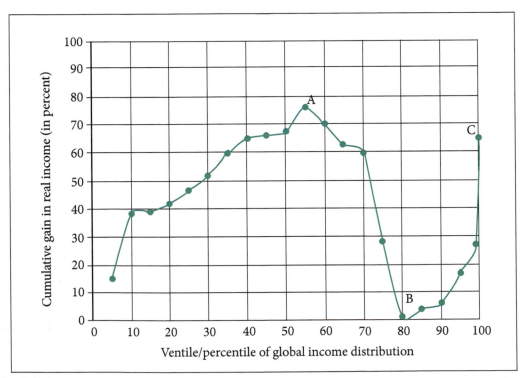

Figure 14.1 The 'Elephant Graph': Relative changes in per capita income by global income level, 1988–2008
Source: Milanovic 2016: 11.

of the centre of gravity of the world economy (Dicken 2015) and underpins growing opposition to globalisation in the US and Europe (Horner and Hulme 2017: 14). In response, a discourse of 'inclusive growth' has emerged amongst policy-makers at the international, national and city-regional scales (see, for example, Lagarde 2014). This raises the question of whether existing growth models can be rendered more inclusive through the greater participation of marginalised groups, or whether existing models and policies need to be fundamentally rethought in an effort to construct more inclusive economies.

Rapid technological change and the explosion of so-called big data in recent years represents another development with profound consequences for the social and geographical organisation of the world economy. Of central importance here are a group of advanced technologies often labelled **Industry 4.0** – the fourth industrial revolution after a first industrial revolution based on steam power and mechanisation, a second one involving electricity and mass production and a third one associated with ICTs and the establishment of the internet. Originating in a German Government-supported research project in 2011, Industry 4.0 emphasises the digital transformation of manufacturing and services, facilitated by a cluster of interrelated technologies, including cloud computing, artificial intelligence (AI), big data, the internet of things, 3D visualisations and **automation** (Gotz and Jankowska 2017). The prospect of intensified automation with robots and AI replacing jobs has prompted widespread concern. Workers in routine occupations and regions which are disproportionately reliant on such work are likely to be particularly exposed to the consequences of automation with the benefits accruing to the owners of technologies and businesses and the highly skilled (Lawrence *et al.* 2017). This underlines the need for new forms of regulation to manage the process of automation and for new models of ownership such as wealth funds and employee ownership trusts to spread the dividends and promote economic democracy (ibid.).

The immediate responses to the economic crisis in 2008–09 appeared to signal a shift away from neoliberalism to something more akin to neo-Keynesianism, with governments lowering interest rates and undertaking fiscal stimulus packages to try to boost economic

recovery. By 2010, however, this had given way to a new round of **austerity** politics in response to rising levels of public debt, much of it caused by propping up the failed financial system. The turn to **austerity** reflects the entrenched institutional power of neoliberalism as a policy discourse that benefits financial and business elites at the expense of other groups in society (Harvey 2005). Accordingly, the underlying causes of the crisis have not been addressed, while the solution of cutting government spending undermined the process of economic recovery from 2011 (Bivens 2016).

More recently, the economic insecurity generated by globalisation and **austerity** has fuelled a new populism, particularly amongst the people and places left behind by deindustrialisation and technological change. This is associated with forms of economic nationalism, expressed in slogans such as 'America first', which take issue with central elements of globalisation such as free trade, capital mobility and immigration (Rodrik 2016). More generally, while the neoliberal agenda of economic integration and liberalisation remains in place at the international level (Peck *et al.* 2010), this may be accompanied by a renewed emphasis on the role of states in managing the consequences of economic change and promoting social cohesion (Cowley 2017).

While this climate of entrenched inequality, insecurity and economic populism might seem depressing, at best, even frighteningly uncertain and unstable, it is important to remind ourselves that the economy itself remains socially constructed and open to economic alternatives. Although the right-wing populist revolt against globalisation has captured much attention since 2016, the mainstream neoliberal model of globalisation has been opposed by a diverse **counter-globalisation movement** since the 1990s (section 1.2.1). At the same time, as highlighted in Chapter 13, a range of alternative economic spaces and initiatives have grown over recent decades, from fair trade schemes to **cooperatives** and **credit unions**. These reflect how the evolution of capitalism tends to be accompanied by elements of resistance, and the articulation of alternative economic models. Interest in alternative and fairer ways of regulating and organising economies may well increase in the absence of meaningful economic and social reforms to tackle inequality. Crucial questions remain, however, about the scope for reforming capitalism in more progressive directions based on an expansion of alternative economic practices, or whether these, instead, provide the beginnings of a new economic model.

References

Bivens, J. (2016) *Why Is Recovery Taking so Long – and Who's to Blame?* Washington, DC: Economic Policy Institute.

Bryson, J.R. (2016) Service economies, spatial divisions of expertise and the second global shift. In Daniels, P., Bradshaw, M., Shaw, D., Sidaway, J. and Hall, T. (eds) *An Introduction to Human Geography*, 5th edition. Harlow: Pearson, pp. 343–64.

Cook, I. and Crang, P. (2016) Consumption and its geographies. In Daniels, P., Bradshaw, M., Shaw, D., Sidaway, J. and Hall, T. (eds) *Introduction to Human Geography*, 5th edition. London: Pearson Education, pp. 379–96.

Cowley, J. (2017) The May doctrine. *New Statesman*, 8 February.

Cumbers, A. (2016) Remunicipalization, the low-carbon transition, and energy democracy. In Worldwatch Institute, *State of the World Report 2016*. Washington, DC: Worldwatch.

Dicken, P. (2015) *Global Shift: Mapping the Changing Contours of the World Economy*, 7th edition. London: Sage.

Economist Intelligence Unit (2018) *Global Outlook*, February. London: Economist Intelligence Unit.

Friedmann, T. (2005) *The World is Flat: A Brief History of the Twenty First Century*. New York: Farrar, Straus & Giroux.

Gotz, M. and Jankowska, B. (2017) Clusters and industry 4.0 – do they fit together? *European Planning Studies* 25: 1633–53.

Harvey, D. (2005) *A Brief History of Neoliberalism*. Oxford: Oxford University Press.

Horner, R. and Hulme, M. (2017) Converging divergence: unpacking the new geography of 21st century global development. *Global Development Institute Working Paper Series*, 2017–010. Manchester: The University of Manchester Global Development Institute.

Johnston, R.J. (1984) The world is our oyster. *Transactions, Institute of British Geographers* NS 9: 443–59.

Lagarde, C. (2014) Economic inclusion and financial integrity. An address to the *Conference on Inclusive Capitalism*. Washington, DC: IMF.

Lawrence, M., Roberts, C. and King, L. (2017) Managing automation: employment, inequality and ethics in the digital age. *Institute for Public Policy Research Discussion Paper*. London: Institute for Public Policy Research.

Massey, D. (1994) A global sense of place. In Massey, D. (ed.) *Place, Space and Gender*. Cambridge: Polity, pp. 146–56.

Milanovic, B. (2016) *Global Inequality: A New Approach for the Age of Globalisation*. Cambridge, MA: Harvard University Press.

Myrdal, G. (1957) *Economic Theory and the Under-developed Regions*. London: Duckworth.

O'Brien, R. (1992) *Global Financial Integration: The End of Geography*. London: Pinter.

Peck, J., Theodore, N. and Brenner, N. (2010) Postneoliberalism and its malcontents. *Antipode* 42: 94–116.

Rodrik, D. (2016) The surprising thing about the backlash against globalization. At www.weforum.org/agenda/2016/07/the-surprising-thing-about-the-backlash-against-globalization?utm_source=feedburner&utm_medium=feed&utm_campaign=Feed%3A+inside-the-world-economic-forum+(Inside+The+World+Economic+Forum), 15 July. Last accessed 9 August 2016.

Smith, N. (1984) *Uneven Development: Nature, Capital and the Production of Space*. Oxford: Blackwell.

Glossary

Absorptive capacity. This refers to a firm's ability to recognise, assimilate and exploit knowledge. It seems to rely upon the existence of a common corporate culture and language, meaning that everybody shares the same broad outlook and sense of the company's overall purpose and objectives.

Agglomeration. The tendency for industries to cluster in particular places, underpinned by the operation of **agglomeration economies**.

Agglomeration economies. A set of economic advantages for individual firms which are derived from the concentration of other firms in the same location – such as the availability of skilled labour, proximity to a large number of customers, knowledge spillovers, and access to specialist suppliers of goods and services. They are comprised of both **localisation** and **urbanisation economies**

Alternative economic spaces. Economic initiatives that seek to create alternative economic identities, practices and sites outside the mainstream capitalist economy.

Anthropocene. A term used to define a proposed epoch of human impact on global change, involving increased CO_2 emissions and other greenhouse gases. Some commentators date this to the industrial revolution in the late eighteenth century and others to the dawn of the atomic age in the 1950s.

Austerity. A form of economic adjustment based on the reduction of prices and wages and public expenditure by the cutting of states' budgets and debts. Usually associated with neoliberal measures to shrink the size of the state, austerity programmes have been widely implemented since 2010 in response to the financial and economic crisis of 2008–09 and the high levels of public debt that ensued, particularly in the Eurozone.

Automation. The use of technology to perform economic processes and tasks without human assistance or labour. As such, it involves the replacement of labour by machines, something that has long been a feature of industrial capitalism. A new round of intensified automation is expected to occur through advances in artificial intelligence and the spread of robots.

Backwash effects. The adverse effects of growth in a core region on a surrounding region in terms of capital and labour being 'sucked out' of the latter region into the former which offers higher profits and wages. In this situation, the virtuous circle of growth in the core is matched by a vicious circle of decline in the periphery, expressed in classic symptoms of underdevelopment such as a lack of capital and depopulation.

Basic needs. An approach to development prominent in the 1960s and 1970s which focused attention on the everyday lives of the poor, reflecting concerns that these were being neglected by orthodox modernisation policies. The concern for identifying and meeting such needs in relation to food, shelter, employment, education, health, etc. is shared by the recent work on grassroots development.

Branch plant economy. A term coined to describe the way particular regions become dominated by 'branch plants', undertaking basic assembly and manufacturing, but controlled from outside. Branch plant economies emerged through economic restructuring processes in the 1960s and 1970s whereby processes of corporate restructuring at national and international levels produced a new **spatial division of labour** taking over from or displacing local ownership.

Brands/brand. A name, design or symbol that distinguishes a firm or product from other firms or products in a market. With the shift to a more consumption-based form of capitalism, the ability of the sellers of a product to differentiate it from competitor products has become a critical source of value.

BRICS. A term originally coined in 2001 as an acronym for the rapidly emerging economic powers of Brazil, China, India and Russia with South Africa added in 2010.

Capital accumulation. The process of investment and reinvestment in production to generate higher profits. Accumulation lies at the heart of the capitalist system, representing the basic driving force for economic growth and innovation.

Capitalism. A particular mode of production, dominant since the eighteenth century, based on the private ownership of the means of production – factories, equipment, etc. – and the associated need for the majority of people to sell their labour power to capitalists in exchange for a wage.

Capital switching. The process by which capital is moved between sectors of the economy and regions, in response to changing investment opportunities. In geographical terms, capital is often transferred from regions dominated by declining sectors to 'new industrial spaces' in distant regions offering more attractive conditions for investment.

Carbon democracy. A term coined by the US-based sociologist Tim Mitchell to refer to the emergence of mass industrial democracies in Europe and North America in the late nineteenth and twentieth centuries which relied heavily on coal as a source of energy, giving considerable power to the workers and unions that extracted and transported it.

Central banks. State institutions that are responsible for controlling the supply of money in a country, usually involving the setting of interest rates and the monitoring of inflation.

Central place theory. Developed by Walter Christaller in the 1930s, this is perhaps the best known of the **German locational models**. Based on the assumption of economic rationality and the existence of certain geographical conditions such as uniform population distribution across an area, central place theory offers an account of the size and distribution of settlements within an urban system. The need for shop owners to select central locations produces a hexagonal network of central places.

Circuit of capital. The basic process of producing commodities for profit under capitalism, involving the combination of the means of production – factories, machines, materials, etc. – and labour power to produce a commodity for sale, generating a profit above the initial money outlay. Part of

this profit is reinvested back into the production process, which recreates the circuit anew and forms the basis of capital accumulation. In addition to the primary circuit of capital in production, secondary and tertiary circuits are also commonly identified, referring to investment in the built environment, and education, health and welfare respectively.

Climate change. A large-scale, long-term shift in the world's average temperature or weather patterns. Whilst subject to great natural variability, there is strong evidence that global temperatures have been rising in recent decades as a result of human activity related to increased emissions of CO_2 and other greenhouse gases.

Clusters. A more popular term for agglomeration involving the concentration of economic activities in a particular location. The term has become associated with the work of Michael Porter, a Harvard business economist, in recent years. Porter (1998: 197) defines clusters as "geographical concentrations of interconnected companies, specialised suppliers, service providers, firms in related industries, and associated institutions (for example universities, standards agencies and trade associations) that compete but also co-operate".

Codified knowledge. Also termed explicit knowledge, this refers to formal, systematic knowledge that can be conveyed in written form through, for example, programmes or operating manuals.

Cold War. The political and ideological conflict between western capitalist countries led by the USA and a communist block led by the Soviet Union, lasting from the mid-1940s to the late 1980s.

Colonialism. A political and economic system based on territorial empires, involving the direct political control of colonies by the colonial powers.

Commercial geography. An early form of economic geography, prominent from the 1880s to the 1930s. Closely linked to European imperialism, it provided knowledge about colonial territories in Africa, Asia and Latin America, identifying and mapping key resources, crops, ports and trade routes, relating these to climate and settlement patterns.

Commodity fetishism. The tendency for the geographies of production and distribution that underpin the goods on sale in shops to be obscured by the emphasis consumers attach to the physical appearance and price of goods.

Community economy. An attempt to encourage a more locally contained and reciprocal economic system in opposition to the profit-centred global economy.

Comparative advantage. A key principle of international trade theory, classically expressed by David Ricardo in 1817. It states that a country should specialise in exporting goods which it can produce more cheaply than other goods and import goods which are more expensive to produce domestically. Through specialisation, both countries gain by focusing on the goods in which they have a comparative or relative cost advantage and importing those in which a country has a comparative disadvantage.

Competence or resource-based theory of the firm. Derived from the work of the economist Edith Penrose (1959), the competence perspective views firms as bundles of assets and competencies that have been built up over time. Competencies can be seen as particular sets of skills, practices and forms of knowledge.

Competitive advantage. The dynamic advantages derived from the active creation of technology, human skills, economies of scale, etc. by firms. This can be contrasted with the rather static and naturalistic notion of **comparative advantage** which relates efficiency to pre-existing endowments of the main factors of production.

Competitiveness. A key concept underpinning economic development policy since the early 1990s. It refers to the underlying capacity of an economy to compete with other countries, based on the assumption that nations and regions compete for global market share in a similar fashion to firms. Key aspects of competitiveness include levels of innovation, enterprise and workforce skills, and governments are charged with fostering these capacities.

Consumer culture. A key feature of contemporary society, which is increasingly organised around the logic of individual choice in the marketplace with shopping and consumption an increasingly central part of people's lives and identities.

Consumption. The processes involved in the purchase and use of commodities by individuals.

Cooperatives. Mutual (as opposed to private) forms of ownership of businesses or other organisations (e.g. housing associations) set up to act collectively in the best interests of particular groups (e.g. employees, consumers, tenants) and usually attempting to pursue more ethical and social goals alongside economic ones.

Counter-globalisation movement. A movement that grew in the 1990s and 2000s to oppose the inequalities brought about by the neoliberal or free market agenda of economic globalisation. It espouses a more open, participative 'bottom-up' model of globalisation, evident in initiatives like the World Social Forum.

Creative classes. A group of highly skilled and educated workers who have distinct lifestyle preferences, according to Richard Florida. Florida argues that cities and regions should actively seek to attract the creative classes who are drawn to open and diverse places.

Creative destruction. A term coined by the Austrian economist Joseph Schumpeter to capture the dynamism of capitalism in terms of innovation and the development of new technologies. As new products and technologies are developed and adopted (creation), they often render existing industries obsolete, unable to compete on the basis of quality or price (destruction).

Credit unions. Financial **cooperatives** that are owned and run by their members, who can borrow from pooled deposits at low interest rates. As such, they provide opportunities for individuals to save or borrow money on better terms than mainstream banks or 'loan sharks' of the underground economy.

Cultural turn. An important development in human geography and the social sciences from the late 1980s, involving a shift of focus from economic to cultural questions.

Cumulative causation. A model of the process of uneven regional development derived from the work of the Swedish economist Gunnar Myrdal. This explains the spatial concentration of industry in terms of a spiral of self-reinforcing advantages that build up in an area and the adverse effect this has on other regions, creating a core-periphery pattern.

Debt crisis. The problem of large-scale indebtedness facing many developing countries since the 1980s, undermining development efforts and threatening the viability of the world financial system at times. The origins of the debt crisis lie in the interactions between three sets of factors: the borrowing of large sums by developing countries from northern banks and institutions in the 1970s; rising interest rates in the late 1970s and early 1980s; and reduced commodity prices and thereby export earnings since the early 1980s.

Deindustrialisation. A decline in the importance of manufacturing industry as a sector of the economy, expressed in terms of its share of employment or output. Generally associated with the growth of service industries and the closure of older heavy industries such as coal, steel and shipbuilding, deindustrialisation has been a common experience across developed countries since the 1960s.

Demand side of the economy. In contrast to the **supply side**, this is concerned with overall demand for goods and services in the economy, emphasising the importance of consumer expenditure. Demand-side theories are associated with the work of the renowned British economist John Maynard Keynes, who argued that output is determined by effective demand. From a regional perspective, demand-side approaches seek to raise the demand for labour and increase income and consumer expenditure in the regional economy.

Dependency theories. The most prominent set of **structuralist theories**, particularly associated with the radical economist Andre Gunder Frank. According to Frank, the metropolitan core exploits its 'satellites', extracting profits (surplus) for investment elsewhere. **Colonialism** was a key force here, creating unequal economic relations which were then perpetuated by the more informal imperialism characteristic of the post-war period.

Deregulation. The reduction of the rules and laws under which business operates, a key component of **neoliberalism**.

Derivatives. New financial instruments, defined as "contracts that specify rights/obligations based on (and hence derived from) the performance of some other currency, commodity or service used to manage risk and volatility in global markets" (Pollard 2005: 347). Key forms of derivatives include futures, swaps and options.

Deskilling of labour. The removal of more rewarding aspects of work such as design, planning and variation, often associated with an increased division of labour. The sub-division and fragmentation of the labour process increases employers' control of production, reducing workers to small cogs in the system and making them vulnerable to replacement by machines.

Development. In the context of economic policy, the term conveys a sense of positive change over time, making a particular country or region more prosperous and advanced. As an economic and social policy, development has been directed at those 'underdeveloped areas' of the world that require economic growth and modernisation.

Developmental state. A particular type of state which is heavily involved in the promotion of economic development. The term is derived from the experience of East Asian countries such as Japan, South Korea and Singapore. Weiss's (2000: 23) definition of the developmental state emphasises three key criteria: the aim of increasing production and closing the economic gap with the industrialised countries; the establishment of a strong government department to coordinate and promote industrial development; and close cooperative ties with business.

Devolution. The transfer of powers and responsibilities downwards from the central states to political authorities at regional or local level (Agranoff 2004). It typically involves the establishment of elected political institutions (governments and/or assemblies) at the regional scale of government. This process is also commonly captured by the term 'political decentralisation'.

Diamond model. Michael Porter's representation of how business clusters actually operate, enhancing competitiveness and productivity by fostering the interaction between four sets of factors: demand conditions; supporting and related industries; factor conditions; and firm strategy, structure and rivalry.

Digital divides. Inequalities in access to information and communication technologies (ICTs) and the internet which tend to reflect broader social divisions of income, class, age, gender and race. As internet access has spread, differences in the speed and quality of connection and usage have become more important.

Digital economy. "The pervasive use of IT (hardware, software, applications and telecommunications) in all aspects of the economy" (Moriset and Malecki 2008: 259). The concept emphasises the combination of digitisation and the internet, allowing information to be collected, packaged and distributed more rapidly and efficiently than before, alongside the spread of digital devices and applications.

Discourses. A key term derived from post-structuralist philosophy which refers to networks of linked concepts, statements and practices that produce distinct bodies of knowledge. Crucially, meaning is generated through particular discourses which, instead of simply reflecting an underlying reality, actively create it.

Diverse economies. A term which signifies an emphasis on the different forms of economic activity and organisation that coexist within capitalism such as domestic work, cooperatives and gift-buying.

Division of labour. A key principle of industrial society, which has technical, social and geographical dimensions. The technical division of labour can be defined as the process of dividing production into a large number of highly specialised parts, so that each worker concentrates on a single task rather than trying to cover several. Under industrialisation, as Adam Smith argued, an increased division of labour results in huge rises in productivity. The social division of labour encompasses the vast array of specialised jobs that people perform in society (see **spatial division of labour**).

Economic geography. A major branch of human geography which addresses questions about the location and distribution of economic activity, the role of uneven geographical development and processes of local and regional economic development.

Economics. An important neighbouring discipline to economic geography which views the economy as governed by market forces which basically operate in the same fashion everywhere, irrespective of time and space. In contrast to the diversity and open-endedness of economic geography, economics is a formal theoretical discipline based on modelling and quantification.

Economies of scale. The tendency for firms' costs for each unit of output to fall when production is carried out on a large scale, reflecting greater efficiency.

Equity finance. Funds raised by investors buying a stake or share in a firm. For large, publicly quoted firms, this occurs through the issuing of shares on the stock market.

Eurozone sovereign debt crisis. A public debt crisis experienced by several peripheral economies within the Eurozone from 2010, namely Portugal, Italy, Ireland, Greece and Spain (christened the PIIGS). Amidst great concern about debt levels and rising bond yields on financial markets, Greece, Ireland, and Portugal were forced to seek bailouts from the EU, ECB and IMF in 2010, which were provided on condition of severe cuts in government spending (**austerity**) and tax rises.

Evolutionary economic geography (EEG). A prominent sub-field of economic geography that emerged in the mid-2000s, focusing on how the economic landscape is transformed over time. Informed by concepts from evolutionary economics and biology, EEG is principally concerned with processes of path dependence and lock-in, the clustering of industries in space and the role of innovation and knowledge in shaping economic development.

Export-oriented industrialisation (EOI). Often regarded as the opposite strategy to **import-substitution industrialisation**, EOI involves countries producing goods and services for selling in external markets. It is compatible with traditional notions of free trade and comparative advantage in contrast to ISI which involves high levels of protection and state intervention.

Export processing zones (EPZs). Special economic zones which offer tax and investment incentives for foreign-owned firms to locate there. Typically these will include up to 100 per cent rebates on local taxation, the provision of all infrastructure and the relaxation of the usual rules governing foreign ownership.

Factors of production. The different elements that are brought together to produce particular goods and services: capital, labour, land and knowledge.

Fair or ethical trade. Alternative principles of trade aimed at giving producers a fair and stable price that allows decent living standards and is not subject to global market fluctuations or cost pressures from dominant multinationals. Fair Trade initiatives link concerned consumers (primarily in richer countries) with farmers and producers in the global South.

Financial crises. Episodes in which the assets of financial institutions and actors lose a large proportion of their value and markets crash. They do not always have a large negative impact upon other parts of the economy, depending upon their nature and the ability of governments to respond, but at various points in time (e.g. the 1930s and 2008–09) they can result in severe global economic recessions.

Financial exclusion. Defined as "processes that serve to prevent certain social groups and individuals from gaining access to the financial system" (Leyshon and Thrift 1995: 314). It is based on income, compounding the difficulties facing disadvantaged individuals and groups.

Financial instability hypothesis. A theory of how unregulated financial markets can create financial crises, developed by the US economist Hyman Minsky in the 1960s, which attracted renewed interest in the wake of the financial crisis of 2008. Minsky's theory identifies three types of borrowers: hedge ones, who are able to pay off the initial loan and interest; speculative units who are able to cover interest payments, but unable to pay off the principal debt; and Ponzi units who are unable to pay either the interest or the principal. Over time, as economic boom conditions develop, financial institutions tend to shift from hedge to the more risky speculative and Ponzi forms. If asset prices fall once an investment bubble bursts, Ponzi borrowers can no longer repay their loans, resulting in a financial crisis if Ponzi-style finance has been widely used across the system.

Financialisation. The increasing role of financial motives, financial markets, financial actors and financial institutions in daily life.

Firm, the. A legal entity involved in the production of goods and services, owned by individual capitalists or, more commonly, a range of shareholders. The standard organisational form of capital.

Fiscal policies. These are based on the raising or lowering of taxes and the level of public expenditure. They play a particularly important role in Keynesian policies of demand management, based on the need for adjustment across the economic cycle.

Flexible production. A new form of production regarded in the 1980s and early 1990s as the successor to **Fordism**. In the sphere of production, flexibility was associated with the widespread use of information and communication technologies (ICTs) which enabled processes and equipment to be continually reprogrammed and reset. In the sphere of consumption, it was rooted in a new emphasis on niche markets and customisation compared to the standardised mass markets of the post-war period.

Forced labour. This is defined by the International Labour Organisation as "situations in which persons are coerced to work, through the use of violence and intimidation or more subtle means such as accumulated debt, retention of identity papers or threats of denunciation to immigration authorities". Forced labour is closely related to contemporary forms of slavery, debt bondage and human trafficking (see www.ilo.org/global/topics/forced-labour/news/WCMS_237569/lang--en/index.htm).

Fordism. A **regime of accumulation**, dominant from the 1940s to the 1970s, based on a crucial link between mass production and **mass consumption**, provided by rising wages for workers and increased productivity in the workplace. This was supported by a Fordist-Keynesian **mode of regulation** where the state adopted highly interventionist policies of demand management, full employment, welfare provision, trade union recognition and national collective bargaining. Fordism is named after the American car manufacturer Henry Ford, who pioneered the use of mass production techniques. As such, the term is also used in a narrower sense to refer to an intensified labour process, based on a highly elaborate division of labour and the introduction of moving assembly lines.

Geographical political economy (GPE). An approach within economic geography that emerged in the 1970s, based upon the application of a **political economy** framework to geographical questions such as regional development and urban restructuring.

Geography of responsibility. A phrase associated with the geographer Doreen Massey whereby people living in one location recognise the interconnections between events and relations in their local economy and distant others as a result of globalisation processes, and take responsibility for them.

German location theory. A body of spatial economic theory from the nineteenth and early twentieth centuries, based on the work of German theorists such as Von Thunen, Weber, Christaller and Losch, which developed models of the economic landscape derived from the assumptions of neoclassical economic theory.

Global commodity chains (GCCs). The network of connections involved in the production, circulation and consumption of a commodity, covering the different stages from the supply of materials to final consumption. They typically incorporate a range of actors, for example farmers, subcontractors, manufacturers, transport operators, distributors, retailers and consumers.

Global consumer culture. A phrase emphasising the creation of a single global market, centred upon brands like McDonald's, Coca-Cola and Nike. For many commentators, this is erasing the distinctiveness of local places and cultures, heralding the 'end of geography'. As a number of studies have shown, however, this cultural homogenisation argument is highly simplistic.

Globalisation. A process of economic integration on a global scale, creating increasingly close connections between people and firms located in different places. Manifested in terms of increased flows of goods, services, money, information and people across national and continental borders.

Global pipelines. Channels of communication that are built by firms in regional clusters with selected partners outside the cluster, enabling them to collaborate and share knowledge. Such strategic partnerships offer access to knowledge and assets not available locally. This concept emphasises the broader relational linkages between clusters and firms located elsewhere, in contrast to earlier approaches which were preoccupied with intra-cluster networks.

Global production networks (GPNs). An approach developed by economic geographers to the study of economic globalisation, incorporating a wide range of actors and relationships, in contrast to the more restrictive global **commodity chain** approach. A GPN is defined as an "organisational arrangement, comprising interconnected economic and noneconomic actors, coordinated by a global firm, and producing goods or services across multiple geographic locations for worldwide markets" (Yeung and Coe 2015: 32).

Global sense of place. An attempt to rethink place in an era of globalisation, associated with the geographer Doreen Massey. This approach rejects the idea of place as isolated and bounded, viewing it as a meeting place, a kind of node where wider social relations and connections come together.

Global value chains. An approach, based to a considerable extent on earlier global commodity chains research, which is concerned with the organisation of global industries, particularly in terms of different stages of the production process being located in different regions. The GVC approach is particularly focused on issues of inter-firm governance. It is closely related to the GPN approach, but is less geographically oriented.

Global warming. A gradual increase in the temperature of the earth's atmosphere that is generally held to be the result of a greenhouse effect associated with increased emissions of CO_2 and other pollutants. Also referred to as **climate change**.

Governance. A term used to refer to the growth of new ways of governing societies in place of the traditional notion of government, incorporating special-purpose agencies, business interests and voluntary organisations alongside government.

Green economy. An economy that aims to reduce environmental damage and promote sustainable development, based on low-carbon and resource-efficient technologies and practices.

Homeshoring. The dispersal of work tasks and activities to workers who work from home.

Human Development Index (HDI). A widely known composite measure of development established by the United Nations Development Programme (UNDP), and published annually since 1990 in its *Human Development Report*. The HDI measures the overall achievement of a country in three basic dimensions of human development – health, education and a decent standard of living.

Hypermobility of capital. A phrase increasingly used by geographers and economists in the context of globalisation and, more specifically, the deregulation and unleashing of financial flows across borders since the 1970s and the growth of electronic forms of money. It refers to the dramatic and potentially crisis-inducing effects that stem from the ability of financial actors to instantaneously move vast amounts of money and finance from one part of the globe to another at the touch of a button.

Import-substitution industrialisation. It involves a country attempting to produce for itself goods that were formerly imported. Newly created '**infant industries**' are protected from outside competition through the erection of high tariff barriers, allowing for the diversification of country's economy and reduced dependence on foreign technology and capital.

Industrial districts. Specialised industrial areas based on networks of small firms and craftsmen. Originally associated with nineteenth-century regions like the Sheffield cutlery district until the 1980s, recent attention has focused on the revival of industrial districts in central and north-eastern Italy in the 1970s and 1980s.

Industry 4.0. The digital transformation of manufacturing and services, facilitated by a cluster of interrelated technologies, including cloud computing, artificial intelligence, big data, the internet of things, 3D visualisations and automation.

Infant industries. Emerging industries regarded as strategically important by the state, which seeks to foster their development. This involves protection from outside competition until such industries are strong enough to compete in global markets.

Innovation. The creation of new products and services or the modification of existing ones to gain a competitive advantage in the market. The commercial exploitation of ideas is crucial, distinguishing innovation from invention.

Institutions. Broad social and organisational conventions, practices and rules that shape economic life. Key institutions include firms, markets, the monetary system, the state and trade unions.

Interactive approach to innovation. A recent perspective which views innovation as a circular process based on cooperation and collaboration between manufacturers or service providers, users (customers), suppliers, research institutes, development agencies, etc. This is often contrasted with the conventional **linear model of innovation**.

International division of labour. A pattern of geographical development which involves different countries specialising in different types of economic activity. The classic 'old' international division of labour of the nineteenth century, associated with colonialism, involved the developed countries of Europe and North America producing manufactured goods whilst the underdeveloped world specialised in the production of raw materials and foodstuffs. This can be contrasted with the **'new' international division of labour (NIDL)** of the 1970s and 1980s.

Internationalisation of retail. The process by which large retail firms expand into international markets, largely through foreign direct investment in stores, sometimes in connection with a local partner or through a franchise model. While retailing has traditionally been a nationally specific activity, it has experienced a dramatic internationalisation since the late 1990s.

Knowledge-based economy. The idea that a new type of economy has developed since the early 1990s in which knowledge has become the key resource and learning the key process for firms and individuals.

Kondratiev cycles. Named after the Soviet economist Kondratieff who first identified such cycles in the 1920s, the term refers to long waves of economic development, based on distinctive systems of technology. Five Kondratiev cycles are usually identified from the late eighteenth century.

Labour control regime. Stable local institutional frameworks where local and even national governments act in concert with businesses to create local environments for the effective regulation and reproduction of a labour force. These local spaces of labour control are based on the creation of reciprocal relations between the spheres of production, reproduction and consumption (see Jonas 1996).

Labour geography. A branch of economic geography that is concerned with how workers and their trade unions influence the changing geography of the economy. A central concern of labour geography is to counter accounts dominated by business and state actors, stressing the agency of workers and their representatives.

Labour market flexibility. An employment regime in which wages, conditions and worker attitudes become more responsive to the pressures of competition and the needs of business, requiring workers to accept varying pay rates and hours of work whilst learning new skills and undertaking new tasks. The creation of labour market flexibility has been a key goal of **neoliberal** policy since the 1980s.

Liberalisation. The opening up of protected sectors of the economy to competition, a key component of **neoliberalism**.

Linear model of innovation. The traditional understanding of innovation as a series of well-defined stages running from the research laboratory to the production line, marketing department and retail outlet.

Livelihoods approach. An influential framework in development studies for understanding how individuals and households in the global South adapt to economic change by developing a range of income-generating and employment strategies, based on their capabilities, assets and activities.

Local currencies. Currencies that can only be spent within a designated local area through participating organisations. These run alongside official national currencies rather than replacing them and often aim to increase spending within a local economy, supporting locally owned businesses and activities.

Localisation economies. A particular type of **agglomeration economy**, stemming from the concentration of firms in the *same* industry.

Local labour markets. A term used to signify the importance of the local scale in the operation of most forms of employment, whereby the majority of the population live and work within relatively small spatially delineated areas. The term 'travel-to-work-area' has been developed to statistically define local labour markets, identifying geographical boundaries that correspond to the limits within which the majority of a population of an area live and work.

Marxism. A set of social and economic theories derived from the writings of Karl Marx. Marxism adopts a materialist view of society, stressing the importance of underlying social relations and forces over ideas. The economy is structured by a capitalist **mode of production** defined by the antagonistic social relations between the capitalist and working classes. The exploitation of workers forms the basis of capitalist profit, but this contradictory relationship will also ensure that capitalism is ultimately overthrown by socialism.

Mass consumption. A form of consumption based on the purchase of standard consumer durables such as automobiles, fridges and washing machines by large numbers of households. This was a key dimension of post-war **Fordism**, linked to mass production by rising wages for workers.

Millennium Development Goals (MDGs). A set of objectives for development to be achieved by 2015 agreed by the United Nations in 2000. They included the eradication of extreme poverty and hunger, the achievement of universal primary education and the promotion of gender equality.

Modernisation theory. The dominant approach to development in 1950s and 1960s. The basic idea is that developing countries undergo a linear process of transformation (modernisation), analogous to the changes experienced by developed countries in the nineteenth century following the industrial revolution. Economic growth is paramount, generating increased income and employment opportunities which, it was assumed, would 'trickle down' to the poorest groups in society. This thinking was famously expressed in US economist Walt Rostow's stages of growth model.

Modes of production. Economic and social systems by which societies are organised, determining how resources are utilised, work is organised and wealth is distributed. Economic historians have identified a number of modes of production, principally subsistence, slavery, feudalism, capitalism and socialism.

Modes of regulation. A concept developed by the French regulationist school of political economy (see **regulation approach**). Modes of regulation are focused on five key aspects of capitalism in particular: labour and the wage relation, forms of competition and business organisation, the monetary system, the state and the international regime.

The post-war period of economic growth was based upon a Fordist-Keynesian mode of regulation (see **Fordism**).

Monetary policies. These are concerned with managing the money supply and its rate of circulation in the form of credit. The control of inflation through interest rate policy is of central importance, a function that is often undertaken by central banks.

Moral economy. A term that has its origins in pre-capitalist forms of social relations where in certain religions (notably Christianity and Islam) there were strong moral (and legal) pressures against greed and usury (e.g. charging interest on loans is still illegal in some Muslim countries). The Marxist historian E.P. Thompson famously refers to the 'moral economy' of the crowd in relation to riots by the poor in response to rising food prices in the eighteenth century. The term has more recently been used as part of the attempt to develop ethical or fair trading systems for global commodities, replacing pure capitalist values with more ethical ones.

Multi-level perspective (MLP). An influential framework for thinking about sustainability transitions which focuses upon how technological change comes about as a result of social as well as technological processes. The MLP identifies three key levels of analysis: niches (micro); **technological regimes** (meso); and wider landscapes (macro).

Neoliberal approach/neoliberalism. A political and economic philosophy and approach to economic policy that seeks to reduce state intervention and embrace the free market, stressing the virtues of enterprise, competition and individual self-reliance.

New economic geography (NEG). An approach to economic geography developed by the economist Paul Krugman and others, involving the application of mathematical modelling techniques to analyse issues of industrial location. The new geographical economics addresses questions such as why, and under what conditions, do industries concentrate? It applies the methods of mainstream economics, devising models based on a number of simplifying assumptions.

New industrial spaces. Areas distinct from the old industrial cores which became centres of flexible production from the 1970s onwards. Three different kinds of 'new industrial spaces' have been identified in Europe and North America: craft-based industrial districts like central and north-eastern Italy; centres of high-technology industries such as Silicon

Valley in California; and clusters of advanced financial and producer services in world cities.

New international division of labour (NIDL). A term that refers to the process by which **TNCs** based in western countries have shifted low-status assembly and processing operations to developing countries where costs are much lower. It is a form of the **spatial division of labour**, operating at the global scale, facilitated by the increasing division of labour in large TNCs, new transport and communications technologies and the creation of a pool of available labour in developing countries. In recent years, a more complex 'new' NIDL has emerged, based upon more complex interrelations between a greater number of countries. Central elements of this include an increased focus on services and growing investment by large emerging economies such as China, Brazil and South Africa (see **BRICS**) in other parts of the global South.

Newly industrialising countries (NICs). Formerly underdeveloped countries that experienced rapid economic development and significant foreign direct investment between 1960 and 1990, including Asian economies such as Hong Kong, Korea, Singapore and Taiwan, but also including Brazil and Mexico.

New regionalism. A collective label used to describe a body of research in contemporary economic geography which stresses the renewed importance of 'the region' as a scale of economic organisation under late capitalism.

New trade theory. An approach developed by the economist Paul Krugman and others since the 1970s which recognises that comparative advantage does not simply reflect pre-existing endowments of the **factors of production**. Rather, it is actively created by firms through the development of technology, skills, economies of scale, etc.

Non-government organisations (NGOs). Organisations, often of a voluntary or charitable nature, which make up the so-called 'third sector' belonging to neither the private nor public sectors.

Offshoring. The relocation of economic activities from developed countries to low-wage economies. A characteristic of manufacturing since the 1960s, creating the **new international division of labour** (NIDL), it has become a feature of service operations in recent years, popularised by the move of some call centres to developing countries like India.

Outsourcing. The organisational process of a firm subcontracting specific tasks and work packages to other firms rather than undertaking them in house. Global outsourcing to firms based in other countries (**offshoring**) has been a key trend reshaping the geography of the global economy in recent decades, whereby many routine tasks are now undertaken by firms based in lower-cost developing countries (See NIDL).

Path dependence. A key evolutionary idea adopted by economic geographers, referring to how past decisions and experiences shape the economic landscape, particularly in terms of structuring and informing economic actors' responses to wider processes of economic change.

Place. A particular area, usually occupied, to which a group of people have become attached, endowing it with meaning and significance. Often associated with notions of family, home and community.

Places of consumption. The particular sites at which goods and services are bought and consumed, including the department store, the mall, the street, the market and the home as well as a host of more inconspicuous sites of consumption (for example, charity shops and car boot sales).

Political ecology. An approach which adopt a more 'political' sense of nature-economy-society relations in contrast to traditional environmental management perspectives. It is concerned with how processes of capitalist accumulation produce uneven spatial effects that are detrimental for the environment and society, and with the articulation of an alternative politics to tackle the inequalities produced by capitalism, linking to notions of ecological, or climate, justice.

Political economy. A broad perspective on economic life which analyses the economy within its wider social and political context, focusing on production and the distribution of wealth between different sections of the population as well as the exchange of commodities through the market.

Positivism. A philosophy of science which states that the goal of science is to generate explanatory laws which explain and predict events and patterns in the real world.

Post-Fordism. A term widely used in the 1980s and 1990s to describe the growth of new production methods. In contrast to the standardised mass production techniques of Fordism, post-Fordism is defined by flexibility in face of more fragmented and customised patterns of market demand (see **flexible production**).

Post-Keynesian economics. A school of heterodox economics which developed in the UK and US from the 1940s onwards. Post-Keynesians place a strong Keynesian emphasis upon the role of aggregate demand in the functioning of a market economy and in the role of government in stabilising the economy during economic downturns. Contrary to the increasingly mainstream orientation of Keynesian economics in the post-Second World War period, post-Keynesians also emphasise the importance of uncertainty, social relations and institutions to the functioning of modern economies.

Postmodernism. Defined as "a movement in philosophy, the arts and social sciences characterised by scepticism towards the grand claims and grand theories of the modern era, and their privileged vantage point, stressing in its place openness to a range of voices in social enquiry, artistic experimentation and political empowerment" (Ley 1994: 466).

Post-neoliberalism. Political and economic projects based upon a rejection of neoliberal free market policies. The term has been applied to a range of left-wing governments that were elected in Latin America in the late 1990s and 2000s. While there are major differences between individual governments, the Latin American form of post-neoliberalism revolves around the dual aims of redirecting a market economy towards social ends and reviving citizenship through a new politics of participation (Yates and Bakker 2014).

Post-structuralism. A set of theories derived from philosophy and literary and cultural studies which stress the fractured identities of individual subjects, seeing these as products of broader social categories and discourses.

Power. The ability or capacity to take decisions that involve or affect other people. Whilst neglected by mainstream economists, alternative approaches such as political economy tend to stress the importance of power in shaping the operation of the economy, particularly in terms of the (social) relations between different economic actors.

Precariat. A new class identified by Guy Standing as part of a sixfold classification of contemporary work. The identification of this global precariat signifies how processes of **globalisation** and flexibility have fostered a generalised shift away from stable and permanent occupations to a rise

in insecure and precarious forms of work. The precariat is defined by its reliance on temporary and contingent forms of employment, representing the fifth of Standing's sixth classes (moving from the most to least privileged), with only an underclass of the unemployed and incapacitated below it.

Privatisation. The policy of transferring state-owned enterprises into private ownership, a key component of **neoliberal** reform programmes since the early 1980s.

Qualitative state. A conception of the state as a dynamic process rather than a fixed 'thing' or object. The term, coined by the Australian economic geographer Philip O'Neill (1997), reflects a general shift of emphasis from a concern with quantitative aspects of state intervention to an interest in its qualitative characteristics.

Regime of accumulation. Another key regulationist term, referring to a relatively stable form of economic organisation which structures particular periods of capitalist development, creating a balance between production and consumption. Regimes of accumulation are supported by particular **modes of regulation**.

Regional geography. This represented the leading approach to human geography between the 1920s and 1950s. It was defined as a project of 'areal differentiation' which describes and interprets the variable character of the earth's surface, expressed through the identification of distinct regions.

Regional policy. A set of programmes and measures established by governments to promote regional growth and development. Conventional regional policy involved governments inducing companies to locate factories and offices in depressed regions by offering grants and financial incentives. At the same time, development in core regions such as south-east England and Paris was restricted. Classical regional policy reached its peak in the 1960s and 1970s, contributing to a narrowing of the income gap between rich and poor regions in Europe.

Regional sectoral specialisation. A pattern of industrial location during the nineteenth and early twentieth centuries where particular regions become specialised in certain sectors of industry. Characteristically, all the main stages of production from resource extraction to final manufacture were carried out within the same region.

Regulation approach. Developed by a group of French economists in the 1970s, the regulation approach stresses the role of wider processes of social regulation in stabilising capitalist development. These wider processes of regulation find expression in specific institutional arrangements which mediate the underlying contradictions of the capitalist system, enabling renewed growth to occur. This occurs through the consolidation of specific **modes of regulation**, underpinning a period of stable growth, known as a **regime of accumulation** (e.g. **Fordism** and **post-Fordism**).

Relational economic geography. This reflects the rise of relational thinking in geography since the early 2000s and is concerned with economic action and interaction which it views as being embedded in wider social and economic relations. It is particularly associated with the study of networks and economic practices.

Rescaling. Changes in the relationships between geographical scales (see **scale**). The term is particularly associated with debates on governance and the restructuring of nation-states, involving shifts of power and responsibilities in two principal directions: from the national to the supranational level; and from the national to the sub-national level of cities and regions (see **devolution**).

Scale. The different geographical levels of human activity: local, regional, national, supranational and global.

Scientific management or Taylorism. An approach to industrial organisation, associated with the Fordist mass production system, named after its founder and principal advocate, F.W. Taylor. Taylorism involved the reorganisation of work according to rational principles designed to maximise productivity, based on an increased division of labour, enhanced coordination and control by management and the close monitoring and analysis of work performance and organisation.

Sharing economy. Sometimes also referred to as the 'digital platform economy' or 'gig economy', the sharing economy refers to forms of economic "exchange that are facilitated through online platforms, encompassing a diversity of for-profit and non-profit activities that all broadly aim to open access to under-utilised resources through what is termed sharing" (Richardson 2015: 121). The use of the term 'sharing' is controversial since it is based on the appropriation of a wider common sense definition of sharing that excludes forms of exchange in which a monetary benefit accrues to one or more parties.

Smart city. This refers to cities harnessing the wealth of 'big data' now available and using the power of ubiquitous computing to address specific urban problems. The concept refers to the harnessing of **ICTs** to support economic development and the use of data derived from software-embedded technologies to augment urban management.

Social economy. A broad term that has been developed to characterise economic activities that are not viewed as part of the capitalist mainstream, but are involved in providing activities to meet social needs rather than produce a profit. A wide range of activities is frequently and somewhat problematically included in this category from privatised welfare services to more radical attempts to develop autonomous activity in opposition to capitalism.

Social networks. The informal social ties between individuals working in different firms, providing a channel for the sharing of information and ideas. The role of such networks has often been cited in accounts of the rise of **new industrial spaces** like Silicon Valley.

Social relations. The sets of relationships between different groups of economic actors. The relations between employers and workers have attracted particular attention, but other relations include those between producers and consumers, manufacturers and suppliers, supervisors and ordinary workers and government agencies and firms.

Social reproduction. The daily processes of feeding, clothing, sheltering and socialising which support and sustain labour, processes which rely on family, friends and the local community, occurring outside the market.

Socio-environmental hybridities. A set of approaches to the study of human-environmental relations, informed by post-structuralism, which give equal weighting to the agency of nature with humans in the co-construction of the economic landscape. Human geographers adopting this approach are concerned with how combinations of the human and non-human come together to create temporary spatial fixes. 'Natural' or environmental entities are viewed as sets of ongoing material processes subject to transformation and change.

Space. An area of the earth's surface, such as that between two particular points or locations. Often thought of in terms of the distance and time it takes to travel or communicate between two points.

Spatial analysis. An approach to economic geography that became influential in the 1960s and 1970s as part of the so-called 'quantitative revolution' in geography. Spatial analysis in economic geography involves the use of statistical and mathematical methods to analyse problems of industrial location, distance and movement.

Spatial dispersal. The opposite process to **agglomeration** where industries or firms move out of existing centres of production into new regions.

Spatial division of labour. A concept developed by Doreen Massey to explain how an increasing division of labour within large corporations produced new spatial patterns. Companies were locating the higher-order functions in cities and regions where there are large pools of highly educated and well-qualified workers, with lower-order functions such as assembly locating increasingly in those regions and places where costs (especially wage rates) are lowest.

Spatial fix. The establishment of relatively stable geographical arrangements which facilitate the expansion of the capitalist economy for a certain period of time. Examples include imperialism during the nineteenth century, **Fordism** in the post-war period and **globalisation** since the 1980s which has involved the **deindustrialisation** of many established centres of production in the 'rustbelts' of North America and Western Europe and the growth of new industry in 'sunbelt' regions and the newly industrialising countries of East Asia.

Spread effects. A set of processes which allow surrounding regions to benefit from increased growth in a core region. One important mechanism here is increased demand in the core region for food, consumer goods and other products, creating opportunities for firms in peripheral regions to supply this growing market. At the same time, rising costs of land, labour and capital in the core region, together with associated problems like congestion, can push investment out into surrounding regions.

State, the. A set of public institutions that exercise authority over a particular territory, including the government, parliament, civil service, judiciary, police, security services and local authorities.

Strategic coupling. The dynamic processes of interaction between **global production networks (GPNs)** and regional assets which underpins regional development. Regional institutions play a key role in attracting and retaining inward investment by shaping and moulding regional assets to fit the needs of lead firms in GPNs.

Strategic decoupling. This term involves a reduction or rupturing of the relationship between a region and a GPN

through disinvestment, the exit of foreign firms and the loss of foreign markets.

Strategic recoupling. This is based upon a renewal of the relationship between regions and **GPNs** through the recombination of existing regional assets with a new round of investment.

Structural adjustment programmes (SAPs). Economic reform packages developed by the IMF and World Bank in the 1980s and 1990s as part of the **Washington Consensus**. SAPs have been adopted by a large number of developing countries in exchange for financial assistance. They encompass a range of measures requiring countries to open up to trade and investment and to reduce public expenditure.

Structuralist theories. A set of theories of development which explained global inequalities in terms of the structure of the world economy, particularly the relationship between developed countries and developing countries. The structuralist approach was particularly associated with a group of theorists (such as Raul Prebisch) and activists based in Latin America in the 1950s and 1960s.

Structured coherence. A term introduced by David Harvey to describe the social, economic and political relations that develop in association with particular forms of production in specific places (Harvey 1982). It has been particularly associated with the main centres of heavy industry in developed countries during the late nineteenth and twentieth centuries which became known as working-class regions with strong socialist and trade union traditions.

Sub-prime housing market. A particular segment of the housing market made up of high-risk borrowers (usually from poorer groups) whose levels of income make it unlikely that they will be able to meet mortgage payments if market conditions deteriorate.

Supply side of the economy. This is defined in terms of the quality of the main factors of production such as labour (training, skills), capital (emphasising enterprise and innovation) and land in terms of sites and infrastructure for investors. Improving these supply-side factors has been the central focus of urban and regional development policy since the 1980s in contrast to the demand-side focus of previous Keynesian approaches.

Sustainable Development Goals (SDGs). A new set of goals established by the UN in 2015 to replace the **Millennium Development Goals**. These 17 goals encompass a far broader programme than the MDGs, reflecting a global agenda for sustainable development, covering all countries not just poor ones.

Sustainability transitions. A set of processes leading to fundamental shifts in the socio-technical systems that underpin specific economic sectors like energy supply, water supply or transportation. A transition involves major changes across multiple dimensions (technological, material, organisational, institutional, political, economic and socio-cultural) and involves a range of actors.

Tacit knowledge. In contrast to **codified knowledge**, this refers to direct experience and expertise, which is not communicable through written documents. It is a form of practical 'know-how' embodied in the skills and work practices of individuals or organisations.

Technological regime. A central concept within sustainability transitions research and the **multi-level perspective** in particular, defined as an established system comprised of institutions, technologies, rules and practices that are clearly associated with prevailing technologies. The regime and incumbent actors linked to it seek to preserve existing arrangements and resist radical innovations, which may emanate from emerging technological niches outside the regime.

Terms of trade. The ratio of the price of a country's export goods relative to the price of its import goods. A deterioration in the terms of trade was experienced by many developing countries in the 1980s and 1990s, reducing their export revenues relative to the price of imported goods and undermining development. An improvement in the terms of trade, as many commodity-producing countries experienced in the 2000s, by contrast, generates additional export revenues to support development.

Territorial embeddedness. A concept which emphasises how certain forms of economic activity are grounded or rooted in particular places.

Territorial innovation models. A group of related theories of innovation and learning in cities and regions. These models include: innovative milieu, **industrial districts**, regional innovation systems, **new industrial spaces, clusters** and learning regions.

Time-space compression. A term that refers to the effects of information and communication technologies in reducing the time and costs of transmitting information and money across space. This reduces the 'friction of distance' which geographers have traditionally stressed.

Time-space convergence. This concept emphasises how "places approach each other in time-space" "as a result of transport innovation[s]" that reduce the travel time between them (Janelle 1969: 351). It overlaps considerably with the associated concept of **time-space compression**.

Trade unions. Collective organisations representing workers which grew in strength in the late nineteenth and early twentieth centuries, and in many countries were associated with the formation of parliamentary labour parties.

Transnational corporations (TNCs). Companies which conduct operations in a number of countries, allowing them to access different markets and to take advantage of geographical differences in conditions of production, such as the skills and costs of labour. TNCs have been key agents of **globalisation** since the 1970s.

Uneven development. The tendency for some countries and regions to be more economically prosperous than others. Uneven development is an inherent feature of the capitalist economy, reflecting the tendency for growth and investment to become concentrated in particular locations which offer profitable opportunities for investment. Over time, patterns of uneven development are periodically restructured as capital moves or 'see-saws' between locations in the search for profit.

Upgrading. The improvement of an actor's or firm's position in a commodity chain or production network, involving the capture of additional value. Four types of upgrading have been identified: process upgrading based on more efficient production, product upgrading (the production of sophisticated high-value goods or services), functional upgrading and intersectoral upgrading whereby a firm uses its existing competences to move into new sectors

Urban economics. A sub-discipline of economics that, in common with the **NEG**, seek to explain the uneven spatial distribution of economic activities with a particular focus on cities and urban regions. It has become closely associated with **agglomeration**, emphasising the economic benefits generated by the scale and density of economic activity and wealth within cities.

Urban entrepreneurialism. A focus on economic development and regeneration which became a key part of urban policy from the early 1980s. The entrepreneurial approach saw cities focus on the need to generate growth and employment, seeking to attract new investment and funds from outside and generate new business and income from within.

Urbanisation economies. A second type of **agglomeration economy**, derived from the concentration of firms in *different* industries in large urban areas.

Value. The economic return (profit) or rent generated by the production of commodities for sale, involving the conversion of labour power into actual labour through the labour process. Firms may create value through: the control of particular product and process technologies; the development of certain organisational and management capabilities; the harnessing of inter-firm relationships; and the prominence of brand names in key markets.

Varieties of capitalism. A concept developed by leading political economists and comparative sociologists to draw attention to the diverse institutional forms that capitalism continues to assume. Hall and Soskice (2001) identify two main national varieties of capitalism: coordinated-market economies which are associated with countries such as Germany, Japan and Sweden; and liberal-market economies, exemplified by the likes of the UK, US and Australia. In practice, however, capitalism is actually far more diverse than the reliance upon these two archetypes would suggest, incorporating regional as well as national differences.

Venture capital. A form of private equity finance provided by outside investors to new or growing firms that are generally not quoted on the stock exchange. Such investment tends to be high risk, focusing on firms with a high growth potential and attracting investors who aim to make high returns by selling their stake at a later date.

Washington Consensus. A set of economic policies, based upon **neoliberal** economic principles, adopted and implemented by the US Treasury and World Bank and IMF in the 1980s and 1990s, all based in Washington, DC. Key elements include reducing public expenditure, economic liberalisation, privatisation and the promotion of exports and foreign direct investment.

References

Agranoff, R. (2004) Autonomy, devolution and intergovernmental relations. *Regional & Federal Studies* 14: 26–65.

Hall, P. and Soskice, D. (eds) (2001) *Varieties of Capitalism: The Institutional Foundations of Comparative Advantage.* Oxford University Press, Oxford.

Harvey, D. (1982) *The Limits to Capital.* Blackwell, Oxford.

Janelle, D. (1969) Spatial reorganisation: a model and concept. *Annals of the Association of American Geographers* 59, 348–64.

Jonas, A. (1996) Local labour control regimes: uneven development and the social regulation of production. *Regional Studies* 30, 323–38.

Ley, D. (1994). 'Postmodernism'. Entry in Johnston, R. J., D. Gregory and D. M. Smith (Eds) *The Dictionary of Human Geography*, 3rd edition, pp. 466–8.

Leyshon, A and Thrift, N (1995) Geographies of financial exclusion: financial abandonment in Britain and the United States *Transactions of the Institute of British Geographers* NS 20, 312–41.

Moriset, B. and Malecki, E. (2008) Organization versus space: the paradoxical geographies of the digital economy. *Geography Compass* 3: 256–74.

O'Neill, P. (1997) 'Bringing the qualitative state into economic geography' in Lee, R and Wills, J (eds) *Geographies of Economies.* Arnold, London, pp. 290–301.

Pollard, J. (2005) The global financial system: worlds of monies. In Daniels, P., Bradshaw, M., Shaw, D. and Sidaway, J. (eds) *Human Geography: Issues for the Twenty First Century,* 2nd edition. Harlow: Pearson, pp. 358–75.

Porter, M.E (1998) Clusters and the new economics of competition. *Harvard Business Review,* December, 77–90.

Richardson, L. (2015) Performing the sharing economy. *Geoforum* 67, 121–29.

Weiss, L. (2000) Developmental states in transition: adapting, dismantling, innovating, not 'normalising'. *Pacific Review* 13: 21–55.

Yates, J.S. and Bakker, K. (2014) Debating the post-neoliberal turn in Latin America. *Progress in Human Geography* 38, 62–90.

Yeung, H.W-C and Coe, N. M. 2015. Towards a dynamic theory of global production networks. *Economic Geography* 91, 29–58.

Index